LONDON MATHEMATICAL SOCIETY LECTURE NOTE SERIES

Managing Editor: Professor N.J. Hitchin, Mathematical Institute,
University of Oxford, 24–29 St. Giles, Oxford OX1 3LB, United Kingdom

The titles below are available from booksellers, or from Cambridge University Press at www.cambridge.org/mathematics

141 Surveys in combinatorics 1989, J. SIEMONS (ed)
144 Introduction to uniform spaces, I.M. JAMES
146 Cohen-Macaulay modules over Cohen-Macaulay rings, Y. YOSHINO
148 Helices and vector bundles, A.N. RUDAKOV *et al*
149 Solitons, nonlinear evolution equations and inverse scattering, M. ABLOWITZ & P. CLARKSON
150 Geometry of low-dimensional manifolds 1, S. DONALDSON & C.B. THOMAS (eds)
151 Geometry of low-dimensional manifolds 2, S. DONALDSON & C.B. THOMAS (eds)
152 Oligomorphic permutation groups, P. CAMERON
153 L-functions and arithmetic, J. COATES & M.J. TAYLOR (eds)
155 Classification theories of polarized varieties, TAKAO FUJITA
158 Geometry of Banach spaces, P.F.X. MÜLLER & W. SCHACHERMAYER (eds)
159 Groups St Andrews 1989 volume 1, C.M. CAMPBELL & E.F. ROBERTSON (eds)
160 Groups St Andrews 1989 volume 2, C.M. CAMPBELL & E.F. ROBERTSON (eds)
161 Lectures on block theory, B. KÜLSHAMMER
163 Topics in varieties of group representations, S.M. VOVSI
164 Quasi-symmetric designs, M.S. SHRIKANDE & S.S. SANE
166 Surveys in combinatorics, 1991, A.D. KEEDWELL (ed)
168 Representations of algebras, H. TACHIKAWA & S. BRENNER (eds)
169 Boolean function complexity, M.S. PATERSON (ed)
170 Manifolds with singularities and the Adams-Novikov spectral sequence, B. BOTVINNIK
171 Squares, A.R. RAJWADE
172 Algebraic varieties, G.R. KEMPF
173 Discrete groups and geometry, W.J. HARVEY & C. MACLACHLAN (eds)
174 Lectures on mechanics, J.E. MARSDEN
175 Adams memorial symposium on algebraic topology 1, N. RAY & G. WALKER (eds)
176 Adams memorial symposium on algebraic topology 2, N. RAY & G. WALKER (eds)
177 Applications of categories in computer science, M. FOURMAN, P. JOHNSTONE & A. PITTS (eds)
178 Lower K- and L-theory, A. RANICKI
179 Complex projective geometry, G. ELLINGSRUD *et al*
180 Lectures on ergodic theory and Pesin theory on compact manifolds, M. POLLICOTT
181 Geometric group theory I, G.A. NIBLO & M.A. ROLLER (eds)
182 Geometric group theory II, G.A. NIBLO & M.A. ROLLER (eds)
183 Shintani zeta functions, A. YUKIE
184 Arithmetical functions, W. SCHWARZ & J. SPILKER
185 Representations of solvable groups, O. MANZ & T.R. WOLF
186 Complexity: knots, colourings and counting, D.J.A. WELSH
187 Surveys in combinatorics, 1993, K. WALKER (ed)
188 Local analysis for the odd order theorem, H. BENDER & G. GLAUBERMAN
189 Locally presentable and accessible categories, J. ADAMEK & J. ROSICKY
190 Polynomial invariants of finite groups, D.J. BENSON
191 Finite geometry and combinatorics, F. DE CLERCK *et al*
192 Symplectic geometry, D. SALAMON (ed)
194 Independent random variables and rearrangement invariant spaces, M. BRAVERMAN
195 Arithmetic of blowup algebras, W. VASCONCELOS
196 Microlocal analysis for differential operators, A. GRIGIS & J. SJÖSTRAND
197 Two-dimensional homotopy and combinatorial group theory, C. HOG-ANGELONI *et al*
198 The algebraic characterization of geometric 4-manifolds, J.A. HILLMAN
199 Invariant potential theory in the unit ball of C^n, M. STOLL
200 The Grothendieck theory of dessins d'enfant, L. SCHNEPS (ed)
201 Singularities, J.-P. BRASSELET (ed)
202 The technique of pseudodifferential operators, H.O. CORDES
203 Hochschild cohomology of von Neumann algebras, A. SINCLAIR & R. SMITH
204 Combinatorial and geometric group theory, A.J. DUNCAN, N.D. GILBERT & J. HOWIE (eds)
205 Ergodic theory and its connections with harmonic analysis, K. PETERSEN & I. SALAMA (eds)
207 Groups of Lie type and their geometries, W.M. KANTOR & L. DI MARTINO (eds)
208 Vector bundles in algebraic geometry, N.J. HITCHIN, P. NEWSTEAD & W.M. OXBURY (eds)
209 Arithmetic of diagonal hypersurfaces over finite fields, F.Q. GOUVÊA & N. YUI
210 Hilbert C*-modules, E.C. LANCE
211 Groups 93 Galway/St Andrews I, C.M. CAMPBELL *et al* (eds)
212 Groups 93 Galway/St Andrews II, C.M. CAMPBELL *et al* (eds)
214 Generalised Euler-Jacobi inversion formula and asymptotics beyond all orders, V. KOWALENKO *et al*
215 Number theory 1992-93, S. DAVID (ed)
216 Stochastic partial differential equations, A. ETHERIDGE (ed)
217 Quadratic forms with applications to algebraic geometry and topology, A. PFISTER
218 Surveys in combinatorics, 1995, P. ROWLINSON (ed)
220 Algebraic set theory, A. JOYAL & I. MOERDIJK
221 Harmonic approximation., S.J. GARDINER
222 Advances in linear logic, J.-Y. GIRARD, Y. LAFONT & L. REGNIER (eds)
223 Analytic semigroups and semilinear initial boundary value problems, KAZUAKI TAIRA
224 Computability, enumerability, unsolvability, S.B. COOPER, T.A. SLAMAN & S.S. WAINER (eds)
225 A mathematical introduction to string theory, S. ALBEVERIO, *et al*
226 Novikov conjectures, index theorems and rigidity I, S. FERRY, A. RANICKI & J. ROSENBERG (eds)
227 Novikov conjectures, index theorems and rigidity II, S. FERRY, A. RANICKI & J. ROSENBERG (eds)
228 Ergodic theory of Z^d actions, M. POLLICOTT & K. SCHMIDT (eds)
229 Ergodicity for infinite dimensional systems, G. DA PRATO & J. ZABCZYK
230 Prolegomena to a middlebrow arithmetic of curves of genus 2, J.W.S. CASSELS & E.V. FLYNN
231 Semigroup theory and its applications, K.H. HOFMANN & M.W. MISLOVE (eds)
232 The descriptive set theory of Polish group actions, H. BECKER & A.S. KECHRIS
233 Finite fields and applications, S. COHEN & H. NIEDERREITER (eds)
234 Introduction to subfactors, V. JONES & V.S. SUNDER
235 Number theory 1993–94, S. DAVID (ed)
236 The James forest, H. FETTER & B. G. DE BUEN
237 Sieve methods, exponential sums, and their applications in number theory, G.R.H. GREAVES *et al*
238 Representation theory and algebraic geometry, A. MARTSINKOVSKY & G. TODOROV (eds)
240 Stable groups, F.O. WAGNER
241 Surveys in combinatorics, 1997, R.A. BAILEY (ed)
242 Geometric Galois actions I, L. SCHNEPS & P. LOCHAK (eds)

243 Geometric Galois actions II, L. SCHNEPS & P. LOCHAK (eds)
244 Model theory of groups and automorphism groups, D. EVANS (ed)
245 Geometry, combinatorial designs and related structures, J.W.P. HIRSCHFELD *et al*
246 *p*-Automorphisms of finite *p*-groups, E.I. KHUKHRO
247 Analytic number theory, Y. MOTOHASHI (ed)
248 Tame topology and o-minimal structures, L. VAN DEN DRIES
249 The atlas of finite groups: ten years on, R. CURTIS & R. WILSON (eds)
250 Characters and blocks of finite groups, G. NAVARRO
251 Gröbner bases and applications, B. BUCHBERGER & F. WINKLER (eds)
252 Geometry and cohomology in group theory, P. KROPHOLLER, G. NIBLO & R. STÖHR (eds)
253 The *q*-Schur algebra, S. DONKIN
254 Galois representations in arithmetic algebraic geometry, A.J. SCHOLL & R.L. TAYLOR (eds)
255 Symmetries and integrability of difference equations, P.A. CLARKSON & F.W. NIJHOFF (eds)
256 Aspects of Galois theory, H. VÖLKLEIN *et al*
257 An introduction to noncommutative differential geometry and its physical applications 2ed, J. MADORE
258 Sets and proofs, S.B. COOPER & J. TRUSS (eds)
259 Models and computability, S.B. COOPER & J. TRUSS (eds)
260 Groups St Andrews 1997 in Bath, I, C.M. CAMPBELL *et al*
261 Groups St Andrews 1997 in Bath, II, C.M. CAMPBELL *et al*
262 Analysis and logic, C.W. HENSON, J. IOVINO, A.S. KECHRIS & E. ODELL
263 Singularity theory, B. BRUCE & D. MOND (eds)
264 New trends in algebraic geometry, K. HULEK, F. CATANESE, C. PETERS & M. REID (eds)
265 Elliptic curves in cryptography, I. BLAKE, G. SEROUSSI & N. SMART
267 Surveys in combinatorics, 1999, J.D. LAMB & D.A. PREECE (eds)
268 Spectral asymptotics in the semi-classical limit, M. DIMASSI & J. SJÖSTRAND
269 Ergodic theory and topological dynamics, M.B. BEKKA & M. MAYER
270 Analysis on Lie groups, N.T. VAROPOULOS & S. MUSTAPHA
271 Singular perturbations of differential operators, S. ALBEVERIO & P. KURASOV
272 Character theory for the odd order theorem, T. PETERFALVI
273 Spectral theory and geometry, E.B. DAVIES & Y. SAFAROV (eds)
274 The Mandlebrot set, theme and variations, TAN LEI (ed)
275 Descriptive set theory and dynamical systems, M. FOREMAN *et al*
276 Singularities of plane curves, E. CASAS-ALVERO
277 Computational and geometric aspects of modern algebra, M.D. ATKINSON *et al*
278 Global attractors in abstract parabolic problems, J.W. CHOLEWA & T. DLOTKO
279 Topics in symbolic dynamics and applications, F. BLANCHARD, A. MAASS & A. NOGUEIRA (eds)
280 Characters and automorphism groups of compact Riemann surfaces, T. BREUER
281 Explicit birational geometry of 3-folds, A. CORTI & M. REID (eds)
282 Auslander-Buchweitz approximations of equivariant modules, M. HASHIMOTO
283 Nonlinear elasticity, Y. FU & R.W. OGDEN (eds)
284 Foundations of computational mathematics, R. DEVORE, A. ISERLES & E. SÜLI (eds)
285 Rational points on curves over finite, fields, H. NIEDERREITER & C. XING
286 Clifford algebras and spinors 2ed, P. LOUNESTO
287 Topics on Riemann surfaces and Fuchsian groups, E. BUJALANCE *et al*
288 Surveys in combinatorics, 2001, J. HIRSCHFELD (ed)
289 Aspects of Sobolev-type inequalities, L. SALOFF-COSTE
290 Quantum groups and Lie theory, A. PRESSLEY (ed)
291 Tits buildings and the model theory of groups, K. TENT (ed)
292 A quantum groups primer, S. MAJID
293 Second order partial differential equations in Hilbert spaces, G. DA PRATO & J. ZABCZYK
294 Introduction to the theory of operator spaces, G. PISIER
295 Geometry and Integrability, L. MASON & YAVUZ NUTKU (eds)
296 Lectures on invariant theory, I. DOLGACHEV
297 The homotopy category of simply connected 4-manifolds, H.-J. BAUES
298 Higher operads, higher categories, T. LEINSTER
299 Kleinian Groups and Hyperbolic 3-Manifolds Y. KOMORI, V. MARKOVIC & C. SERIES (eds)
300 Introduction to Möbius Differential Geometry, U. HERTRICH-JEROMIN
301 Stable Modules and the D(2)-Problem, F.E.A. JOHNSON
302 Discrete and Continuous Nonlinear Schrödinger Systems, M.J. ABLORWITZ, B. PRINARI & A.D. TRUBATCH
303 Number Theory and Algebraic Geometry, M. REID & A. SKOROBOGATOV (eds)
304 Groups St Andrews 2001 in Oxford Vol. 1, C.M. CAMPBELL, E.F. ROBERTSON & G.C. SMITH (eds)
305 Groups St Andrews 2001 in Oxford Vol. 2, C.M. CAMPBELL, E.F. ROBERTSON & G.C. SMITH (eds)
306 Peyresq lectures on geometric mechanics and symmetry, J. MONTALDI & T. RATIU (eds)
307 Surveys in Combinatorics 2003, C.D. WENSLEY (ed.)
308 Topology, geometry and quantum field theory, U.L. TILLMANN (ed)
309 Corings and Comodules, T. BRZEZINSKI & R. WISBAUER
310 Topics in Dynamics and Ergodic Theory, S. BEZUGLYI & S. KOLYADA (eds)
311 Groups: topological, combinatorial and arithmetic aspects, T.W. MÜLLER (ed)
312 Foundations of Computational Mathematics, Minneapolis 2002, FELIPE CUCKER *et al* (eds)
313 Transcendental aspects of algebraic cycles, S. MÜLLER-STACH & C. PETERS (eds)
314 Spectral generalizations of line graphs, D. CVETKOVIC, P. ROWLINSON & S. SIMIC
315 Structured ring spectra, A. BAKER & B. RICHTER (eds)
316 Linear Logic in Computer Science, T. EHRHARD *et al* (eds)
317 Advances in elliptic curve cryptography, I.F. BLAKE, G. SEROUSSI & N. SMART
318 Perturbation of the boundary in boundary-value problems of Partial Differential Equations, D. HENRY
319 Double Affine Hecke Algebras, I. CHEREDNIK
321 Surveys in Modern Mathematics, V. PRASOLOV & Y. ILYASHENKO (eds)
322 Recent perspectives in random matrix theory and number theory, F. MEZZADRI & N.C. SNAITH (eds)
323 Poisson geometry, deformation quantisation and group representations, S. GUTT *et al* (eds)
324 Singularities and Computer Algebra, C. LOSSEN & G. PFISTER (eds)
325 Lectures on the Ricci Flow, P. TOPPING
326 Modular Representations of Finite Groups of Lie Type, J. E. HUMPHREYS
328 Fundamentals of Hyperbolic Manifolds, R.D. CANARY, A. MARDEN & D.B.A. EPSTEIN (eds)
329 Spaces of Kleinian Groups, Y. MINSKY, M. SAKUMA & C. SERIES (eds)
330 Noncommutative Localization in Algebra and Topology, A. RANICKI (ed)
331 Foundations of Computational Mathematics, Santander 2005, L. PARDO, A. PINKUS, E. SULI & M. TODD (eds)
332 Handbook of Tilting Theory, L. ANGELERI HÜGEL, D. HAPPEL & H. KRAUSE (eds)
333 Synthetic Differential Geometry 2ed, A. KOCK
334 The Navier-Stokes Equations, P.G. DRAZIN & N. RILEY
335 Lectures on the Combinatorics of Free Probability, A. NICU & R. SPEICHER
336 Integral Closure of Ideals, Rings, and Modules, I. SWANSON & C. HUNEKE
337 Methods in Banach Space Theory, J.M.F. CASTILLO & W.B. JOHNSON (eds)
338 Surveys in Geometry and Number Theory N. YOUNG (ed)

London Mathematical Society Lecture Note Series. 332

Handbook of Tilting Theory

Edited by
LIDIA ANGELERI HÜGEL
Università degli Studi dell' Insubria

DIETER HAPPEL
Technische Universität Chemnitz

HENNING KRAUSE
Universität Paderborn

CAMBRIDGE
UNIVERSITY PRESS

CAMBRIDGE UNIVERSITY PRESS
Cambridge, New York, Melbourne, Madrid, Cape Town, Singapore, São Paulo, Delhi

Cambridge University Press
The Edinburgh Building, Cambridge CB2 8RU, UK

Published in the United States of America by Cambridge University Press, New York

www.cambridge.org
Information on this title: www.cambridge.org/9780521680455

© Cambridge University Press 2007

First published 2007

A catalogue record for this publication is available from the British Library

ISBN 978-0-521-68045-5 paperback

Transferred to digital printing 2009

Contents

1	Introduction	*page* 1
2	**Basic results of classical tilting theory**	
	L. Angeleri Hügel, D. Happel, and H. Krause	9
	REFERENCES	12
3	**Classification of representation-finite algebras and their modules**	
	T. Brüstle	15
	1 Introduction	15
	2 Notation	16
	3 Representation-finite algebras	18
	4 Critical algebras	24
	5 Tame algebras	26
	REFERENCES	28
4	**A spectral sequence analysis of classical tilting functors**	
	S. Brenner and M. C. R. Butler	31
	1 Introduction	31
	2 Tilting modules	32
	3 Tilting functors, spectral sequences and filtrations	35
	4 Applications	43
	5 Edge effects, and the case $t = 2$	46
	REFERENCES	47
5	**Derived categories and tilting**	
	B. Keller	49
	1 Introduction	49
	2 Derived categories	51

3   Derived functors   63
4   Tilting and derived equivalences   66
5   Triangulated categories   72
6   Morita theory for derived categories   78
7   Comparison of t-structures, spectral sequences   83
8   Algebraic triangulated categories and dg algebras   90
REFERENCES   97

6   Hereditary categories
        H. Lenzing   105
1   Fundamental concepts   106
2   Examples of hereditary categories   108
3   Repetitive shape of the derived category   112
4   Perpendicular categories   114
5   Exceptional objects   115
6   Piecewise hereditary algebras and Happel's theorem   117
7   Derived equivalence of hereditary categories   121
8   Modules over hereditary algebras   121
9   Spectral properties of hereditary categories   124
10   Weighted projective lines   125
11   Quasitilted algebras   142
REFERENCES   143

7   Fourier-Mukai transforms
        L. Hille and M. Van den Bergh   147
1   Some background   147
2   Notations and conventions   149
3   Basics on Fourier-Mukai transforms   149
4   The reconstruction theorem   155
5   Curves and surfaces   159
6   Threefolds and higher dimensional varieties   166
7   Non-commutative rings in algebraic geometry   170
REFERENCES   173

**8   Tilting theory and homologically finite subcategories
        with applications to quasihereditary algebras**
        I. Reiten   179
1   The Basic Ingredients   181
2   The Correspondence Theorem   191
3   Quasihereditary algebras and their generalizations   200
4   Generalizations   207
REFERENCES   211

9 **Tilting modules for algebraic groups and finite dimensional algebras**
  S. Donkin 215
  1 Quasi-hereditary algebras 217
  2 Coalgebras and Comodules 220
  3 Linear Algebraic Groups 225
  4 Reductive Groups 228
  5 Infinitesimal Methods 233
  6 Some support for tilting modules 238
  7 Invariant theory 239
  8 General Linear Groups 241
  9 Connections with symmetric groups and Hecke algebras 244
  10 Some recent applications to Hecke algebras 247
  REFERENCES 254

10 **Combinatorial aspects of the set of tilting modules**
  L. Unger 259
  1 Introduction 259
  2 The partial order of tilting modules 260
  3 The quiver of tilting modules 261
  4 The simplicial complex of tilting modules 270
  REFERENCES 275

11 **Infinite dimensional tilting modules and cotorsion pairs**
  J. Trlifaj 279
  1 Cotorsion pairs and approximations of modules 281
  2 Tilting cotorsion pairs 292
  3 Cotilting cotorsion pairs 298
  4 Finite type, duality, and some examples 304
  5 Tilting modules and the finitistic dimension conjectures 312
  REFERENCES 316

12 **Infinite dimensional tilting modules over finite dimensional algebras**
  Ø. Solberg 323
  1 Definitions and preliminaries 324
  2 The subcategory correspondence 327
  3 The finitistic dimension conjectures 332
  4 Complements of tilting and cotilting modules 336
  5 Classification of all cotilting modules 340
  REFERENCES 341

13   Cotilting dualities
       R. Colpi and K. R. Fuller   345
       1   Generalized Morita Duality and Finitistic Cotilting
            Modules   348
       2   Cotilting Modules and Bimodules   350
       3   Weak Morita Duality   353
       4   Pure Injectivity of Cotilting Modules and Reflexivity   355
       REFERENCES   356

14   Representations of finite groups and tilting
       J. Chuang and J. Rickard   359
       1   A brief introduction to modular representation theory   359
       2   The abelian defect group conjecture   360
       3   Symmetric algebras   361
       4   Characters and derived equivalence   366
       5   Splendid equivalences   370
       6   Derived equivalence and stable equivalence   373
       7   Lifting stable equivalences   375
       8   Clifford theory   376
       9   Cases for which the Abelian Defect Group Conjecture
            has been verified   378
       10   Nonabelian defect groups   383
       REFERENCES   384

15   Morita theory in stable homotopy theory
       B. Shipley   393
       1   Introduction   393
       2   Spectral Algebra   396
       3   Quillen model categories   399
       4   Differential graded algebras   403
       5   Two topologically equivalent DGAs   406
       REFERENCES   409

Appendix   **Some remarks concerning tilting modules and
       tilted algebras. Origin. Relevance. Future.**
       C. M. Ringel   413
       1   Basic Setting   414
       2   Connections   423
       3   The new cluster tilting approach   446
       REFERENCES   470

1

Introduction

Tilting theory arises as a universal method for constructing equivalences between categories. Originally introduced in the context of module categories over finite dimensional algebras, tilting theory is now considered an essential tool in the study of many areas of mathematics, including finite and algebraic group theory, commutative and non-commutative algebraic geometry, and algebraic topology. In particular, tilting complexes were shown by Rickard to be the necessary ingredient in developing a Morita theory for derived categories. The aim of this handbook is to present both the basic concepts of tilting theory together with a variety of applications.

Tilting theory can trace its history back to 1973 and two articles by Bernstein, Gelfand and Ponomarev. It had recently been shown by Gabriel (1972) that the path algebra $k\Delta$ of a finite quiver Δ over an algebraically closed field k admits only finitely many isomorphism classes of indecomposable modules precisely when the underlying graph of Δ is a disjoint union of Dynkin diagrams of type $\mathbb{A}_n$, $\mathbb{D}_n$, $\mathbb{E}_6$, $\mathbb{E}_7$ or $\mathbb{E}_8$. In this case, the isomorphism classes are in bijection with the positive roots of the corresponding semisimple Lie algebra. The insight provided by Bernstein, Gelfand and Ponomarev was that these indecomposable modules can be constructed recursively from the simple modules via reflection functors in the same manner as the positive roots are constructed from the simple roots via the action of the Weyl group. As a consequence it was noted that changing the orientation of the quiver does not greatly change the module category.

It is known that any finite dimensional hereditary k-algebra is Morita equivalent to such a path algebra $k\Delta$, and that by the Krull-Remak-

Schmidt Theorem, every finite dimensional module decomposes in an essentially unique way into a direct sum of indecomposable modules. The above theorems therefore provide a partial answer to the central problem in the representation theory of finite dimensional algebras: to describe the category $\operatorname{mod}\Lambda$ of finite dimensional Λ-modules for a finite dimensional algebra Λ as completely as possible.

This procedure of constructing indecomposable modules via reflection functors was generalised in a paper by Auslander, Platzeck and Reiten (1979), in which the first tilting modules were considered, although not under that name. Suppose that Λ admits a simple projective non-injective module $S = \Lambda e$ for some primitive idempotent e. Define the Λ-module $T = \tau^- S \oplus \Lambda(1-e)$ and its endomorphism algebra $\Gamma = \operatorname{End} T$, where τ^- denotes the inverse of the Auslander-Reiten translation. Then the functor $\operatorname{Hom}(T, -)$ can be used to compare the two module categories $\operatorname{mod}\Lambda$ and $\operatorname{mod}\Gamma$.

The major milestone in the development of tilting theory was the article by Brenner and Butler (1979). It was here that the notion of a tilting module T for Λ was axiomatised, and the equivalence induced by $\operatorname{Hom}(T, -)$ between certain subcategories of $\operatorname{mod}\Lambda$ and $\operatorname{mod}\Gamma$ for $\Gamma = \operatorname{End} T$ proved in general. Central results like the Brenner-Butler Theorem and the behaviour of the associated quadratic forms are contained in this article. In fact it is due to the latter considerations that we have the name tilting, for in passing from Λ to Γ, the coordinate axes in the two Grothendieck groups are tilted. The approach to tilting presented there is still the approach taken up in the various forms of generalisations considered today. For details of the beginning of tilting theory and its use in determining the representation type of an algebra we refer to the first article and to the subsequent article of Brüstle in this volume.

The set of axioms was relaxed and simplified by Happel and Ringel (1982). They considered additional functors such as $\operatorname{Ext}^1(T, -)$ in order to obtain a much more complete picture. This was further generalised by Miyashita (1986) and Happel (1987), who also introduced a new concept to the subject of tilting theory. Let T be a tilting module for Λ and let $\Gamma = \operatorname{End} T$. Then, whereas the categories $\operatorname{mod}\Lambda$ and $\operatorname{mod}\Gamma$ are similar but equivalent only in trivial situations, Happel showed that the bounded derived categories of both algebras are always equivalent as triangulated categories. Derived categories were introduced in the

1960s by Grothendieck and Verdier in the study of derived functors and spectral sequences, and have since proved to be of fundamental importance in mathematics. In particular, they have been shown to be the correct setting for tilting theory, for not only do they allow the main results to be easily formulated and proved, they also offer new insights concerning homological properties shared by the algebras involved. The general use of derived categories and tilting is explained in the article by Keller. The article by Butler contains a spectral sequence approach to tilting theory.

It was soon realised by Rickard (1989) that a Morita theory for derived categories was possible using tilting complexes. Besides giving a beautiful answer to the problem of deciding when two algebras have equivalent derived categories, Rickard's approach opened the way to applications of tilting theory to selfinjective algebras, in particular to applications in the modular representation theory of finite groups. For details we refer to the article of Chuang and Rickard in this volume.

Another important point of view in tilting theory was introduced by Auslander and Reiten (1991). Usually tilting theory deals with the problem of comparing a fixed module category with the category of modules over some endomorphism algebra. Auslander and Reiten instead used tilting theory to investigate subcategories of the initial category and showed that there is a bijection between certain subcategories and isomorphism classes of multiplicity-free tilting modules. This approach is discussed in the article by Reiten in this volume. The correspondence between tilting modules and subcategories initiated several important developments in representation theory: it was realised by Ringel (1991) that this can be successfully applied in the theory of quasi-hereditary algebras, which in turn inspired Donkin (1993) to translate the theory into the language of algebraic groups and Lie theory. We refer to the articles by Donkin and Reiten.

The correspondence above naturally led to a combinatorial study of the set of isomorphism classes of multiplicity-free tilting modules, an approach first taken up by Riedtmann and Schofield (1991), and then by Unger (1993), who investigated a partial order on the set of all tilting modules together with an associated simplicial complex. The combinatorial aspect of tilting theory is contained in the article by Unger in this volume.

The work of Auslander and Reiten also opened the way to generalising tilting theory to infinite dimensions. Various approaches have been proposed in order to apply the theory to arbitrary rings on one side and to infinite dimensional modules over finite dimensional algebras on the other. Both points of view and their developments are covered by the articles of Trlifaj and Solberg in this volume.

Another aspect of tilting theory is the study of cotilting modules and the dualities which they induce. In this way the classical concept of Morita duality is generalised. This theory is of interest for modules over general rings; it is discussed in the article by Colpi and Fuller.

Tilting theory also appears in both commutative and non-commutative algebraic geometry as a means of relating different categories of coherent sheaves, as well as comparing these to module categories over finite dimensional algebras. As an example we mention the work of Beilinson (1978), which established a derived equivalence between the category of coherent sheaves over projective n-space and the category of finite dimensional modules over a non-commutative algebra. This approach is explained in the article of Hille and van den Bergh.

The category of finite dimensional modules over a finite dimensional hereditary algebra and the category of coherent sheaves on a weighted projective line are standard examples of hereditary categories with a tilting object. In fact, up to derived equivalence, these are the only two examples of such categories. The article by Lenzing is devoted to this topic.

The article by Shipley discusses some of the occurrences of tilting theory in algebraic topology. More specifically, the Morita theory for derived categories is extended from associative rings to ring spectra, which arise as basic structures in stable homotopy theory.

Further aspects of tilting theory and historical references are contained in the individual articles in this handbook. Finally, the appendix by Ringel provides a guideline to the various features of tilting theory, focusing on the historical development of the subject as well as on new perspectives opened up by the recent important results in cluster tilting theory.

The idea of publishing this collection of articles emerged during the meeting Twenty Years of Tilting Theory, which took place in November 2002 on the island Fraueninsel in the South of Germany. We are most grateful for the substantial support provided by the Stiftung Volkswagenwerk for this conference. We also wish to express our thanks to the authors and referees of the articles in this volume, to our TeXpert Marc Jesse and to the staff of the Cambridge University Press for their help in preparing this handbook for publication.

Varese, Chemnitz, and Paderborn, April 2006

Lidia Angeleri Hügel, Dieter Happel, and Henning Krause

Leitfaden

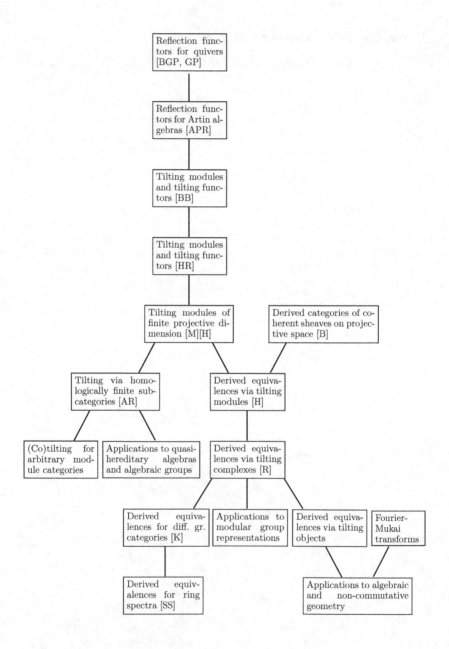

References

[APR] Maurice Auslander, María Inés Platzeck, and Idun Reiten, *Coxeter functors without diagrams*, Trans. Amer. Math. Soc. **250** (1979), 1–46.

[AR] Maurice Auslander and Idun Reiten, *Applications to contravariantly finite subcategories*, Adv. Math. **86** (1991), 111–152.

[B] A. A. Beĭlinson, *Coherent sheaves on $\mathbf{P}^n$ and problems in linear algebra*, Funktsional. Anal. i Prilozhen. **12** (1978), no. 3, 68–69.

[BB] Sheila Brenner and M. C. R. Butler, *Generalizations of the Bernstein-Gel'fand-Ponomarev reflection functors*, Representation theory, II (Proc. Second Internat. Conf., Carleton Univ., Ottawa, Ont., 1979), Lecture Notes in Math., vol. 832, Springer, Berlin, 1980, pp. 103–169.

[BGP] I. N. Bernšteĭn, I. M. Gel'fand, and V. A. Ponomarev, *Coxeter functors, and Gabriel's theorem*, Uspehi Mat. Nauk **28** (1973), no. 2(170), 19–33.

[GP] I. M. Gel'fand, and V. A. Ponomarev, *Problems of linear algebra and classification of quadruples of subspaces in a finite-dimensional vector space*, Hilbert Space Operators. (Proc. Internat. Conf., Tihany, 1970), Colloquia Mathematica Societatis Janos Bolyai **5** (1970), 163-237.

[H] Dieter Happel, *On the derived category of a finite-dimensional algebra*, Comment. Math. Helv. **62** (1987), no. 3, 339–389.

[HR] Dieter Happel and Claus Michael Ringel, *Tilted algebras*, Trans. Amer. Math. Soc. **274** (1982), no. 2, 399–443.

[K] Bernhard Keller, *Deriving DG categories*, Ann. Sci. École Norm. Sup. (4) **27** (1994), no. 1, 63–102.

[M] Yoichi Miyashita, *Tilting modules of finite projective dimension*, Math. Z. **193** (1986), 113–146.

[R] Jeremy Rickard, *Morita theory for derived categories*, J. London Math. Soc. **39** (1989), 436–456.

[SS] Stefan Schwede and Brooke Shipley, *Stable model categories are categories of modules*, Topology **42** (2003), no. 1, 103–153.

2

Basic results of classical tilting theory

Lidia Angeleri Hügel, Dieter Happel, and Henning Krause

The aim of this introductory note is to present the main features of classical tilting theory, as originally developed in the early 1980s. We will state here the main definitions and properties, but refer for the proofs to the original articles, or leave it as an exercise to obtain those from the later developments on derived equivalences. We will follow here the account given in [3] or [4].

Let Λ be a finite dimensional algebra over a field k and let $\operatorname{mod}\Lambda$ be the category of finitely generated left Λ-modules. We denote by $D = \operatorname{Hom}_k(-, k)$ the duality between left and right Λ-modules.

Definition 1. A module $T \in \operatorname{mod}\Lambda$ is called a *tilting module* provided the following three conditions are satisfied:

(T1) the projective dimension of T is at most one,

(T2) $\operatorname{Ext}_\Lambda^1(T, T) = 0$, and

(T3) there exists an exact sequence $0 \to {}_\Lambda\Lambda \to T^0 \to T^1 \to 0$ such that each T^i is a direct summand of a direct sum of copies of T.

Note that cotilting modules are defined dually. Given a tilting module $T \in \operatorname{mod}\Lambda$, we can consider the endomorphism algebra $\Gamma = \operatorname{End}_\Lambda(T)$. One of the main objectives of tilting theory is to compare $\operatorname{mod}\Lambda$ with $\operatorname{mod}\Gamma$.

The basic general properties are summarized in the Theorem of Brenner and Butler. Before stating this explicitly we need some further notation and terminology. Given a tilting module $T \in \operatorname{mod}\Lambda$ with $\Gamma = \operatorname{End}_\Lambda(T)$, we consider the following torsion pairs in $\mathrm{mod}\,\Lambda$ and $\mathrm{mod}\,\Gamma$. We denote by $\mathcal{T}$ the full subcategory of $\mathrm{mod}\,\Lambda$ whose objects are generated by T and by $\mathcal{F}$ the full subcateory of $\mathrm{mod}\,\Lambda$ whose objects X satisfy $\mathrm{Hom}_\Lambda(T,X) = 0$. Then the pair $(\mathcal{T},\mathcal{F})$ is a torsion pair in $\mathrm{mod}\,\Lambda$. It follows easily that $\mathcal{T}$ coincides with the full subcategory of $\mathrm{mod}\,\Lambda$ whose objects X satisfy $\mathrm{Ext}^1_\Lambda(T,X) = 0$. We denote by $\mathcal{X}$ the full subcategory of $\mathrm{mod}\,\Gamma$ whose objects Y satisfy $T \otimes_\Gamma Y = 0$ and by $\mathcal{Y}$ the full subcategory of $\mathrm{mod}\,\Gamma$ whose objects Y satisfy $\mathrm{Tor}^\Gamma_1(T,Y) = 0$. Then the pair $(\mathcal{X},\mathcal{Y})$ is a torsion pair in $\mathrm{mod}\,\Gamma$.

Given a tilting module $T \in \mathrm{mod}\,\Lambda$ with $\Gamma = \mathrm{End}_\Lambda(T)$, we have two pairs of adjoint funcors between $\mathrm{mod}\,\Lambda$ and $\mathrm{mod}\,\Gamma$. The first pair is given by $F = \mathrm{Hom}_\Lambda(T,-)$ and $G = T \otimes_\Gamma -$ while the second is given by $F' = \mathrm{Ext}^1_\Lambda(T,-)$ and $G' = \mathrm{Tor}^\Gamma_1(T,-)$.

Theorem 1 (Brenner-Butler). *Let $T \in \mathrm{mod}\,\Lambda$ be a tilting module with $\Gamma = \mathrm{End}_\Lambda(T)$. Then the following holds.*

(1) *$_\Gamma D(T)$ is a cotilting module and $\Lambda \cong \mathrm{End}_\Gamma(D(T))$.*

(2) *The restriction of F and G to $\mathcal{T}$ and $\mathcal{Y}$ is a pair of inverse equivalences.*

(3) *The restriction of F' and G' to $\mathcal{F}$ and $\mathcal{X}$ is a pair of inverse equivalences.*

Further properties of the tilting situation can be obtained if additional assumptions are imposed. The first result we are going to discuss is related to the name 'tilting'. Recall that for an algebra Λ of finite global dimension the Grothendieck group $K_0(\Lambda)$ is endowed with a bilinear form. Viewing modules $X,Y \in \mathrm{mod}\,\Lambda$ as elements in $K_0(\Lambda)$, we set

$$\langle X,Y \rangle = \sum_{i \geq 0}(-1)^i \dim_k \mathrm{Ext}^i_\Lambda(X,Y).$$

Given a tilting module $T \in \mathrm{mod}\,\Lambda$ with $\Gamma = \mathrm{End}_\Lambda(T)$, we obtain a map $f\colon K_0(\Lambda) \to K_0(\Gamma)$ given by

$$f(X) = \dim_k \mathrm{Hom}_\Lambda(T,X) - \dim_k \mathrm{Ext}^1_\Lambda(T,X).$$

Theorem 2. *Let $T \in \mathrm{mod}\,\Lambda$ be a tilting module with $\Gamma = \mathrm{End}_\Lambda(T)$. If Λ has finite global dimension, then the following holds.*

(1) *The global dimension of Γ is finite.*

(2) *The map f is an isometry.*

In fact it can be shown that the absolute difference between the global dimensions of Λ and Γ is at most one.

As already mentioned in the introduction to this volume, the name 'tilting' comes from this observation. The positivity cone in the Grothendieck group $K_0(\Lambda)$ given by the standard coordinate axes is tilted when passing to $K_0(\Gamma)$ by the isometry given above.

The most important applications of tilting theory to the representation theory of finite dimensional algebras are obtained when assuming that Λ is a finite dimensional hereditary algebra. In this case, given a tilting module $T \in \text{mod} \, \Lambda$ with $\Gamma = \text{End}_\Lambda(T)$, we call Γ a *tilted algebra*.

Theorem 3. *Let T be a tilting module over a finite dimensional hereditary algebra Λ and $\Gamma = \text{End}_\Lambda(T)$. Then the following holds.*

(1) *The global dimension of Γ is at most two.*

(2) *The torsion pair $(\mathcal{X}, \mathcal{Y})$ on $\text{mod} \, \Gamma$ is a split pair.*

Let T be a tilting module over a finite dimensional hereditary algebra Λ with $\Gamma = \text{End}_\Lambda(T)$. We consider the injective cogenerator $D(\Lambda_\Lambda)$. Clearly it is contained in $\mathcal{T}$. It follows that $F(D(\Lambda_\Lambda))$ is a complete slice in $\text{mod} \, \Gamma$. We refer to [4] or to [2, Section 3] for a definition. But also the converse holds.

Theorem 4. *Let Γ be a finite dimensional algebra and let Σ be a complete slice in $\text{mod} \, \Gamma$, then there exists a finite dimensional hereditary algebra Λ and a tilting module T such that $\Gamma = \text{End}_\Lambda(T)$ and $\Sigma = F(D(\Lambda_\Lambda))$.*

Particular examples of tilted algebras can be obtained from the last theorem. We first recall some terminology. For a finite dimensional algebra Λ, a *path* in $\text{mod} \, \Lambda$ is a finite sequence of morphisms $X_0 \to X_1 \to \ldots \to X_m$ between indecomposable Λ-modules $X_0, \ldots, X_m$ such that each morphism is non-zero and not invertible. Further, an *oriented cycle* is a path $X_0 \to X_1 \to \ldots \to X_m$ where $X_0 \cong X_m$. If $\text{mod} \, \Lambda$ does not contain an oriented cycle, then the algebra Λ is called *representation directed*. Note that such an algebra is necessarily representation finite

[4]. Moreover, a Λ-module X is called *sincere* provided each simple Λ-module occurs as a composition factor of X. As an application of the characterization theorem above, one obtains the following.

Corollary 1. *Let Γ be a representation directed algebra with an indecomposable sincere Λ-module X. Then Γ is a tilted algebra.*

As a consequence, we see that the support algebra of an indecomposable module over a representation directed algebra is tilted.

Further applications of tilting theory to the representation theory of finite dimensional algebras are obtained for critical algebras with a preprojective component. It turns out that these are precisely the tame concealed algebras, that is, the tilted algebras which arise by considering endomorphism algebras of preprojective or of preinjective tilting modules over tame hereditary algebras. For more details and further applications we refer to [2].

References

[1] S. Brenner, M. C. R. Butler, *Generalizations of the Bernstein-Gelfand-Ponomarev reflection functors*, in Proceedings ICRA II, Ottawa 1979, Lectures Notes in Math. 832 (1980), 103–169.

[2] Th. Brüstle, *Classification of representation-finite algebras and their modules*, this volume.

[3] D. Happel, C. M. Ringel, *Tilted algebras*, Trans. Amer. Math. Soc. 274 (1982), 399–443.

[4] C. M. Ringel, *Tame algebras and integral quadratic forms*, Springer Lecture Notes in Mathematics 1099, Heidelberg (1984).

Lidia Angeleri Hügel
DICOM
Università degli Studi dell'Insubria
Via Mazzini 5
I-21100 Varese
Italy
Email: lidia.angeleri@uninsubria.it

Dieter Happel
Fakultät für Mathematik
Technische Universität Chemnitz

D-09107 Chemnitz
Germany
E-mail: happel@mathematik.tu-chemnitz.de

Henning Krause
Institut für Mathematik
Universität Paderborn
D-33095 Paderborn
Germany
E-mail: hkrause@math.uni-paderborn.de

3

Classification of representation-finite algebras and their modules

Thomas Brüstle

Abstract

We describe how tilting modules are used to classify the representation-finite algebras and their indecomposable modules.

1 Introduction

Probably the first appearance of tilting modules in representation theory of finite-dimensional algebras was in 1973 the use of reflection functors when Bernstein, Gelfand and Ponomarev [5] reproved Gabriel's classification of the representation-finite hereditary quiver algebras. Dlab and Ringel [18] extended in 1976 the use of reflection functors to arbitrary representation-finite hereditary algebras. Next, the concept of reflection functors has been generalized in 1979 by Auslander, Platzeck and Reiten [2] (they called it "Coxeter functors without diagrams"), and finally in 1980 by Brenner and Butler [13], who coined the term tilting and gave the first general definition of a tilting module, together with basic properties of tilting functors.

In a time where most people working with representation-finite algebras were knitting Auslander-Reiten sequences, this was a new approach: To study a class of modules which are given by abstract properties. Tilting modules have then been used very successfully by Bongartz [9] and by Happel and Vossieck [26] to find a far-reaching generalization of Gabriel's Theorem to representation-directed algebras, see Theorems 6 and 7 below.

The aim of these notes is to describe this powerful application of tilting theory in the classification of representation-finite algebras, and to add some more details on the representation-infinite case and the classification of tame algebras. These notes arose from two lectures which I gave at the meeting on 'Tilting Theory' at the Fraueninsel near Munich. I would like to thank the organizers of this conference for the opportunity to present these lectures.

2 Notation

Throughout, we fix an algebraically closed field k. We deal with finite-dimensional, associative algebras A over k and study the category $\mod A$ of finitely generated left $A-$modules. By a fundamental observation due to Gabriel, it is sufficient (up to Morita-equivalence) to consider algebras presented in the form $A = kQ/I$, where Q is a finite quiver, that is, an oriented graph, and I is an admissible ideal in the path algebra kQ. Here, an ideal I of kQ is said to be admissible if there is some number m such that $kQ_1^m \subseteq I \subseteq kQ_1^2$ where kQ_1 denotes the ideal generated by the arrows of Q.

We freely use the language of quivers and admissible ideals (generated by certain relations) here and refer to the textbooks [3, 24, 28] for more details. The set of vertices of Q is denoted by Q_0, and the set of arrows by Q_1. The vertices of Q are in bijection with the isomorphism classes of simple $A-$modules, and we denote by S_i a simple $A-$module corresponding to the vertex $i \in Q_0$. The arrows of Q encode extensions between the simple modules, in the way that $\dim_k \operatorname{Ext}_A^1(S_i, S_j)$ is the number of arrows in Q from i to j.

We consider in this note only algebras of the form $A = kQ/I$ with an admissible ideal I. These algebras are hereditary precisely when the ideal I is zero, thus A is the path algebra of a quiver Q. Each module M in $\mod A$ is given by a family $(M_i)_{i \in Q_0}$ of vector spaces and a family $(M_\alpha)_{\alpha \in Q_1}$ of linear maps such that the relations generating the ideal I of $A = kQ/I$ are satisfied. We recall from [10] that the Tits form of A is the quadratic form $q_A \colon \mathbb{Z}^{Q_0} \to \mathbb{Z}$ defined by the following formula:

$$q_A(x) = \sum_{i \in Q_0} x_i^2 - \sum_{(i \to j) \in Q_1} x_i x_j + \sum_{(i,j) \in Q_0} r(i,j) x_i x_j,$$

where $r(i,j) = |R \cap I(i,j)|$ for a minimal set of generators $R \subset \bigcup_{i,j \in Q_0} I(i,j)$ of the ideal I. Note that these numbers can also be interpreted as $r(i,j) = \dim_k \operatorname{Ext}_A^2(S_i, S_j)$. A positive root of q_A is a vector $x \in \mathbb{N}^{Q_0}$ such that $q_A(x) = 1$. An algebra A is said to be representation-finite if it admits only finitely many indecomposable modules up to isomorphism. The relation between the representation type of A and properties of q_A has been studied intensively in representation theory of finite-dimensional algebras. One of the starting points was the class of hereditary algebras.

We say that the quiver Q is a Dynkin quiver if the underlying graph of Q is a Dynkin graph, likewise for extended Dynkin quivers. In our context (working over an algebraically closed field), only the simply laced Dynkin diagrams of type $\mathbb{A}_n, \mathbb{D}_n$ and $\mathbb{E}_6, \mathbb{E}_7$ and $\mathbb{E}_8$ occur. The following theorem of Gabriel determines the precise relationship between the representation type of a quiver and its quadratic form:

Theorem 2.1 ([22]). *Let H be a hereditary algebra of the form $H = kQ$ where Q is a connected quiver. Then the following are equivalent:*

  (i) *H is representation-finite.*

  (ii) *The quiver Q is a Dynkin quiver of type $\mathbb{A}_n, \mathbb{D}_n, \mathbb{E}_6, \mathbb{E}_7$ or $\mathbb{E}_8$.*

  (iii) *The Tits form q_H is positive definite.*

Moreover, if H is representation-finite, then there is a bijection between the isomorphism classes of the indecomposable $H-$modules and the positive roots of the quadratic form q_H.

This result motivated Bernstein, Gelfand and Ponomarev [5] to use ideas from Lie theory: the Weyl group W associated to the Dynkin diagram is generated by the simple reflections. The roots may be viewed as the union of the $W-$orbits of the simple roots. With their reflection functors, they copied this behavior for representations. Auslander, Platzeck and Reiten [2] showed that these reflection functors are of the form $\operatorname{Hom}_A(T, -)$ where the $A-$module T is a direct sum $T = \oplus_{i \in Q_0} T_i$ where all but one T_i are indecomposable projective (and non-isomorphic), and the remaining summand T_j is obtained from the remaining indecomposable projective by a shift with the inverse Auslander-Reiten translation.

This fits into the general concept of a tilting functor $\mathrm{Hom}_A(T, -)$ where T is a tilting module. Among various versions of the notion of a tilting module, we are using here the original definition from [25]: a finitely generated A−module T is a tilting module if it satisfies the following conditions:

(i) $\mathrm{pd}\, T \leqslant 1$

(ii) $\mathrm{Ext}^1_A(T, T) = 0$

(iii) There is an exact sequence $0 \rightarrow A \rightarrow X_0 \rightarrow \cdots \rightarrow X_n \rightarrow 0$ with $X_i \in \mathrm{add}\, T$.

If $A = kQ$ is a hereditary algebra with a tilting module T, then the endomorphism ring $\mathrm{End}_A(T)$ is called a tilted algebra (of type Q). We will see below that the tilted algebras play a major role in the classification of representation-finite algebras.

We recall the definition of the Auslander-Reiten quiver Γ_A of the category $\mathrm{mod}\, A$ (see [3, 24, 28] for more details): It has as set of vertices the isoclasses $[X]$ of indecomposable modules X in $\mathrm{mod}\, A$. The number of arrows from $[X]$ to $[Y]$ is defined as $\dim_k \mathrm{rad}\,(X, Y)/\mathrm{rad}^2(X, Y)$, where rad denotes the radical of the category $\mathrm{mod}\, A$ (i.e., the ideal generated by the non-invertible maps between indecomposables). Note that Γ_A is a translation quiver with Auslander-Reiten translation τ.

3 Representation-finite algebras

Let $A = kQ/I$ be a representation-finite algebra. We are aiming at a generalization of Gabriel's theorem to algebras of arbitrary global dimension. An argument of Tits (see [10]) shows that for every representation-finite algebra A, the form q_A is weakly positive, that is, positive when evaluated on positive vectors.

The converse, however, is not true in general. Consider for example the algebra $A = kQ/I$ given by the quiver Q below and the ideal I which is generated by the relation $\xi\eta = 0$:

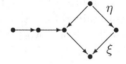

Then the algebra A has the same Tits form as a tilted algebra of type $\mathbb{D}_6$, thus q_A is weakly positive. But the universal covering of A contains a critical subcategory of type $\tilde{\mathbb{E}}_6$, therefore A is known to be representation-infinite.

So the representation-finite algebras are not characterized by the weak positivity of the Tits form, and it needs considerable effort to detect the representation-finite algebras. The main technique one uses to determine the representation type of a general algebra is covering theory. As this note is mainly concerned with tilting, we do not recall the concept of coverings and its use in representation theory here, but rather refer to [23, 7] for the details. We do however describe the effect that the use of covering spaces has to our question:

An A−module M is said to be sincere when $M_i \neq 0$ for all $i \in Q_0$. And the algebra A is called sincere if it admits a sincere indecomposable module. Moreover, a representation-finite algebra A is said to be representation-directed if there is no cycle $X_1 \rightarrow \cdots \rightarrow X_n \rightarrow X_1$ in the Auslander-Reiter quiver of A (see [28, 2.4(6) and (9)] for a more general definition). The main result which we need from covering theory is the following:

Theorem 3.1. *Each indecomposable module over every representation-finite algebra is obtained as a push-down of an indecomposable module over some sincere representation-directed algebra.*

The major part of this result is shown in the work of Bautista, Gabriel, Roiter and Salmeron [4] on multiplicative bases. They show there that a representation-finite algebra admits a universal covering $\tilde{A} \rightarrow A$ by a simply connected algebra $\tilde{A}$ (see [1, 30] for a discussion of simply connected algebras) except when char $k = 2$ and A contains a so-called penny-farthing. These are algebras whose quiver P has the following shape:

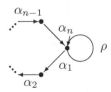

The ideal J of a penny-farthing kP/J is generated by one of the following two systems of relations (see also [8]) :

$$0 = \alpha_n \cdots \alpha_2 \alpha_1 - \rho^2 = \alpha_1 \alpha_n = \alpha_{f(i)} \cdots \alpha_1 \rho \alpha_n \cdots \alpha_{i+1} \text{ or}$$

$$0 = \alpha_n \cdots \alpha_2 \alpha_1 - \rho^2 = \alpha_1 \alpha_n - \alpha_1 \rho \alpha_n = \alpha_{f(i)} \cdots \alpha_1 \rho \alpha_n \cdots \alpha_{i+1},$$

where $f\colon \{1, 2, \ldots, n-1\} \to \{1, 2, \ldots, n\}$ is a non-decreasing function. However, a penny-farthing is not sincere: Every indecomposable module vanishes at some vertex. When studying indecomposable modules over an algebra $A = kQ/I$, however, it is sufficient to consider the case when A is sincere: Given any indecomposable module M, we may replace A by the algebra $B = eAe$ where $e = \sum_{M_i \neq 0} e_i$ (remember that e_i denotes the primitive idempotent of the algebra A associated to the vertex $i \in Q_0$). This algebra B is also called the support of the $A-$module M. By construction, the module M is non-zero at all vertices from B, hence sincere when considered as a $B-$module.

Thus, when studying indecomposable modules, one does not need to consider the penny-farthing. In fact, refining the methods from [4], Bongartz could show in [11] that no representation-finite algebra which contains a penny-farthing is sincere, so one can avoid any possible problem with covering theory. As a consequence, when studying indecomposable modules over representation-finite algebras, up to covering techniques it is sufficient to study modules over sincere representation-directed algebras. This is the point where tilting modules come into play:

Theorem 3.2 ([25]). *Let A be a sincere representation-directed algebra. Then $A = \operatorname{End}_H(T)$ where T is a tilting module over a hereditary algebra $H = kQ$ whose quiver Q is a tree.*

This theorem is one of the main contributions of tilting theory to the classification of representation-finite algebras, and we would like to recall the major steps of its proof: Denote by Γ_A the Auslander-Reiten quiver of A. By assumption, there exists a sincere indecomposable $A-$module M. We consider the set Σ of all vertices $X \in \Gamma_A$ such that there exists a path $X \to \cdots \to M$ in Γ_A and such that there is no path $X \to \cdots X_i \to X_{i+1} \to X_{i+2} \to \cdots \to M$ in Γ_A with $\tau X_{i+2} = X_i$ (remember that τ denotes the Auslander-Reiten translation). A typical example of such a set $\Sigma = \{X_1, \ldots, X_5, M\}$ is depicted below:

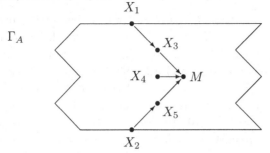

It is clear from the construction that Σ satisfies the properties of a slice in Γ_A, that is to say:

(a) $T = \bigoplus_{X \in \Sigma} X$ is a sincere module

(b) Σ is path-closed: Given any path $X_i \to \cdots \to Y \to \cdots \to X_j$ in Γ_A with X_i and X_j in Σ, then also Y belongs to Σ.

(c) Let $0 \to X \to Y \to Z \to 0$ be an Auslander-Reiten sequence. If Y has an indecomposable direct summand in Σ, then one of X and Z, but not both, belong to Σ.

Note that there exist different versions of the definition of a slice in the literature, and various names such as (complete) slice or (complete) section. The definition we are using here is equivalent to the one in [28, 4.2], and the last condition underlines the symmetry of the concept. If Σ is any slice in Γ_A, then $T = \bigoplus_{X \in \Sigma} X$ is a tilting module, and the algebra $H = \operatorname{End}_A(T)$ is hereditary where the quiver Q of the algebra $H = kQ$ is a tree. In the example for the set Σ above, the quiver of H is of type $\mathbb{E}_6$, its vertices correspond to the summands of Σ and all the arrows are reversed. For the rest of the proof we refer to [28, 4.2 (3)] where it is shown that every slice can be realized as the set of indecomposable injective modules over a hereditray algebra as above, thus the given algebra A is tilted.

The results discussed so far open the door to a classification of all sincere representation-directed algebras: Take a hereditary algebra $H = kQ$ (it is sufficient to consider the cases where Q is a tree), compute all tilting modules T of H and decide if $A = \operatorname{End}_H(T)$ is representation-finite. Of course, it is very difficult in general to compute all tilting modules. Moreover, to determine the representation type, we distinguish several cases: If H is representation-finite, then A is so as well ([28, 4.2 (1)]).

Consider now the algebras $H = kQ$ where Q is one of the extended Dynkin quivers $\tilde{\mathbb{A}}_n, \tilde{\mathbb{D}}_n, \tilde{\mathbb{E}}_6, \tilde{\mathbb{E}}_7$ or $\tilde{\mathbb{E}}_8$. Then the Auslander-Reiten Γ_H is well understood (see [28, 3.6]): It consists of a preprojective component $\mathcal{P}$, a preinjective component $\mathcal{I}$ and a $\mathbb{P}_1(k)$–family $\mathcal{T} = (\mathcal{T}_\lambda)_{\lambda \in k \cup \infty}$ of tubes $\mathcal{T}_\lambda$ of rank n_λ. Here a component of Γ_H is called preprojective if it contains no cyclic paths, has only a finite number of τ–orbits and each of those orbits contains one projective module. Preinjective components are defined likewise, requiring an injective instead of a projective module in every τ–orbit. Note that we are considering only finite-dimensional algebras here, so the cyclic orientation is excluded in case $Q = \tilde{\mathbb{A}}_n$. As a concrete example we sketch here the Auslander-Reiten quiver of the algebra $A = kQ$ where Q is an extended Dynkin diagram of type $\tilde{\mathbb{D}}_4$ (it is the four subspace quiver). Then Γ_H has the following form:

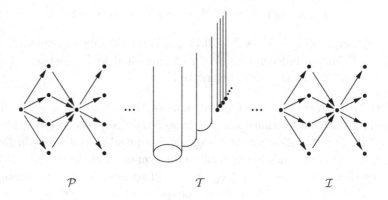

$$\mathcal{P} \qquad\qquad\qquad \mathcal{T} \qquad\qquad\qquad \mathcal{I}$$

Proposition 3.3. *Let $H = kQ$ where Q is an extended Dynkin quiver, and let T be a tilting module of H. Then $A = \text{End}_H(T)$ is representation-finite if and only if the module T contains at least one summand from $\mathcal{P}$ and at least one summand from $\mathcal{I}$.*

Unfortunately, the situation is rather complicated for quivers Q which are neither Dynkin nor extended Dynkin. Bongartz [8] showed that every sincere representation-directed algebra with more than a certain number n_v of vertices occurs in a list of 24 infinite families of sincere representation-directed algebras (such as the families $\mathbb{A}_n, \mathbb{D}_n$ for $n \in \mathbb{N}$). In fact, the optimal bound for n_v is 13, see [28]. The small sincere representation-directed algebras with at most 13 vertices were then classified by computer (see [19]), there are 16344 of them (sorted in

2242 diagrams) which do not belong to one of the 24 infinite families.

The list of the 16344 exceptional algebras is quite impossible to read, of course, even if they are presented very well in graphical form in [29]. The production of this list seems to be one of the starting points of the use of computers in representation theory, and Bongartz's first list has been checked and verified by Dräxler and others.

One should note that, even if the production of all sincere directing algebras is an almost impossible task, it follows at least a systematic approach given by tilting theory. Through Theorem 3.2 one also obtains information about the sincere representation-directed algebras which would be difficult to show directly, for instance the global dimension for all these algebras is bounded by 2 (as for all tilted algebras).

We illustrate by another example how useful the concept of Theorem 3.2 is: An algebra $A = kQ/I$ is called a tree algebra if the underlying graph of the quiver Q is a tree. A module M over a tree algebra is said to have a peak if there exists some vertex i of Q such that every arrow of Q pointing towards i is represented in M by an injective map, and every arrow of Q pointing away from i is represented in M by a surjection. Bongartz and Ringel show in [6] the following result:

Theorem 3.4. *Let $A = kQ/I$ be a representation-finite tree algebra. Then every indecomposable A−module has a peak.*

Of course, this theorem is proven using the approach of writing A as endomorphism ring of a tilting module over a hereditary algebra (while assuming that A is sincere). In fact, Bongartz was first working with lengthy lists of tree algebras, or as the authors say in their paper about their result: "This ... was conjectured and partially proved by the first author using a quite technical inductive argument", and later on in the paper: "The original proof of the theorem used this list. To convince the reader of the arising combinatorial difficulties, we give the list of all representation-finite tree algebras kT/R, such that $\dim U(x) \leqslant 4$ for all $x \in T$ and U indecomposable, and such there exists at least one sincere indecomposable V." What follows is a list of tree algebras which is longer than the proof of the above theorem. We believe this convinces every reader how useful tilting modules are.

4 Critical algebras

We did not discuss so far how one can test if a given algebra is representation-finite. The lengthy list of sincere representation-directed algebras is clearly not useful. It turns out that a good test is established by the minimal representation-infinite algebras: these are representation-infinite algebras such that every "smaller" algebra is representation-finite (we specify below how the term "small" is defined in this context). Surprisingly, tilting modules play an important role again: Let Q be an extended Dynkin quiver of type $\tilde{\mathbb{A}}_n, \tilde{\mathbb{D}}_n, \tilde{\mathbb{E}}_6, \tilde{\mathbb{E}}_7$ or $\tilde{\mathbb{E}}_8$, and let T be a tilting module over the algebra $H = kQ$. The algebra $A = \operatorname{End}_H(T)$ is called tame concealed (or critical) if all summands of T belong to the preprojective component $\mathcal{P}$ of Γ_H. The tame concealed algebras have been classified by Happel and Vossieck [26]. Their list is rather handy, there are 5 infinite families (such as $\tilde{\mathbb{A}}_n$ and $\tilde{\mathbb{D}}_n$, $n \in \mathbb{N}$) and a number of exceptional algebras, sorted in 134 frames.

In the representation-infinite situation, many algebras do not admit a simply connected covering, thus one has to restrict to certain well-behaved classes of algebras. The first appraoch is to consider those algebras which admit a preprojective component. The hereditary algebras given by an extended Dynkin quiver admit a preprojective component, as we discussed above, and also all tame concealed algebras enjoy this property.

Furthermore, we need to make precise which kind of minimality we are using: An algebra $A = kQ/I$ can also be viewed as a k–category whose set of objects is the set of idempotents e_i where i runs through the vertices of the quiver Q. The morphism spaces are identified with $e_j A e_i$. A (full) subcategory of A is then of the form $C = eAe$ where $e = \sum_{i \in J} e_i$ for some subset J of the vertices of Q. The following theorem explains why the tame concealed algebras are also called critical:

Theorem 4.1 ([26]). *Let A be an algebra which admits a preprojective component. Then A is representation-finite if and only if A does not contain a subcategory C which is a tame concealed algebra.*

The subcategory $C = eAe$ is said to be convex if the set of vertices J is path-closed (as in part (b) in the definition of a slice above). It is much easier in practice to search for subcategories of a given quiver algebra which are convex. The following theorem provides a criterion for representation-finiteness using this kind of minimality (while working with a slightly smaller class of algebras which excludes the algebras of the form $\widetilde{\mathbb{A}}_n$):

Theorem 4.2 ([9]). *Let A be an algebra which admits a simply connected preprojective component. Then A is representation-finite if and only if A does not contain a critical algebra C as convex subcategory.*

The two theorems above were obtained independently, and it turned out a posteriori that the corresponding lists of minimal representation-infinite algebras coincide (excluding $\widetilde{\mathbb{A}}_n$). Thus, it turns out that Bongartz's critical algebras are tame concealed (hence tilted), and that the tame concealed algebras classified by Happel and Vossieck are minimal representation-infinite with respect to convex subcategories.

It is this criterion together with covering theory which is mostly used when one likes to decide if a given algebra is representation-finite or not. In fact, the restriction of working with algebras that admit a preprojective component is justified by the previous use of covering theory: Given an algebra A, one tries to find a covering with directed quiver Q. If this does not exist, then either A is not sincere, or representation-infinite (see our previous discussions). If such a covering exists, one restricts to finite subcategories of the covering and continues as follows: In [20] there is an algorithm to decide if the given algebra admits a preprojective component. If not, the algebra is representation-infinite, if yes, the theorems above can be applied.

Besides the algorithmic method described in [20], several sufficient conditions have been found when an algebra admits a preprojective component. For instance, every directed, Schurian algebra admits a preprojective component if it satisfies a certain separation property (an algebra $A = kQ/I$ is said to be directed if Q has no oriented cycles, and it is said to be Schurian provided $\dim_k \operatorname{Hom}_A(P(x), P(y)) \leqslant 1$ for all $x, y \in Q_0$ where $P(x)$ denotes the indecomposable projective A–module associated to the vertex $x \in Q_0$).

Unfortunately, most of these concepts are motivated by techniques developed for studying representation-finite algebras. We consider finally a smaller class of algebras which is intensively studied in the representation-infinite cases: An algebra $A = kQ/I$ is strongly simply connected if it is directed and the first Hochschild cohomology $HH^1(C, C)$ vanishes for every convex subcategory C of A (see [30]). For instance, all tree algebras are strongly simply connected. One can show that every strongly simply connected algebra admits a preprojective component, thus the minimality criterion above applies to it, as well as the following generalization of Gabriels theorem on the representation-finite hereditary algebras:

Theorem 4.3 ([10]). *Let A be an algebra which admits a preprojective component. Then A is representation-finite precisely when the Tits form q_A is weakly positive and in this case, there is a bijection between the isomorphism classes of the indecomposable $A-$modules and the positive roots of q_A.*

5 Tame algebras

We finally discuss which of the previous results can be generalized to representation-infinite algebras. By Drozd's fundamental theorem, these algebras are divided into two disjoint classes, called tame and wild.

Theorem 5.1 ([21], [14]). *Every finite-dimensional algebra is either tame or wild, but not both.*

Here, an algebra A is called tame if there exists, for each dimension d, a parametrization of the indecomposable $d-$dimensional $A-$modules by a finite number of one-parameter families. The module category of a wild algebra, on the other hand, contains information about *all* indecomposable modules over *all* finite-dimensional algebras. Consequently, wild algebras admit families of indecomposable $A-$modules of a fixed dimension which depend on an arbitrarily high number of parameters. One of the main aims in representation theory of finite-dimensional algebras is to determine the borderline between the tame and wild algebras.

In contrast to the representation-finite case, it is no longer sufficient to apply covering theory and to study modules which are obtained via tilting from hereditary algebras: At first, the covering theory is much less applicable, and secondly, the sincere directed tame algebras are far from tilted. Consider, for instance, the algebras Δ_n (named "dancing girls" by S. Brenner) which are given by the quiver Q below, modulo the ideal I generated by all products $\alpha_{i+1}\alpha_i$:

These algebras Δ_n are tree algebras, thus covering theory does not apply to them and they also admit a preprojective component. Moreover, they are tame and sincere (they belong to a class called clannish algebras which is studied in [15]). But, the algebra Δ_n is of global dimension $n-1$, whereas tilted algebras have global dimension at most two.

However, one result still holds under certain restrictive conditions which are dictated by the weakness of covering theory: The minimal non-tame algebras are tilted. Consider the minimal wild hereditary tree algebras, given in the following figure, where in case $\tilde{\tilde{\mathbb{D}}}_n$, the graph has $n+2$ vertices and $4 \leqslant n \leqslant 8$.

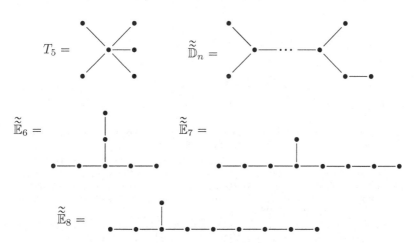

Analogous to the definition of the critical algebras, an algebra A is said to be hypercritical if it is of the form $A = \operatorname{End}_H(T)$ where H is a minimal wild hereditary tree algebra and T is a tilting $H-$module such that all summands of T belong to the preprojective component $\mathcal{P}$ of Γ_H. The hypercritical algebras have been completely classified ([27, 32, 31]), they are given by a list of 176 frames. Up to now, only for some narrow clases of algebras it could be shown that the hypercritical algebras serve as the minimal wild algebras. The theorem below is one of these results (it will be generalized in a forthcoming paper with Skowronski to the class of strongly simply connected algebras):

Theorem 5.2 ([12]). *Let A be a tree algebra. Then A is tame if and only if A contains no hypercritical convex subalgebra.*

The corresponding property for the Tits form of an algebra is the following: The quadratic form $q_A \colon \mathbb{Z}^{Q_0} \to \mathbb{Z}$ is called weakly non-negative if $q_A(x) \geqslant 0$ for all $x \in \mathbb{N}^{Q_0}$. Just as weak positivity in the representation-finite case, the weak non-negativity is a necessary condition for tameness, but it does not characterize the tame algebras:

Theorem 5.3 ([16]). *Let A be a tame algebra. Then the Tits form of A is weakly non-negative.*

For the class of strongly simply connected algebras, however, de la Peña [17] showed that weak non-negativity of q_A is characterized by the fact that the algebra A contains no hypercritical algebra. This is where tilting theory comes into play again.

References

[1] I. ASSEM, A. SKOWROŃSKI, On some classes of simply connected algebras, Proc. London Math. Soc. 56, (1988), 417–450.

[2] M. AUSLANDER, M.I. PLATZECK, I. REITEN, Coxeter functors without diagrams. Trans. Amer. Math. Soc. 250 (1979), 1–46.

[3] M. AUSLANDER, I. REITEN, S.O. SMALØ, Representation theory of Artin algebras. Cambridge University Press, Cambridge, 1995.

[4] R. BAUTISTA, P. GABRIEL, A.V. ROĬTER, L. SALMERON, Representation-finite algebras and multiplicative bases. Invent. Math. 81 (1985), no. 2, 217–285.

[5] I. N. BERNSTEIN, I. M. GELFAND, V. A. PONOMAREV, Coxeter functors, and Gabriel's theorem. (Russian) Uspehi Mat. Nauk 28 (1973), no. 2(170), 19–33.

[6] K. BONGARTZ, C.M. RINGEL, Representation-finite tree algebras. Representations of algebras (Puebla, 1980), pp. 39–54, Lecture Notes in Math., 903, Springer, Berlin-New York, 1981.

[7] K. BONGARTZ, P. GABRIEL, Covering spaces in representation-theory. Invent. Math. 65 (1981/82), no. 3, 331–378.

[8] K. BONGARTZ, Treue einfach zusammenhängende Algebren I, Comment. Math. Helvet. 57 (1982), 282-330.

[9] K. BONGARTZ, Critical simply connected algebras, Manuscripta Math. **46** (1984), 117–136.

[10] K. BONGARTZ, Algebras and quadratic forms. J. London Math. Soc. (2) 28 (1983), no. 3, 461–469.

[11] K. BONGARTZ, Indecomposables are standard, Comment. Math. Helv. 60 (1985), no. 3, 400–410.

[12] TH. BRÜSTLE, Tame tree algebras, J. reine angew. Math. 567 (2004), 51–98.

[13] S. BRENNER, M.C.R. BUTLER, Generalizations of the Bernstein-Gelfand-Ponomarev reflection functors. Representation theory, II (Proc. Second Internat. Conf., Carleton Univ., Ottawa, Ont., 1979), pp. 103–169, Lecture Notes in Math., 832, Springer, Berlin-New York, 1980.

[14] W. W. CRAWLEY-BOEVEY, On tame algebras and bocses, Proc. London Math. Soc. (3), 56(3):451–483, 1988.

[15] W. W. CRAWLEY-BOEVEY, Functorial filtrations II. Clans and the Gelfand problem, J. London Math. Soc. (2), 40 (1989), no.1, 9–30.

[16] J.A. DE LA PEÑA, On the representation type of one point extensions of tame concealed algebras, Manuscripta Math. 61 (1988), 183–194.

[17] J.A. DE LA PEÑA, Algebras with hypercritical Tits form, in: Topics in algebra, Part 1 (Warsaw, 1988), 353–369, PWN, Warsaw, 1990.

[18] V. DLAB, C.M. RINGEL, Indecomposable representations of graphs and algebras. Mem. Amer. Math. Soc. 6 (1976), no. 173.

[19] P. DRÄXLER, Aufrichtige gerichtete Ausnahmealgebren, Bayreuth. Math. Schr. No. 24 (1989), 191 pp.

[20] P. DRÄXLER, K. KÖGLER, An algorithm for finding all preprojective components of the Auslander-Reiten quiver, Math. Comp. 71 (2002), no. 238, 743–759.

[21] JU. A. DROZD, Tame and wild matrix problems, In *Representation theory, II (Proc. Second Internat. Conf., Carleton Univ., Ottawa, Ont., 1979)*, pages 242–258. Springer, Berlin, 1980.

[22] P. GABRIEL, Unzerlegbare Darstellungen I, Manuscripta Math. **6** (1972), 71–103.

[23] P. GABRIEL, The universal cover of a representation-finite algebra. In *Representations of algebras (Puebla, 1980)*, 68–105, Lecture Notes Math. 903, Springer,Berlin, 1981.

[24] P. GABRIEL, A. V. ROĬTER, Representations of finite-dimensional algebras. In Algebra, VIII, pages 1–177. Springer, Berlin, 1992. With a chapter by B. Keller.

[25] D. HAPPEL, C.M. RINGEL, Tilted algebras. Trans. Amer. Math. Soc. 274 (1982), no. 2, 399–443.

[26] D. HAPPEL, D. VOSSIECK, Minimal algebras of infinite representation type with preprojective component, Manuscripta Math. **42** (1983), no. 2-3, 221–243.

[27] M. LERSCH, Minimal wilde Algebren, Diplomarbeit, Düsseldorf (1987).

[28] C.M. RINGEL, Tame algebras and integral quadratic forms, Springer-Verlag, Berlin, 1984.

[29] A. ROGAT, TH. TESCHE, The Gabriel quivers of the sincere simply connected algebras, SFB-preprint, Ergänzungsreihe 93-005, Bielefeld 1993.

[30] A. SKOWROŃSKI, Simply connected algebras and Hochschild cohomologies, in: Representations of Algebras, CMS Conference Proc. 31, (1993), 431–447.

[31] L. UNGER, The concealed algebras of the minimal wild, hereditary algebras, Bayreuth. Math. Schr. No. 31, (1990), 145–154.

[32] J. WITTMANN, *Verkleidete zahme und minimal wilde Algebren*, Diplomarbeit, Bayreuth (1990).

Thomas Brüstle

Bishop's University and Université de Sherbrooke

4

A spectral sequence analysis of classical tilting functors

Sheila Brenner[1] and M. C. R. Butler

1 Introduction

Let A be a ring, $T = T_A$ be a tilting module of finite projective dimension, t, in the sense of [5] and [9] - spelt out in detail in Section 2, below - and $B = \mathrm{End}_A(T)$. This article contains some spectral sequences which provide a systematic framework for studying the tilting functors

$$F = \mathrm{Hom}_A(T, -) \qquad \text{and} \qquad G = - \otimes_B T,$$

their derived functors $R^n F = \mathrm{Ext}_A^n(T, -)$ and $L_n G = \mathrm{Tor}_n^B(-, T)$, and the associated filtrations of finite length $t+1$ on A- and B-modules. The context is strictly that of classical homological algebra of modules. For a far more general construction covering, for example, equivalences of derived categories given by tilting complexes, see Section 7 of Bernhard Keller's article [7] in this volume.

The spectral sequences in question seem first to have been written down, but nowhere published, by Dieter Vossieck in the mid-1980's, and were re-discovered by the authors during the summer of 2002 whilst preparing the talk for the conference 'Twenty Years of Tilting Theory' at Chiemsee in November 2002 on which this article is based. After that talk, Helmut Lenzing mentioned Vossieck's work, and kindly supplied a copy of his notes of a lecture in July 1986 by Vossieck at the University of Paderborn entitled *Tilting theory, derived categories and spectral sequences*. The main part of this lecture gave an account of the then recently proved theorem of Happel, [5], and Cline-Parshall-Scott, [4], that F and G

---

1 Sheila Brenner sadly died on October 10, 2002, so the composition of this article has been the responsibility of the second author.

induce inverse equivalences of the derived categories of A and B, but in the last two sections Vossieck briefly described the spectral sequences and filtration formulae which are stated and proved in the main part, Section 3, of this article.

Section 3 is preceded by a brief derivation in Section 2 of the main properties of tilting modules, and followed by two short sections of older and newer applications.

Conventions   Throughout this note, rings are associative and have identities and modules are unital. For a ring A, the notations Mod-A, mod-A, Proj-A and proj-A denote the categories of all right A-modules, all finitely presented right A-modules, all projective right A-modules and all finitely generated projective right A-modules, respectively; the corresponding categories of left A-modules have A as a prefix. For a right A-module X (left A-module Y), we write add-X (Y-add) for the category of all modules isomorphic to direct summands of *finite* direct sums of copies of X (Y). Finally, for any category $\mathcal{X}$ of modules, we say that a *bounded* complex

$$X^\bullet : \qquad \cdots \to X^n \to X^{n+1} \to \cdots ,$$

which has terms in $\mathcal{X}$ and is acyclic except at $n = 0$, is *a finite resolution of $H^0(X^\bullet)$ in $\mathcal{X}$* if $X^n = 0$ for all $n > 0$, and *a finite coresolution of $H^0(X^\bullet)$ in $\mathcal{X}$* if $X^n = 0$ for all $n < 0$.

2 Tilting modules

This section contains the definition and basic properties of the *tilting modules* to be used throughout this article for the construction of spectral sequences. For a fuller more analytic discussion, we refer to the original papers [4, 5, 9].

Definition 2.1. Let A be a ring. The right A-module, T, is called *a tilting module* if it satisfies the following *tilting conditions*:

TC1   there is a finite resolution, $P^\bullet$, of T_A in proj-A;

TC2   there is a finite coresolution, $T^\bullet$, of A_A in add-T;

TC3 T_A has no self-extensions, that is,

$$\operatorname{Ext}_A^n(T, T) = 0 \qquad \text{for all} \qquad n \geqslant 1.$$

Given such a tilting module, $T = T_A$, let

$$B = \mathrm{End}_A(T).$$

Then $T = {}_BT_A$ is a B, A-bimodule, which may be used to define, for each right A-module X, a left B-module $^*X = \mathrm{Hom}_A(X, T)$, and for each left B-module Y, a right A-module $Y^* = \mathrm{Hom}_B(Y, T)$. We call $^*X$ and Y^* the *T-duals of X and Y*, respectively, and shall make use of the usual double duality map

$$\mathrm{e}_X \colon X \to (^*X)^* = \mathrm{Hom}_B(\mathrm{Hom}_A(X, T_A), {}_BT)$$

in Mod-A. Clearly e_T is an isomorphism, and so also then is e_X for every X in add-T. The following proposition asserts *interalia* that e_X is an isomorphism for every X in proj-A.

Proposition 2.2. *Let the right A-module T_A be a tilting module, with finite resolution $P^\bullet$ in proj-A and finite coresolution $T^\bullet$ in add-T. Then:*

(1) *The left B-module ${}_BT$ is a tilting module, the T-dual complexes $^*(T^\bullet)$ and $^*(P^\bullet)$ being, respectively, a finite resolution of ${}_BT$ in B-proj and a finite coresolution of ${}_BB$ in T-add.*

(2) *The duality map $\mathrm{e}_X \colon X \to (^*X)^*$ is an isomorphism for modules X in add-$(T \oplus A)_A$. In particular, the complex morphisms $\mathrm{e}_{T^\bullet}$ and $\mathrm{e}_{P^\bullet}$ are isomorphisms, as also is the natural ring homomorphism of A into $\mathrm{End}_B(T)$.*

The proof makes use of the following easy lemma.

Lemma 2.3. *Let $X^\bullet$ be an acyclic bounded above complex and Y be a module such that, for all $m \geqslant 1$, $\mathrm{Ext}^m(X^\bullet, Y) = 0$. Then $\mathrm{Hom}(X^\bullet, Y)$ is acyclic.*

Proof the lemma. We may assume that $X^n \neq 0$ implies $n \leqslant 0$. Let $Z^n = \mathrm{Im}\,(X^{n-1} \to X^n)$, so that $Z^0 = X^0$. For all $m \geqslant 1$ and $n \leqslant 0$, we have $\mathrm{Ext}^m(Z^{n-1}, Y) \cong \mathrm{Ext}^{m+1}(Z^n, Y)$, from which it follows that $\mathrm{Ext}^m(Z^n, Y) = 0$ for all such m and n. Hence, for each $n \leqslant 0$, the sequence $0 \to \mathrm{Hom}(Z^n, Y) \to \mathrm{Hom}(X^{n-1}, Y) \to \mathrm{Hom}(Z^{n-1}, Y) \to 0$ is exact, and the lemma follows by splicing these sequences together. $\square$

Proof of the proposition. We use the lemma to show that ${}_BT$ satisfies the conditions TC1 and TC2 for being a tilting module. First, apply the lemma with $Y = T$ and $X^{\bullet}$ the coaugmented complex

$$\cdots 0 \to A_A \to T^0 \to T^1 \to \cdots$$

determined by $T^{\bullet}$. By TC3 for T_A and the projectivity of A_A, the conditions of the lemma are satisfied, so since $^*A_A = {}_BT$ and $^*(\text{add}-T) = B-\text{proj}$, it follows that $^*(T^{\bullet})$ is a finite resolution of $_BT$ in B-proj. Next, the conditions of the lemma are again satisfied on taking $Y = T$ but $X^{\bullet}$ to be the augmented complex

$$\cdots \to P^{-1} \to P^0 \to T_A \to 0 \to \cdots$$

determined by $P^{\bullet}$. Since $^*T_A = {}_BB$ and $^*(\text{proj-}A) = T$-add, it follows that $^*(P^{\bullet})$ is a finite coresolution of $_BB$ in T-add.

To verify TC3 for $_BT$, we use the projective resolution $^*(T^{\bullet})$ to calculate the self-extensions of $_BT$ as the homology of the complex $(^*(T^{\bullet}))^* = \text{Hom}_B(^*(T^{\bullet}), {}_BT)$. As already noted, it is clear that for $X = T_A$ and, hence for any X in add-T, the duality map $e_X \colon X \to (^*X)^*$ is an isomorphism. Thus $e_{T^{\bullet}} \colon T^{\bullet} \to (^*(T^{\bullet}))^*$ is an isomorphism of complexes, so induces isomorphisms of homology. However, in degree $n \geqslant 1$, $H^n(T^{\bullet}) = 0$, so that $\text{Ext}_B^n(T, T) = 0$. This verifies TC3 and completes the proof of part (1) of the proposition. In degree 0, it shows that the map $e_A \colon A = H^0(T^{\bullet}) \to \text{End}_B(T)$ is an isomorphism – so that T is indeed a balanced B, A-bimodule – and it then follows that e_X is an isomorphism for any X in proj-A, as also then is the complex morphism $e_{P^{\bullet}}$. This completes the proof of the second part of the proposition. $\quad\square$

Remark 2.4. The following lemma and corollary - given by Dieter Happel in [5] - show that the finite resolutions and coresolutions involved in Proposition 2.2 may all be chosen to have the same length.

Lemma 2.5. *Let T be a module of projective dimension at most t, with no self-extensions. Then any module X possessing a finite coresolution in* add-T *has such a coresolution $X^{\bullet}$ in which $X^n = 0$ for all $n > t$.*

Corollary 2.6. *Let T_A be a tilting module over the ring A, and $B = \text{End}_A(T)$. Then T_A and $_BT$ have the same finite projective dimensions.*

3 Tilting functors, spectral sequences and filtrations

From now on, fix a ring A, a tilting module T_A, and the bounded complexes $P^\bullet$ and $T^\bullet$ required by TC1 and TC2, respectively. Set $B = \mathrm{End}_A(T)$. By Proposition 2.2, $_BT$ is a tilting module and the complexes $^*(T^\bullet)$ and $^*(P^\bullet)$ are those required by the left B-module versions of TC1 and TC2. We shall further assume, as Remark 2.4 allows, that

$$P^{-n} = 0 = T^n \qquad \text{for all} \qquad n > t,$$

where $t = \mathrm{pd}\, T_A$ denotes the finite projective dimension of T_A.

The bimodule $_BT_A$ determines a pair of adjoint functors

$$F = \mathrm{Hom}_A(T, -)\colon \mathrm{Mod}\text{-}A \to \mathrm{Mod}\text{-}B$$

and

$$G = - \otimes_B T \colon \mathrm{Mod}\text{-}B \to \mathrm{Mod}\text{-}A$$

and in this section we set up spectral sequences analysing the homological properties of the composite functors GF and FG, and study the filtrations of length $t+1$ on both A- and B-modules canonically induced by the tilting theory. The use of spectral sequences for studying composite functors goes back a long time. It is the last topic in the famous book [3] of Cartan and Eilenberg, and is further developed in Section 5.8, entitled Grothendieck Spectral Sequences, of Weibel's book [13], with several examples. We conclude this section with a Remark briefly summarising the Grothendieck-Roos duality theory, [11], for finitely generated modules over regular rings, which was explicitly discussed in parallel with tilting theory in Vossieck's Paderborn lecture and, in the derived category setting, in his joint paper [8] with Keller.

The analysis of course involves the *derived functors*

$$\mathrm{R}^n F = \mathrm{Ext}_A^n(T, -) \qquad \text{and} \qquad \mathrm{L}_n G = \mathrm{Tor}_n^B(-, T)$$

of F and G, as was already the case in the early papers [1, 2, 6] (in the case $t = 1$) and [5, 9]. Although the term *tilting functor* was first used in [2] for F and for G, it seems appropriate to extend it to include any of their derived functors. Their number is essentially finite since

$$\mathrm{R}^n F = 0 = \mathrm{L}_n G \quad \text{for all} \quad n > t.$$

We start by noting a trivial but important lemma.

Lemma 3.1.   (1) *For modules M_A and M'_A there is a natural map*

$$F(M) \otimes_B {}^*M' \to \mathrm{Hom}_A(M', M), \qquad f \otimes_B \phi \mapsto (x' \mapsto f(\phi(x'))),$$

which is an isomorphism when M'_A is in add-T.

(2) *For modules M_A and N_B, there is a natural map*

$$N \otimes_B {}^*M \to \mathrm{Hom}_A(M, G(N)), \qquad y \otimes_B \theta \mapsto (x \mapsto y \otimes_B \theta(x)),$$

which is an isomorphism when M_A is in proj-A.

In the next two theorems, use is made of the standard notations $\mathrm{I}^{\bullet\bullet}_{\bullet}$ and $\mathrm{II}^{\bullet\bullet}_{\bullet}$ for the terms of the spectral sequences determined by the column and row filtrations (respectively) of the single complex associated with a double complex. Also, we use the term *support of a spectral sequence* $E^{\bullet\bullet}_{\bullet}$ to mean the set of points $(p, q) \in \mathbb{Z} \times \mathbb{Z}$ such that $E^{pq}_r \neq 0$ for some $r \geq 2$.

Theorem 3.2. *For each right A-module M, there is a convergent spectral sequence*

$$\mathrm{II}^{p,q}_2 = \mathrm{II}^{p,q}_2(M) \overset{q}{\Longrightarrow} \mathrm{H}^{\bullet}(M)$$

in Mod-A *in which*

$$\mathrm{H}^0(M) = M \qquad and \qquad \mathrm{H}^n(M) = 0 \quad for \quad n \neq 0,$$

and

$$\mathrm{II}^{p,q}_2 = (\mathrm{L}_{-q}G \circ \mathrm{R}^p F)(M) = \mathrm{Tor}^B_{-q}(\mathrm{Ext}^p_A(T, M), T),$$

so that its support is in the 4-th quadrant square in which $0 \leq p \leq t$ and $-t \leq q \leq 0$. Hence

$$\mathrm{II}_{t+1} = \mathrm{II}_\infty,$$

and there is an induced filtration on $M = \mathrm{H}^0(M)$,

$$M = M(t) \supset M(t-1) \supset \cdots \supset M(0) \supset M(-1) = 0,$$

with filtration factors

$$M(q)/M(q-1) = \mathrm{II}^{q,-q}_{t+1},$$

whilst

$$\mathrm{II}^{p,q}_{t+1} = 0 \quad for \quad p \neq -q.$$

The formulae for the terms $M(n)$ of the filtration require some more notation.

Notation 3.3. Let $A^n = \mathrm{Ker}\,(T^n \to T^{n+1})$, so that $A^n = 0$ unless $0 \leqslant n \leqslant t$; also $A^0 = A$ and $A^t = T^t$. Let $M \in$ Mod-A. From the short exact sequences $0 \to A^n \to T^n \to A^{n+1} \to 0$, we obtain connecting homomorphisms,

$$\sigma^n \colon \mathrm{Ext}_A^n(A^n, M) \;\to\; \mathrm{Ext}_A^{n+1}(A^{n+1}, M),$$

whose composite $\sigma^n \cdots \sigma^1 \sigma^0$ is a map

$$\pi^n \colon M \longrightarrow \mathrm{Ext}_A^{n+1}(A^{n+1}, M).$$

Remark 3.4. The connecting homomorphism σ^n may be viewed as splicing n-fold extensions of M by A^n with the short exact sequence $0 \to A^n \to T^n \to A^{n+1} \to 0$. It then follows by an easy induction that, for each element $m \in M$, the Yoneda class of $\pi^n(m)$ in $\mathrm{Ext}_A^{n+1}(A^{n+1}, M)$ is represented by the pushout of the exact sequence

$$0 \to A \to T^0 \to \cdots \to T^n \to A^{n+1} \to 0$$

along the map $a \mapsto ma$ of A into M.

Proposition 3.5. *The terms of the filtration of M in Theorem 3.2 are given by*

$$M(n) \;=\; \mathrm{Ker}\,(\pi^n).$$

Proof of Theorem 3.2.   Choose any injective resolution,

$$I^\bullet : \quad \cdots \to 0 \to I^0 \to I^1 \to I^2 \to \cdots ,$$

of the right A-module M. Then Lemma 3.1, (1), yields two different formulae for the double complex

$$C^{\bullet,\bullet} = F(I^\bullet) \otimes_B {}^*(T^\bullet) \cong \mathrm{Hom}_A(T^\bullet, I^\bullet),$$

and its terms,

$$C^{p,q} = F(I^p) \otimes_B {}^*(T^{-q}) \cong \mathrm{Hom}_A(T^{-q}, I^p),$$

are 0 except when $p \geqslant 0$ and $-t \leqslant q \leqslant 0$, thus when (p, q) is in the fourth quadrant. The associated single complex $C^\bullet$ has general term

$$C^n \;=\; \oplus_{p+q=n} C^{p,q},$$

which is 0 unless $n \geqslant -t$, and is always a direct sum of at most $t+1$ terms. Hence its two spectral sequences do converge to $\mathrm{H}^\bullet(M) := \mathrm{H}^\bullet(C^\bullet)$.

The first spectral sequence collapses. Indeed, the injectivity of each I^p shows that

$$\mathrm{I}_1^{p,q} = \mathrm{H}_{\mathrm{II}}^q(C^{p\bullet}) = \mathrm{H}_{\mathrm{II}}^q(\mathrm{Hom}_A(T^\bullet, I^p)) = \mathrm{Hom}_A(\mathrm{H}^q(T^\bullet), I^p),$$

and then the fact that $T^\bullet$ is a coresolution of A implies that $\mathrm{I}_1^{\bullet q} = 0$ for $q \neq 0$ and that $\mathrm{I}_1^{\bullet 0} = I^\bullet$. It further follows that $\mathrm{I}_2^{\bullet q} = 0$ for $q \neq 0$ and for $p \neq 0$, whereas $\mathrm{I}_2^{00} = M$ because $I^\bullet$ is an injective resolution of M. In particular, this implies that $\mathrm{H}^n(M) = 0$ for $n \neq 0$, whereas $\mathrm{H}^0(M) = M$.

We next calculate the first two terms of the second spectral sequence. Since the terms of $^*(T^\bullet)$ are projective left B-modules, the term $\mathrm{II}_1^{p\bullet}$ is given by

$$\mathrm{H}_\mathrm{I}^p(F(I^\bullet) \otimes_B {}^*(T^\bullet)) = \mathrm{H}_\mathrm{I}^p(F(I^\bullet)) \otimes_B {}^*(T^\bullet) = \mathrm{R}^p F(M) \otimes_B {}^*(T^\bullet).$$

But $^*(T^\bullet)$ is a projective resolution of $_B T$, $G = - \otimes_B T$, and the differentials in II_1 are induced by the second differential in $C^{\bullet\bullet}$, that is, by the differential in $^*(T^\bullet)$. Therefore,

$$\mathrm{II}_2^{p,q} = (\mathrm{L}_{-q}G)((\mathrm{R}^p F)(M)),$$

and this formula shows that the spectral sequence has its support inside the given square of side length $t+1$. Since the differentials in the r-th page of the spectral sequence increase the filtration index, q, by r, they all vanish when $r \geqslant t+1$, so that $\mathrm{II}_{t+1} = \mathrm{II}_\infty$. Now the remaining statements in the theorem follow from the fact that the terms $\mathrm{II}_\infty^{p,q}$ are the factors of the filtration of $\mathrm{H}^{p+q}(M)$ induced by the row filtration of the complex $C^{\bullet,\bullet}$. □

For the proof of Proposition 3.5 we need to recall the definition of the filtration of $\mathrm{H}^\bullet(M) = \mathrm{H}^\bullet(C^\bullet)$. The r-th row filtration term of $C^\bullet$ is

$$\mathrm{F}^r(C^\bullet) := \bigoplus_{p,q:q \geqslant r} C^{p,q},$$

– which we often abbreviate to $\mathrm{F}^r C$ – so that

$$\cdots = \mathrm{F}^{-t}C = C^\bullet \supset \mathrm{F}^{-t+1}C \supset \cdots \supset \mathrm{F}^0 C \supset 0 = \mathrm{F}^1 C = \cdots$$

is a decreasing sequence of subcomplexes of $C^\bullet$ – often abbreviated to

C in the next few lines. The images

$$\mathrm{F}^r\mathrm{H}^\bullet(M) := \operatorname{Im}\left[\mathrm{H}^\bullet(\mathrm{F}^r C) \to \mathrm{H}^\bullet(C)\right]$$

of the induced maps of homology are the terms of the required decreasing filtration of $\mathrm{H}^\bullet(M)$, though we shall actually use the alternative formula

$$\mathrm{F}^r\mathrm{H}^\bullet(M) = \operatorname{Ker}\left[H^\bullet(C) \to \mathrm{H}^\bullet(C/\mathrm{F}^r C)\right]$$

obtained from the exactness of the homology sequence of the exact sequence of complexes

$$0 \to \mathrm{F}^r C \to C \to C/\mathrm{F}^r C \to 0.$$

The proposition is an easy consequence of the following lemma.

Lemma 3.6. *For all $p \geqslant 0$ and $n \geqslant 0$, there is a commutative square*

$$
\begin{array}{ccc}
\mathrm{H}^p(C/\mathrm{F}^{-(n-1)}C) & \xrightarrow{\ \theta^{n-1}\ } & \mathrm{Ext}_A^{p+n}(A^n, M) \\
\ \downarrow{\nu^n} & & \ \downarrow{\sigma^n} \\
\mathrm{H}^p(C/\mathrm{F}^{-n}C) & \xrightarrow{\ \theta^n\ } & \mathrm{Ext}_A^{p+n}(A^{n+1}, M),
\end{array}
$$

in which ν^n is induced by the natural projection $p^n \colon C/\mathrm{F}^{-(n-1)}C \to C/\mathrm{F}^{-n}C$, σ^n is the connecting homomorphism induced by the exact sequence $0 \to A^n \to T^n \to A^{n+1} \to 0$, and the maps θ^{n-1} and θ^n are natural isomorphisms.

Proof. First, observe that a p-cocycle in $C/\mathrm{F}^{-(n-1)}C$ is a sequence $f^{(n)} = (f^i)_{i \geqslant n}$ of maps $f^i \in \operatorname{Hom}_A(T^i, I^{p+i})$ such that the diagram

$$
\begin{array}{ccccccccc}
T^n & \to & T^{n+1} & \to & \cdots & \to & T^t & \to & 0 & \to \cdots \\
\downarrow{f^n} & & \downarrow{f^{n+1}} & & & & \downarrow{f^t} & & \downarrow{f^{t+1}} \\
I^{p+n} & \to & I^{p+n+1} & \to & \cdots & \to & I^{p+t} & \to & I^{p+t+1} & \to \cdots
\end{array}
$$

commutes. Its natural image under p^n in $C/\mathrm{F}^n C$ is the sequence $f^{(n+1)} = (f^i)_{i \geqslant n+1}$, which is again a p-cocycle; this induces the map ν^n in the lemma. On the other hand, the commutativity of the above diagram implies that the restriction $f^n|A^n \colon A^n \to I^{p+n}$ of f^n to A^n gives a $p + n$-cocycle in the complex $\operatorname{Hom}_A(A^n, I^\bullet)$, and so determines an element of $\mathrm{Ext}_A^{p+n}(A^n, M)$ which - abusing notation - we denote by $\theta^{n-1}(f^{(n)})$. Similarly, $\theta^n(f^{(n+1)})$ in $\mathrm{Ext}_A^{p+n+1}(A^{n+1}, M)$ is represented by the $p + n + 1$-cocycle $f^{n+1}|A^{n+1}$ in the complex $\operatorname{Hom}_A(A^{n+1}, I^\bullet)$.

Now the proof of commutativity of the square in the lemma follows immediately from the commutativity of the square in the diagram

$$
\begin{array}{ccccccc}
0 & \to & A^n & \to & T^n & \to & A^{n+1} & \to & 0 \\
& & & & f^n \downarrow & & \downarrow f^{n+1}|A^{n+1} & & \\
& & & & I^{p+n} & \to & I^{p+n+1} & &
\end{array}
$$

There are two ways of showing that the horizontal maps are isomorphisms. The formula given above for $\theta^{n-1}(f^{(n)})$ can easily be used to prove "by hand" that θ^{n-1} is bijective. Alternatively, just as was done for the complex C in the proof of Theorem 3.2, one can show that the *first* spectral sequence for the complex $C/F^{-(n-1)}C$ collapses at the I_2 page, on which the only non-zero terms are $I_2^{p+n,-n} \cong \operatorname{Ext}_A^p(A^n, M)$, in the $(-n)$-th row, the isomorphism being induced by θ^{n-1}.   □

Proof of Proposition 3.5. Recall that, for $n \geqslant 0$, the $(-n)$-th filtration term of $\mathrm{H}^0(C) \cong M$ is given by

$$
\mathrm{F}^{-n}\mathrm{H}^0(C) = \operatorname{Ker}\left[\mathrm{H}^0(C) \to \mathrm{H}^0(C/\mathrm{F}^{-n}C)\right]
$$

where, since $C = C/\mathrm{F}^{-1}C$, the map is the composite of the maps $\nu^0, \ldots, \nu^n$ in the lemma, with $p = 0$. Since π^n is the composite of the maps $\sigma^0, \ldots, \sigma^n$ in the lemma, with $p = 0$, the proposition follows immediately.   □

We now formulate the corresponding theorem and proposition obtained by an appropriate spectral sequence analysis of the endofunctor FG of Mod-B, but shall only briefly discuss their proofs.

Theorem 3.7. *For each right B-module N there is a convergent spectral sequence*

$$
I_2^{p,q} \stackrel{p}{\Longrightarrow} \mathrm{H}^\bullet(N)
$$

in Mod-B *in which*

$$
\mathrm{H}^0(N) = N \qquad and \qquad \mathrm{H}^n(N) = 0 \quad for \quad n \neq 0,
$$

and

$$
I_2^{p,q} = (\mathrm{R}^p F \circ \mathrm{L}_{-q}G)(N) = \operatorname{Ext}_A^p(T, \operatorname{Tor}_{-q}^B(N,T)),
$$

so that its support is in the 4-th quadrant square in which $0 \leqslant p \leqslant t$ and $-t \leqslant q \leqslant 0$. Hence

$$
I_{t+1} = I_\infty,
$$

and there is an induced filtration on $N = \mathrm{H}^0(N)$,

$$N = N(0) \supset N(1) \cdots N(t) \supset N(t+1) = 0,$$

with filtration factors

$$N(p)/N(p+1) = \mathrm{I}_{t+1}^{p,-p},$$

whilst

$$\mathrm{I}_{t+1}^{p,q} = 0 \quad \text{for} \quad q \neq -p.$$

Notation 3.8. Let $B^n = \mathrm{Ker}\,[{}^*(P^{-n}) \to {}^*(P^{-n-1})]$ be the n-th kernel in the coresolution ${}^*(P^\bullet)$ of ${}_B B$ in add-${}_B T$, so that $B^n = 0$ unless $0 \leqslant n \leqslant t$, $B^0 = B$ and $B^t = {}^*(P^{-t})$. Let $N \in \mathrm{Mod} - B$. From the short exact sequences $0 \to B^{n-1} \to {}^*(P^{-n+1}) \to B^n \to 0$ with $n \geqslant 1$, we obtain connecting homomorphisms

$$\rho_{n-1} : \mathrm{Tor}_n^B(N, B^n) \to \mathrm{Tor}_{n-1}^B(N, B^{n-1}),$$

whose composite $\rho_0 \cdots \rho_{n-1}$ is a map

$$\tau_n : \mathrm{Tor}_n^B(N, B^n) \to N.$$

Proposition 3.9. *The terms of the filtration of N in Theorem 3.7 are given by*

$$N(n) = \mathrm{Im}\,(\tau_n).$$

The proofs of Theorem 3.7 and Proposition 3.9 are quite obvious variants of those of Theorem 3.2 and Proposition 3.5, and we just indicate their starting point. Choose any projective resolution, $Q^\bullet$, of N_B. By Lemma 3.1, (2), there are two different formulae for the double complex

$$D^{\bullet,\bullet} = \mathrm{Hom}_A(P^\bullet, G(Q^\bullet)) \cong Q^\bullet \otimes_B \mathrm{Hom}_A(P^\bullet, T) \cong Q^\bullet \otimes_B {}^*(P^\bullet),$$

and its terms, $D^{p,q}$, are 0 except when $0 \leqslant p \leqslant t$ and $q \leqslant 0$, thus when (p,q) is in the fourth quadrant. The term D^n of the single complex is 0 unless $n \leqslant t$, and is a direct sum of at most $t+1$ terms, so its two spectral sequences do converge to $\mathrm{H}^\bullet(N) := \mathrm{H}^\bullet(D^\bullet)$. For this complex, it is the second spectral sequence which collapses, to show that $\mathrm{H}^n(D) = 0$ for $n \neq 0$ and $\mathrm{H}^0(D) = N$, whereas it is the first spectral sequence which contains non-trivial imformation, including the formulae for the induced filtrations on B-modules.

Remark 3.10. In the original case considered in [1, 2, 6] of a tilting module of projective dimension 1, the second pages of the spectral sequences in both theorems are already the E_∞ pages, and the induced filtrations of modules are just those arising from the well-known torsion theories considered in the versions of the Brenner- Butler Theorem given by Bongartz, [1], and Happel and Ringel, [6]. For example, in an A-module M, $M(0) = \text{II}_2^{0,0} = GF(M)$ is the torsion submodule, its torsionfree quotient is $\text{II}_2^{1,1} = (L_1G)((R^1F)(M))$, and the last statement in Theorem 3.2 gives the 'orthogonality' relations $G \circ R^1 F = 0 = L_1 G \circ F$.

Remark 3.11. The Grothendieck-Roos duality theory in [11] starts with the construction, which we now briefly describe, of a spectral sequence. Let R be a regular ring, that is, a commutative noetherian ring of finite global dimension, which number we denote by t. The category mod-R admits an obvious duality functor

$$D: \quad M \mapsto D(M) = \text{Hom}_R(M, R),$$

which is a contravariant equivalence on proj-R. The spectral sequence provides a homological analysis of the natural double duality morphism $M \to D^2(M)$, this of course being an isomorphism exactly when M is a finitely generated projective R-module. Roos starts his paper with a theorem which includes all the statements in the following analogue of the two tilting theorems above.

Theorem 3.12.   *For each finitely generated R-module M there is a convergent spectral sequence*

$$E_2^{p,q} \overset{p}{\Longrightarrow} \text{H}^\bullet(M)$$

in mod-R *in which*

$$\text{H}^0(M) = M \qquad and \qquad \text{H}^n(M) = 0 \quad for \quad n \neq 0,$$

and

$$E_2^{p,q} = \text{Ext}_R^p(\text{Ext}_R^{-q}(M, R), R),$$

so that its support is in the 4-th quadrant square in which $0 \leqslant p \leqslant t$ and $-t \leqslant q \leqslant 0$. Hence

$$E_{t+1} = E_\infty$$

and the filtration induced on $M = \text{H}^0(M)$ has the form

$$M = M(0) \supset M(1) \supset \cdots \supset M(t) \supset M(t+1) = 0,$$

with filtration factors

$$M(p)/M(p+1) = E_{t+1}^{p,-p},$$

whilst

$$E_{t+1}^{p,q} = 0 \quad for \quad q \neq -p.$$

A proof analogous to the tilting theorem proofs may be given by choosing a finite resolution, $P^\bullet$, of M in proj-R, and an injective resolution $I^\bullet$ of R, and studying the single complex associated with the double complex with terms

$$\mathrm{Hom}_R(D(P^q), I^p).$$

Since the $D(P^q)$'s are projective, the spectral sequence determined by the filtration by columns collapses to give the stated homology. The spectral sequence determined by the filtration by rows is the Grothendieck- Roos spectral sequence.

4 Applications

We start with short proofs of two basic properties of tilting functors given in [5] and [9].

Corollary 4.1.   (1) *If M_A satisfies $R^p F(M) = 0$ for each $p \geqslant 0$, then $M = 0$.*

(2) *If N_B satisfies $L_q(N) = 0$ for each $q \geqslant 0$, then $N = 0$.*

Proof. If $R^p F(M) = 0$ for each $p \geqslant 0$, then every term on the second page, II_2, and hence on all subsequent pages, of the spectral sequence in Theorem 3.2 is 0; therefore, $M = 0$. This proves (1), and (2) is proved similarly using Theorem 3.7   □

Next, both Happel and Miyashita consider, for each n with $0 \leqslant n \leqslant t$, the subcategories

$$K^n(A) = \bigcap_{p \neq n} \mathrm{Ker}\, R^p F \quad \text{and} \quad K_n(B) = \bigcap_{q \neq n} \mathrm{Ker}\, L_q G$$

of Mod-A and Mod-B, respectively.

Theorem 4.2.   *For each n with $0 \leqslant n \leqslant t$, the tilting functor $R^n F$ maps $K^n(A)$ isomorphically onto $K_n(B)$ and has $L_n G$ as inverse.*

Proof.   Let $M \in K^n(A)$. In Theorem 3.2 for this module M, the only non-zero column on the page II_2 is the nth column. So $d_2 = 0$, and $\text{II}_2^{n,-q} = H^{n-q}(M)$ for each q. Hence $L_n G(R^n F(M)) = M$, and $L_q G(R^n F(M)) = 0$ for $q \neq n$, so that $R^n F(M) \in K_n(B)$. Using Theorem 3.7 similarly, we see that if $N \in K_n(B)$ then $L_n G(N) \in K^n(A)$ and $R^n F(L_n G(N)) = N$. The theorem follows. $\square$

Next, we use the spectral sequences to prove an interesting result of Alberto Tonolo's in [12].

Proposition 4.3.   (1) *Let M be a right A-module such that*

$$L_n G(R^{n+1} F(M)) = 0 = L_{n+1} G(R^n F(M))$$

for all $n \geqslant 0$. Then, the filtration factors for M defined in Theorem 3.2 are given by the formulae

$$M(n)/M(n-1) \cong L_n G(R^n F(M)) \quad for \ \ 0 \leqslant n \leqslant t.$$

   (2) *Let N be a right B-module such that*

$$R^{n+1} F(L_n G(N)) = 0 = R^n F(L_{n+1} G(N))$$

for all $n \geqslant 0$. Then, the filtration factors for N defined in Theorem 3.7 are given by the formulae

$$N(n)/N(n+1) \cong R^n F(L_n G(N)) \quad for \ \ 0 \leqslant n \leqslant t.$$

Proof. For (1), it suffices to prove that $\text{II}_2^{n,-n} = \text{II}_3^{n,-n} = \ldots = \text{II}_{t+1}^{n,-n}$ for each n. The hypotheses on M imply, for all $m \geqslant 0$ and all $r \geqslant 2$, that

$$\text{II}_r^{m+1,-m} = 0 = \text{II}_r^{m,-m-1}.$$

Now $\text{II}_{r+1}^{n,-n}$ is the homology of a complex

$$\text{II}_r^{n+r-1,-n-r} \longrightarrow \text{II}_r^{n,-n} \longrightarrow \text{II}_r^{n-r+1,-n+r},$$

and since the end terms are trivial, it follows as required that $\text{II}_{r+1}^{n,-n} = \text{II}_r^{n,-n}$. The proof of (2) is similar. $\square$

Tonolo calls the modules of $K^n(A)$ and of $K_n(B)$ the *n-static A-modules* and the *n-costatic B-modules*, respectively. Then, pursuing an analogy with the idea of sequentially Cohen-Macaulay modules, he defines a right A-module, M, to be *sequentially static* if $L_m G(R^n F(M)) = 0$ for all $m \neq n$, and a right B-module N to be *sequentially costatic* if $R^n F(L_m G(N)) = 0$ for all $m \neq n$. These conditions are just that the E_2 pages of the spectral sequences for these modules of Theorems 3.2 and 3.7 are trivial except on the 'diagonal' $p + q = 0$, so that $E_2 = E_3 = \cdots = E_{t+1}$. In particular, the filtration factors of the modules are given by the terms $E_2^{n,-n}$ of the appropriate spectral sequences, which proves the *only if* parts of the following characterisation by Tonolo of his sequentially static and costatic modules.

Theorem 4.4 ([12], Theorems 1.10 and 1.11).   (1) *The   right A-module, M, is sequentially static if, and only if, it possesses a filtration*

$$M = M_t \supset M_{t-1} \supset \cdots \supset M_0 \supset M_{-1} = 0$$

in which, for each n, the filtration factor M_n/M_{n-1} is an n-static module, in which case this factor is isomorphic to $L_n G(R^n F(M))$.

(2) *The right B-module, N, is sequentially costatic if, and only if, it possesses a filtration*

$$N = N_0 \supset N_1 \supset \cdots \supset N_t \supset N_{t+1} = 0$$

in which, for each n, the filtration factor N_n/N_{n+1} is an n-costatic module, in which case this factor is isomorphic to $R^n F(L_n G(N))$.

Proof. We have noted that the *only if* parts are immediate consequences of the properties of the spectral sequences. For completeness, we conclude with Tonolo's elegant proof of the *if* part of (1). Assume therefore that M_A has a filtration as in (1), with n-th factor M_n/M_{n-1} an n-static module for each n. We first show that $R^m F(M_n) = 0$ for all $m > n$. This is true for $n = 0$ since $M_0 = M_0/M_{-1}$ is 0-static. Suppose $m > n > 0$ and that the result holds for smaller n, so in particular, $R^m F(M_{n-1}) = 0$. Then, in the portion

$$\cdots \to R^{m-1} F(M_n/M_{n-1}) \to R^m F(M_{n-1}) \to$$

$$\to R^m F(M_n) \to R^m F(M_n/M_{n-1}) \to R^{m+1}F(M_{n-1}) \to \cdots$$

of the long exact connected sequence associated with the n-static n-th factor, the second and fourth displayed terms are trivial, and so therefore is $R^m F(M_n)$.

Now, fix $m \geqslant 0$ and consider these exact sequences for $n = m, m + 1, \ldots, t$. For $n = m$, the preceding discussion shows that the second and last terms are trivial, so we obtain an isomorphism of $R^m F(M_m)$ with $R^m F(M_m/M_{m-1})$. For $n > m$, the first and fourth terms are trivial because M_n/M_{n-1} is n-static, so we obtain isomorphisms

$$R^m F(M_m) \cong R^m F(M_{m+1}) \cong \cdots \cong R^m F(M_t) = R^m F(M).$$

Thus $R^m F(M_m/M_{m-1}) \cong R^m F(M)$, and so

$$L_n(R^m F(M)) \cong L_n G(R^m F(M_m/M_{m-1}))$$

for all n. Now M_m/M_{m-1} is m-static, so the RHS is trivial for $n \neq m$, and we conclude that M is sequentially static. Also, by Theorem 4.2, the functor $L_m G \circ R^m F$ is equivalent to the identity on m-static modules, so the above formula with $n = m$ shows that

$$M_n/M_{n-1} \cong L_m(R^m F(M)).$$

This completes the proof of Tonolo's theorem. $\qquad\qquad\qquad\square$

5 Edge effects, and the case $t = 2$

Given a tilting module of arbitrary projective dimension t, we note some edge effects stemming from the specific features of the spectral sequences relating the tilting functors. In the case $t = 2$, we show that rather complete imformation about the filtrations induced on modules is given by the spectral sequences.

Proposition 5.1.   (1) *Let M be a right A-module.*
     Then $F(M) \in \operatorname{Ker} L_t G$ and $R^t F(M) \in \operatorname{Ker} G$; if also $t \geqslant 2$, then $F(M) \in \operatorname{Ker} L_{t-1}G$ and $R^t F(M) \in \operatorname{Ker} L_1 G$.

     (2) *Let N be a right B-module.*
     Then $G(N) \in \operatorname{Ker} R^t F$ and $L_t G(N) \in \operatorname{Ker} F$; if also $t \geqslant 2$, then $G(N) \in \operatorname{Ker} R^{t-1}F$ and $L_t G(N) \in \operatorname{Ker} R^1 F$.

Proof. In Theorem 3.2, d_2 has bidegree $(-1, 2)$, so $\mathrm{II}_{t+1}^{p,q} = \mathrm{II}_2^{p,q}$ if (p, q) is any one of the 'edge' pairs $(0, -t)$, $(0, 1 - t)$, $(t, 0)$ or $(t, -1)$. The first assertion then follows from the last part of Theorem 3.2. The proof of (2) uses the last part of Theorem 3.7.   □

One can formulate other edge effects. In Theorem 3.2, for example, since d_3 has bidegree $(-2, 3)$, we have $\mathrm{II}_3^{1,-t} = \mathrm{II}_{t+1}^{1,-t} = 0$ for $t \geqslant 2$. Hence $d_2 : \mathrm{II}_2^{1,-t} \to \mathrm{II}_2^{0,2-t}$ is injective, and we obtain an exact sequence

$$0 \to \mathrm{II}_2^{1,-t} \to \mathrm{II}_2^{0,2-t} \to \mathrm{II}_{t+1}^{0,2-t} \to 0,$$

where the last 0 depends on the fact that $d_2 = 0$ on the first column. For $t \geqslant 3$, this shows that

$$\mathrm{II}_2^{1,-t} \cong \mathrm{II}_2^{0,2-t},$$

whereas, for $t = 2$, it reduces to a presentation of the smallest filtration term $M(0)$ in M as a quotient of $GF(M)$. Rather than list further such formulae, we gather below all that we know about the filtrations of A-modules in the case $t = 2$.

When $t = 2$, an arbitrary A-module has a filtration of the form

$$0 \subset M(0) \subset M(1) \subset M(2) = M,$$

and $\mathrm{II}_3 = \mathrm{II}_\infty$ is given by Theorem 3.2. We have already noted that there is an exact sequence presenting $M(0)$,

$$0 \to L_2 G(R^1 F(M)) \to GF(M) \to M(0) \to 0.$$

A similar argument gives a presentation of the top quotient of the filtration via the exact sequence

$$0 \to M(1) \to M \to L_2 G(R^2 F(M)) \to G(R^1 F(M)) \to 0,$$

and similarly there is an exact sequence

$$0 \to M(0) \to M(1) \to L_1 G(R^1 F(M)) \to 0$$

presenting the middle quotient. Finally, there are four 'orthogonality relations'

$$0 = L_1 G \circ F = L_2 G \circ F = G \circ R^2 F = L_1 G \circ R^2 F.$$

References

[1] K. Bongartz, *Tilted algebras*, Lecture Notes in Mathematics 903 (1982) 26-38.

[2] Sheila Brenner and M.C.R.Butler, *Generalisations of the Bernstein-Gelfand-Ponomarev reflection functors*, Representation Theory II, Lecture Notes in Mathematics, 832, Springer, 1980.

[3] H. Cartan and S. Eilenberg, *Homological algebra*, Princeton University Press, 1956.

[4] E. Cline, B. Parshall and L. Scott, *Derived categories and Morita theory*, J. Algebra 104(1986) 397-409.

[5] D. Happel. *Triangulated categories in the representation theory of finite dimensional algebras*, London Mathematical Society Lecture Note Series 119, Cambridge University Press, 1988.

[6] D. Happel and C. M. Ringel, *Tilted algebras*, Transactions Amer. Math. Soc. 274 (1982) 339-443.

[7] B. Keller, *Derived categories and tilting*, this volume.

[8] B. Keller and D. Vossieck, *Dualité de Grothendieck-Roos et basculement*, C. R. Acad. Sci. Paris, 307, Série I, 1988, 543-546.

[9] Y. Miyashita, *Tilting modules of finite projective dimension*, Math. Zeit., 193 (1986), 113-146.

[10] J. Rickard *Morita theory for derived categories*, J. London Math. Soc.(2)39 (1989) 436-456.

[11] J-E. Roos, *Bidualité et structure des foncteurs dérivées de* $\varprojlim$ *dans la catégorie des modules sur un anneau régulier*, C.R. Acad.Sci.Paris, 280, Série I, 1962, 1556-1558.

[12] A. Tonolo, *Tilting modules of finite projective dimension: sequentially static and costatic modules*, J. Algebra Appl. 1 (2002) 295-305.

[13] C. A. Weibel, *An introduction to homological algebra*, Cambridge University Press, 1994.

Department of Mathematical Sciences

The University of Liverpool

Liverpool, L69 7ZL

UK

email: mcrb@liv.ac.uk

5
Derived categories and tilting

Bernhard Keller

Abstract

We review the basic definitions of derived categories and derived functors. We illustrate them on simple but non trivial examples. Then
we explain Happel's theorem which states that each tilting triple yields
an equivalence between derived categories. We establish its link with
Rickard's theorem which characterizes derived equivalent algebras. We
then examine invariants under derived equivalences. Using t-structures
we compare two abelian categories having equivalent derived categories.
Finally, we briefly sketch a generalization of the tilting setup to differential graded algebras.

1 Introduction

1.1 Motivation: Derived categories as higher invariants

Let k be a field and A a k-algebra (associative, with 1). We are especially
interested in the case where A is a non commutative algebra. In order
to study A, one often looks at various invariants associated with A, for
example its Grothendieck group $K_0(A)$, its center $Z(A)$, its higher K-
groups $K_i(A)$, its Hochschild cohomology groups $HH^*(A, A)$, its cyclic
cohomology groups .... Of course, each isomorphism of algebras $A \to B$
induces an isomorphism in each of these invariants. More generally, for
each of them, there is a fundamental theorem stating that the invariant
is preserved not only under isomorphism but also under passage from A
to a matrix ring $M_n(A)$, and, more generally, that it is preserved under

49

Morita equivalence. This means that it only depends on the category Mod A of (right) A-modules so that one can say that the map taking A to any of the above invariants factors through the map which takes A to its module category:

$$A \longmapsto K_0(A), Z(A), K_i(A), HH^*(A, A), HC_*(A), \ldots$$

Now it turns out that for each of these invariants, there is a second fundamental theorem to the effect that the invariant does not depend so much on the module category but only on its *derived category* D Mod A in the sense that each (triangle) equivalence between derived categories allows us to construct an isomorphism in the invariant. So we obtain a second factorization:

$$A \longmapsto K_0(A), Z(A), K_i(A), HH^*(A, A), HC_*(A), \ldots$$

In this picture, the derived category appears as a kind of *higher invariant*, an invariant which, as we will see, is much coarser than the module category (at least in the non commutative case) but which is still fine enough to determine all of the classical homological and homotopical invariants associated with A.

Tilting theory enters the picture as a rich source of derived equivalences. Indeed, according to a theorem by D. Happel, if B is an algebra and T a tilting module for B with endomorphism ring A, then the total derived tensor product by T is an equivalence from D Mod A to D Mod B. In particular, A and B then share all the above-mentioned invariants. But an equivalence between the derived categories of Mod A and Mod B also yields strong links between the abelian categories Mod A and Mod B themselves: often, it allows one to identify suitable 'pieces' of Mod A with 'pieces' of Mod B. This has proved to be an extremely useful method in representation theory.

1.2 Contents

We will recall the definition of the derived category of an abelian category. We will make this abstract construction more intuitive by considering the quivers of module categories and their derived categories in several examples. These examples will suggest the existence of certain equivalences between derived categories. We will construct these equivalences using D. Happel's theorem: the derived functor of tensoring by a tilting module is an equivalence. We will then proceed to a first crude analysis of the relations between module categories with equivalent derived categories and examine some examples. In the next section, we generalize Happel's theorem to Rickard's Morita theorem for derived categories. Here, the key notion is that of a tilting complex. This generalizes the notion of a tilting module. Tilting modules over selfinjective algebras are always projective, but there may exist non trivial tilting complexes. We illustrate this by exhibiting the action of a braid group on the derived category of a selfinjective algebra following Rouquier-Zimmermann. Then we proceed to a more sophisticated analysis of the links between two abelian categories with equivalent derived categories. We use aisles (=t-structures) and also sketch the link with the spectral sequence approach due to Vossieck and Brenner-Butler. Finally, we show how the notion of a tilting complex can be weakened even more if, instead of algebras, we consider differential graded algebras. We present the description of suitable 'algebraic' triangulated categories via derived categories of differential graded algebras. As an illustration, we present D. Happel's description of the derived category of a finite-dimensional algebra via the category of graded modules over its trivial extension.

2 Derived categories

2.1 First definition

Let $\mathcal{A}$ be an additive category. For example, $\mathcal{A}$ could be the category $\mathsf{Mod}\,A$ of (right) modules over a ring A or the category $\mathsf{Mod}\,O_X$ of sheaves of O_X-modules on a ringed space (X, O_X). A *complex over* $\mathcal{A}$ is a diagram

$$\cdots \longrightarrow M^p \xrightarrow{\;d^p\;} M^{p+1} \longrightarrow \cdots \,, \; p \in \mathbb{Z}\,,$$

such that $d^p \circ d^{p-1} = 0$ for all $p \in \mathbb{Z}$. A *morphism of complexes* is a morphism of diagrams. We obtain the *category of complexes* $\mathsf{C}\,\mathcal{A}$.

Now suppose that $\mathcal{A}$ is abelian. This is the case for the above examples. For $p \in \mathbb{Z}$, the *pth homology* $H^p M$ of a complex M is $\ker d^p / \operatorname{im} d^{p-1}$.

A morphism of complexes is a *quasi-isomorphism* if it induces isomorphisms in all homology groups. Ignoring a set-theoretic problem, we define the *derived category* $\mathsf{D}\,\mathcal{A}$ as the localization of the category of complexes with respect to the class of quasi-isomorphisms. This means that the objects of the derived category are all complexes. And morphisms in the derived category between two complexes are given by paths composed of morphisms of complexes and formal inverses of quasi-isomorphisms, modulo a suitable equivalence relation *cf.* [28].

This quick definition is not very explicit but it immediately yields an important universal property of the derived category: The canonical functor $\mathsf{C}\,\mathcal{A} \to \mathsf{D}\,\mathcal{A}$ makes all quasi-isomorphisms invertible and is universal among all functors $F \colon \mathsf{C}\,\mathcal{A} \to \mathcal{C}$ with this property. More precisely, for each category $\mathcal{C}$, the canonical functor $\mathsf{C}\,\mathcal{A} \to \mathsf{D}\,\mathcal{A}$ yields an isomorphism of functor categories

$$\mathsf{Fun}(\mathsf{D}\,\mathcal{A}, \mathcal{C}) \xrightarrow{\sim} \mathsf{Fun}_{qis}(\mathsf{C}\,\mathcal{A}, \mathcal{C})\,,$$

where the category on the right is the full subcategory on the functors making all quasi-isomorphisms invertible. We deduce that a pair of exact adjoint functors between two abelian categories induces a pair of adjoint functors between their derived categories.

2.2 Second definition

We keep the notations of paragraph 2.1. A morphism of complexes $f \colon L \to M$ is *null-homotopic* if there are morphisms $r^p \colon L^p \to M^{p-1}$ such that $f^p = d^{p-1} \circ r^p + r^{p+1} d^p$ for all $p \in \mathbb{Z}$. Null-homotopic morphisms form an *ideal* in the category of complexes. We define the *homotopy category* $\mathsf{H}\,\mathcal{A}$ to be the quotient of $\mathsf{C}\,\mathcal{A}$ by this ideal. Thus, the objects of $\mathsf{H}\,\mathcal{A}$ are all complexes, and morphisms between two objects are classes of morphisms of complexes modulo null-homotopic morphisms. Note that the homology functors $M \mapsto H^p M$ descend to functors defined on the homotopy category. A *quasi-isomorphism in* $\mathsf{H}\,\mathcal{A}$ is a morphism whose image under the homology functors is invertible. Let Σ be the class of quasi-isomorphisms in $\mathsf{H}\,\mathcal{A}$. The following lemma states that the analogues of the Ore conditions in the localization theory of rings hold for the class Σ (the assumption that the elements to be made invertible be non-zero divisors is weakened into condition c).

Lemma 2.1.   a) *Identities are quasi-isomorphisms and composi-tions of quasi-isomorphisms are quasi-isomorphisms.*

b) *Each diagram*

$$L' \xleftarrow{\;s\;} L \xrightarrow{\;f\;} M \quad (resp. \;\; L' \xrightarrow{\;f'\;} M' \xleftarrow{\;s'\;} M)$$

of H $\mathcal{A}$*, where s (resp. s′) is a quasi-isomorphism, may be em-bedded into a square*

$$
\begin{array}{ccc}
L & \xrightarrow{\;f\;} & M \\
{\scriptstyle s}\downarrow & & \downarrow{\scriptstyle s'} \\
L' & \xrightarrow{\;f'\;} & M'
\end{array}
$$

which commutes in H $\mathcal{A}$.

c) *Let f be a morphism of* H $\mathcal{A}$*. Then there is a quasi-isomorphism s such that sf = 0 in* H $\mathcal{A}$ *if and only if there is a quasi-isomorphism t such that ft = 0 in* H $\mathcal{A}$.

The lemma is proved for example in [44, 1.6.7]. Clearly condition a) would also be true for the pre-image of Σ in the category of complexes. However, for b) and c) to hold, it is essential to pass to the homotopy category. Historically [39], this observation was the main reason for in-serting the homotopy category between the category of complexes and the derived category. We now obtain a second, equivalent, definition [82] of the *derived category* D $\mathcal{A}$: it is the category of fractions of the ho-motopy category with respect to the class of quasi-isomorphisms. This means that the derived category has the same objects as the homotopy category (namely all complexes) and that morphisms in the derived cat-egory from L to M are given by 'left fractions' "$s^{-1}f$", *i.e.* equivalence classes of diagrams

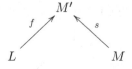

where s is a quasi-isomorphism, and a pair (f, s) is equivalent to (f', s')

iff there is a commutative diagram of $H\mathcal{A}$

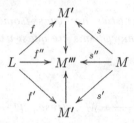

where s'' is a quasi-isomorphism. Composition is defined by

$$\text{``}t^{-1}g\text{''} \circ \text{``}s^{-1}f\text{''} = \text{``}(s't)^{-1} \circ g'f\text{''} ,$$

where $s' \in \Sigma$ and g' are constructed using condition b) as in the following commutative diagram of $H\mathcal{A}$

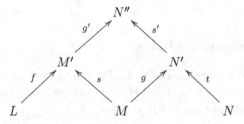

One can then check that composition is associative and admits the obvious morphisms as identities. Using 'right fractions' instead of left fractions we would have obtained an isomorphic category (use lemma 2.1 b). The universal functor $C\mathcal{A} \to D\mathcal{A}$ of paragraph 2.1 descends to a canonical functor $H\mathcal{A} \to D\mathcal{A}$. It sends a morphism $f\colon L \to M$ to the fraction $\text{``}1_M^{-1}f\text{''}$. It makes all quasi-isomorphisms invertible and is universal among functors with this property.

2.3 Cofinal subcategories

A subcategory $\mathcal{U} \subset H\mathcal{A}$ is *left cofinal* if, for each quasi-isomorphism $s\colon U \to V$ with $U \in \mathcal{U}$ and $V \in H\mathcal{A}$, there is a quasi-isomorphism $s'\colon U \to U'$ with $U' \in \mathcal{U}$ and a commutative diagram

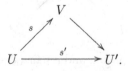

Dually, one defines the notion of a right cofinal subcategory.

Lemma 2.2. *If $\mathcal{U} \subset \mathcal{H}\mathcal{A}$ is left or right cofinal, then the essential image of $\mathcal{U}$ in $\mathsf{D}\mathcal{A}$ is equivalent to the localization of $\mathcal{U}$ at the class of quasi-isomorphisms $s\colon U \to U'$ with $U, U' \in \mathcal{U}$.*

For example, the category $\mathsf{H}^-(\mathcal{A})$ of complexes U with $U^n = 0$ for all $n \gg 0$ is easily seen to be left cofinal in $\mathcal{H}\mathcal{A}$. The essential image of $\mathsf{H}^-(\mathcal{A})$ in $\mathsf{D}\mathcal{A}$ is the *right bounded derived category* $\mathsf{D}^- \mathcal{A}$, whose objects are all complexes U with $H^n U = 0$ for all $n \gg 0$. According to the lemma, it is equivalent to the localization of the category $\mathsf{H}^- \mathcal{A}$ with respect to the class of quasi-isomorphisms it contains. Similarly, the category $\mathsf{H}^+ \mathcal{A}$ of all complexes U with $U^n = 0$ for all $n \ll 0$ is right cofinal in $\mathcal{H}\mathcal{A}$ and we obtain an analogous description of the *left bounded derived category* $\mathsf{D}^+ \mathcal{A}$. Finally, the category $\mathsf{H}^b \mathcal{A}$ formed by the complexes U with $U^n = 0$ for all $|n| \gg 0$ is left cofinal in $\mathsf{H}^+ \mathcal{A}$ and right cofinal in $\mathsf{H}^- \mathcal{A}$. We infer that the *bounded derived category* $\mathsf{D}^b \mathcal{A}$, whose objects are the U with $H^n U = 0$ for all $|n| \gg 0$, is equivalent to the localization of $\mathsf{H}^b(\mathcal{A})$ with respect to its quasi-isomorphisms.

2.4 Morphisms and extension groups

The following lemma yields a more concrete description of some morphisms of the derived category. We use the following notation: An object $A \in \mathcal{A}$ is identified with the complex

$$\ldots \to 0 \to A \to 0 \to \ldots$$

having A in degree 0. If M is an arbitrary complex, we denote by $S^n M$ or $M[n]$ the complex with components $(S^n M)^p = M^{n+p}$ and differential $d_{S^n M} = (-1)^n d_M$. A complex I (resp. P) is *fibrant* (resp. *cofibrant*) if the canonical map

$$\mathsf{Hom}_{\mathsf{H}\mathcal{A}}(L, I) \to \mathsf{Hom}_{\mathsf{D}\mathcal{A}}(L, I) \quad \text{resp.} \quad \mathsf{Hom}_{\mathsf{H}\mathcal{A}}(P, L) \to \mathsf{Hom}_{\mathsf{D}\mathcal{A}}(P, L)$$

is bijective for each complex L.

Lemma 2.3.   a) *The category $\mathsf{D}\mathcal{A}$ is additive and the canonical functors $\mathsf{C}\mathcal{A} \to \mathsf{H}\mathcal{A} \to \mathsf{D}\mathcal{A}$ are additive.*

    b) *If I is a left bounded complex (i.e. $I^n = 0$ for all $n \ll 0$) with*

> *injective components then I is fibrant. Dually, if P is a right*
> *bounded complex with projective components, then P is cofibrant.*

c) *For all $L, M \in \mathcal{A}$, there is a canonical isomorphism*

$$\partial\colon \operatorname{Ext}^n_{\mathcal{A}}(L, M) \xrightarrow{\sim} \operatorname{Hom}_{\mathsf{D}\mathcal{A}}(L, S^n M)$$

valid for all $n \in \mathbb{Z}$ if we adopt the convention that Ext^n vanishes for $n < 0$. In particular, the canonical functor $\mathcal{A} \to \mathsf{D}\mathcal{A}$ is fully faithful.

The calculus of fractions yields part a) of the lemma (cf. [28]). Part b) follows from [38, I, Lemma 4.5]. Part c) is in [38, I, §6]. Let us prove c) in the case where $\mathcal{A}$ has enough injectives (*i.e.* each object admits a monomorphism into an injective). In this case, the object M admits an injective resolution, i.e. a quasi-isomorphism $s\colon M \to I$ of the form

$$
\begin{array}{ccccccccc}
\cdots & \to & 0 & \to & M & \to & 0 & \to & 0 & \to & \cdots \\
 & & \downarrow & & \downarrow & & \downarrow & & \downarrow & & \\
\cdots & \to & 0 & \to & I^0 & \to & I^1 & \to & I^2 & \to & \cdots
\end{array}
$$

where the I^p are injective. Then, since s becomes invertible in $\mathsf{D}\mathcal{A}$, it induces an isomorphism

$$\operatorname{Hom}_{\mathsf{D}\mathcal{A}}(L, S^n M) \xrightarrow{\sim} \operatorname{Hom}_{\mathsf{D}\mathcal{A}}(L, S^n I).$$

By part b) of the lemma, we have the isomorphism

$$\operatorname{Hom}_{\mathsf{D}\mathcal{A}}(L, S^n I) \xleftarrow{\sim} \operatorname{Hom}_{\mathsf{H}\mathcal{A}}(L, S^n M).$$

Finally, the last group is exactly the nth homology of the complex

$$\operatorname{Hom}_{\mathcal{A}}(L, I),$$

which identifies with $\operatorname{Ext}_{\mathcal{A}}(L, M)$ by (the most common) definition.

2.5 Derived categories of semi-simple or hereditary categories

In two very special cases, we can directly describe the derived category in terms of the module category: First suppose that $\mathcal{A}$ is *semi-simple*, *i.e.* $\operatorname{Ext}^1_{\mathcal{A}}(A, B) = 0$ for all $A, B \in \mathcal{A}$. For example, this holds for the category of vector spaces over a field. Then it is not hard to show that the functor $M \mapsto H^* M$ establishes an equivalence between $\mathsf{D}\mathcal{A}$ and the category of $\mathbb{Z}$-graded k-vector spaces. In the second case, suppose that $\mathcal{A}$ is *hereditary* (*i.e.* $\operatorname{Ext}^2_{\mathcal{A}}(A, B) = 0$ for all $A, B \in \mathcal{A}$). We claim that each object M of $\mathsf{D}\,\mathcal{A}$ is quasi-isomorphic to the sum of the $(H^n M)[-n]$, $n \in \mathbb{Z}$. To prove this, let us denote by Z^n the kernel of $d^n \colon M^n \to M^{n+1}$, and put $H^n = H^n(M)$. Then we have an exact sequence

$$0 \longrightarrow Z^{n-1} \longrightarrow M^{n-1} \xrightarrow{\;\delta\;} Z^n \longrightarrow H^n \longrightarrow 0$$

for each $n \in \mathbb{Z}$, where δ is induced by d. Its class in $\mathsf{Ext}_{\mathcal{A}}^2$ vanishes by the assumption on $\mathcal{A}$. Therefore, there is a factorization

$$M^{n-1} \xrightarrow{\;\;\varepsilon\;\;} E^n \xrightarrow{\;\zeta\;} Z^n$$

of δ where ε is a monomorphism, ζ an epimorphism, Z^{n-1} identifies with the kernel of ζ and H^n with the cokernel of ε. It follows that the direct sum H of the complexes $H^n[-n]$ is quasi-isomorphic to the direct sum S of the complexes

$$\cdots \longrightarrow 0 \longrightarrow M^{n-1} \xrightarrow{\;\varepsilon\;} E^n \longrightarrow 0 \longrightarrow \cdots .$$

There is an obvious quasi-isomorphism $S \to M$. Thus we have a diagram of quasi-isomorphisms

$$M \longleftarrow S \longrightarrow H$$

and the claim follows. Note that the direct sum of the $(H^n M)[-n]$, $n \in \mathbb{Z}$, identifies with their direct product. Therefore, if L and M are two complexes, then the morphisms from L to M in $\mathsf{D}\,\mathcal{A}$ are in bijection with the families (f_n, ε_n), $n \in \mathbb{Z}$, of morphisms $f_n \colon H^n L \to H^n M$ and extensions $\varepsilon_n \in \mathsf{Ext}_{\mathcal{A}}^1(H^n L, H^{n-1} M)$.

2.6 The quiver of a k-linear category

We briefly sketch the definition of this important invariant (*cf.* [27, Ch. 9] and [1, Ch. VII] for thorough treatments). It will enable us to visualize the abelian categories and derived categories appearing in the examples below. Let k be a field and $\mathcal{A}$ a k-linear category such that all morphism spaces $\mathcal{A}(A, B)$, $A, B \in \mathcal{A}$, are finite-dimensional. Recall that an object U of $\mathcal{A}$ is *indecomposable* if it is non zero and is not the direct sum of two non zero objects. We suppose that $\mathcal{A}$ *is multilocular [27, 3.1]*, i.e.

   a) each object of $\mathcal{A}$ decomposes into a finite sum of indecomposables and

   b) the endomorphism ring of each indecomposable object is local.

Thanks to condition b), the decomposition in a) is then unique up to isomorphism and permutation of the factors [27, 3.3].

For example, the category mod A of finite-dimensional modules over a finite-dimensional algebra A is multilocular, cf. [27, 3.1], and so is the category coh X of coherent sheaves on a projective variety X, cf. [72]. The bounded derived categories of these abelian categories are also multilocular.

A multilocular category $\mathcal{A}$ is determined by its full subcategory ind $\mathcal{A}$ formed by the indecomposable objects. Condition b) implies that the sets

$$\mathsf{rad}(U, V) = \{f \colon U \to V \mid f \text{ is not invertible}\}$$

form an ideal in ind $\mathcal{A}$. For $U, V \in$ ind $\mathcal{A}$, we define the *space of irreducible maps* to be

$$\mathsf{irr}(U, V) = \mathsf{rad}(U, V)/\mathsf{rad}^2(U, V).$$

The *quiver* $\Gamma(\mathcal{A})$ is the quiver (=oriented graph) whose vertices are the isomorphism classes $[U]$ of indecomposable objects U of $\mathcal{A}$ and where, for two vertices $[U]$ and $[V]$, the number of arrows from $[U]$ to $[V]$ equals the dimension of the space of irreducible maps $\mathsf{irr}(U, V)$.

For example, the quiver of the category of finite dimensional vector spaces mod k has a single vertex (corresponding to the one-dimensional vector space) and no arrows. The quiver of the bounded derived category D^b mod k has vertex set $\mathbb{Z}$, where $n \in \mathbb{Z}$ corresponds to the isoclass of $k[n]$, and has no arrows. The quiver of the category of finite-dimensional modules over the algebra of lower triangular 5×5-matrices is depicted in the top part of figure 5.1. This example and several others are discussed below in section 2.8.

2.7 Algebras given by quivers with relations

Interesting but accessible examples of abelian categories arise as categories of modules over non semi-simple algebras. To describe a large class of such algebras, we use quivers with relations. We briefly recall the main construction: A *quiver* is an oriented graph. It is thus given by a set Q_0 of points, a set Q_1 of arrows, and two maps $s, t \colon Q_1 \to Q_0$ associating with each arrow its source and its target. A simple example is the quiver

$$\vec{A}_{10} : \quad 1 \xrightarrow{\alpha_1} 2 \xrightarrow{\alpha_2} 3 \to \ldots \to 8 \xrightarrow{\alpha_8} 9 \xrightarrow{\alpha_9} 10.$$

A *path* in a quiver Q is a sequence $(y|\beta_r|\beta_{r-1}|\ldots|\beta_1|x)$ of composable arrows β_i with $s(\beta_1) = x$, $s(\beta_i) = t(\beta_{i-1})$, $2 \leqslant i \leqslant r$, $t(\beta_r) = y$. In particular, for each point $x \in Q_0$, we have the *lazy path* $(x|x)$. It is neutral for the obvious *composition* of paths. The *quiver algebra* kQ has as its basis all paths of Q. The product of two basis elements equals the composition of the two paths if they are composable and 0 otherwise. For example, the quiver algebra of $Q = \vec{A}_{10}$ is isomorphic to the algebra of lower triangular 10×10 matrices.

The construction of the quiver algebra kQ is motivated by the (easy) fact that the category of left kQ-modules is equivalent to the category of all diagrams of vector spaces of the shape given by Q. It is not hard to show that each quiver algebra is hereditary. It is finite-dimensional iff the quiver has no oriented cycles. Gabriel [26] has shown that the quiver algebra of a finite quiver has only a finite number of k–finite-dimensional indecomposable modules (up to isomorphism) iff the underlying graph of the quiver is a disjoint union of Dynkin diagrams of type A, D, E.

The above example has underlying graph of Dynkin type A_{10} and thus its quiver algebra has only a finite number of finite-dimensional indecomposable modules.

An ideal I of a finite quiver Q is *admissible* if, for some N, we have

$$(kQ_1)^N \subseteq I \subseteq (kQ_1)^2,$$

where (kQ_1) is the two-sided ideal generated by all paths of length 1. A *quiver Q with relations R* is a quiver Q with a set R of generators for an admissible ideal I of kQ. The algebra kQ/I is then the *algebra associated with* (Q, R). Its category of left modules is equivalent to the category of diagrams of vector spaces of shape Q obeying the relations in R. The algebra kQ/I is finite-dimensional (since I contains all paths of length at least N), hence artinian and noetherian. By induction on the number of points one can show that if the quiver Q contains no oriented cycle, then the algebra kQ/I is of finite global dimension.

One can show that every finite-dimensional algebra over an algebraically closed field is Morita equivalent to the algebra associated with a quiver with relations and that the quiver is unique (up to isomorphism).

2.8 Example: Quiver algebras of type A_n

Let k be a field, $n \geqslant 1$ an integer and $\mathcal{A}$ the category of k-finite-dimensional (right) modules over the quiver algebra A of the quiver $\vec{A}_n$ given by

$$\vec{A}_n : 1 \longrightarrow 2 \longrightarrow \cdots \longrightarrow n-1 \longrightarrow n \, .$$

The quiver $\Gamma(\mathcal{A})$ is triangle-shaped with $n(n+1)/2$ vertices. For $n = 5$, it is given in the top part of figure 5.1: There are n (isomorphism classes of) indecomposable projective modules given by the $P_i = e_{ii}A$, $1 \leqslant i \leqslant n$. They occur in increasing order on the left rim of the triangle. There are n simple modules $S_i = P_i/P_{i-1}$, $1 \leqslant i \leqslant n$ (where $P_0 := 0$). They are represented in increasing order by the vertices at the bottom. There are n injective modules $I_i = \mathsf{Hom}_k(Ae_{ii}, k)$, $1 \leqslant i \leqslant n$. They are represented in decreasing order by the vertices on the right rim. Note that each simple module has a projective resolution of length 1 which confirms that $\mathcal{A}$ is hereditary.

Using 2.5 we see that the indecomposable objects of $\mathsf{D}^b \mathcal{A}$ are precisely the $U[n]$, $n \in \mathbb{Z}$, $U \in \mathrm{ind}(\mathcal{A})$. Thus the quiver $\mathsf{D}^b \mathcal{A}$ has the vertices $[S^n U]$, $n \in \mathbb{Z}$, where $[U]$ is a vertex of $\Gamma(\mathcal{A})$. Arrows from $[S^n U]$ to $[S^m V]$ can occur only if m equals n or $n+1$, again by 2.5. Now Lemma 2.3 shows that the functor

$$\mathrm{ind}\,\mathcal{A} \to \mathrm{ind}\,\mathsf{D}\,\mathcal{A} \, , \ M \mapsto S^n U$$

preserves the spaces of irreducible maps. So the arrows $[S^n U] \to [S^n V]$, where U and V are indecomposable in $\mathcal{A}$, are in bijection with the arrows $[U] \to [V]$ in $\Gamma(\mathcal{A})$. The additional arrows $[S^n U] \to [S^{n+1} V]$ are described in [34, 5.5] for $\mathcal{A} = \mathrm{mod}\,A$, where A is an arbitrary finite-dimensional path algebra of a quiver. For $A = k\vec{A}_n$, the quiver $\Gamma(\mathsf{D}^b \mathcal{A})$ is isomorphic to the infinite stripe $\mathbb{Z}\vec{A}_n$ depicted in the middle part of figure 5.1. The objects $[U]$, $U \in \mathrm{ind}\,\mathcal{A}$, correspond to the vertices (g, h) in the triangle

$$g \geqslant 0 \, , \ h \geqslant 1 \, , \ g+h \leqslant n.$$

The translation $U \mapsto SU$ corresponds to the glide-reflection $(g, h) \mapsto (g+h, n+1-h)$. Remarkably, this quiver was actually discovered twenty years before D. Happel's work appeared in R. Street's Ph. D. Thesis [77] [76], *cf.* also [78] [79] [75].

The quiver

$$Q : 1 \longleftarrow 2 \longleftarrow 3 \longrightarrow 4 \longrightarrow 5$$

is obtained from $\vec{A}_5$ by changing the orientation of certain arrows. The quiver of $\mathsf{mod}\,kQ$ is depicted in the lower part of figure 5.1. The quiver of $\mathsf{D}^b(\mathsf{mod}\,kQ)$ turns out to be isomorphic to that of $\mathsf{D}^b(\mathsf{mod}\,k\vec{A}_5)$! The isomorphism commutes with the shift functor $U \mapsto SU$. In fact, the isomorphism between the quivers of $\mathsf{D}^b(\mathsf{mod}\,\vec{A}_5)$ and $\mathsf{D}^b(\mathsf{mod}\,kQ)$ comes from an equivalence between the derived categories themselves, as we will see below. However, this equivalence does not respect the module categories embedded in the derived categories. This is also visible in figure 5.1: Some modules for $k\vec{A}_5$ correspond to *shifted* modules for kQ and vice versa. Note that the module categories of kQ and $k\vec{A}_5$ cannot be equivalent, since the quivers of the module categories are not isomorphic.

2.9 Example: Commutative squares and representations of $\vec{D}_4$

Let A be the algebra given by the quiver with relations

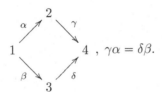

$$, \ \gamma\alpha = \delta\beta.$$

A (right) A-module is the same as a commutative diagram of vector spaces

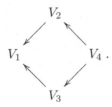

The quivers of $\mathsf{mod}\,A$ and $\mathsf{D}^b(\mathsf{mod}\,A)$ are depicted in figure 5.2. Their computation is due to D. Happel [33] and, independently, R. Street, *cf.* p. 118 of [75]. The shift functor $U \mapsto SU$ corresponds to the map $(g,h) \mapsto (g+3,h)$. Note that the algebra A is not hereditary. Therefore, some indecomposable objects of the derived category are not isomorphic to shifted modules. In the notations of the figure, these are the translates of Y. Let Q be the quiver $\vec{D}_4$:

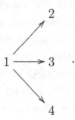

The quiver of $\operatorname{mod} kQ$ is depicted in the lower part of figure 5.2. The quiver of the derived category $\mathsf{D}^b(\operatorname{mod} kQ)$ turns out to be isomorphic to that of $\mathsf{D}^b(\operatorname{mod} A)$! Moreover, the isomorphism respects the shift functors. The isomorphism between the quivers of the bounded derived categories of A and kQ comes from an equivalence between the categories themselves, as we will see below.

2.10 Example: Kronecker modules and coherent sheaves on the projective line

Let Q be the *Kronecker quiver*

$$1 \rightrightarrows 2 \; .$$

The quiver of the category $\operatorname{mod} kQ$ is depicted in the top part of figure 5.3, *cf.* [67]. It is a disjoint union of infinitely many connected components: one *postprojective* component containing the two (isoclasses of) indecomposable projective modules P_1, P_2, one *preinjective* component containing the two indecomposable injective modules I_1, I_2 and an infinity of components containin the *regular* modules $R(t, n)$ indexed by a point $(t_0 : t_1)$ of the projective line $\mathbb{P}^1(k)$ and an integer $n \geqslant 1$. Explicitly, the module $R(x, n)$ is given by the diagram

$$V_{n+1} \underset{x_1}{\overset{x_0}{\rightleftarrows}} V_n \; ,$$

where V_n is the nth homogeneous component of the graded space $k[x_0, x_1]/((t_1 x_0 - t_0 x_1)^n)$. The category $\operatorname{mod} kQ$ is hereditary. Thus the indecomposables in its derived category are simply shifted copies of indecomposable modules. The quiver of the derived category is depicted in the middle part of figure 5.3. Remarkably, it is isomorphic to the quiver of the derived category of the category $\operatorname{coh} \mathbb{P}^1$ of coherent sheaves on the projective line. The quiver of the category $\operatorname{coh} \mathbb{P}^1$ is depicted in the bottom section of figure 5.3. It contains one component whose vertices are the line bundles $O(n)$, $n \in \mathbb{Z}$, and an infinity of components containing the skyscraper sheaves O_{nx} of length $n \geq 1$ concentrated at a point $x \in \mathbb{P}^1$. Note that via the isomorphism of the quivers of the derived categories, these correspond to the indecomposable regular modules over kQ, while the line bundles correspond to postprojective modules and to preinjective modules shifted by one degree. We will see that the isomorphism between the quivers of the derived categories of the categories $\operatorname{mod} kQ$ and $\operatorname{coh} \mathbb{P}^1$ comes from an equivalence between the derived categories themselves.

3 Derived functors

3.1 Deligne's definition

The difficulty in finding a general definition of derived functors is to establish a framework which allows one to prove, in full generality, as many as possible of the pleasant properties found in the examples. This seems to be best achieved by Deligne's definition [23], which we will give in this section (compare with Grothendieck-Verdier's definition in [82]).

Let $\mathcal{A}$ and $\mathcal{B}$ be abelian categories and $F \colon \mathcal{A} \to \mathcal{B}$ an additive functor. A typical example is the fixed point functor

$$\operatorname{Fix}_G \colon \operatorname{Mod} \mathbb{Z} G \to \operatorname{Mod} \mathbb{Z}$$

which takes a module M over a group G to the abelian group of G-fixed points in M. The additive functor F clearly induces a functor $\mathsf{C}\mathcal{A} \to \mathsf{C}\mathcal{B}$ between the categories of complexes (obtained by applying F componentwise) and a functor $\mathsf{H}\mathcal{A} \to \mathsf{H}\mathcal{B}$ between the homotopy categories. By abuse of notation, both will be denoted by F as well. We are looking for a functor $? \colon \mathsf{D}\mathcal{A} \to \mathsf{D}\mathcal{B}$ which should make the following square commutative

$$
\begin{array}{ccc}
\mathsf{H}\mathcal{A} & \xrightarrow{F} & \mathsf{H}\mathcal{B} \\
\downarrow & & \downarrow \\
\mathsf{D}\mathcal{A} & \xrightarrow{?} & \mathsf{D}\mathcal{B}
\end{array}
$$

However, if F is not exact, it will not transform quasi-isomorphisms to quasi-isomorphisms and the functor in question cannot exist. What we will define then is a functor $\mathbf{R}F$ called the 'total right derived functor', which will be a 'right approximation' to an induced functor. More precisely, for a given $M \in \mathsf{D}\,\mathcal{A}$, we will not define $\mathbf{R}F(M)$ directly but only a functor

$$(\mathbf{r}F)(?, M)\colon (\mathsf{D}\,\mathcal{B})^{op} \to \mathsf{Mod}\,\mathbb{Z}$$

which, if representable, will be represented by $\mathbf{R}F(M)$. For $L \in \mathsf{D}\,\mathcal{B}$, we define $(\mathbf{r}F)(L, M)$ to be the set of 'left F-fractions', i.e. equivalence classes of diagrams

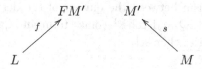

where f is a morphism of $\mathsf{D}\,\mathcal{B}$ and s a quasi-isomorphism of $\mathsf{H}\,\mathcal{A}$. Equivalence is defined in complete analogy with section 2.1. We say that $\mathbf{R}F$ is *defined at* $M \in \mathsf{D}\,\mathcal{A}$ if the functor $(\mathbf{r}F)(?, M)$ is representable and if this is the case, then the value $\mathbf{R}FM$ is defined by the isomorphism

$$\mathsf{Hom}_{\mathsf{D}\,\mathcal{B}}(?, (\mathbf{R}F)(M)) \xrightarrow{\sim} (\mathbf{r}F)(?, M).$$

The link between this definition and more classical constructions is established by the

Proposition 3.1. *Suppose that $\mathcal{A}$ has enough injectives and M is left bounded. Then $\mathbf{R}F$ is defined at M and we have*

$$\mathbf{R}FM = FI$$

where $M \to I$ is a quasi-isomorphism with a left bounded complex with injective components.

Under the hypotheses of the proposition, the quasi-isomorphism $M \to I$ always exists [44, 1.7.7]. Viewed in the homotopy category $\mathsf{H}\,\mathcal{A}$ it is functorial in M since it is in fact the universal morphism from M to a fibrant (2.4) complex. For example, if M is concentrated in degree 0, then I may be chosen to be an injective resolution of M and we find that

$$H^n \mathbf{R}FM = (\mathbf{R}^n F)(M)\,, \tag{3.1.1}$$

the nth right derived functor of F in the sense of Cartan-Eilenberg [19].

The above definition works not only for functors induced from functors

$F\colon \mathcal{A} \to \mathcal{B}$ but can also be applied without any changes to arbitrary functors $F\colon \mathsf{H}\mathcal{A} \to \mathsf{H}\mathcal{B}$. One obtains $\mathbf{R}F$ (defined in general only on a subcategory). Dually, one defines the functor $\mathbf{L}F\colon \mathsf{D}\mathcal{A} \to \mathsf{D}\mathcal{B}$: For each $M \in \mathsf{D}\mathcal{A}$, where $\mathbf{L}F(M)$ is defined, it represents the functor

$$L \mapsto \mathbb{IF}(M, L),$$

where $\mathbb{IF}(M, L)$ is the set of equivalence classes of diagrams

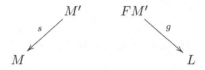

As an exercise, the reader can prove the isomorphism of functors

$$\mathbf{R}\mathsf{Fix}_{G/H} \circ \mathbf{R}\mathsf{Fix}_H = \mathbf{R}\mathsf{Fix}_G$$

for a group G and a normal subgroup H of G. Here, all derived functors are defined on the full subcategory of left bounded complexes $\mathsf{D}^+ \mathsf{Mod}\,\mathbb{Z}G$ of $\mathsf{D}\,\mathsf{Mod}\,\mathbb{Z}G$. This isomorphism replaces the traditional Lyndon-Hochschild-Serre spectral sequence:

$$E_2^{pq} = H^p(G/H, H^q(H, M)) \Rightarrow H^{p+q}(G, M) \qquad (3.1.2)$$

for a G-module M. In fact, using the methods of section 7 one can obtain the spectral sequence from the isomorphism of functors.

Equation 3.1.1 shows that in general, derived functors defined on $\mathsf{D}^b(\mathcal{A})$ will take values in the unbounded derived categories. It is therefore useful to work with unbounded derived categories from the start. The following theorem ensures the existence of derived functors in all the cases we will need: Let A be a k-algebra and $\mathcal{B}$ a Grothendieck category (*i.e.* an abelian category having a generator, such that all set-indexed colimits exist and all filtered colimits are exact).

Theorem 3.2.   a) *Every functor with domain* $\mathsf{H}(\mathcal{B})$ *admits a total right derived functor.*

   b) *Every functor with domain* $\mathsf{H}(\mathsf{Mod}\,A)$ *admits a total right derived functor and a total left derived functor.*

   c) *If* (F, G) *is a pair of adjoint functors from* $\mathsf{H}(\mathsf{Mod}\,A)$ *to* $\mathsf{H}(\mathcal{B})$, *then* $(\mathbf{L}F, \mathbf{R}G)$ *is a pair of adjoint functors from* $\mathsf{D}(\mathsf{Mod}\,A)$ *to* $\mathsf{D}(\mathcal{B})$.

We refer to [25] and [80] for a) and to [74] and [49] for b). Statement c) is a special case of the following easy

Lemma 3.3. *Let (F, G) be an adjoint pair of functors between the homotopy categories $\mathsf{H}(\mathcal{A})$ and $\mathsf{H}(\mathcal{B})$ of two abelian categories $\mathcal{A}$ and $\mathcal{B}$. Suppose that $\mathbf{L}F$ and $\mathbf{R}G$ are defined everywhere. Then $(\mathbf{L}F, \mathbf{R}G)$ is an adjoint pair between $\mathsf{D}(\mathcal{A})$ and $\mathsf{D}(\mathcal{B})$.*

4 Tilting and derived equivalences

4.1 Tilting between algebras

Let A and B be associative unital k-algebras and T an A-B-bimodule. Then we have adjoint functors

$$? \otimes_A T : \operatorname{Mod} A \underset{\longleftarrow}{\overset{\longrightarrow}{\rule{1.5cm}{0pt}}} \operatorname{Mod} B : \operatorname{Hom}_B(T, ?)$$

(and in fact each pair of adjoint functors between module categories is of this form). One variant of Morita's theorem states that these functors are quasi-inverse equivalences iff

a) T_B is finitely generated projective,

b) the canonical map $A \to \operatorname{Hom}_B(T_B, T_B)$ is an isomorphism, and

c) the free B-module of rank one B_B is a direct factor of a finite direct sum of copies of T.

If, in this statement, we replace the module categories by their derived categories, and adapt the conditions accordingly, we obtain the statement of the

Theorem 4.1 (Happel [33]). *The total derived functors*

$$\mathbf{L}(? \otimes_A T) : \mathsf{D}(\operatorname{Mod} A) \underset{\longleftarrow}{\overset{\longrightarrow}{\rule{1.5cm}{0pt}}} \mathsf{D}(\operatorname{Mod} B) : \mathbf{R}\operatorname{Hom}_B(T, ?)$$

are quasi-inverse equivalences iff

a) *As a B-module, T admits a finite resolution*

$$0 \to P_n \to P_{n-1} \to \ldots \to P_1 \to P_0 \to T \to 0$$

by finitely generated projective B-modules P_i,

b) *the canonical map*

$$A \to \mathsf{Hom}_B(T, T)$$

is an isomorphism and for each $i > 0$, we have $\mathsf{Ext}_B^i(T, T) = 0$, and

c) *there is a long exact sequence*

$$0 \to B \to T^0 \to T^1 \to \ldots \to T^{m-1} \to T^m \to 0$$

where B is considered as a right B-module over itself and the T^i are direct factors of finite direct sums of copies of T.

If these conditions hold and, moreover, A and B are right noetherian, then the derived functors restrict to quasi-inverse equivalences

$$\mathsf{D}^b(\mathsf{mod}\, A) \; \underset{\longleftarrow}{\overset{\longrightarrow}{\rule{0pt}{0pt}}} \; \mathsf{D}^b(\mathsf{mod}\, B) \;,$$

where $\mathsf{mod}\, A$ denotes the category of finitely generated A-modules.

4.2 First links between the module categories

Now assume that (A, T, B) is a *tilting triple*, *i.e.* that the conditions of Happel's theorem 4.1 hold. Note that we make no assumption on the dimensions over k of A, B, or T. Let w be the maximum of the two integers n and m occuring in conditions a) and c). Put

$$F = ? \otimes_A^{\mathsf{L}} T \,, \quad G = \mathbf{R}\mathsf{Hom}_B(T, ?)$$

and, for $n \in \mathbb{Z}$, put

$$F_n = H^{-n} \circ F | \, \mathsf{Mod}\, A \,, \quad G^n = H^n \circ G | \, \mathsf{Mod}\, B.$$

These functors are homological, *i.e.* each short exact sequence of modules will give rise to a long exact sequence in these functors. This makes it clear that the subcategories

$$\mathcal{A}_n = \{ M \in \mathsf{Mod}\, A \mid F_i(M) = 0 \,, \; \forall i \neq n \}$$
$$\mathcal{B}_n = \{ N \in \mathsf{Mod}\, B \mid G^i(N) = 0 \,, \; \forall i \neq n \}$$

are closed under extensions, that they vanish for $n < 0$ or $n > w$ and that $\mathcal{A}_w$ and $\mathcal{B}_0$ are closed under submodules and $\mathcal{A}_0$ and $\mathcal{B}_w$ under quotients. Moreover, since

$$F_n | \mathcal{A}_n \overset{\sim}{\to} S^{-n} F | \mathcal{A}_n \quad \text{and} \quad G^n | \mathcal{B}_n \overset{\sim}{\to} S^n G | \mathcal{B}_n$$

the functors F_n and G_n induce quasi-inverse equivalences between $\mathcal{A}_n$ and $\mathcal{B}_n$. Let us now assume that Mod A is hereditary. Then each indecomposable of $\mathsf{D}(\mathrm{Mod}\,A)$ is concentrated in precisely one degree. Thus, for each indecomposable N of Mod A, FN will have non-vanishing homology in exactly one degree and so N will lie in precisely one of the $\mathcal{B}_n$. Thus, as an additive category, Mod B is made up of 'pieces' of the hereditary category Mod A. Whence the terminology that Mod B is *piecewise hereditary*. The algebras in the examples below are all hereditary or piecewise hereditary.

This first analysis of the relations between abelian categories with equivalent derived categories will be refined in section 7.

4.3 Example: $k\vec{A}_5$

We continue example 2.8. In $\mathrm{mod}\,k\vec{A}_5$, we consider the module T given by the sum of the indecomposables T_i, $1 \leqslant i \leqslant 5$, marked in the top part of figure 5.1. The endomorphism ring of T over $k\vec{A}_5$ is isomorphic to kQ so that T becomes a kQ-$k\vec{A}_5$-bimodule. It is not hard to check that $(kQ, T, k\vec{A}_5)$ is a tilting triple. The resulting equivalence between the derived categories gives rise to the identification of their quivers depicted in figure 5.1. For two indecomposables U and V, let us write $U \leqslant V$ if there is a path from U to V in the quiver of the module category. Then we can describe the indecomposables of the subcategories $\mathcal{A}_n$ and $\mathcal{B}_n$ of 4.2 as follows:

$$\mathcal{B}_0 : U \in \mathrm{ind}(k\vec{A}_5) \text{ such that } U \geqslant T_i \text{ for some i}$$
$$\mathcal{A}_0 : U \in \mathrm{ind}(kQ) \text{ such that } U \leqslant GI_i \text{ for some i}$$
$$\mathcal{B}_1 : P_1, P_2, S_2 \in \mathrm{ind}(k\vec{A}_5)$$
$$\mathcal{A}_1 : S_2', I_2', I_1' \in \mathrm{ind}(kQ)$$

In terms of representations of Q and $\vec{A}_5$, the functor $G_0 = \mathrm{Hom}_{k\vec{A}_5}(T, ?)$ corresponds to the 'reflection functor' [10] which sends a representation

$$V_1 \xleftarrow{\alpha_1} V_2 \xleftarrow{\alpha_2} V_3 \longleftarrow V_4 \longleftarrow V_5$$

to

$$\mathrm{ker}(\alpha_2) \longrightarrow \mathrm{ker}(\alpha_1\alpha_2) \longrightarrow V_3 \longleftarrow V_4 \longleftarrow V_5 \quad .$$

The functor $G_1 = \mathsf{Ext}_{k\vec{A}_5}(T, ?)$ corresponds to the functor which sends a representation

$$V_1 \xleftarrow{\ \alpha_1\ } V_2 \xleftarrow{\ \alpha_2\ } V_3 \longleftarrow V_4 \longleftarrow V_5$$

to

$$\mathsf{cok}(\alpha_2) \longrightarrow \mathsf{cok}(\alpha_1\alpha_2) \longrightarrow 0 \longleftarrow 0 \longleftarrow 0 \ .$$

To describe the total right derived functor $G = \mathbf{R}\mathsf{Hom}_{k\vec{A}_5}(T, ?)$, we need the *mapping cone*: recall that if $f: K \to L$ is a morphism of complexes, the mapping cone $C(f)$ is the complex with components $L^p \oplus K^{p+1}$ and with the differential

$$\begin{bmatrix} d_L & f \\ 0 & -d_K \end{bmatrix}.$$

We view complexes of $k\vec{A}_5$-modules as representations of $\vec{A}_5$ in the category of complexes of vector spaces and similarly for complexes of kQ-modules. Then the functor $\mathbf{R}\mathsf{Hom}_{k\vec{A}_5}(T, ?)$ is induced by the exact functor which sends

$$V_1 \xleftarrow{\ \alpha_1\ } V_2 \xleftarrow{\ \alpha_2\ } V_3 \longleftarrow V_4 \longleftarrow V_5$$

(where the V_i are complexes of vector spaces) to

$$C(\alpha_2) \longrightarrow C(\alpha_1\alpha_2) \longrightarrow V_3 \longleftarrow V_4 \longleftarrow V_5 \ .$$

4.4 Example: Commutative squares and representations of $\vec{D}_4$

We continue example 2.9. Let T be the $k\vec{D}_4$-module which is the direct sum of the indecomposables T_i, $1 \leqslant i \leqslant 4$, marked in the lower part of figure 5.2. It is not hard to see that the endomorphism ring of T is isomorphic to A and that $(A, T, k\vec{D}_4)$ is a tilting triple. The resulting equivalence of derived categories leads to the identification of their quivers depicted in figure 5.2. The indecomposables of the subcategories $\mathcal{A}_n$ and $\mathcal{B}_n$ of 4.2 are as follows

$$\mathcal{A}_0 : U \in \mathrm{ind}\, A \text{ such that } U < \tau I_4$$
$$\mathcal{B}_0 : U \in \mathrm{ind}\, kQ \text{ such that } U \neq Y, Z \text{ and } U \geqslant T_1$$
$$\mathcal{A}_1 : U \in \mathrm{ind}\, A \text{ such that } U \geqslant \tau I_4$$
$$\mathcal{B}_1 : P_1', P_2', P_4', Z \in \mathrm{ind}\, A.$$

Note that GY has homology in degrees 0 and 1 so that Y belongs neither to $\mathcal{B}_0$ nor to $\mathcal{B}_1$. In terms of representations of quivers, the functor $G_0 = \mathrm{Hom}_{k\vec{D}_4}(T, ?)$ is constructed as follows: Given a diagram V, we form

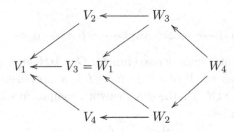

where all 'squares' are cartesian. Then the image of V is the commutative square W.

4.5 Historical remarks

Happel's theorem 4.1, the links between module categories described in section 4.2 and examples like the above form the theory of tilting as it was developped in the representation theory of finite-dimensional algebras in the 1970s and 80s. Important precursors to the theory were: Gelfand-Ponomarev [31] [30], Bernstein-Gelfand-Ponomarev [10], Auslander-Platzeck-Reiten [2], Marmaridis [56], ... The now classical theory (based on homological algebra but avoiding derived categories) is due to: Brenner-Butler [14], who first proved the 'tilting theorem', Happel-Ringel [37], who improved the theorem and defined tilted algebras, Bongartz [13], who streamlined the theory, and Miyashita [58], who generalized it to tilting modules of projective dimension > 1. The use of derived categories goes back to D. Happel [33]. Via the work of Parshall-Scott [21], this lead to J. Rickard's Morita theory for derived categories [65], which we present below.

4.6 Tilting from abelian categories to module categories

Let $\mathcal{B}$ be a k-linear abelian Grothendieck category. An object T of $\mathcal{B}$ is a *tilting object* if the functor

$$\mathbf{R}\mathrm{Hom}(T, ?)\colon\ \mathsf{D}(\mathcal{B}) \to \mathsf{D}(\mathrm{Mod}\,\mathrm{End}(T))$$

is an equivalence.

Proposition 4.2 ([8], [12]). *Suppose that $\mathcal{B}$ is locally noetherian of finite homological dimension and that $T \in \mathcal{B}$ has the following properties:*

a) *T is noetherian.*

b) *We have $\mathsf{Ext}^n(T, T) = 0$ for all $n > 0$.*

c) *Let $\mathsf{add}(T)$ be the closure of T under forming finite direct sums and direct summands. The closure of $\mathsf{add}(T)$ under kernels of epimorphisms contains a set of generators for $\mathcal{B}$.*

Then T is a tilting object. If, moreover, $\mathsf{End}(T)$ is noetherian, the functor $\mathbf{R}\mathsf{Hom}(T, ?)$ induces an equivalence

$$\mathsf{D}^b(\mathcal{B}_{noe}) \to \mathsf{D}^b(\mathrm{mod}\,\mathsf{End}(T)),$$

where $\mathcal{B}_{noe}$ is the subcategory of noetherian objects of $\mathcal{B}$.

An analysis of the links between $\mathcal{B}$ and $\mathrm{mod}\,\mathsf{End}(T)$ analogous to 4.2 can be carried out. The more refined results of section 7 also apply in this situation.

4.7 Example: Coherent sheaves on the projective line

We continue example 2.10. Let $\mathcal{A}$ be the category of quasi-coherent sheaves on the projective line $\mathbb{P}^1(k)$. Let T be the sum of $O(-1)$ with $O(0)$. Then the conditions of the above proposition hold: Indeed, $\mathcal{A}$ is locally noetherian and hereditary and T is noetherian. So condition a) holds. Condition b) is a well-known computation. The sheaves $O(-n)$, $n \in \mathbb{N}$, form a system of generators for $\mathcal{A}$. Therefore condition c) follows from the existence of the short exact sequences

$$0 \to O(-n-1) \to O(-n) \oplus O(-n) \to O(-n+1) \to 0, \; n \in \mathbb{Z}.$$

The endomorphism ring of T is isomorphic to the Kronecker algebra of example 2.10. The resulting equivalence

$$\mathbf{R}\mathsf{Hom}(T, ?) \colon \; \mathsf{D}^b(\mathsf{coh}(\mathbb{P}^1)) \to \mathsf{D}^b(\mathsf{End}(T))$$

induces the identification of the quivers depicted in figure 5.3. With notations analogous to 4.2, the indecomposables of $\mathcal{A}_0$ are those of the postprojective and the regular components. Those of $\mathcal{A}_1$ are the ones

in the preinjective component. The indecomposables in $\mathcal{B}_0$ are the line
bundles $O(n)$ with $n \geqslant 0$ and the skyscraper sheaves. Those of $\mathcal{B}_1$ are
the line bundles $O(n)$ with $n < 0$.

This example is a special case of Beilinson's [11] description of the derived category of coherent sheaves on $\mathbb{P}^n(k)$. It was generalized to other
homogeneous varieties by Kapranov [40] [41] [42] [43] and to weighted
projective lines by Geigle and Lenzing [29] and Baer [8].

5 Triangulated categories

5.1 Definition and examples

Let $\mathcal{A}$ be an abelian category (for example, the category $\mathsf{Mod}\,R$ of modules over a ring R). One can show that the derived category $\mathsf{D}\,\mathcal{A}$ is
abelian only if all short exact sequences of $\mathcal{A}$ split. This deficiency is
partly compensated by the so-called triangulated structure of $\mathsf{D}\,\mathcal{A}$, which
we are about to define. Most of the material of this section first appears
in [82].

A *standard triangle* of $\mathsf{D}\,\mathcal{A}$ is a sequence

$$X \xrightarrow{Qi} Y \xrightarrow{Qp} Z \xrightarrow{\partial\varepsilon} X[1]\,,$$

where $Q\colon \mathsf{C}\,\mathcal{A} \to \mathsf{D}\,\mathcal{A}$ is the canonical functor,

$$\varepsilon\colon 0 \to X \xrightarrow{i} Y \xrightarrow{p} Z \to 0$$

a short exact sequence of complexes, and $\partial\varepsilon$ a certain morphism of $\mathsf{D}\,\mathcal{A}$,
functorial in ε, and which lifts the connecting morphism $H^*Z \to H^{*+1}X$
of the long exact homology sequence associated with ε. More precisely,
$\partial\varepsilon$ is the fraction $"s^{-1} \circ j"$ where j is the inclusion of the subcomplex Z
into the complex $X'[1]$ with components $Z^n \oplus Y^{n+1}$ and differential

$$\begin{bmatrix} d_Z & p \\ 0 & -d_Y \end{bmatrix},$$

and $s\colon X[1] \to X'[1]$ is the morphism $[0, i]^t$. A *triangle* of $\mathsf{D}\,\mathcal{A}$ is a
sequence (u', v', w') of $\mathsf{D}\,\mathcal{A}$ *isomorphic* to a standard triangle, i.e. such

that we have a commutative diagram

$$
\begin{array}{ccccccc}
X' & \xrightarrow{u'} & Y' & \xrightarrow{v'} & Z' & \xrightarrow{w'} & X'[1] \\
{\scriptstyle x}\downarrow & & \downarrow & & \downarrow & & \downarrow{\scriptstyle x[1]} \\
X & \longrightarrow & Y & \longrightarrow & Z & \longrightarrow & X[1]
\end{array} \quad ,
$$

where the vertical arrows are isomorphisms of $\mathsf{D}\mathcal{A}$ and the bottom row is a standard triangle.

Lemma 5.1. T1 *For each object X, the sequence*

$$
0 \to X \xrightarrow{1} X \to S0
$$

is a triangle.

T2 *If (u, v, w) is a triangle, then so is $(v, w, -Su)$.*

T3 *If (u, v, w) and (u', v', w') are triangles and x, y morphisms such that $yu = u'x$, then there is a morphism z such that $zv = v'y$ and $(Sx)w = w'z$.*

$$
\begin{array}{ccccccc}
X & \xrightarrow{u} & Y & \xrightarrow{v} & Z & \xrightarrow{w} & SX \\
{\scriptstyle x}\downarrow & & {\scriptstyle y}\downarrow & & {\scriptstyle z}\downarrow & & \downarrow{\scriptstyle Sx} \\
X' & \xrightarrow{u'} & Y' & \xrightarrow{v'} & Z' & \xrightarrow{w'} & SX'
\end{array} \quad .
$$

T4 *For each pair of morphisms*

$$
X \xrightarrow{u} Y \xrightarrow{v} Z
$$

there is a commutative diagram

$$
\begin{array}{ccccccc}
X & \xrightarrow{u} & Y & \xrightarrow{x} & Z' & \longrightarrow & SX \\
\| & & {\scriptstyle v}\downarrow & & {\scriptstyle w}\downarrow & & \| \\
X & \longrightarrow & Z & \xrightarrow{y} & Y' & \xrightarrow{s} & SX \\
& & \downarrow & & {\scriptstyle t}\downarrow & & \downarrow{\scriptstyle Su} \\
& & X' & = & X' & \xrightarrow{r} & SY \\
& & {\scriptstyle r}\downarrow & & \downarrow & & \\
& & SY & \xrightarrow{Sx} & SZ' & &
\end{array} \quad ,
$$

where the first two rows and the two central columns are triangles.

Property T4 can be given a more symmetric form if we represent a
morphism $X \to SY$ by the symbol $X \xrightarrow{+} Y$ and write a triangle in the
form

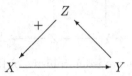

With this notation, the diagram of T4 can be written as an octahedron
in which 4 faces represent triangles. The other 4 as well as two of the 3
squares 'containing the center' are commutative.

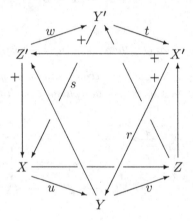

A *triangulated category* is an additive category $\mathcal{T}$ endowed with an auto-
equivalence $X \mapsto X[1]$ and a class of sequences (called triangles) of the
form

$$X \to Y \to Z \to X[1]$$

which is stable under isomorphisms and satisfies properties T1–T4.

Note that 'being abelian' is a property of an additive category, whereas
'being triangulated' is the datum of extra structure.

A whole little theory can be deduced from the axioms of triangulated
categories. This theory is nevertheless much poorer than that of abelian
categories. The main reason for this is the non-uniqueness of the morphism z in axiom T3.

We mention only two consequences of the axioms: a) They are actually self-dual, in the sense that the opposite category $\mathcal{T}^{op}$ also carries a canonical triangulated structure. b) For each $U \in \mathcal{T}$, the functor $\mathsf{Hom}_{\mathcal{T}}(U, ?)$ is *homological*, *i.e.* it takes triangles to long exact sequences. Dually, the functor $\mathsf{Hom}_{\mathcal{T}}(?, V)$ is *cohomological* for each V of $\mathcal{T}$. By the 5-lemma, this implies for example that if in axiom T3, two of the three vertical morphisms are invertible, then so is the third.

For later use, we record a number of examples of triangulated categories: If $\mathcal{A}$ is abelian, then not only the derived category $\mathsf{D}\,\mathcal{A}$ is triangulated but also the homotopy category $\mathsf{H}\,\mathcal{A}$. Here the triangles are constructed from componentwise split short exact sequences of complexes.

If $\mathcal{T}$ is a triangulated category, a *full triangulated subcategory of* $\mathcal{T}$ is a full subcategory $\mathcal{S} \subset \mathcal{T}$ such that $\mathcal{S}[1] = \mathcal{S}$ and that whenever we have a triangle (X, Y, Z) of $\mathcal{T}$ such that X and Z belong to $\mathcal{T}$ there is an object Y' of $\mathcal{S}$ isomorphic to Y. For example, the full subcategories $\mathsf{H}^*\,\mathcal{A}$, $* \in \{-, +, b\}$, of $\mathsf{H}\,\mathcal{A}$ are full triangulated subcategories. Note that the categories $\mathsf{H}^*\,\mathcal{A}$, $* \in \{\varnothing, +, -, b\}$, are in fact defined for any additive category $\mathcal{A}$.

If $\mathcal{T}$ is a triangulated category and $\mathcal{X}$ a class of objects of $\mathcal{T}$, there is a smallest *strictly* (=closed under isomorphism) full triangulated subcategory $\mathsf{tria}(\mathcal{X})$ of $\mathcal{T}$ containing $\mathcal{X}$. It is called the *triangulated subcategory generated by* $\mathcal{X}$. For example, the category $\mathsf{D}^b\,\mathcal{A}$ is generated by $\mathcal{A}$ (identified with the category of complexes concentrated in degree 0).

If R is a ring, a very important triangulated subcategory is the full subcategory $\mathsf{per}\,R \subset \mathsf{D}\,\mathsf{Mod}\,R$ formed by the *perfect* complexes, i.e. the complexes quasi-isomorphic to bounded complexes with components in $\mathsf{proj}\,R$, the *category* of finitely generated projective R-modules. The subcategory $\mathsf{per}\,R$ may be intrinsically characterized [65, 6.3] as the subcategory of *compact objects* of $\mathsf{D}\,\mathsf{Mod}\,R$, i.e. objects X whose associated functor $\mathsf{Hom}(X, ?)$ commutes with arbitrary set-indexed coproducts. Note that by lemma 2.3, the canonical functor

$$\mathsf{H}^b\,\mathsf{proj}\,R \to \mathsf{per}\,R$$

is an equivalence so that the category $\mathsf{per}\,R$ is relatively accessible to explicit computations.

5.2 Auslander-Reiten sequences and triangles

How are short exact sequences or triangles reflected in the quiver of a
multilocular abelian or triangulated category ? The problem is that the
three terms of a triangle, like that of a short exact sequence, are only
very rarely all indecomposable. The solution to this problem is provided
by Auslander-Reiten theory, developed in [3] [4] [5] [6] [7] and presented,
for example, in [27] and [1]. The typical 'mesh structure' which we
observe in the quivers in figures 5.1 to 5.3 is produced by the 'minimal
non split' exact sequences (resp. triangles), *i.e.* the Auslander-Reiten
sequences (resp. triangles).

Let $\mathcal{A}$ be a multilocular abelian category and let X and Z be indecomposable objects of $\mathcal{A}$. An *almost split sequence (or Auslander-Reiten
sequence) from Z to X* is a non-split exact sequence

$$0 \to X \xrightarrow{i} Y \xrightarrow{p} Z \to 0$$

having the two equivalent properties

   i) each non isomorphism $U \to Z$ with indecomposable U factors
      through p;

   ii) each non isomorphism $X \to V$ with indecomposable V factors
       through i.

In this case, the sequence is determined up to isomorphism by Z (as well
as by X) and X is the *translate* of Z (resp. Z the *cotranslate* of X).
Moreover, if an indecomposable U occurs in Y with multiplicity μ, then
there are μ arrows from X to U and μ arrows from U to Z in the quiver
of $\mathcal{A}$. We write $X = \tau Z$ resp. $Z = \tau^- X$. This yields the following
additional structure on the quiver $\Gamma(\mathcal{A})$:

   - a bijection τ from set of 'non-projective' vertices to the set of
     'non-injective' vertices;

   - for each non projective vertex $[Z]$ and each indecomposable U, a
     bijection σ from the set of arrows from $[U]$ to $[Z]$ to the set of
     arrows from $\tau[Z]$ to $[U]$.

Auslander-Reiten have shown, *cf.* [27] or [1], that if $\mathcal{A}$ is the category
mod A of finite-dimensional modules over a finite-dimensional algebra,
each non-projective indecomposable Z occurs as the right hand term of

an almost split sequence and each non-injective indecomposable X as the left hand term.

Analogously, if $\mathcal{A}$ is a multilocular triangulated category, an *almost split triangle (or Auslander-Reiten triangle)* is defined as a triangle

$$X \xrightarrow{i} Y \xrightarrow{p} Z \xrightarrow{e} X[1]$$

such that X and Z are indecomposable and the equivalent conditions i) and ii) above hold. Almost split triangles have properties which are completely analogous to those of almost split sequences. D. Happel [33] has shown that in the derived category of the category of finite-dimensional modules over a finite-dimensional algebra, an object Z occurs as the third term of an almost split triangle iff it is isomorphic to a bounded complex of finitely generated projectives. For example, for the quiver of $\mathsf{D}^b(\text{mod } k\vec{A}_n)$ in the middle part of figure 5.1, the translation $X \mapsto \tau X$ is given by $(g, h) \mapsto (g-1, h)$. In the quivers of the two module categories, it is the 'restriction' (where defined) of this map. Similarly, in the middle part of figure 5.2, the translation τ is given by $(g, h) \mapsto (g-1, h)$ and in the lower part of the figure by the restriction (where defined) of this map. The analogous statement is true for the category of 'commutative squares' in the top part of the figure except for $\tau^- P_1$, whose translate is P_1 and I_4, whose translate is τI_4 (such exceptions are to be expected because the category of commutative squares is not hereditary).

5.3 Grothendieck groups

Then *Grothendieck group* $K_0(\mathcal{T})$ of a triangulated category $\mathcal{T}$ is defined [32] as the quotient of the free abelian group on the isomorphism classes $[X]$ of objects of $\mathcal{T}$ divided by the subgroup generated by the relators

$$[X] - [Y] + [Z]$$

where (X, Y, Z) runs through the triangles of $\mathcal{T}$.

For example, if R is a right coherent ring, then the category $\text{mod } R$ of finitely presented R-modules is abelian and the K_0-group of the triangulated category $\mathsf{D}^b \text{mod } R$ is isomorphic to $G_0 R = K_0(\text{mod } R)$ via the Euler characteristic:

$$[M] \mapsto \sum_{i \in \mathbb{Z}} (-1)^i [H^i M].$$

If R is any ring, the K_0-group of the triangulated category per R is isomorphic to $K_0 R$ via the map

$$[P] \mapsto \sum_{i \in \mathbb{Z}} (-1)^i [P^i], \; P \in \mathsf{H}^b \operatorname{proj} R.$$

Note that this shows that any two rings with the 'same' derived category, will have isomorphic K_0-groups. To make this more precise, we need the notion of a triangle equivalence (cf. below).

5.4 Triangle functors

Let $\mathcal{S}, \mathcal{T}$ be triangulated categories. A *triangle functor* $\mathcal{S} \to \mathcal{T}$ is a pair (F, φ) formed by an additive functor $F \colon \mathcal{S} \to \mathcal{T}$ and a functorial isomorphism

$$\varphi X \colon F(X[1]) \xrightarrow{\sim} (FX)[1],$$

such that the sequence

$$FX \xrightarrow{Fu} FY \xrightarrow{Fv} FZ \xrightarrow{(\varphi X) Fw} (FX)[1]$$

is a triangle of $\mathcal{T}$ for each triangle (u, v, w) of $\mathcal{S}$.

For example, if $\mathcal{A}$ and $\mathcal{B}$ are abelian categories and $F \colon \mathcal{A} \to \mathcal{B}$ is an additive functor, one can show [23] that the domain of definition of the right derived functor $\mathbf{R}F$ is a strictly full triangulated subcategory $\mathcal{S}$ of $\mathsf{D}\mathcal{A}$ and that $\mathbf{R}F \colon \mathcal{S} \to \mathsf{D}\mathcal{B}$ becomes a triangle functor in a canonical way.

A triangle functor (F, φ) is a *triangle equivalence* if the functor F is an equivalence. We leave it to the reader as an exercise to define 'morphisms of triangle functors', and 'quasi-inverse triangle functors', and to show that a triangle functor admits a 'quasi-inverse triangle functor' if and only if it is a triangle equivalence [53].

6 Morita theory for derived categories

6.1 Rickard's theorem

Let k be a commutative ring. One version of the Morita theorem states that for two k-algebras A and B the following statements are equivalent:

(i) There is a k-linear equivalence $F\colon \operatorname{Mod} A \to \operatorname{Mod} B$.

(ii) There is an A-B-bimodule X (with central k-action) such that the tensor product $? \otimes_A X$ is an equivalence from $\operatorname{Mod} A$ to $\operatorname{Mod} B$.

(iii) There is a finitely generated projective B-module P which generates $\operatorname{Mod} B$ and whose endomorphism ring is isomorphic to A.

This form of the Morita theorem carries over to the context of derived categories. The following theorem is due to J. Rickard [65] [66]. A direct proof can be found in [49] (with a more didactical version in [52]).

Theorem 6.1 (Rickard). *Let A and B be k-algebras which are flat as modules over k. The following are equivalent*

i) *There is a k-linear triangle equivalence $(F, \varphi)\colon \operatorname{D}\operatorname{Mod} A \to \operatorname{D}\operatorname{Mod} B$.*

ii) *There is a complex of A-B-modules X such that the total left derived functor*

$$\mathbf{L}(? \otimes_A X)\colon \operatorname{D}\operatorname{Mod} A \to \operatorname{D}\operatorname{Mod} B$$

is an equivalence.

iii) *There is a complex T of B-modules such that the following conditions hold*

    a) *T is perfect,*

    b) *T generates $\operatorname{D}\operatorname{Mod} B$ as a triangulated category with infinite direct sums,*

    c) *we have*

$$\operatorname{Hom}_{\operatorname{D} B}(T, T[n]) = 0 \text{ for } n \neq 0 \text{ and } \operatorname{Hom}_{\operatorname{D} B}(T, T) \cong A \ ;$$

Condition b) in iii) means that $\operatorname{D}\operatorname{Mod} B$ coincides with its smallest strictly full triangulated subcategory stable under forming arbitrary (set-indexed) coproducts. The implication from ii) to i) is clear. To prove the implication from i) to iii), one puts $T = FA$ (where A is regarded as the free right A-module of rank one concentrated in degree 0). Since F is a triangle equivalence, it is then enough to check that the analogues of a), b), and c) hold for the object A of $\operatorname{D}\operatorname{Mod} A$. Properties a) and c) are clear. Checking property b) is non-trivial [51]. The hard part of the proof is the implication from iii) to ii). Indeed, motivated by the proof of the classical Morita theorem we would like to put $X = T$. The problem is that although A acts on T as an object of the derived category, it does not act on the individual components of T, so that T is not a complex of bimodules as required in ii). We refer to [48] for a direct solution of this problem.

Condition b) of iii) may be replaced by the condition that the direct summands of T generate per B as a triangulated category, which is easier to check in practice.

If the algebras A and B are even projective as modules over k, then the complex X may be chosen to be bounded and with components which are projective from both sides. In this case, the tensor product functor $? \otimes_A X$ is exact and the total left derived functor $? \otimes_A^{\mathbf{L}} X$ is isomorphic to the one induced by $? \otimes_A X$.

By definition [66], the algebra A is *derived equivalent* to B if the conditions of the theorem hold. In this case, T is called a *tilting complex*, X a *two-sided tilting complex* and $\mathbf{L}(? \otimes_A X)$ a *standard equivalence*.

We know that any equivalence between module categories is given by the tensor product with a bimodule. Strangely enough, in the setting of derived categories, it is an open question whether all k-linear triangle equivalences are (isomorphic to) standard equivalences.

An important special case of the theorem is the one where the equivalence F in (i) takes the free A-module A_A to an object $T = F(A_A)$ whose homology is concentrated in degree 0. Then T becomes an A-B-bimodule in a natural way and we can take $X = T$ in (ii). The equivalence between (ii) and (iii) then specializes to Happel's theorem (4.1). In particular, this yields many non-trivial examples of derived equivalent algebras which are not Morita equivalent.

Derived equivalence is an equivalence relation, and if two algebras A and B are related by a tilting triple, then they are derived equivalent. One may wonder whether derived equivalence coincides with the smallest equivalence relation containing all pairs of algebras related by a tilting triple. Let us call this equivalence relation *T-equivalence*. It turns out that T-equivalence is strictly stronger than derived equivalence. For example, any T-equivalence between self-injective algebras comes from a Morita-equivalence but there are many derived equivalent self-injective algebras which are not Morita equivalent, *cf.* below. On the other hand, two *hereditary* finite-dimensional algebras are T-equivalent iff they are derived equivalent, by a result of Happel-Rickard-Schofield [36].

In the presence of an equivalence $D(A) \to D(B)$, strong links exist between the abelian categories $\mathsf{Mod}\,A$ and $\mathsf{Mod}\,B$. They can be analyzed in analogy with 4.2 (where w now becomes the width of an interval containing all non-zero homology groups of T). The more refined results of section 7 also apply in this situation.

6.2 Example: A braid group action

To illustrate theorem 6.1, let $n \geqslant 2$ and consider the algebra A given by the quiver

with the relators

$$\alpha_{i+1}\alpha_i \,,\; \beta_i\beta_{i+1} \,,\; \alpha_i\beta_i - \beta_{i+1}\alpha_{i+1} \text{ for } 1 \leqslant i < n-1$$

and

$$\alpha_1\beta_1\alpha_1 \,,\; \beta_{n-1}\alpha_{n-1}\beta_{n-1}.$$

Note that the bilinear form

$$< [P],[Q] >= \dim \mathsf{Hom}(P,Q)$$

defined on $K_0(A)$ is symmetric and non degenerate. In fact, its matrix in the basis given by the $P_i = e_i A$, $1 \leqslant i \leqslant n$, is the Cartan matrix of the root system of type A_n. For $1 \leqslant i \leqslant n$, let X_i be the complex of A-A-bimodules

$$0 \to Ae_i A \to A \to 0 \,,$$

where A is concentrated in degree 0. It is not very hard to show that X_i is a two-sided tilting complex. Note that the automorphism σ_i of $K_0(A)$ induced by $? \otimes_A^{\mathsf{L}} X_i$ is the reflection at the ith simple root $[P_i]$ so that the group generated by these automorphisms is the Weyl group of A_n (*i.e.* the symmetric group of degree $n+1$). Rouquier-Zimmermann [69] (*cf.* also [55]) have shown that the functors $F_i =? \otimes^{\mathsf{L}} X_i$ themselves satisfy the braid relations (up to isomorphism of functors) so that we obtain a (weak) action of the braid group on the derived category $\mathsf{D}\,A$.

6.3 The simplest form of Broué's conjecture

A large number of derived equivalent (and Morita non equivalent) algebras is provided by Broué's conjecture [17], [16], which, in its simplest form, is the following statement

Conjecture 6.2 (Broué). *Let k be an algebraically closed field of characteristic p and let G be a finite group with abelian p-Sylow subgroups. Then $B_{pr}(G)$ (the principal block of of kG) is derived equivalent to $B_{pr}(N_G(P))$, where P is a p-Sylow subgroup.*

We refer to [64] for a proof of the conjecture for blocks of group algebras with cyclic p-Sylows and to J. Chuang and J. Rickard's contribution to this volume [20] for much more information on the conjecture.

6.4 Rickard's theorem for bounded derived categories

Often, it makes sense to consider subcategories of the derived category defined by suitable finiteness conditions. The following theorem shows, among other things, that this yields the same derived equivalence relation:

Theorem 6.3 ([65]). *If A is derived equivalent to B, then*

a) *there is a triangle equivalence* per $A \xrightarrow{\sim}$ per B *(and conversely, if there is such an equivalence, then A is derived equivalent to B);*

b) *if A and B are right coherent, there is a triangle equivalence* $\mathsf{D}^b \operatorname{mod} A \xrightarrow{\sim} \mathsf{D}^b \operatorname{mod} B$ *(and conversely, if A and B are right coherent and there is such an equivalence, then A is derived equivalent to B).*

6.5 Subordinate invariants

One of the main motivations for considering derived categories is the fact that they contain a large amount of information about classical homological invariants. Suppose that A and B are k-algebras, projective as modules over k and that there is a complex of A-B-bimodules X such that $? \otimes^{\mathsf{L}} X$ is an equivalence.

a) The algebra A is of finite global dimension iff this holds for the algebra B and in this case, the difference of their global dimensions is bounded by $r - s + 1$ where $[r, s]$ is the smallest interval containing the indices of all non vanishing homology groups of X, *cf.* [27, 12.5]. Note that the homological dimensions may actually differ, as we see from example 4.4.

b) There is a canonical isomorphism $K_0 A \xrightarrow{\sim} K_0 B$ and, if A and B are right coherent, an isomorphism $G_0 A \xrightarrow{\sim} G_0 B$, *cf.* [65].

c) There is a canonical isomorphism between the centers of A and of B, *cf.* [65]. More generally, there is a canonical isomorphism between the Hochschild cohomology algebras of A and B, *cf.* [35] [66]. Moreover, this isomorphism is compatible with the Gerstenhaber brackets, *cf.* [47].

d) There is a canonical isomorphism between the Hochschild homologies of A and B, *cf.* [66], as well as between all variants of their cyclic homologies (in fact, the mixed complexes associated with A and B are linked by a quasi-isomorphism of mixed complexes, *cf.* [50]).

e) There is a canonical isomorphism between $K_i(A)$ and $K_i(B)$ for all $i \geqslant 0$. In fact, Thomason-Trobaugh have shown [81] how to deduce this from Waldhausen's results [84], *cf.* [22] or [24]. If A and B are right noetherian of finite global dimension, so that $K_i(A) = G_i(A)$, $i \geqslant 0$, it also follows from Neeman's description of the K-theory of an abelian category $\mathcal{A}$ purely in terms of the *triangulated category* $\mathsf{D}^b(\mathcal{A})$, *cf.* [60] [61] [62].

f) The topological Hochschild homologies and the topological cyclic homologies of A and B are canonically isomorphic. This follows from work of Schwede-Shipley, *cf.* [70].

7 Comparison of t-structures, spectral sequences

The reader is advised to skip this section on a first reading.

7.1 Motivation

Let $\mathcal{A}$ and $\mathcal{B}$ be abelian categories and suppose that there is a triangle equivalence

$$\Phi \colon \mathsf{D}^b(\mathcal{A}) \to \mathsf{D}^b(\mathcal{B})$$

between their derived categories. Our aim is to obtain relations between the categories $\mathcal{A}$ and $\mathcal{B}$ themselves. We will refine the analysis which we performed in section 4.2. For this, we will use the fact that the derived category $\mathsf{D}^b(\mathcal{A})$ is 'glued together' from countably many copies of $\mathcal{A}$. The gluing data are encoded in the natural t-structure on $\mathsf{D}^b(\mathcal{A})$. On the other hand, thanks to the equivalence Φ, we may also view $\mathsf{D}^b(\mathcal{A})$ as glued together from copies of $\mathcal{B}$. This is encoded in a second t-structure on $\mathsf{D}^b(\mathcal{A})$, the pre-image under Φ of the natural t-structure on $\mathcal{B}$. We now have two t-structures on $\mathsf{D}^b(\mathcal{A})$. The sought for relations between $\mathcal{A}$ and $\mathcal{B}$ will be obtained by comparing the two t-structures. We will see how spectral sequences arise naturally in this comparison. This generalizes an idea first used in tilting theory by Vossieck [83] and developped in this volume by Brenner-Butler [15]. Finally, we will review the relatively subtle results [54] which are obtained by imposing compatibility conditions between the two t-structures. These compatibility conditions (strictly) imply the vanishing of 'half' the E_2-terms of the spectral sequences involved.

Note that to obtain the second t-structure on $\mathsf{D}^b(\mathcal{A})$ we could equally well have started from a duality

$$\Psi \colon \mathsf{D}^b(\mathcal{A}) \to (\mathsf{D}^b(\mathcal{B}))^{op}.$$

Indeed, both tilting theory and Grothendieck-Roos duality theory [68] yield examples which fit into the framework which we are about to sketch.

7.2 Aisles and t-structures

Let $\mathcal{T}$ be a triangulated category with suspension functor S. A full additive subcategory $\mathcal{U}$ of $\mathcal{T}$ is called an *aisle* in $\mathcal{T}$ if

a) $S\mathcal{U} \subset \mathcal{U}$,

b) $\mathcal{U}$ is stable under extensions, i.e. for each triangle $X \to Y \to Z \to SX$ of $\mathcal{T}$ we have $Y \in \mathcal{U}$ whenever $X, Z \in \mathcal{U}$,

c) the inclusion $\mathcal{U} \to \mathcal{T}$ admits a right adjoint $\mathcal{T} \to \mathcal{U}$, $X \mapsto X_{\mathcal{U}}$.

For each full subcategory $\mathcal{V}$ of $\mathcal{T}$ we denote by $\mathcal{V}^{\perp}$ (resp. $^{\perp}\mathcal{V}$) the full additive subcategory consisting of the objects $Y \in \mathcal{T}$ satisfying $\mathsf{Hom}(X, Y) = 0$ (resp. $\mathsf{Hom}(Y, X) = 0$) for all $X \in \mathcal{V}$.

Proposition 7.1 ([46]). *A strictly (=closed under isomorphisms) full subcategory $\mathcal{U}$ of $\mathcal{T}$ is an aisle iff it satisfies a) and c')*

c') *for each object X of $\mathcal{T}$ there is a triangle $X_{\mathcal{U}} \to X \to X^{\mathcal{U}^{\perp}} \to S(X_{\mathcal{U}})$ with $X_{\mathcal{U}} \in \mathcal{U}$ and $X^{\mathcal{U}^{\perp}} \in \mathcal{U}^{\perp}$.*

Moreover, a triangle as in c') is unique.

Given an aisle $\mathcal{U} \subset \mathcal{T}$ and $n \in \mathbb{Z}$, we define

$$\mathcal{U}_{\leqslant n} = \mathcal{U}_{<n+1} = S^{n}\mathcal{U} \,,\; \mathcal{U}_{>n} = \mathcal{U}_{\geqslant n+1} = (\mathcal{U}_{\leqslant n})^{\perp} \,,$$
$$\tau_{\leqslant n}^{\mathcal{U}} X = \tau_{<n+1}^{\mathcal{U}} X = X_{\mathcal{U}_{\leqslant n}} \,,\; \tau_{>n}^{\mathcal{U}} X = \tau_{\geqslant n+1}^{\mathcal{U}} X = X^{\mathcal{U}_{>n}}.$$

Then the proposition above shows that $(\tau_{\leqslant n}^{\mathcal{U}}, \tau_{>n}^{\mathcal{U}})_{n \in \mathbb{Z}}$ is a *t-structure* [9] on $\mathcal{T}$ and that we have a bijection between aisles in $\mathcal{T}$ and *t*-structures on $\mathcal{T}$.

For example, let $\mathcal{T}$ be the derived category $\mathsf{D}(\mathcal{A})$ of an abelian category $\mathcal{A}$. Then the full subcategory $\mathcal{U}$ formed by the complexes X such that $H^{n}(X) = 0$ for all $n > 0$ is the *natural aisle* on $\mathsf{D}(\mathcal{A})$. Its right orthogonal is formed by the complexes Y with $H^{n}(Y) = 0$ for all $n \leqslant 0$. The corresponding truncation functors $\tau_{\leqslant 0}$ and $\tau_{>0}$ are given by

$$\tau_{\leqslant 0}(X) = (\ldots \to X^{-1} \to Z^{0}(X) \to 0 \to \ldots)$$
$$\tau_{>0}(X) = (\ldots 0 \to X^{0}/Z^{0}(X) \to X^{1} \to \ldots).$$

The corresponding *t*-structure is the *natural t-structure* on $\mathsf{D}(\mathcal{A})$. Let $\mathcal{U} \subset \mathcal{T}$ be an aisle. Its *heart* is the full subcategory

$$\mathcal{U} \cap S(\mathcal{U}^{\perp}) = \mathcal{U}_{\leqslant 0} \cap \mathcal{U}_{\geqslant 0}.$$

It equals the heart of the corresponding *t*-structure [9]. For example, the heart of the natural *t*-structure on $\mathsf{D}(\mathcal{A})$ equals $\mathcal{A}$ (identified with the full subcategory of the complexes with homology concentrated in degree 0). In general, the heart $\mathcal{H}$ of an aisle $\mathcal{U}$ is always abelian, each short exact sequence (i, p) of $\mathcal{H}$ fits into a unique triangle

$$A \xrightarrow{i} B \xrightarrow{p} C \xrightarrow{e} SA$$

and the functor $H_\mathcal{U}^0 = \tau_{\leq 0}\tau_{\geq 0}$ is a homological functor. We put $H_\mathcal{U}^n = H_\mathcal{U}^0 \circ S^n$.

7.3 Example: classical tilting theory

Let (A, T, B) be a tilting triple. In $\mathcal{T} = \mathsf{D}^b(\mathsf{Mod}\, A)$, we consider the natural aisle $\mathcal{U}$ and the aisle $\mathcal{V}$ which is the image of the natural aisle of $\mathsf{D}^b(\mathsf{Mod}\, A)$ under the functor $\mathbf{R}\mathsf{Hom}_B(T, ?)$. Then the heart $\mathcal{A}$ of $\mathcal{U}$ identifies with $\mathsf{Mod}\, A$, the heart $\mathcal{B}$ of $\mathcal{V}$ with $\mathsf{Mod}\, B$, the functor $H_\mathcal{V}^n|\mathcal{A}$ with $\mathsf{Tor}_{-n}^A(?, T)$ and the functor $H_\mathcal{U}^n|\mathcal{B}$ with $\mathsf{Ext}_B^n(T, ?)$.

7.4 Example: duality theory

Let R be a commutative ring which is noetherian and regular, *i.e.* of finite homological dimension. Recall [57, 17.4] that

a) For each finitely generated R-module M, the codimension

$$c(M) = \inf\{\dim R_p : p \in \mathsf{Spec}\,(R)\,, M_p \neq 0\}$$

equals the grade

$$g(M) = \inf\{i : \mathsf{Ext}_R^i(M, R) \neq 0\}.$$

b) We have $c(\mathsf{Ext}_R^n(M, R)) \geq n$ for all finitely generated R-modules M and N and each n.

The derived functor $D = \mathbf{R}\mathsf{Hom}_R(?, R)$ induces a duality

$$\mathsf{D}^b(\mathsf{mod}\, R) \xrightarrow{\sim} (\mathsf{D}^b(\mathsf{mod}\, R))^{op}.$$

In $\mathcal{T} = \mathsf{D}^b(\mathsf{mod}\, R)$, we consider the natural aisle $\mathcal{V}$ and the aisle $\mathcal{U}$ which is the image of the natural co-aisle under D. The heart $\mathcal{B}$ of $\mathcal{V}$ identifies with $\mathsf{mod}\, R$ and the heart $\mathcal{A}$ of $\mathcal{U}$ with $(\mathsf{mod}\, R)^{op}$. The functors $H_\mathcal{U}^n|\mathcal{B}$ and $H_\mathcal{V}^n|\mathcal{A}$ are given by $\mathsf{Ext}_R^{-n}(?, R)$ and $\mathsf{Ext}_R^n(?, R)$.

7.5 Spectral sequences

Let T be a triangulated category and H^0 a homological functor defined on T with values in an abelian category. Put $H^n = H^0 \circ S^n$, $n \in \mathbb{Z}$. Let

$$\ldots \to X_{q-1} \xrightarrow{i_q} X_q \to \ldots \,, \ q \in \mathbb{Z}$$

be a diagram in T such that $X_q = 0$ for all $q \ll 0$ and i_q is invertible for all $q \gg 0$. Let X be the colimit (=direct limit) of this diagram. Let us choose a triangle

$$X_{q-1} \xrightarrow{i_q} X_q \to X_{q-1}^q \to SX_q$$

for each $q \in \mathbb{Z}$. Then the sequences

$$\ldots \to H^{p+q}(X_{q-1}) \to H^{p+q}(X_q) \to H^{p+q}(X_{q-1}^q) \to \ldots \,, \ p, q \in \mathbb{Z},$$

combine into an exact couple

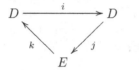

where $D^{pq} = H^{p+q}(X_q)$ and $E^{pq} = H^{p+q}(X_{q-1}^q)$. The associated spectral sequence has $E_2^{pq} = E^{pq}$ and its rth differential is of degree $(r, 1-r)$. It converges after finitely many pages to $H^{p+q}(X)$

$$E_2^{pq} = H^{p+q}(X_q) \Longrightarrow H^{p+q}(X).$$

The qth term of the corresponding filtration of $H^{p+q}(X)$ is the image of $H^{p+q}(X_q)$ under the canonical map ι so that we have canonical isomorphisms

$$E_\infty^{pq} \xrightarrow{\sim} \iota(H^{p+q}(X_q))/\iota(H^{p+q}(X_{q-1})).$$

Now suppose that in T, we are given two aisles $\mathcal{U}$ and $\mathcal{V}$ with hearts $\mathcal{A}$ and $\mathcal{B}$. We suppose that $\mathcal{A}$ generates $\mathcal{U}$ as a triangulated category and that the same holds for $\mathcal{B}$. This entails that for each $X \in T$, the sequence

$$\ldots \to \tau_{\leqslant q-1}^{\mathcal{V}} X \to \tau_{\leqslant q}^{\mathcal{V}} X \to \ldots$$

satisfies the assumptions made above. We choose the canonical triangles

$$\tau_{\leqslant q-1}^{\mathcal{V}} X \to \tau_{\leqslant q}^{\mathcal{V}} X \to S^{-q} H_{\mathcal{V}}^q(X) \to S\tau_{\leqslant q-1}^{\mathcal{V}} X.$$

If we apply the above reasoning to these data and to the homological functor $H_{\mathcal{U}}^0$, we obtain a spectral sequence, convergent after finitely many pages, with

$$E_2^{pq} = H_{\mathcal{U}}^{p+q}(S^{-q}H_{\mathcal{V}}^q(X)) = H_{\mathcal{U}}^p H_{\mathcal{V}}^q(X) \Longrightarrow H^{p+q}(X). \qquad (7.5.1)$$

Of course, if we exchange $\mathcal{U}$ and $\mathcal{V}$, we also obtain a spectral sequence

$$E_2^{pq} = H_{\mathcal{V}}^p H_{\mathcal{U}}^q(X) \Longrightarrow H^{p+q}(X). \qquad (7.5.2)$$

In example 7.3, the two sequences become

$$E_2^{pq} = \mathsf{Ext}_B^p(T, \mathsf{Tor}_{-q}^A(M, T)) \Longrightarrow M$$

and

$$E_2^{pq} = \mathsf{Tor}_{-p}^A(\mathsf{Ext}_B^q(T, N)) \Longrightarrow N,$$

where we suppose that $M \in \mathsf{Mod}\, A$ and $N \in \mathsf{Mod}\, B$. They lie respectively in the second and in the fourth quadrant and have their non zero terms inside a square of width equal to the projective dimension of T. Thus, we have $E_\infty = E_{r+1}$ if r is the projective dimension of T. In particular, if $r = 1$, then $E_2 = E_\infty$. For the first sequence, the qth term of the corresponding filtration of

$$M = \mathbf{R}\mathsf{Hom}_B(T, M \otimes_A^{\mathbf{L}} T)$$

equals the image of the map $H_{\mathcal{U}}^0(\tau_{\leqslant q}^{\mathcal{V}} M) \to M$, *i.e.* of

$$H^0(\mathbf{R}\mathsf{Hom}_B(T, \tau_{\leqslant q}(M \otimes_A^{\mathbf{L}} T)) \to M.$$

In example 7.4, the first sequence becomes

$$E_2^{pq} = \mathsf{Ext}_R^p(\mathsf{Ext}_R^{-q}(M, R), R) \Longrightarrow M,$$

where we suppose that $M \in \mathsf{mod}\, R$.

7.6 Compatibility of t-structures

Let $\mathcal{T}$ be a triangulated category with suspension functor S. Let $\mathcal{U}$ and $\mathcal{V}$ be aisles in $\mathcal{T}$. We use the notations of 7.5 for the t-structures associated with the aisles. Moreover, we put $\mathcal{A}_{\geqslant n} = \mathcal{A} \cap \mathcal{V}_{\geqslant n}$ and $\mathcal{B}_{\leqslant n} = \mathcal{B} \cap \mathcal{U}_{\leqslant n}$. Thus we have filtrations

$$\mathcal{A} \supset \dots \supset \mathcal{A}_{\geqslant n} \supset \mathcal{A}_{\geqslant n+1} \supset \dots \quad \text{and} \quad \dots \subset \mathcal{B}_{\leqslant n} \subset \mathcal{B}_{\leqslant n+1} \subset \dots \subset \mathcal{B}.$$

Note that $M \in \mathcal{A}$ belongs to $\mathcal{A}_{\geqslant n}$ iff $H_{\mathcal{V}}^q(M) = 0$ for all $q < n$. This occurs iff all the lines below $q = n$ vanish in the spectral sequence 7.5.1.

Similarly, $N \in \mathcal{B}$ belongs to $\mathcal{B}_{\leqslant n}$ iff all the lines above $q = n$ vanish in the spectral sequence 7.5.2.

The co-aisle $\mathcal{U}^\perp$ is *compatible* with the aisle $\mathcal{V}$ if $\mathcal{U}^\perp$ is stable under the truncation functors $\tau^\mathcal{V}_{\geqslant n}$ for all $n \in \mathbb{Z}$. Dually, the aisle $\mathcal{U}$ is *compatible* with the co-aisle $\mathcal{V}^\perp$, if $\mathcal{U}$ is stable under the truncation functors $\tau^\mathcal{V}_{\leqslant n}$ for all $n \in \mathbb{Z}$, cf. [54]. If $\mathcal{U}$ is compatible with $\mathcal{V}^\perp$, it is not hard to check that $\mathcal{U}$ is also stable under $\tau^\mathcal{V}_{>n}$ and we have $H^n_\mathcal{V}(\mathcal{U}_{\leqslant 0}) \subset \mathcal{B}_{\leqslant -n}$. Thus we obtain

$$H^p_\mathcal{U} H^q_\mathcal{V} | \mathcal{U}_{\leqslant 0} = 0$$

for all $p + q > 0$. Thus, if X belongs to $\mathcal{U}_{\leqslant 0}$, then in the spectral sequence 7.5.1, all terms above the codiagonal $p + q = 0$ vanish. The following proposition shows that the converse often holds:

Proposition 7.2 ([54]). *Suppose that $\mathcal{A}$ generates $\mathcal{T}$ as a triangulated category and that the same holds for $\mathcal{B}$. Then the following are equivalent*

(i) *$\mathcal{U}$ is compatible with $\mathcal{V}^\perp$.*

(ii) *$\mathcal{U} = \{X \in \mathcal{T} \mid H^n_\mathcal{V}(X) \in \mathcal{B}_{\leqslant -n}$ for all $n \in \mathbb{Z}\}$.*

(iii) *We have*

    a) *$H^p_\mathcal{U} H^q_\mathcal{V} | \mathcal{A} = 0$ for all $p + q > 0$ and*

    b) *for each morphism $g\colon N \to N'$ of $\mathcal{B}$ with $N \in \mathcal{B}_{\leqslant n}$ and $N' \in \mathcal{B}_{\leqslant n+1}$, we have $\ker(g) \in \mathcal{B}_{\leqslant n}$ and $\operatorname{cok}(g) \in \mathcal{B}_{\leqslant n+1}$.*

It is not hard to show that in example 7.4, the aisle $\mathcal{U}$ is compatible with $\mathcal{V}^\perp$ and $\mathcal{V}^\perp$ is compatible with $\mathcal{U}$. In example 7.3 these properties are satisfied if T is of projective dimension 1. They are not always satisfied for tilting modules T of higher projective dimension.

7.7 Links between the hearts of compatible t-structures

Keep the notations of 7.6 and suppose moreover that the *t-structures* defined by $\mathcal{U}$ and $\mathcal{V}$ are compatible, i.e. that $\mathcal{U}$ is compatible with $\mathcal{V}^\perp$ and $\mathcal{V}^\perp$ compatible with $\mathcal{U}$. Then for each object $N \in \mathcal{B}$, one obtains a short exact sequence

$$0 \to H^0_\mathcal{V}(\tau^\mathcal{U}_{\leqslant q} N) \to N \to H^0_\mathcal{V}(\tau^\mathcal{U}_{>q} N) \to H^1_\mathcal{V}(\tau^\mathcal{U}_{\leqslant q} N) \to 0.$$

Its terms admit intrinsic descriptions: First consider $N_{\leq q} = H^0_{\mathcal{V}}(\tau^{\mathcal{U}}_{\leq q} N)$. One shows that for each $N \in \mathcal{B}$, the morphism $N_{\leq q} \to N$ is the largest subobject of N contained in $\mathcal{B}_{\leq q}$. It follows that $\mathcal{B}_{\leq q}$ is stable under quotients. Note that $N_{\leq q}$ is also the qth term of the filtration on N given by the spectral sequence

$$E^{pq}_2 = H^p_{\mathcal{V}} H^q_{\mathcal{U}}(N) \Longrightarrow N.$$

Now consider $N_{>q} = H^0_{\mathcal{V}}(\tau^{\mathcal{U}}_{>q} N)$. Call a morphism $t \colon N \to N'$ of $\mathcal{B}$ a q-*quasi-isomorphism* if its kernel belongs to $\mathcal{B}_{\leq q}$ and its cokernel to $\mathcal{B}_{\leq q-1}$; call an object of B of $\mathcal{B}$ q-*closed* if the map

$$\mathsf{Hom}(t, B) \colon \mathsf{Hom}(N', B) \to \mathsf{Hom}(N, B)$$

is bijective for each q-quasi-isomorphism $t \colon N \to N'$. It is easy to see that the morphism $N \to N_{>q}$ is a q-quasi-isomorphism. Moreover, one shows that $N_{>q}$ is q-closed. Thus the functor $N \mapsto N_{>q}$ is left adjoint to the inclusion of the subcategory of q-closed objects in $\mathcal{B}$.

Dually, one defines q-co-quasi-isomorphisms and q-co-closed objects in $\mathcal{A}$. Let $\underline{\mathcal{A}}_q \subset \mathcal{A}_{\geq q}$ be the full subcategory of $(q+1)$-co-closed objects and $\underline{\mathcal{B}}_q \subset \mathcal{B}_{\leq q}$ the full subcategory of $(q+1)$-closed objects. Then we have the

Proposition 7.3. *The functors $H^q_{\mathcal{U}}$ and $H^q_{\mathcal{V}}$ induce a pair of adjoint functors*

$$H^q_{\mathcal{U}} : \mathcal{B}_{\leq -q} \xrightarrow{\quad\quad} \mathcal{A}_{\geq q} : H^q_{\mathcal{V}}$$

and inverse equivalences

$$\underline{\mathcal{B}}_q \xrightarrow{\quad\quad} \underline{\mathcal{A}}_{-q} \, .$$

8 Algebraic triangulated categories and dg algebras

8.1 Motivation

One form of Morita's theorem characterizes module categories among abelian categories: if $\mathcal{A}$ is an abelian category admitting all set-indexed coproducts and P is a compact (*i.e.* $\mathsf{Hom}(P, ?)$ commutes with all set-indexed coproducts) projective generator of $\mathcal{A}$, then the functor

$$\mathsf{Hom}(P, ?) \colon \mathcal{A} \to \mathsf{Mod}\,\mathsf{End}(P)$$

is an equivalence. Is there an analogue of this theorem for triangulated categories? Presently, it is not known whether such an analogue exists for arbitrary triangulated categories. However, for triangulated categories obtained as homotopy categories of Quillen model categories, there are such analogues. The most far-reaching ones are due to Schwede-Shipley, *cf.* [71] and [73]. The simplest, and historically first [49], case is the one where the triangulated category is the stable category of a Frobenius category. It turns out that all triangulated categories arising in algebra are actually of this form. In this case, the rôle of the module category $\mathsf{Mod}\,\mathsf{End}(P)$ is played by the derived category of a differential graded algebra. In this section, we will review the definition of differential graded algebras and their derived categories, state the equivalence theorem and illustrate it with Happel's description of the derived category of an ordinary algebra.

8.2 Differential graded algebras

Let k be a commutative ring. Following Cartan [18] a *differential graded (=dg) k-algebra* is a $\mathbb{Z}$-graded associative k-algebra

$$A = \bigoplus_{p \in \mathbb{Z}} A^p$$

endowed with a differential, *i.e.* a homogeneous k-linear endomorphism $d \colon A \to A$ of degree $+1$ such that $d^2 = 0$ and the Leibniz rule holds: we have

$$d(ab) = d(a)\,b + (-1)^p a\,d(b)$$

for all $a \in A^p$ and all $b \in A$. Let A be a dg algebra. A *differential graded A-module* is a $\mathbb{Z}$-graded A-module M endowed with a differential $d \colon M \to M$ homogeneous of degree $+1$ such that the Leibniz rule holds:

$$d(ma) = d(m)\,a + (-1)^p m\,d(a)$$

for all $m \in M^p$ and all $a \in A$. Note that the homology $H^*(A)$ is a $\mathbb{Z}$-graded algebra and that $H^*(M)$ becomes a graded $H^*(A)$-module for each dg A-module M.

If $A^p = 0$ for all $p \neq 0$, then A is given by the ordinary algebra A^0. In this case, a dg A-module is nothing but a complex of A^0-modules. In the general case, A becomes a dg module over itself: the free A-module of rank one. If M is an arbitrary A-module and n is an integer, then the shifted complex $M[n]$ carries a natural dg A-module structure (no additional sign changes here).

To give a more interesting example of a dg algebra, let us recall the morphism complex: Let B be an ordinary associative k-algebra. For two complexes

$$M = (\ldots \to M^p \overset{d_M^p}{\to} M^{p+1} \to \ldots)$$

and N of B-modules, the *morphism complex* $\text{Hom}^\bullet_B(M, N)$ has as its nth component the k-module of B-linear maps $f\colon M \to N$, homogeneous of degree n (which need not satisfy any compatibility condition with the differential). The differential of the morphism complex is defined by

$$d(f) = d_N \circ f - (-1)^n f \circ d_M \,,$$

where f is of degree n. Note that the zero cycles of the morphism complex identify with the morphisms of complexes $M \to N$ and that its 0th homology identifies with the set of homotopy classes of such morphisms. Then the composition of graded maps yields a natural structure of dg algebra on the endomorphism complex $\text{Hom}^\bullet_B(M, M)$ and for each complex N, the morphism complex $\text{Hom}^\bullet_B(M, N)$ becomes a natural dg module over $\text{Hom}^\bullet(M, M)$. Note that even if M is concentrated in degrees ≥ 0, the dg algebra $\text{Hom}^\bullet_B(M, M)$ may have non-zero components in positive and negative degrees.

8.3 The derived category

Let A be a dg algebra. A morphism $s\colon L \to M$ of dg A-modules is a *quasi-isomorphism* if it induces a quasi-isomorphism in the underlying complexes. By definition, the *derived category* $\text{D}(A)$ is the localization of the category of dg A-modules at the class of quasi-isomorphisms. If A is concentrated in degree 0, *i.e.* $A = A^0$, then $\text{D}(A)$ equals the ordinary derived category $\text{D}(A^0)$. Note that, for arbitrary dg algebras A, homology yields a well defined functor

$$H^*\colon \text{D}(A) \to \text{Grmod}(H^*A) \,, \quad M \mapsto H^*(M) \,,$$

where $\text{Grmod}(H^*A)$ denotes the category of graded $H^*(A)$-modules. To compute morphism spaces in the derived category, it is useful to introduce the homotopy category: a morphism of dg modules $f\colon L \to M$ is *nullhomotopic* if there is a morphism $r: L \to M$ of *graded* A-modules (not compatible with the differential) such that

$$f = d_M \circ r + r \circ d_L.$$

The nullhomotopic morphisms form an ideal in the category of dg A-modules and the quotient by this ideal is the homotopy category $\mathsf{H}(A)$. We have a canonical functor $\mathsf{H}(A) \to \mathsf{D}(A)$. A dg module M is *cofibrant* (resp. *fibrant*) if the map

$$\mathsf{Hom}_{\mathsf{H}(A)}(M, L) \to \mathsf{Hom}_{\mathsf{D}(A)}(M, L)$$

resp.

$$\mathsf{Hom}_{\mathsf{H}(A)}(L, M) \to \mathsf{Hom}_{\mathsf{D}(A)}(L, M)$$

is bijective for all dg A-modules L. We have the

Proposition 8.1 ([49]).   a) *The derived category* $\mathsf{D}(A)$ *admits a canonical triangulated structure whose suspension functor is* $M \to M[1]$. *Moreover, it admits all set-indexed coproducts and these are computed as coproducts of dg A-modules.*

b) *For each dg A-module M, there are quasi-isomorphisms*

$$\mathbf{p}M \to M \quad and \quad M \to \mathbf{i}M$$

such that $\mathbf{p}M$ *is cofibrant and* $\mathbf{i}M$ *is fibrant.*

c) *The free A-module A_A is cofibrant. We have*

$$\mathsf{Hom}_{\mathsf{D}(A)}(A, M[n]) \xrightarrow{\sim} H^n(M)$$

for all dg A-modules M. In particular, the functor $\mathsf{Hom}_{\mathsf{D}(A)}(A, ?)$ *commutes with coproducts and we have*

$$H^n(A) \xrightarrow{\sim} \mathsf{Hom}_{\mathsf{H}(A)}(A, A[n]) \xrightarrow{\sim} \mathsf{Hom}_{\mathsf{D}(A)}(A, A[n]).$$

Part b) of the proposition shows in particular that

$$\mathsf{Hom}_{\mathsf{D}(A)}(L, M)$$

is actually a set (not just a class) for all dg A-modules L and M. We deduce from the proposition that the object $A \in \mathsf{D}(A)$ is *compact* (*i.e.* its covariant Hom-functor commutes with coproducts) and *generates* $\mathsf{D}(A)$, in the sense that an object M vanishes iff we have $\mathsf{Hom}_{\mathsf{D}(A)}(A, M[n]) = 0$ for all $n \in \mathbb{Z}$.

The objects $\mathbf{p}M$ and $\mathbf{i}M$ are functorial in $M \in \mathsf{D}(A)$. They yield a left and a right adjoint of the canonical functor $\mathsf{H}(A) \to \mathsf{D}(A)$. For each functor $F \colon \mathsf{H}(A) \to \mathcal{C}$, one defines the total right and left derived functors via

$$\mathbf{R}F = F \circ \mathbf{i} \quad \text{and} \quad \mathbf{L}F = F \circ \mathbf{p}.$$

The *perfect derived category* $\mathsf{per}(A)$ is the full subcategory of $\mathsf{D}(A)$ whose objects are obtained from the free A-module of rank one by forming extensions, shifts (in both directions) and direct factors. Clearly it is a triangulated subcategory consisting of compact objects. We have the following important

Proposition 8.2 ([59]). *The perfect derived category* $\mathsf{per}(A)$ *equals the subcategory of compact objects of* $\mathsf{D}(A)$.

An explicit proof, based on [59], can be found in [49]. If A is an ordinary algebra, $\mathsf{per}(A)$ is equivalent to $\mathsf{H}^b(\operatorname{proj} A)$, the homotopy category of bounded complexes with finitely generated projective components.

8.4 Stalk algebras

Proposition 8.3. *Let $f \colon A \to B$ be a morphism of dg algebras which is a quasi-isomorphism of the underlying complexes. Then the restriction functor*

$$\mathsf{D}(B) \to \mathsf{D}(A)$$

is an equivalence.

It A is a dg algebra, then the complex

$$\tau_{\leqslant 0}(A) = (\ldots \to A^{-2} \to A^{-1} \to Z^0(A) \to 0 \to \ldots)$$

becomes a dg subalgebra and the canonical map $\tau_{\leqslant 0}(A) \to H^0 A$ a morphism of dg algebras (where we consider $H^0 A$ as a dg algebra concentrated in degree 0). Thus, if $H^*(A)$ is concentrated in degree 0, then A is linked to $H^0(A)$ by two quasi-isomorphisms. Thus we get the

Corollary 8.4. *If A is a dg algebra such that $H^*(A)$ is concentrated in degree 0, then there is a canonical triangle equivalence*

$$\mathsf{D}(A) \xrightarrow{\sim} \mathsf{D}(H^0 A)$$

8.5 Example: mixed complexes

Let Λ be the exterior k-algebra on one generator x of degree -1. Endow Λ with the zero differential. Then a dg Λ-module is given by a $\mathbb{Z}$-graded k-module M endowed with $b = d_M$ and with the map

$$B \colon M \to M \,, \; m \mapsto (-1)^{\deg(m)} m.x \,,$$

which is homogeneous of degree -1. We have

$$b^2 = 0 \,, \; B^2 = 0 \,, \; bB + Bb = 0.$$

By definition, the datum of the $\mathbb{Z}$-graded k-module M together with b and B satisfying these relations is a *mixed complex*, cf. [45]. The augmentation of Λ yields the Λ-bimodule k. The tensor product over Λ by k yields a functor

$$? \otimes_\Lambda k \colon \; \mathsf{H}(\Lambda) \to \mathsf{H}(k)$$

and if we compose its derived functor $? \otimes_\Lambda^{\mathsf{L}} k \colon \mathsf{D}(\Lambda) \to \mathsf{D}(k)$ with H^{-n}, we obtain the cyclic homology:

$$HC_n(M) = H^{-n}(M \otimes_\Lambda^{\mathsf{L}} k).$$

Moreover, the negative cyclic homology groups identify with morphism spaces in the derived category:

$$HN_n(M) = \mathsf{Hom}_{\mathsf{D}(\Lambda)}(k, M[n]).$$

8.6 Frobenius categories

A *Frobenius category* is an exact category in the sense of Quillen [63] which has enough injectives, enough projectives and where the class of the injectives coincides with that of the projectives. Let $\mathcal{E}$ be a Frobenius category. The morphisms factoring through a projective-injective form an ideal and the quotient by this ideal is the associated *stable category* $\underline{\mathcal{E}}$. The stable category admits a canonical structure of triangulated category [33] whose suspension functor S is defined by choosing admissible short exact sequences

$$0 \to L \to I \to S(L) \to 0$$

with projective-injective I for each object L. The triangles are constructed from the admissible exact sequences of $\mathcal{E}$.

For example, let A be a dg algebra (*e.g.* an ordinary algebra). Then the category of dg A-modules becomes a Frobenius category if we define a short exact sequence

$$0 \to L \to M \to N \to 0$$

of dg A-modules to be admissible exact if it splits in the category of graded A-modules. Then the morphisms factoring through projective-injectives are precisely the nullhomotopic morphisms and the associated stable category is the category $H(A)$. Now let $\mathcal{C}_c(A)$ be the full subcategory of the category of dg A-modules whose objects are the cofibrant dg A-modules. It is not hard to see that it inherits the structure of a Frobenius category and that its associated stable category is equivalent to $D(A)$ as a triangulated category.

8.7 Algebraic triangulated categories and dg algebras

Let $\mathcal{T}$ be an algebraic triangulated category, *i.e.* a triangulated category which is triangle equivalent to the stable category of some Frobenius category. As we have seen at the end of section 8.6, all derived categories of dg algebras are of this form.

Theorem 8.5 ([49]). *Let T be an object of $\mathcal{T}$.*

a) *There is a dg algebra* $\mathbf{R}\mathsf{Hom}(T,T)$ *with homology*

$$H^*(\mathbf{R}\mathsf{Hom}(T,T)) = \bigoplus_{p \in \mathbb{Z}} \mathsf{Hom}_{\mathcal{T}}(T, T[p])$$

and a k-linear triangle functor

$$F : \mathcal{T} \to \mathsf{D}(\mathbf{R}\mathsf{Hom}(T,T))$$

which takes T to the free module of rank one and whose composition with homology is given by

$$\mathcal{T} \to \mathsf{Grmod}(H^*(\mathbf{R}\mathsf{Hom}(T,T))), \quad X \mapsto \bigoplus_{p \in \mathbb{Z}} (T, X[p]).$$

b) *Suppose that $\mathcal{T}$ admits all set-indexed coproducts and that T is a compact generator for $\mathcal{T}$. Then the functor F is a k-linear triangle equivalence*

$$\mathcal{T} \xrightarrow{\sim} \mathsf{D}(\mathbf{R}\mathsf{Hom}(T,T)).$$

c) *Suppose that $\mathcal{T}$ is the closure of T under forming extensions, shifts (in both directions) and direct factors. Then F is a k-linear triangle equivalence*

$$\mathcal{T} \xrightarrow{\sim} \mathrm{per}(\mathbf{R}\mathrm{Hom}(T,T)).$$

If we take $\mathcal{T}$ to be the derived category of a k-algebra B and T a tilting complex, we can deduce the implication from iii) to i) in Rickard's theorem 6.1.

8.8 Illustration: Happel's theorem

Let k be a field and A a finite-dimensional k-algebra. Put $DA = \mathrm{Hom}_k(A,k)$. We view DA as an A-A-bimodule. Let B be the graded algebra with $B^p = 0$ for $p \neq 0,1$, $B^0 = A$ and $B^1 = DA$. Consider the category $\mathsf{Grmod}\,B$ of $\mathbb{Z}$-graded B-modules and its subcategory $\mathsf{grmod}\,B$ of graded B-modules of total finite dimension. If we endow them with all exact sequences, both become abelian Frobenius categories. We would like to apply theorem 8.5 to the stable category $\mathcal{T} = \underline{\mathsf{Grmod}}B$ and the B-module T given by A considered as a graded B-module concentrated in degree 0. A straightforward computation shows that T is compact in $\mathcal{T}$, that

$$\mathrm{Hom}_{\mathcal{T}}(T, T[n]) = 0$$

for all $n \neq 0$ and that $\mathrm{Hom}_{\mathcal{T}}(T,T)$ is canonically isomorphic to A (beware that the suspension in $\mathcal{T}$ has nothing to do with the shift functor of $\mathsf{Grmod}\,B$). By theorem 8.5 and corollary 8.4, we get a triangle functor

$$F\colon \underline{\mathsf{Grmod}}B \to \mathrm{D}(A).$$

Now by proving the hypotheses of b) and c) of theorem 8.5, one obtains the

Theorem 8.6 (Happel [33]). *If A is of finite global dimension, then F is a triangle equivalence*

$$\underline{\mathsf{Grmod}}B \xrightarrow{\sim} \mathrm{D}(A).$$

and induces a triangle equivalence

$$\underline{\mathsf{grmod}}B \xrightarrow{\sim} \mathrm{per}(A) \xrightarrow{\sim} \mathrm{D}^b(\mathrm{mod}\,A).$$

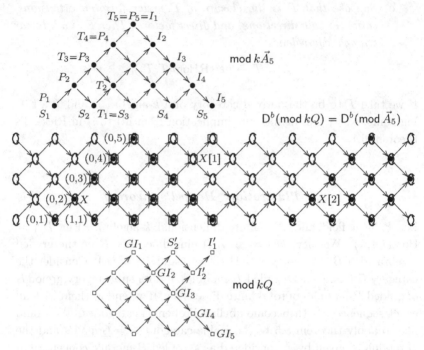

Fig. 5.1. Quivers of categories associated with algebras of type A_n

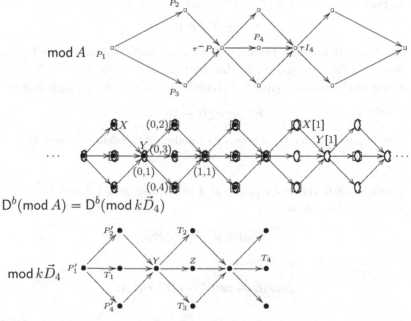

Fig. 5.2. Two module categories with the same derived category of type D_4

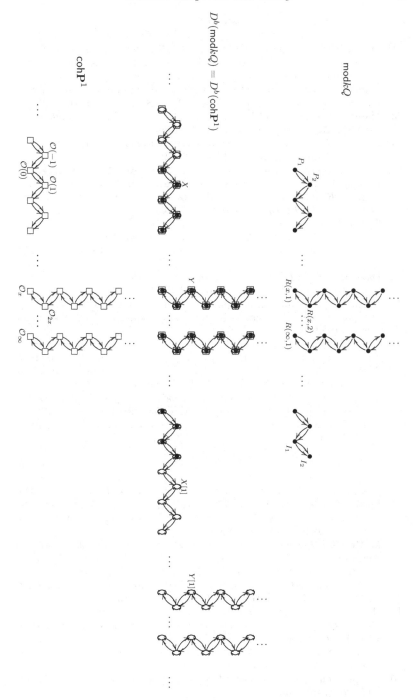

Fig. 5.3. Modules over the Kronecker quiver and coherent sheaves on $\mathbb{P}^1$

References

[1] M. Auslander, I. Reiten, and S. Smalø, *Representation theory of Artin algebras*, Cambridge Studies in Advanced Mathematics, vol. 36, Cambridge University Press, 1995 (English).

[2] Maurice Auslander, María Inés Platzeck, and Idun Reiten, *Coxeter functors without diagrams*, Trans. Amer. Math. Soc. **250** (1979), 1–46.

[3] Maurice Auslander and Idun Reiten, *Stable equivalence of dualizing R-varieties*, Advances in Math. **12** (1974), 306–366.

[4] _____, *Representation theory of Artin algebras. III. Almost split sequences*, Comm. Algebra **3** (1975), 239–294.

[5] _____, *Representation theory of Artin algebras. IV. Invariants given by almost split sequences*, Comm. Algebra **5** (1977), no. 5, 443–518.

[6] _____, *Representation theory of Artin algebras. V. Methods for computing almost split sequences and irreducible morphisms*, Comm. Algebra **5** (1977), no. 5, 519–554.

[7] _____, *Representation theory of Artin algebras. VI. A functorial approach to almost split sequences*, Comm. Algebra **6** (1978), no. 3, 257–300.

[8] Dagmar Baer, *Tilting sheaves in representation theory of algebras*, Manuscripta Math. **60** (1988), no. 3, 323–347.

[9] Alexander A. Beilinson, Joseph Bernstein, and Pierre Deligne, *Analyse et topologie sur les éspaces singuliers*, Astérisque, vol. 100, Soc. Math. France, 1982 (French).

[10] I. N. Bernšteĭn, I. M. Gel'fand, and V. A. Ponomarev, *Coxeter functors, and Gabriel's theorem*, Uspehi Mat. Nauk **28** (1973), no. 2(170), 19–33.

[11] A. A. Beĭlinson, *Coherent sheaves on $\mathbf{P}^n$ and problems in linear algebra*, Funktsional. Anal. i Prilozhen. **12** (1978), no. 3, 68–69.

[12] A. I. Bondal, *Representations of associative algebras and coherent sheaves*, Izv. Akad. Nauk SSSR Ser. Mat. **53** (1989), no. 1, 25–44.

[13] Klaus Bongartz, *Tilted algebras*, Representations of algebras (Puebla, 1980), Lecture Notes in Math., vol. 903, Springer, Berlin, 1981, pp. 26–38.

[14] Sheila Brenner and M. C. R. Butler, *Generalizations of the Bernstein-Gel'fand-Ponomarev reflection functors*, Representation theory, II (Proc. Second Internat. Conf., Carleton Univ., Ottawa, Ont., 1979), Lecture Notes in Math., vol. 832, Springer, Berlin, 1980, pp. 103–169.

[15] Sheila Brenner and Michael C. R. Butler, *A spectral sequence analysis of classical tilting functors*, this volume.

[16] M. Broué, *Rickard equivalences and block theory*, Groups '93 Galway/St. Andrews, Vol. 1 (Galway, 1993), London Math. Soc. Lecture Note Ser., vol. 211, Cambridge Univ. Press, Cambridge, 1995, pp. 58–79.

[17] Michel Broué, *Blocs, isométries parfaites, catégories dérivées*, C. R. Acad. Sci. Paris Sér. I Math. **307** (1988), no. 1, 13–18.

[18] H. Cartan, *Séminaire Henri Cartan de l'Ecole Normale Supérieure, 1954/1955. Algèbres d'Eilenberg-Maclane*, Secrétariat mathématique, 1956, Exposés 2 à 11.

[19] Henri Cartan and Samuel Eilenberg, *Homological algebra*, Princeton University Press, Princeton, N. J., 1956.

[20] Joseph Chuang and Rickard Jeremy, *Representations of finite groups and tilting*, this volume.

[21] E. Cline, B. Parshall, and L. Scott, *Derived categories and Morita theory*, J. Algebra **104** (1986), no. 2, 397–409.

[22] Julien Dalpayrat-Glutron, *Equivalences dérivées et k-théorie (d'après Thomason-Trobaugh)*, Mémoire de DEA, Université Denis Diderot – Paris 7, 1999.

[23] Pierre Deligne, *Cohomologie à supports propres*, pp. vi+640, Springer-Verlag, Berlin, 1973, Séminaire de Géométrie Algébrique du Bois-Marie 1963–1964 (SGA 4), Dirigé par M. Artin, A. Grothendieck et J. L. Verdier. Avec la collaboration de P. Deligne et B. Saint-Donat, Lecture Notes in Mathematics, Vol. 305.

[24] Daniel Dugger and Brooke Shipley, *K-theory and derived equivalences*, math.kt/0209084, 2002.

[25] Jens Franke, *On the Brown representability theorem for triangulated categories*, Topology **40** (2001), no. 4, 667–680.

[26] P. Gabriel, *Unzerlegbare Darstellungen I*, Manuscripta Math. **6** (1972), 71–103.

[27] P. Gabriel and A.V. Roiter, *Representations of finite-dimensional algebras*, Encyclopaedia Math. Sci., vol. 73, Springer–Verlag, 1992.

[28] P. Gabriel and M. Zisman, *Calculus of fractions and homotopy theory*, Ergebnisse der Mathematik und ihrer Grenzgebiete, Band 35, Springer-Verlag New York, Inc., New York, 1967.

[29] Werner Geigle and Helmut Lenzing, *A class of weighted projective curves arising in representation theory of finite-dimensional algebras*, Singularities, representation of algebras, and vector bundles (Lambrecht, 1985), Lecture Notes in Math., vol. 1273, Springer, Berlin, 1987, pp. 265–297.

[30] I. M. Gel'fand and V. A. Ponomarev, *Quadruples of subspaces of finite-dimensional vector space*, Dokl. Akad. Nauk SSSR **197** (1971), 762–765.

[31] _____, *Problems of linear algebra and classification of quadruples of subspaces in a finite-dimensional vector space*, Hilbert space operators and operator algebras (Proc. Internat. Conf., Tihany, 1970), North-Holland, Amsterdam, 1972, pp. 163–237. Colloq. Math. Soc. János Bolyai, 5.

[32] Alexandre Grothendieck, *Groupes de classes des catégories abéliennes et triangulées, complexes parfaits*, pp. 351–371, Springer-Verlag, Berlin, 1977, SGA 5, Exposé VIII, Lecture Notes in Mathematics, Vol. 589.

[33] Dieter Happel, *On the derived category of a finite-dimensional algebra*, Comment. Math. Helv. **62** (1987), no. 3, 339–389.

[34] _____, *Triangulated categories in the representation theory of finite-dimensional algebras*, Cambridge University Press, Cambridge, 1988.

[35] Dieter Happel, *Hochschild cohomology of finite-dimensional algebras*, Séminaire d'Algèbre Paul Dubreil et Marie-Paul Malliavin, 39ème Année (Paris, 1987/1988), Lecture Notes in Math., vol. 1404, Springer, Berlin, 1989, pp. 108–126.

[36] Dieter Happel, Jeremy Rickard, and Aidan Schofield, *Piecewise hereditary algebras*, Bull. London Math. Soc. **20** (1988), no. 1, 23–28.

[37] Dieter Happel and Claus Michael Ringel, *Tilted algebras*, Trans. Amer. Math. Soc. **274** (1982), no. 2, 399–443.

[38] Robin Hartshorne, *Residues and duality*, Lecture Notes in Mathematics, vol. 20, Springer–Verlag, 1966.

[39] Luc Illusie, *Catégories dérivées et dualité: travaux de J.-L. Verdier*, Enseign. Math. (2) **36** (1990), no. 3-4, 369–391.

[40] M. M. Kapranov, *The derived category of coherent sheaves on Grassmann varieties*, Funktsional. Anal. i Prilozhen. **17** (1983), no. 2, 78–79.

[41] ———, *Derived category of coherent sheaves on Grassmann manifolds*, Izv. Akad. Nauk SSSR Ser. Mat. **48** (1984), no. 1, 192–202.

[42] ———, *Derived category of coherent bundles on a quadric*, Funktsional. Anal. i Prilozhen. **20** (1986), no. 2, 67.

[43] ———, *On the derived categories of coherent sheaves on some homogeneous spaces*, Invent. Math. **92** (1988), no. 3, 479–508.

[44] Masaki Kashiwara and Pierre Schapira, *Sheaves on manifolds*, Grundlehren der Mathematischen Wissenschaften [Fundamental Principles of Mathematical Sciences], vol. 292, Springer-Verlag, Berlin, 1994, With a chapter in French by Christian Houzel, Corrected reprint of the 1990 original.

[45] Christian Kassel, *Cyclic homology, comodules and mixed complexes*, J. Alg. **107** (1987), 195–216.

[46] B. Keller and D. Vossieck, *Aisles in derived categories*, Bull. Soc. Math. Belg. Sér. A **40** (1988), no. 2, 239–253.

[47] Bernhard Keller, *Hochschild cohomology and derived Picard groups*, J. Pure Appl. Algebra **190** (2004), no. 1-3, 177–196.

[48] ———, *A remark on tilting theory and DG algebras*, Manuscripta Math. **79** (1993), no. 3-4, 247–252.

[49] ———, *Deriving DG categories*, Ann. Sci. École Norm. Sup. (4) **27** (1994), no. 1, 63–102.

[50] ———, *Basculement et homologie cyclique*, Algèbre non commutative, groupes quantiques et invariants (Reims, 1995), Sémin. Congr., vol. 2, Soc. Math. France, Paris, 1997, pp. 13–33.

[51] ———, *Invariance and localization for cyclic homology of DG algebras*, J. Pure Appl. Algebra **123** (1998), no. 1-3, 223–273.

[52] ———, *On the construction of triangle equivalences*, Derived equivalences for group rings, Lecture Notes in Math., vol. 1685, Springer, Berlin, 1998, pp. 155–176.

[53] Bernhard Keller and Dieter Vossieck, *Sous les catégories dérivées*, C. R. Acad. Sci. Paris Sér. I Math. **305** (1987), no. 6, 225–228.

[54] ———, *Dualité de Grothendieck-Roos et basculement*, C. R. Acad. Sci. Paris Sér. I Math. **307** (1988), no. 11, 543–546.

[55] Mikhail Khovanov and Paul Seidel, *Quivers, Floer cohomology, and braid group actions*, J. Amer. Math. Soc. **15** (2002), no. 1, 203–271 (electronic).

[56] Nikolaos Marmaridis, *Reflection functors*, Representation theory, II (Proc. Second Internat. Conf., Carleton Univ., Ottawa, Ont., 1979), Lecture Notes in Math., vol. 832, Springer, Berlin, 1980, pp. 382–395.

[57] H. Matsumura, *Commutative ring theory*, Cambridge University Press, 1986.

[58] Yoichi Miyashita, *Tilting modules of finite projective dimension*, Math. Z. **193** (1986), no. 1, 113–146.

[59] Amnon Neeman, *The connection between the K–theory localisation theorem of Thomason, Trobaugh and Yao, and the smashing subcategories of Bousfield and Ravenel*, Ann. Sci. École Normale Supérieure **25** (1992), 547–566.

[60] ———, *K–theory for triangulated categories I(A): homological functors*, Asian Journal of Mathematics **1** (1997), 330–417.

[61] _____, *K-theory for triangulated categories I(B): homological functors,* Asian Journal of Mathematics **1** (1997), 435–519.

[62] _____, *K-theory for triangulated categories II: the subtlety of the theory and potential pitfalls,* Asian Journal of Mathematics **2** (1998), 1–125.

[63] Daniel Quillen, *Higher algebraic K-theory. I,* Algebraic *K*-theory, I: Higher *K*-theories (Proc. Conf., Battelle Memorial Inst., Seattle, Wash., 1972), Lecture Notes in Math., vol. 341, Springer verlag, 1973, pp. 85–147.

[64] Jeremy Rickard, *Derived categories and stable equivalence,* J. Pure and Appl. Algebra **61** (1989), 303–317.

[65] Jeremy Rickard, *Morita theory for derived categories,* J. London Math. Soc. **39** (1989), 436–456.

[66] _____, *Derived equivalences as derived functors,* J. London Math. Soc. (2) **43** (1991), no. 1, 37–48.

[67] C.M. Ringel, *Tame algebras and integral quadratic forms,* Lecture Notes in Mathematics, vol. 1099, Springer Verlag, 1984.

[68] Jan-Erik Roos, *Bidualité et structure des foncteurs dérivés de* lim, C. R. Acad. Sci. Paris **280** (1962), 1556–1558.

[69] Raphaël Rouquier and Alexander Zimmermann, *Picard groups for derived module categories,* Proc. London Math. Soc. (3) **87** (2003), no. 1, 197–225.

[70] Stefan Schwede, *Morita theory in abelian, derived and stable model categories,* Notes of lectures at the workshop "Structured ring spectra and their applications" held in Glasgow in January 2002, 2002.

[71] Stefan Schwede and Brooke Shipley, *Stable model categories are categories of modules,* Topology **42** (2003), no. 1, 103–153.

[72] Jean-Pierre Serre, *Faisceaux algébriques cohérents,* Ann. of Math. (2) **61** (1955), 197–278.

[73] Brooke Shipley, *Morita theory in stable homotopy theory,* this volume.

[74] N. Spaltenstein, *Resolutions of unbounded complexes,* Compositio Math. **65** (1988), no. 2, 121–154.

[75] Ross Street, *Homotopy classification by diagrams of interlocking sequences,* Math. Colloq. Univ. Cape Town **13** (1984), 83–120.

[76] Ross H. Street, *Complete invariants for diagrams of short-exact sequences of free-abelian-group complexes,* Handwritten manuscript, 1968.

[77] _____, *Homotopy classification of filtered complexes,* Ph. D. Thesis, University of Sydney, 1968.

[78] _____, *Projective diagrams of interlocking sequences,* Illinois J. Math. **15** (1971), 429–441.

[79] _____, *Homotopy classification of filtered complexes,* J. Australian Math. Soc. **15** (1973), 298–318.

[80] Leovigildo Alonso Tarrío, Ana Jeremías López, and María José Souto Salorio, *Localization in categories of complexes and unbounded resolutions,* Canad. J. Math. **52** (2000), no. 2, 225–247.

[81] Robert W. Thomason and Thomas F. Trobaugh, *Higher algebraic K-theory of schemes and of derived categories,* The Grothendieck Festschrift (a collection of papers to honor Grothendieck's 60'th birthday), vol. 3, Birkhäuser, 1990, pp. 247–435.

[82] Jean-Louis Verdier, *Catégories dérivées, état 0,* SGA 4.5, Lec. Notes in Math., vol. 569, Springer–Verlag, 1977, pp. 262–308 (French).

[83] Dieter Vossieck, *Tilting theory, derived categories and spectral sequences,*

talk at the University of Paderborn, notes taken by H. Lenzing, 1986.

[84] Friedhelm Waldhausen, *Algebraic K-theory of spaces*, Algebraic and geometric topology (New Brunswick, N.J., 1983), Springer Verlag, Berlin, 1985, pp. 318–419.

Bernhard Keller
UFR de Mathématiques
UMR 7586 du CNRS
Case 7012
Université Paris 7
2, place Jussieu
75251 Paris Cedex 05
France

6

Hereditary categories

Helmut Lenzing

Hereditary categories serve as prototypes for many phenomena of representation theory. For instance the concepts of representation-finite, tame and wild algebras are most easily illustrated in the framework of finite quivers and the hereditary categories of their representations. Standard types of Auslander-Reiten components (preprojective, preinjective, tubes, type $\mathbb{Z}\mathbb{A}_\infty$) show up naturally in the hereditary context. Hereditary categories further allow the paradigmatic study of one-parameter families, a central concept in the study of tame algebras. Most important, due to the simple form of their derived category, hereditary categories are the natural domain to study derived equivalence and tilting for quite large classes of algebras. The classes of tilted, quasitilted, iterated tilted, tubular, canonical, concealed hereditary, concealed canonical algebras, among others, all owe their existence and properties to hereditary categories. Categories of coherent sheaves over smooth projective curves provide another main source for hereditary categories. A main aim of this survey is therefore to show the ubiquity of hereditary categories and the variety of effects covered by them.

Let $\mathcal{H}$ be a small abelian, connected k-category where k is a field. We assume that $\mathcal{H}$ is Ext-finite, that is, has all morphism and extension spaces $\mathrm{Ext}^i(X, Y)$ finite dimensional over k. We call $\mathcal{H}$ hereditary if $\mathrm{Ext}^2(-, -)$ vanishes. If k is algebraically closed and the hereditary category $\mathcal{H}$ has a tilting object then Happel's theorem [17] states that — up to derived equivalence — there are only two standard types to consider, namely the category mod-H of finite dimensional modules over a finite dimensional hereditary k-algebra H and the category $\mathrm{coh}\,\mathbb{X}$ of coherent sheaves on a weighted projective line $\mathbb{X}$. Starting from this fundamental result it is then an easy task to describe all hereditary categories with a tilting object, see [16] and [37]. This survey therefore focus on the study of the two standard types of hereditary categories with a tilting object and their main properties. Particular attention is given to a gentle axiomatic introduction of categories of coherent sheaves on a weighted projective line, establishing all their main properties but bypassing the technicalities of the actual construction of sheaves. A particular highlight is the classification of indecomposable sheaves on weighted projective lines of tubular type and the implied structure of the module category for a canonical algebra of tubular type.

1 Fundamental concepts

Let k be a field, supposed to be algebraically closed in order to simplify the exposition. Most results of this survey will be true also for arbitrary base fields, however proofs will often need to be modified and will sometimes even rely on more sophisticated concepts and arguments. The interested reader is referred to [7, 35, 20, 29, 36].

A category is said to be *k-linear* (or a *k-category*) if its morphism spaces are vector spaces over k and, moreover, composition is k-bilinear. We say that an abelian k-category $\mathcal{A}$ is *Ext-finite* if all extension spaces $\mathrm{Ext}^n(X,Y)$, defined in terms of Yoneda-extensions, are finite dimensional over k. Our main interest is the study of small (the isomorphism classes of objects form a set) abelian Ext-finite k-categories $\mathcal{H}$ which are additionally *hereditary*, that is, have vanishing $\mathrm{Ext}^2(-,-)$, and then vanishing $\mathrm{Ext}^n(-,-)$ for all $n \geqslant 2$. We refer to [40] for basic information on abelian categories and Yoneda extensions. Mostly we will deal with categories $\mathcal{H}$ that are *connected*, that is, do not decompose into a coproduct $\mathcal{H}_1 \coprod \mathcal{H}_2$ of nonzero categories $\mathcal{H}_1$ and $\mathcal{H}_2$. We say that $\mathcal{H}$ satisfies *weak Serre duality* if there exists a functor $\tau : \mathcal{H} \to \mathcal{H}$ such that for all X, Y from $\mathcal{H}$ there are isomorphisms $\mathrm{D}\,\mathrm{Ext}^1(X,Y) \xrightarrow{\cong} \mathrm{Hom}(Y, \tau X)$ which are functorial in X and Y. Here, D refers to the k-dual $\mathrm{Hom}(-,k)$. If additionally τ is an equivalence we say that $\mathcal{H}$ satisfies *Serre duality*. In this case $\mathcal{H}$ has no non-zero projectives or injectives, it has almost-split sequences and τ acts (on objects) as the Auslander-Reiten translation for $D^b \mathcal{H}$. Note that Serre duality for $\mathcal{H}$ — in the form specified here — implies that $\mathcal{H}$ is hereditary, see Section 10.2. Such hereditary categories arise in quite different mathematical contexts.

By $D^b\mathcal{H}$ we denote the derived category (of bounded complexes) of $\mathcal{H}$. Since $\mathcal{H}$ is hereditary, $D^b\mathcal{H}$ is the additive closure of the disjoint union $\bigcup_{n\in\mathbb{Z}}\mathcal{H}[n]$, where each $\mathcal{H}[n]$ is a copy of $\mathcal{H}$ with objects denoted $X[n]$, $X \in \mathcal{H}$, see Section 3. Morphisms are determined by $\operatorname{Hom}(X[m], Y[n]) = \operatorname{Ext}_{\mathcal{H}}^{n-m}(X, Y)$, composition is given by the Yoneda-composition of Ext, and translation of $D^b\mathcal{H}$ acts as $X[n] \mapsto X[n + 1]$, where $X \in \mathcal{H}$. Generally, we use the notation $Z \mapsto Z[n]$, $Z \in D^b\mathcal{H}$, for the n-th iterate of the translation functor. We write $D^b\mathcal{H} = \bigvee_{n\in\mathbb{Z}}\mathcal{H}[n]$ in order to indicate that there are no nonzero morphisms backwards, that is from $\mathcal{H}[n]$ to $\mathcal{H}[m]$ for $n > m$.

Assume $\mathcal{H}$ as above satisfies Serre duality or is of the form mod-H with H finite dimensional hereditary. In $D^b\mathcal{H}$ we have Serre duality in the form $\operatorname{D}\operatorname{Hom}(X,Y) = \operatorname{Hom}(Y[-1], \tau X)$, for some self-equivalence τ of $D^b\mathcal{H}$. Accordingly $D^b\mathcal{H}$ has Auslander-Reiten triangles, see [27] or [15] for this concept, and τ serves as the Auslander-Reiten translation. Note in this context a related not so obvious result proved by Happel [15]: D^bmod-A has Serre duality in the above form whenever A is a finite dimensional algebra of finite global dimension. Within this survey we will mostly encounter algebras whose module category is derived-equivalent to a hereditary category.

Let $\mathcal{A}$ be a small Ext-finite abelian k-category. We say that $T \in \mathcal{A}$ is a *tilting object* of $\mathcal{H}$ if

(i) T has no self-extensions, that is, $\operatorname{Ext}^n(T,T) = 0$ for all $n > 0$,

(ii) $\operatorname{Ext}^n(T, -) = 0$ for large n, and

(iii) T generates $\mathcal{H}$ in a homological sense: for every $X \in \mathcal{A}$ the condition $\operatorname{Ext}^n(T, X) = 0$ for every integer $n \geqslant 0$ implies that $X = 0$.

Note that (ii) is automatically satisfied if $\mathcal{A}$ is hereditary. Recall further that an abelian category $\mathcal{G}$ is called a *Grothendieck category* if it has a generator G in the categorical sense, that is $\operatorname{Hom}(G, u) = 0$ for any morphism u implies $u = 0$, and further $\mathcal{G}$ has arbitrary coproducts and exact direct limits. Each Grothendieck category automatically has injective envelopes, see [9], underlining the importance of this concept. Main examples of Grothendieck categories are the category Mod-R of *all* modules over a ring R and the category Qcoh X of quasi-coherent sheaves on a scheme X. An object E of $\mathcal{G}$ is called *finitely presented* if the functor $\operatorname{Hom}(E, -)$ from $\mathcal{G}$ to the category **Ab** of abelian groups commutes with direct limits. In case of the category Mod-R exactly those modules E admitting an exact sequence $R^n \to R^m \to E \to 0$, with integers m and n, are finitely presented. Finally, the author wishes to thank Henning Krause and the referees for for specific suggestions and valuable advice.

2 Examples of hereditary categories

We illustrate the ubiquity of hereditary categories by a list of examples. Each example is accompanied by comments, some of these rely on concepts explained only later.

Example 2.1. Let H be a finite dimensional hereditary k-algebra, for instance the path algebra $H = k[\vec{\Delta}]$ of a finite quiver $\vec{\Delta}$ without oriented cycles. Then the module category $\mathcal{H} = \text{mod-}H$ of finite dimensional (right) H-modules is a small abelian hereditary Ext-finite category. Since H is hereditary, the Auslander-Reiten translation 'is' a functor $\tau : \text{mod-}H \to \text{mod-}H$, take $\tau = \text{DExt}^1(-, H)$. Due to Auslander-Reiten theory we further have weak Serre (or Auslander-Reiten) duality. The category $\mathcal{H} = \text{mod-}H$ has a tilting object (or tilting module) T, we may take $T = H_H$ as a trivial example. Usually, however, there will be lots of further tilting modules, and it is of great interest to study them. The endomorphism rings of tilting modules over hereditary algebras are called *tilted algebras*.

Hereditary categories also arise in algebraic geometry, preferably related to geometric objects of dimension one.

Example 2.2. Let $R = \bigoplus_{n \geqslant 0} R_n$ denote the polynomial algebra $k[x, y]$ in two indeterminates, positively $\mathbb{Z}$-graded by total degree, where hence R_n consists of all homogeneous polynomials of total degree n. By $\text{mod}^{\mathbb{Z}}\text{-}R$ we denote the category of all finitely presented $\mathbb{Z}$-graded R-modules $M = \bigoplus_{n \in \mathbb{Z}} M_n$ with $M_l R_h \subseteq M_{l+h}$. Similarly $\text{mod}_0^{\mathbb{Z}}\text{-}R$ denotes the full category of all graded modules of finite length. Since $\text{mod}_0^{\mathbb{Z}}\text{-}R$ is a Serre subcategory of the abelian category $\text{mod}^{\mathbb{Z}}\text{-}R$ we may form the *quotient category* $\mathcal{H} = \text{mod}^{\mathbb{Z}}\text{-}R/\text{mod}_0^{\mathbb{Z}}\text{-}R$ in the sense of Serre-Grothendieck (see [9], [42]) which is an Ext-finite hereditary abelian connected k-category. Clearly $\mathbb{Z}$ acts on $\text{mod}^{\mathbb{Z}}\text{-}R$, hence on $\mathcal{H}$, by grading shift $M \mapsto M(l)$ where $M(l)$ is the graded R-module with underlying R-module M whose n-th component $M(l)_n$ equals M_{l+n}. We write $\widetilde{M}$

for the image of a graded R-module M in the quotient category $\mathcal{H}$, and in particular $\mathcal{O}$ for $\tilde{R}$. $\mathcal{H}$ satisfies Serre duality, where $\tau : \mathcal{H} \to \mathcal{H}$, $X \mapsto X(-2)$, is the equivalence induced by grading shift $M \mapsto M(-2)$. Using additionally that $\mathrm{Hom}(\mathcal{O}(n), \mathcal{O}(m)) = R_{m-n}$, it is easily checked that $T = \mathcal{O} \oplus \mathcal{O}(1)$ is a tilting object in $\mathcal{H}$ whose endomorphism ring is the *Kronecker algebra* $k[\circ \rightrightarrows \circ]$. It follows that the derived categories of $\mathcal{H}$ and of mod-$k[\circ \rightrightarrows \circ]$ are equivalent as triangulated categories. It is classical, and in fact a special case of Serre's theorem [49], that $\mathcal{H}$ is equivalent to the category of coherent sheaves on the projective line $\mathbb{P}^1(k)$.

This example allows many variations, we mention three of them explicitly.

Example 2.3. Let $R = k[x, y, z]/(h)$, $h = x^2 + y^3 + z^5$. Attaching degrees 15, 10 and 6 to x, y and z, respectively, turns R into a positively $\mathbb{Z}$-graded k-algebra. Again, the category $\mathcal{H}(2, 3, 5) = \mathrm{mod}^{\mathbb{Z}}$-$R/\mathrm{mod}_0^{\mathbb{Z}}$-$R$ is a connected hereditary abelian category with Serre duality $\mathrm{D}\,\mathrm{Ext}^1(X, Y) = \mathrm{Hom}(Y, \tau X)$, where $\tau : \mathcal{H} \to \mathcal{H}$ is the equivalence induced by the grading shift $M \mapsto M(-1)$. Writing $\mathcal{O} = \tilde{R}$ as in the preceding example, the direct sum of all objects from the configuration

$$\mathcal{O}(15)$$

$$\mathcal{O}(0) \quad \begin{matrix} \nearrow \\ \to \\ \searrow \end{matrix} \quad \mathcal{O}(10) \qquad \to \qquad \mathcal{O}(20) \qquad \to \qquad \begin{matrix} \searrow \\ \\ \nearrow \end{matrix} \quad \mathcal{O}(30)$$

$$\mathcal{O}(6) \quad \to \quad \mathcal{O}(12) \quad \to \quad \mathcal{O}(18) \quad \to \quad \mathcal{O}(24)$$

is a tilting object in $\mathcal{H}(2, 3, 5)$ whose endomorphism ring is the *canonical algebra* $\Lambda(2, 3, 5)$. Changing the relation to $h = x^2 + y^3 + z^7$ yields another hereditary abelian category $\mathcal{H}(2, 3, 7)$ with Serre duality, where now τ is induced by the shift $M \mapsto M(1)$. In this case the direct sum of the objects $\mathcal{O}$, $\mathcal{O}(21)$, $\mathcal{O}(14)$, $\mathcal{O}(28)$, $\mathcal{O}(6)$, $\mathcal{O}(12)$, $\mathcal{O}(18)$, $\mathcal{O}(24)$, $\mathcal{O}(30)$, $\mathcal{O}(36)$ and $\mathcal{O}(42)$ is a tilting object whose endomorphism ring is the canonical algebra $\Lambda(2, 3, 7)$. The categories $\mathcal{H}(2, 3, 5)$ and $\mathcal{H}(2, 3, 7)$ are equivalent to categories of coherent sheaves over *weighted projective lines* of weight type $(2, 3, 5)$ and $(2, 3, 7)$, respectively, compare [12]. From a different perspective we will investigate such categories in Section 10.

Example 2.4. Let $R = k[x, y]$ be the polynomial algebra. Consider R as a $\mathbb{Z} \times \mathbb{Z}$-graded algebra, where x gets degree $(1, 0)$ and y gets degree $(0, 1)$. The quotient category $\mathcal{H} = \mathrm{mod}^{\mathbb{Z} \times \mathbb{Z}}$-$R/\mathrm{mod}_0^{\mathbb{Z} \times \mathbb{Z}}$-$R$ is abelian hereditary with Serre duality, where τ is induced by grading shift $M \mapsto M(-1, -1)$. $\mathcal{H}$ does not have a tilting object, but does have an infinite *tilting system* $\mathcal{O}(p,q)$, where p, q runs through the integers satisfying $p + q = 0$ or $p + q = 1$. The full subcategory formed by this tilting system is equivalent to the path category of the infinite zig-zag-quiver

Within this survey we will not investigate such hereditary categories with an infinite tilting system. The interested reader is referred to [43] and [36].

Example 2.5. Let C be a smooth projective curve over an algebraically closed field k. Then the category $\mathcal{H} = \operatorname{coh} C$ of coherent (algebraic) sheaves on C is a small abelian hereditary Ext-finite category with Serre duality. This category has a tilting object if and only if C has genus zero, that is, C is isomorphic to the projective line over k. In a different language, coherent sheaves on the projective line are treated in example 2.2.

Example 2.6. For the base field of complex numbers such categories also occur in a different context. It is classical that there is a bijection between isomorphism classes of compact Riemann surfaces and isomorphism classes of smooth projective curves over $\mathbb{C}$, yielding equivalences of the respective categories of holomorphic and algebraic coherent sheaves, respectively. Let X be a compact Riemann surface. The category $\mathcal{H}$ of holomorphic coherent sheaves on X is an abelian, hereditary, Ext-finite $\mathbb{C}$-category with Serre duality. We have a tilting object in $\mathcal{H}$ if and only if X is isomorphic to the Riemann sphere.

Example 2.7. Let k be any field, and $R = k[[x]]$ denote the power series ring in one indeterminate. The category $\mathcal{T} = \operatorname{mod}_0\text{-}R$ of all R-modules of finite length (=finite k-dimension) is a hereditary k-category with Serre duality $\operatorname{D}\operatorname{Ext}^1(X, Y) = \operatorname{Hom}(Y, \tau X)$ with τ the identity functor. Each indecomposable in $\mathcal{T}$ is *uniserial*, that is, has a unique composition series. We state this by saying that $\mathcal{T}$ is a uniserial category. Clearly $\mathcal{T}$ is connected with just one simple object — up to isomorphism. Its indecomposables form a *tube* of rank (or τ-period) one. An easy variation leads to (uniserial) tubes of rank $p \geqslant 1$. Consider R to be $\mathbb{Z}_p$-graded, putting $R_{[k]} = x^k R$ for $0 \leqslant k < p$. The category $\mathcal{T}_p = \operatorname{mod}_0^{\mathbb{Z}_p}\text{-}R$ of finite dimensional $\mathbb{Z}_p$-graded R-modules has p simples $k([n])$, $0 \leqslant n < p$, arising from the simple R-module $k = R/(x)$ by grading shift. Again, we have Serre duality, where τ is given by grading shift $M \mapsto M(-[1])$. In particular, τ has period p. Note that $\mathcal{T}_p$ has no tilting object.

Example 2.8. There are further hereditary categories with Serre duality not covered by the present framework. Such categories arise for instance in the investigation of surface singularities. Here, we deal only with the most basic example. A detailed investigation is beyond the scope of this survey. Let $R = k[[x,y]]$ be the ring of formal power series in two indeterminates. Let mod-R (resp. mod$_0$-R) denote the category of all finitely presented (resp. all finite length) modules over R. Then the quotient category $\mathcal{H} = $ mod-R/mod$_0$-R is a hereditary abelian category which is noetherian and satisfies Serre duality $\mathrm{D}\,\mathrm{Ext}^1(X,Y) = \mathrm{Hom}(Y,X)$. Here D refers to Matlis-duality $\mathrm{Hom}_R(-,E(k))$ with $E(k)$ the injective hull of the simple R-module k. Note for this example that morphism and extension spaces are reflexive R-modules, that is, the canonical morphism into the Matlis-bidual is an isomorphism, but they are usually not finitely generated R-modules.

All the categories $\mathcal{H}$, mentioned so far, are additionally *noetherian*, meaning that each ascending sequence of subobjects (U_n) of an object E of $\mathcal{H}$ becomes stationary. Noetherianness is however not typical for a hereditary abelian category $\mathcal{H}$ with Serre duality. Just note that the category $\mathcal{H}^{\mathrm{op}}$, opposite to $\mathcal{H}$, is again hereditary abelian with Serre duality and, for the examples above, therefore an artinian category, that is, descending chains of subobjects become stationary. With the exception of Example 2.7 none of these opposite categories is noetherian. A hereditary abelian category with Serre duality not having any simple object, hence not being noetherian nor artinian, is obtained as follows.

Example 2.9. Let Λ be the path algebra of a wild connected quiver $\vec{\Delta}$, that is, we assume that the underlying connected graph Δ is neither Dynkin nor extended Dynkin. Then mod-$\Lambda = (\mathcal{P}(\Lambda) \vee \mathcal{R}(\Lambda)) \vee \mathcal{I}(\Lambda)$ is a cut in mod-Λ (see Section 7), where $\mathcal{P}(\Lambda)$, $\mathcal{R}(\Lambda)$ and $\mathcal{I}(\Lambda)$ respectively are the full subcategories consisting of all preprojective, regular or preinjective Λ-modules. Within the derived category we form the full subcategory

$$\mathcal{H}(\Lambda) = \mathcal{I}(\Lambda)[-1] \vee \mathcal{P}(\Lambda) \vee \mathcal{R}(\Lambda).$$

The category $\mathcal{H}(\Lambda)$ is abelian hereditary with Serre duality. There does not exist any simple object in $\mathcal{H}(\Lambda)$: Assume that S is simple in $\mathcal{H}(\Lambda)$. Since Λ is wild and τ is an equivalence of $\mathcal{H}(\Lambda)$, the objects $\tau^n S$, $n \geqslant 0$, are pairwise non-isomorphic simple objects. Since $\mathrm{Hom}(S, \tau^n S)$ is nonzero for large n, implying $S \cong \tau^n S$ by Schur's lemma, we get a contradiction.

3 Repetitive shape of the derived category

A main reason for the interest in hereditary categories is the very simple form of its derived category. By definition, the *bounded derived category* $D^b \mathcal{H}$ is obtained from the category of bounded complexes in $\mathcal{H}$ by formally inverting all quasi-isomorphisms, that is, those isomorphisms inducing isomorphisms for (co)-homology in each degree. $D^b \mathcal{H}$ carries the structure of a *triangulated category*. We refer to [27] and [14] for further information on derived and triangulated categories. For any abelian category $\mathcal{A}$ its *repetitive category* $\operatorname{Rep} \mathcal{H} = \bigvee_{n \in \mathbb{Z}} \mathcal{A}[n]$ is the additive closure of the union of disjoint copies $\mathcal{A}[n]$ of $\mathcal{A}$, objects written in the form $A[n]$ (A in $\mathcal{A}$), with morphisms given by $\operatorname{Hom}(A[m], B[n]) = \operatorname{Ext}^{n-m}(A, B)$ and composition given by the Yoneda product of extensions. Identifying $\mathcal{A}[n]$ with the complexes with cohomology concentrated in degree n, the repetitive category is a full subcategory of the bounded derived category $D^b \mathcal{A}$ of $\mathcal{A}$. Moreover, with the present notation, the translation functor for the derived category sends $A[m]$ to $A[m+1]$. Each object X in the repetitive category of $\mathcal{A}$ has the form $X = \bigoplus_{n \in \mathbb{Z}} X_n[n]$ where $X_n \in \mathcal{A}$ and only finitely many X_n's are non-zero.

Theorem 3.1. *Let $\mathcal{H}$ be an abelian hereditary category. Then the repetitive category $\operatorname{Rep} \mathcal{H}$ and the bounded derived category $D^b \mathcal{H}$ are naturally equivalent.*

Proof. Let X be a bounded complex, being zero in degrees $> n$. As usual let B^n (resp. Z^{n-1}) be the image (resp. the kernel) of $X^{n-1} \xrightarrow{d} X^n$. Let X' denote the complex obtained from X by replacing X^n by 0 and X^{n-1} by Z^{n-1}. We are going to show that X is quasi-isomorphic to $X' \oplus \mathrm{H}^n X[n]$ which by induction implies the claim. Since $\mathcal{H}$ is hereditary, the epimorphism $c : X^{n-1} \to B^n$, induced by d, induces an epimorphism $\operatorname{Ext}^1(\mathrm{H}^n X, X^{n-1}) \to \operatorname{Ext}^1(\mathrm{H}^n X, B^n)$. We thus obtain a commutative

diagram with exact diagonals

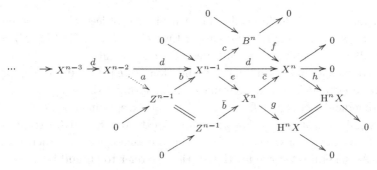

hence an induced diagram of complexes

with quasi-isomorphisms α and β. The claim follows. $\square$

For any abelian category $\mathcal{A}$ the *Grothendieck group* $\mathrm{K}_0(\mathcal{A})$ of $\mathcal{A}$ with respect to short exact sequences is canonically isomorphic to the Grothendieck group of the bounded derived category $D^b\mathcal{A}$ with respect to exact or distinguished triangles: Since short exact sequences in $\mathcal{A}$ yield exact triangles in $D^b\mathcal{A}$, identification of $\mathcal{A}$ with complexes in $D^b\mathcal{A}$ concentrated in degree zero induces a morphism $i_* : \mathrm{K}_0(\mathcal{A}) \to \mathrm{K}_0(D^b\mathcal{A})$ sending the class of X in $\mathcal{A}$ to the class of X in $\mathrm{K}_0(D^b\mathcal{A})$. Conversely, to a complex $C = (C_n)$ in $D^b\mathcal{A}$ we assign the element $\sum_{n\in\mathbb{Z}}(-1)^n[C_n]$ of $\mathrm{K}_0(\mathcal{A})$. Due to the shape of mapping cones this assignment is additive on exact triangles in $D^b\mathcal{A}$, and yields the desired inverse of i_*. For a hereditary category $\mathcal{H}$, where $\mathrm{Rep}\,\mathcal{H} = D^b\mathcal{H}$, the class of an object $X = \bigoplus_{n\in\mathbb{Z}} X_n[n]$, with $X_n \in \mathcal{H}$, is identified with $\sum_{n\in\mathbb{Z}}(-1)^n[X_n]$ in $\mathrm{K}_0(\mathcal{H})$.

Assume now that $\mathcal{H}$ is an Ext-finite hereditary category with Serre duality. In the sequel we will need the *Euler form* $\langle -, - \rangle : \mathrm{K}_0(\mathcal{H}) \times \mathrm{K}_0(\mathcal{H}) \to \mathbb{Z}$, which is the bilinear form on $\mathrm{K}_0(\mathcal{H})$ given on classes of objects by the expression $\langle [X], [Y] \rangle = \dim_k \mathrm{Hom}(X, Y) - \dim_k \mathrm{Ext}^1(X, Y)$. Because of Serre duality the Euler form satisfies $\langle x, y \rangle = -\langle y, \tau x \rangle$ for all $x, y \in \mathrm{K}_0(\mathcal{H})$. Note that here and later we denote the isomorphism induced by τ on $\mathrm{K}_0(\mathcal{H})$ by the same letter.

4 Perpendicular categories

Assume k is an algebraically closed field. An object E of an abelian k-category $\mathcal{C}$ is called *exceptional* if $\operatorname{End}(E) = k$ and $\operatorname{Ext}^n(E,E) = 0$ for all $n \geqslant 1$. Similarly, an object E of a triangulated k-category is called exceptional if $\operatorname{End}(E) = k$ and $\operatorname{Hom}(E, E[n]) = 0$ for each non-zero integer n. The proper framework for forming the left (resp. right) perpendicular category $E^\perp$ (resp. $^\perp E$) of an exceptional object E are the triangulated categories. For a hereditary category $\mathcal{H}$ and E exceptional in $D^b\mathcal{H}$, these constructions reduce to the analogous constructions for $\mathcal{H}$. As general reference for this section we refer to [4] and [13].

We start with some remarks on canonical morphisms in k-categories, respectively triangulated k-categories. Fix an object E. For each finite dimensional vector space V and object E of $\mathcal{H}$ we write $V \otimes E$ for $E^{\dim V}$. Since, obviously, $V \otimes E$ represents the functor $X \mapsto \operatorname{Hom}_k(V, \operatorname{Hom}(E, X))$, the tensor product $V \otimes E$ is functorial in V. It is then easily checked that there are natural isomorphisms $\operatorname{Hom}(X, V \otimes E) = V \otimes \operatorname{Hom}(X, E)$, where all tensor products are taken over k.

Lemma 4.1. *For each pair of objects E and X in $\mathcal{H}$ there is a canonical morphism $\kappa : \operatorname{Hom}(E, X) \otimes E \to X$ such that for any $F \in \mathcal{H}$ the application of $\operatorname{Hom}(F, -)$ induces the composition map $\operatorname{Hom}(E, X) \otimes \operatorname{Hom}(F, E) \to \operatorname{Hom}(F, X)$, $u \otimes v \mapsto u \circ v$.*

Proof. Let $u_1, \ldots, u_n$ be a k-basis of $\operatorname{Hom}(E, X)$, and let κ be the morphism $(u_1, \ldots, u_n) : E^n \to X$. $\square$

We need a variant for triangulated categories. Let E and X be objects of $D^b\mathcal{H}$. Define $\operatorname{Hom}^\bullet(E, X) \otimes E$ as the direct sum $\bigoplus_{n \in \mathbb{Z}} \operatorname{Hom}(E[n], X) \otimes E[n]$, yielding a canonical morphism $\kappa : \operatorname{Hom}^\bullet(E, X) \otimes E \to X$ which combines the canonical morphisms $\kappa_n : \operatorname{Hom}(E[n], X) \otimes E[n] \to X$ to a single morphism. Note, that in the present context the direct sum above has only finitely many non-zero terms. Let E be an object of $D^b\mathcal{H}$. The full subcategory $E^\perp$ of all objects X of $D^b\mathcal{H}$ such that $\operatorname{Hom}(E[n], X) = 0$ for each integer n is a triangulated subcategory of $D^b\mathcal{H}$, that is, if two terms of an exact triangle of $D^b\mathcal{H}$ belong to $E^\perp$ then so does the third. More can be said in the case when E is exceptional. Up to translation we may then assume that E belongs to $\mathcal{H}$ and then form the category $\mathcal{H}' = \mathcal{H} \cap E^\perp$ of all objects

X of $\mathcal{H}$ satisfying the conditions $\mathrm{Hom}(E, X) = 0 = \mathrm{Ext}^1(E, X)$. Since $\mathcal{H}$ is hereditary, $\mathcal{H}'$ is a full exact subcategory of $\mathcal{H}$, hence an abelian category in its own right and moreover hereditary. It is then easy to check that $E^\perp = D^b\mathcal{H}'$. For each X in $D^b\mathcal{H}$ put $E_X = \mathrm{Hom}^\bullet(E, X) \otimes E$, and let $\kappa : E_X \to X$ be the canonical morphism. Since E is exceptional, we have $\mathrm{Hom}(E[n], E_X) = \mathrm{Hom}(E[n], X) \otimes \mathrm{End}(E[n])$, hence the map $\mathrm{Hom}(E[n], \kappa) : \mathrm{Hom}(E[n], X) \otimes \mathrm{End}(E[n]) \to \mathrm{Hom}(E[n], X)$, $u \otimes v \mapsto u \circ v$, is an isomorphism for each integer n. Hence the third term ℓX of the triangle $E_X \xrightarrow{\kappa} X \xrightarrow{\alpha} \ell X \to E_X[1]$ belongs to the right perpendicular category $E^\perp$. Moreover, for any object Y in $E^\perp$ application of $\mathrm{Hom}(-, Y)$ to the triangle above shows that α induces an isomorphism $\mathrm{Hom}(\ell X, Y) \xrightarrow{\approx} \mathrm{Hom}(X, Y)$ which is clearly functorial in Y. Therefore the inclusion $E^\perp \hookrightarrow D^b\mathcal{H}$ admits a left adjoint $\ell : D^b\mathcal{H} \to E^\perp$. It is then not difficult to show that ℓ is an exact functor, yielding the following result due to [4].

Proposition 4.2. *Let E be an exceptional object of $D^b\mathcal{H}$. Then $E^\perp$ is a triangulated subcategory of $D^b\mathcal{H}$ and the inclusion $i : E^\perp \to D^b\mathcal{H}$ admits a left adjoint ℓ which is an exact functor.* □

The following consequence is the basis for "perpendicular induction" with respect to the rank of the Grothendieck group.

Corollary 4.3. *Let E be an exceptional object of $\mathcal{H}$. Let $\mathcal{H}'$ be the right perpendicular category of E formed in $\mathcal{H}$. Then the inclusion functor admits a left adjoint $\ell : \mathcal{H} \to \mathcal{H}'$. Moreover, $\mathrm{K}_0(\mathcal{H}) = \mathbb{Z}[E] \oplus \mathrm{K}_0(\mathcal{H}')$.* □

5 Exceptional objects

Throughout we assume that $\mathcal{H}$ is a hereditary abelian k-category with finite-dimensional morphism and extension spaces spaces. As an abelian category with finite dimensional morphism spaces, $\mathcal{H}$ is a Krull-Schmidt category, that is, each object of $\mathcal{H}$ is a finite direct sum of indecomposable objects with local endomorphism ring. Due to Proposition 3.1 the Krull-Schmidt property also holds in $D^b\mathcal{H}$. The next result is fundamental and due to Happel and Ringel [23].

Proposition 5.1. *Let E and F be indecomposable objects of $\mathcal{H}$ such* *that* $\text{Ext}^1(F, E) = 0$. *Then each nonzero morphism* $f : E \to F$ *is a monomorphism or an epimorphism. In particular, each indecomposable object E without self-extensions is exceptional.*

Proof. Represent f as the composition $f = [E \xrightarrow{p} F' \xhookrightarrow{i} F]$ of an epimorphism p and a monomorphism i, and form the short exact sequence $\eta : 0 \to F' \xrightarrow{i} F \to F'' \to 0$. Since $\mathcal{H}$ is hereditary, p induces an epimorphism $\text{Ext}^1(F'', E) \twoheadrightarrow \text{Ext}^1(F'', F')$. Hence η is the push-out along p of a short exact sequence $\mu \in \text{Ext}^1(F'', E)$. The push-out diagram

$$
\begin{array}{ccccccccc}
\mu : & 0 & \longrightarrow & E & \xrightarrow{j} & X & \longrightarrow & F'' & \longrightarrow & 0 \\
 & & & {\scriptstyle p}\downarrow & & \downarrow{\scriptstyle q} & & \| & & \\
\eta : & 0 & \longrightarrow & F' & \xrightarrow{i} & F & \longrightarrow & F'' & \to & 0
\end{array}
$$

yields a short exact sequence $0 \to E \xrightarrow{\binom{j}{p}} X \oplus F' \xrightarrow{(q,-i)} F \to 0$. By the assumption $\text{Ext}^1(F, E) = 0$ this sequence splits such that $E \oplus F \cong X \oplus F'$. The Krull-Schmidt property then implies that X and F' are also indecomposable. Moreover, we obtain a splitting morphism $(\alpha, \beta) : X \oplus F' \to E$ such that $\alpha j + \beta p = 1_E$. Since E has a local endomorphism ring, αj or βp is an isomorphism. It follows that p is an isomorphism, and then f is a monomorphism, or else that j is an isomorphism (use that X is indecomposable), and then $F'' = 0$ implying that f is an epimorphism. $\qquad \square$

Corollary 5.2. *Let E and F be exceptional objects of $\mathcal{H}$ and assume $\text{Ext}^1(F, E) = 0$. Then at most one of the terms $\text{Hom}(E, F)$ and $\text{Ext}^1(E, F)$ is non-zero.*

Proof. We assume that there is a non-zero morphism $f : E \to F$. By the proposition f is a monomorphism or an epimorphism. In the first case f induces an epimorphism $0 = \text{Ext}^1(F, F) \to \text{Ext}^1(E, F)$ whereas in the second case we obtain an epimorphism $0 = \text{Ext}^1(E, E) \to \text{Ext}^1(E, F)$. In either case therefore $\text{Ext}^1(E, F) = 0$. $\qquad \square$

Exceptional objects for hereditary categories have a somehow combinatorial flavor due to the next proposition. Note that — different from the module case — it may happen that a non-zero object of $\mathcal{H}$ has a trivial class in the Grothendieck group. The next result is due to [25, Lemma 4.2], see also [39, Prop. 4.4.1].

Proposition 5.3. *Each exceptional object E of $\mathcal{H}$ is determined by its class $[E]$ in the Grothendieck group $K_0(\mathcal{H})$.*

Proof. Let E and F be exceptional objects with $[E] = [F]$. Then $0 < \langle [E], [E] \rangle = \langle [E], [F] \rangle$, and hence there exists a non-zero morphism $f : E \to F$ whose kernel and image we denote by E' and F', respectively. We claim that f is an isomorphism and assume, for contradiction, that E' is non-zero. The assumption implies that $\mathrm{Hom}(F', E) = 0$: Otherwise there exists a nonzero morphism $g : F' \to E$ and the composition $E \twoheadrightarrow F' \xrightarrow{g} E$ yields a non-trivial endomorphism, hence an automorphism, of E. It then follows that the morphism $E \twoheadrightarrow F'$, induced by f, is an isomorphism, contradicting $E' \neq 0$. We have shown that the assumption $E' \neq 0$ implies $\mathrm{Hom}(F', E) = 0$. Further, since $\mathcal{H}$ is hereditary the embedding $F' \hookrightarrow F$ induces an epimorphism $0 = \mathrm{Ext}^1(F, F) \twoheadrightarrow \mathrm{Ext}^1(F', F)$ implying $\mathrm{Ext}^1(F', F) = 0$. We thus obtain $\langle [F'], [E] \rangle = -\dim \mathrm{Ext}^1(F', E) \leqslant 0$ and $\langle [F'], [F] \rangle = \dim \mathrm{Hom}(F', F) \geqslant 0$. Since the classes of E and F agree, we get $0 = \langle [F'], [E] \rangle = \langle [F'], [F] \rangle = \dim \mathrm{Hom}(F', F)$, hence $\mathrm{Hom}(F', F) = 0$, contradicting $f \neq 0$. We have thus shown that there exists a monomorphism $f : E \to F$. Similarly, there exists a monomorphism $g : F \to E$ yielding non-zero endomorphisms, hence automorphisms, $f \circ g$ and $g \circ f$ of F resp. E. Therefore f and g are isomorphisms and the claim follows. $\qquad\square$

6 Piecewise hereditary algebras and Happel's theorem

We say that $T \in D^b \mathcal{H}$ is a *tilting complex for* a hereditary category $\mathcal{H}$ if

(i) $\mathrm{Hom}(T, T[n]) = 0$ for each non-zero integer n and,

(ii) for each $X \in D^b \mathcal{H}$ the condition $\mathrm{Hom}(T, X[n]) = 0$ for each integer n implies that $X = 0$.

The following result is fundamental for the study of piecewise hereditary algebras.

Theorem 6.1. *Let $\mathcal{H}$ be a small hereditary abelian k-category with finite dimensional morphism and extension spaces. We assume that T is a tilting complex for $\mathcal{H}$. Then the full subcategory $\mathcal{M}$ of all subobjects X of $D^b \mathcal{H}$, satisfying $\mathrm{Hom}(T, X[n]) = 0$ for each non-zero integer n, is an abelian category, equivalent to the category of finite dimensional modules over $\Lambda = \mathrm{End}(T)$. Moreover, a sequence $\eta : 0 \to M' \xrightarrow{u} M \xrightarrow{v} M'' \to 0$ is exact in $\mathcal{M}$ if and only if $M' \xrightarrow{u} M \xrightarrow{v} M'' \xrightarrow{\eta} M'[1]$ is*

an exact triangle in $D^b\mathcal{H}$ and, further, we have natural isomorphisms
$\mathrm{Ext}^n_{\mathcal{M}}(M, X) = \mathrm{Hom}_{D^b\mathcal{H}}(M, X[n])$ *for all $M, X \in \mathcal{M}$ and integers n.*

Further, Λ has finite global dimension, and there is an equivalence $D^b\mathcal{H} \to D^b\mathrm{mod}\text{-}\Lambda$ of triangulated categories. Also $D^b\mathcal{H}$ satisfies Serre duality.

Proof. For M in $\mathcal{M}$ let $\kappa : T^a \to M$ denote a morphism whose components $\kappa_1, \ldots, \kappa_a$ form a k-basis of $\mathrm{Hom}(T, M)$, implying that $\mathrm{Hom}(T, \kappa) : \mathrm{Hom}(T, T^a) \to \mathrm{Hom}(T, M)$ is surjective. We extend κ to an exact triangle $N \to T^a \xrightarrow{\kappa} M \to N[1]$. *We claim that N also belongs to $\mathcal{M}$ and that, moreover, the sequence $0 \to \mathrm{Hom}(T, N) \to \mathrm{Hom}(T, T^a) \to \mathrm{Hom}(T, M) \to 0$ is exact.* The assertion follows immediately by applying $\mathrm{Hom}(T[n], -)$ to the above triangle, keeping in mind that M belongs to $\mathcal{M}$ and that $\mathrm{Hom}(T, \kappa)$ is surjective. Applying the same argument to N yields a sequence $T^b \to T^a \to M$ inducing an exact sequence $\mathrm{Hom}(T, T^b) \to \mathrm{Hom}(T, T^a) \to \mathrm{Hom}(T, M) \to 0$.

Starting with two such sequences $T^{b'} \to T^{a'} \to M'$ and $T^b \to T^a \to M$ (for M and M' from $\mathcal{M}$) it follows that each morphism $f : M' \to M$ extends to a commutative diagram

$$
\begin{array}{ccccc}
T^{b'} & \xrightarrow{u'} & T^{a'} & \xrightarrow{v'} & M' \\
\downarrow h & & \downarrow g & & \downarrow f \\
T^b & \xrightarrow{u} & T^a & \xrightarrow{v} & M
\end{array}
$$

where f is zero if and only if g lifts to T^b via u.

To sum up: Let $\mathcal{T}$ be the set of all T^n, $n \geqslant 0$. Then $\mathcal{M}$ can be viewed as the category of morphisms in $\mathcal{T}$ where a morphism from $u' : X' \to Y'$ to $u : X \to Y$ (all terms in $\mathcal{T}$) is given by a commutative square

$$
\begin{array}{ccc}
X' & \xrightarrow{u'} & Y' \\
\downarrow h & & \downarrow g \\
X & \xrightarrow{u} & Y
\end{array}
$$

where we identify two such squares if they differ by a morphism from Y' to Y lifting to X. That is, $\mathcal{M}$ is naturally equivalent to the *category of homotopy squares* in $\mathcal{T}$. Note that $\mathcal{T}$ is equivalent to the category $\mathcal{F}$ of all finitely generated free Λ-modules. It is well known, and easy to prove, that the category of all homotopy squares in $\mathcal{F}$ is equivalent to the category of all finite dimensional Λ-modules. It follows that the functor

$\text{Hom}(T, -)$ yields an equivalence $\mathcal{M} \xrightarrow{\approx} \text{mod-}\Lambda$ sending the projective generator T of $\mathcal{M}$ to Λ.

We now deal with the relationship between exact sequences in $\mathcal{M}$ and exact triangles in $D^b\mathcal{H}$. Assume first that $M' \xrightarrow{u} M \xrightarrow{v} M'' \xrightarrow{w} M'[1]$ is an exact triangle in $D^b\mathcal{H}$ where M', M and M'' belong to $\mathcal{M}$. Applying $\text{Hom}(T, -)$ yields an exact sequence $0 \to \text{Hom}(T, M') \to \text{Hom}(T, M) \to \text{Hom}(T, M'') \to 0$ in mod-Λ, proving that $0 \to M' \to M \to M'' \to 0$ is exact in $\mathcal{M}$. Next, assume that $\eta : 0 \to M' \xrightarrow{u'} M \xrightarrow{v} M'' \to 0$ is a short exact sequence in $\mathcal{M}$. We claim that $M' \xrightarrow{u'} M \xrightarrow{v} M'' \xrightarrow{\eta} M'[1]$ is an exact triangle in $D^b\mathcal{H}$. To show this we complete $M \xrightarrow{v} M''$ to an exact triangle $K \xrightarrow{u} M \xrightarrow{v} M'' \xrightarrow{w} K[1]$ in $D^b\mathcal{H}$. Invoking that T is a projective generator in $\mathcal{M}$ and applying $\text{Hom}(T, -)$ to the above triangle shows that K belongs to $\mathcal{M}$ and that, moreover, the sequence $0 \to K \xrightarrow{u} M \xrightarrow{v} M'' \to 0$ is exact in $\mathcal{M}$, hence isomorphic to the sequence η. Since Yoneda composition agrees with the composition in the derived category this proves the relationship between short exact sequences in $\mathcal{M}$ and exact triangles with members from $\mathcal{M}$.

By means of short exact sequences $0 \to N \to P \to M \to 0$ with $M \in \mathcal{M}$ and P projective in $\mathcal{M}$, that is, P lying in the additive closure of T, we obtain natural isomorphisms $\text{Ext}^1_{\mathcal{M}}(M, X) = \text{Hom}_{D^b\mathcal{H}}(M, X[1])$ for each $X \in \mathcal{M}$. Invoking dimension shift $\text{Ext}^1_{\mathcal{M}}(M, X) = \text{Hom}_{D^b\mathcal{H}}(N, X)$ for $X \in \mathcal{M}$ we obtain inductively natural isomorphisms $\text{Ext}^n_{\mathcal{M}}(M, X) = \text{Hom}_{D^b\mathcal{H}}(M, X[n])$ for all $M, X \in \mathcal{M}$. The finitely many simple objects $S_1, \dots, S_n$ from $\mathcal{M} \cong \text{mod-}\Lambda$ belong to a finite number r of consecutive copies $\mathcal{H}[m], \mathcal{H}[m+1], \dots, \mathcal{H}[m+r-1]$. It follows that $\text{Ext}^{r+1}(S_i, S_j) = \text{Hom}(S_i, S_j[r+1]) = 0$ for all i, j, hence gl.dim $\Lambda \leqslant r$.

We finally show that $\mathcal{M}$ and $D^b\mathcal{H}$ are derived-equivalent. By a result of [18] $\mathcal{H}$ has a tilting object T_1, hence is derived-equivalent to mod-Λ_1, where $\Lambda_1 = \text{End}(T_1)$ by an argument of Beilinson [2], compare [12, Theorem 3.2]. Hence T becomes a tilting complex in $D^b\text{mod-}\Lambda_1$, and Rickard's theorem [44] implies that mod-Λ and mod-Λ_1 are derived equivalent, implying the derived equivalence of $\mathcal{H}$ and mod-Λ. Since Λ has finite global dimension and $D^b\text{mod-}\Lambda$ has Auslander-Reiten triangles and Serre duality by a theorem of Happel [15, chap. I, Prop. 4.10], this proves the last claim. □

A tilting complex for $\mathcal{H}$ whose members belong to $\mathcal{H}$ is just a tilting ob-

ject of $\mathcal{H}$. This situation is particularly easy to describe and historically
was the starting point of tilting theory. For a tilting object T in $\mathcal{H}$ we
form the full subcategories $\mathcal{T}$ (resp. $\mathcal{F}$) of $\mathcal{H}$ consisting of all objects X
(resp. Y) of $\mathcal{H}$ satisfying $\mathrm{Ext}^1(T, X) = 0$ (resp. $\mathrm{Hom}(T, Y) = 0$). Clearly
we have $\mathrm{Hom}(\mathcal{T}, \mathcal{F}) = 0$. Obviously, $\mathcal{T} = \mathcal{M} \cap \mathcal{H}$, $\mathcal{F}[1] = \mathcal{M} \cap \mathcal{H}[1]$ and
$\mathcal{M} \cap \mathcal{H}[n] = 0$ for $n \neq 0, 1$. We thus obtain a transparent interpretation
of the module category mod-Λ.

Corollary 6.2 ([23, 5]). *Let T be a tilting object of a small hereditary
abelian Ext-finite k-category $\mathcal{H}$. Then $\mathcal{M} = \mathcal{T} \vee \mathcal{F}[1]$. Moreover, for
each X in $\mathcal{H}$ there is a short exact sequence $0 \to X_\mathcal{T} \to X \to X_\mathcal{F} \to 0$
with $X_\mathcal{T} \in \mathcal{T}$ and $X_\mathcal{F} \in \mathcal{F}$.*

Proof. Only the last assertion needs a proof. Let $X_\mathcal{T} = \mathrm{Hom}(T, X) \otimes_\Lambda$
X. (This makes sense since $\mathcal{H}$ is abelian.) The natural morphism $\kappa :$
$X_\mathcal{T} \to X$ induces an isomorphism $\mathrm{Hom}(T, \kappa)$. Applying $\mathrm{Hom}(T, -)$ to
the triangle $X_\mathcal{T} \xrightarrow{\kappa} X \to Y \to X_\mathcal{T}[1]$ proves that $\mathrm{Hom}(T[n], Y[1]) = 0$
for each $n \neq 0$ such that $Y[1]$ belongs to $\mathcal{M}$, and thus $Y = U[-1] \oplus$
V where $U \in \mathcal{T}$ and $V \in \mathcal{F}$. Since $\mathrm{Hom}(T[-1], Y) = 0$ we obtain
$\mathrm{Hom}(T, U) = 0$, and hence $U = 0$. We conclude that Y belongs to
$\mathcal{F}$; in particular the three terms $X_\mathcal{T}$, X and Y belong to $\mathcal{H}$. Hence
$0 \to X_\mathcal{T} \xrightarrow{\kappa} X \to Y \to 0$ is exact in $\mathcal{H}$ and satisfies the claim. $\square$

A finite dimensional algebra Λ is called *piecewise hereditary* if there
exists a hereditary k-category $\mathcal{H}$ with finite dimensional morphism and
extension spaces and a tilting complex T for $\mathcal{H}$ such that Λ is isomorphic
to the endomorphism algebra of T. Λ is called *quasitilted* if, moreover,
T is a tilting object of $\mathcal{H}$. In this context, Happel's theorem [17], which
has a difficult proof, shows that, up to derived equivalence, there are
only two classes of hereditary categories.

Theorem 6.3 (Happel's theorem). *Let k be an algebraically closed
field and $\mathcal{H}$ be a connected, Ext-finite, hereditary, abelian k-category with
a tilting complex. Then $\mathcal{H}$ is derived equivalent to the category mod-H
for some finite dimensional hereditary k-algebra or to the category $\mathrm{coh}\,\mathbb{X}$
of coherent sheaves over a weighted projective line.* $\square$

7 Derived equivalence of hereditary categories

We assume that $\mathcal{H}$ is a hereditary category such that $\mathcal{H}$, and hence $D^b\mathcal{H}$, has Serre duality. (We allow that $\mathcal{H}$ has projectives or injectives.) By τ we denote the Auslander-Reiten translation in $D^b\mathcal{H}$, where Serre duality takes the form $D\operatorname{Hom}(X, Y[1]) = \operatorname{Hom}(Y, \tau X)$ with τ a self-equivalence of $D^b\mathcal{H}$.

A decomposition $\mathcal{H} = \mathcal{A} \vee \mathcal{B}$ is called a *cut* of $\mathcal{H}$ if $\operatorname{Hom}(\mathcal{B}, \mathcal{A}) = 0$ and $\operatorname{Hom}(\mathcal{B}, \tau\mathcal{A}) = 0$. The second assumption follows from the first if $\mathcal{A}$ is closed under τ or $\mathcal{B}$ is closed under τ^{-1}. The next proposition is taken from [21].

Proposition 7.1. *Let $\mathcal{H} = \mathcal{A} \vee \mathcal{B}$ be a cut in a hereditary category $\mathcal{H}$ with Serre duality. Then the full subcategory $\bar{\mathcal{H}} = \mathcal{B} \vee \mathcal{A}[1]$ of $D^b\mathcal{H}$ is an abelian hereditary category which is derived-equivalent to $\mathcal{H}$.* $\square$

There is a partial converse with an obvious proof.

Proposition 7.2. *Assume $\mathcal{H}$ and $\bar{\mathcal{H}}$ are hereditary categories which are derived equivalent allowing us to identify $\operatorname{Rep}\mathcal{H}$ and $\operatorname{Rep}\bar{\mathcal{H}}$. We assume that the equivalence is* normal *in the sense that $\bar{\mathcal{H}}$ is lying in two consecutive copies of $\mathcal{H}$, say $\bar{\mathcal{H}} \subseteq \mathcal{H} \vee \mathcal{H}[1]$. We put $\mathcal{A} = \mathcal{H} \cap \bar{\mathcal{H}}$ and $\mathcal{B} = \mathcal{H} \cap \bar{\mathcal{H}}[-1]$. Then $\mathcal{H} = \mathcal{B} \vee \mathcal{A}$ is a cut of $\mathcal{H}$, and $\bar{\mathcal{H}} = \mathcal{A} \vee \mathcal{B}[1]$.* $\square$

The hypothesis for the equivalence to be normal is automatically true if $\mathcal{H}$ (or $\bar{\mathcal{H}}$) is hereditary noetherian with Serre duality. On the other side it is easily seen that there are selfequivalences of $D^b\mathcal{H}$, where $\mathcal{H} = \operatorname{mod-}k[\circ \to \circ \to \circ]$, which are not normal.

8 Modules over hereditary algebras

Let A be a finite dimensional k-algebra. By Hölder's theorem the classes of simple modules $S_1, S_2, \ldots, S_n$ form a $\mathbb{Z}$-basis of the Grothendieck group $\operatorname{K}_0(A) = \operatorname{K}_0(\operatorname{mod-}A)$. We thus identify $\operatorname{K}_0(A)$ and $\mathbb{Z}^n$, and also speak of the dimension vector $\underline{\dim} M$ instead of the class $[M]$ of an A-module M. If A is hereditary $\operatorname{K}_0(A)$ is equipped with the *Euler form* which induces the *Tits quadratic form* q_A on $\operatorname{K}_0(A)$ with $q_A(x) = \langle x, x \rangle$. We call $x \in \operatorname{K}_0(A) = \mathbb{Z}^n$ a root of q_A if $q_A(x) = 1$. We call x a positive root if, moreover, $x \in \mathbb{N}^n$.

8.1 The representation-finite case

A finite dimensional k-algebra A is called *representation-finite* if — up to isomorphism — there is only a finite number of finite-dimensional indecomposable modules. By a theorem of Auslander, then *any* A-module is a possibly infinite direct sum of indecomposables. The next theorem [10] is famous, in particular, since it links representation theory of finite dimensional algebras with Lie theory. For the proof we refer to [3].

Theorem 8.1 (Gabriel's theorem). *Let $H = k[\vec{\Delta}]$ be the path algebra of a finite connected quiver $\vec{\Delta}$ without oriented cycles. Then the following holds:*

(i) H is representation-finite if and only if the underlying graph Δ is a Dynkin diagram, that is, is of type $\mathbb{A}_n$, $\mathbb{D}_n$ $(n \geqslant 4)$, $\mathbb{E}_6$, $\mathbb{E}_7$ or $\mathbb{E}_8$.

(ii) In this case, each indecomposable H-module E has the form $\tau^{-n}P$, with P indecomposable projective, in particular E is exceptional.

(iii) The Tits quadratic form q_H associated to H is positive definite. Moreover, the mapping $E \mapsto \underline{\dim}\, E$, sending a module to its dimension vector, establishes a bijection between the set of isomorphism classes of indecomposable H-modules and the set of positive roots of the Tits form q_H. $\qquad\qquad\qquad\qquad\qquad\qquad\qquad\qquad\qquad\qquad\qquad\qquad\qquad$ $\square$

8.2 The tame case

By definition, a connected quiver $\vec{\Delta}$ is called *tame* if the quadratic form q_H associated to $H = k[\vec{\Delta}]$ is positive semidefinite but not positive definite. This holds true exactly if Δ is an extended Dynkin diagram. Recall that the associated Dynkin diagram Δ' is a star $[p, q, r]$ satisfying $1/p + 1/q + 1/r > 1$, called the *Dynkin type* of $\vec{\Delta}$. Thus $\tilde{\mathbb{A}}_{pq}$ has type $[1, p, q]$ or just type $[p, q]$, $\tilde{\mathbb{D}}_n$ $(n \geqslant 4)$ has type $[2, 2, n - 2]$, and $\tilde{\mathbb{E}}_6$, $\tilde{\mathbb{E}}_7$ or $\tilde{\mathbb{E}}_8$ have types $[2, 3, 3]$, $[2, 3, 4]$ and $[2, 3, 5]$, respectively. If Δ is extended Dynkin there exists a unique positive integral valued function λ on the set Δ_0 of vertices of Δ which is *additive*, that is, for $p \in \Delta_0$ the value $2\lambda(p)$ agrees with the sum $\sum_{q-p} \lambda(q)$, extended over all neighbors q of p, and *normalized*, that is, $\lambda(p) = 1$ for some vertex p. For each $p \in \Delta_0$ let $P(p)$ be the indecomposable projective right H-module associated to the vertex p. The unique linear form $r : K_0(H) \to \mathbb{Z}$ satisfying $r([P(p)]) = \lambda(p)$ is called *rank*. The rank is invariant under the Coxeter transformation Φ, the unique endomorphism of $K_0(H)$ with

$\langle y, x \rangle = -\langle x, \Phi y \rangle$ for all x, y.

Theorem 8.2. *Let $\vec{\Delta}$ be a tame quiver of type $[p, q, r]$ and $H = k[\vec{\Delta}]$. Let $\mathcal{P}$, $\mathcal{R}$ and $\mathcal{I}$ denote the additive closure of all indecomposable H-modules of rank > 0, $= 0$ or < 0, respectively. Then the following holds:*
(i) There is a trisection mod-$H = \mathcal{P} \vee \mathcal{R} \vee \mathcal{I}$ *in the subcategories of preprojective, regular and preinjective modules.*
(ii) The indecomposables of $\mathcal{P}$, called preprojective, *are exactly the modules $\tau^{-m} P$, where P is indecomposable projective and $m \geqslant 0$. The indecomposable preprojective modules form a single Auslander-Reiten component.*
(ii) The indecomposables of $\mathcal{I}$, called preinjective, *are exactly the modules $\tau^m I$, where I is indecomposable injective and $m \geqslant 0$. The indecomposable preinjective modules form a single Auslander-Reiten component.*
(iii) The indecomposables of $\mathcal{R}$, called regular, *form a 1-parameter family $(\mathcal{T}_x)$ of tubes, naturally indexed by the points of the projective line $\mathbb{P}^1(k) = k \cup \{\infty\}$, and such that $\mathcal{T}_x$ is homogeneous, that is, fixed under τ, for $x \notin \{0, 1, \infty\}$, and $\mathcal{T}_x$ has τ-period p, q or r, according as $x = 0, 1$ or ∞.*
(iv) Each morphism $f : P \to Q$ with $P \in \mathcal{P}$ and $Q \in \mathcal{I}$ factors through any given tube $\mathcal{T}_x$. □

For the proof we refer to [46], an alternative proof, using weighted projective lines, can be based on Theorem 10.14.

8.3 The wild case

A connected quiver $\vec{\Delta}$ and the path algebra $H = k[\vec{\Delta}]$ are called *wild* if the quadratic form q_H is indefinite. As in the tame case an indecomposable module is called *preprojective* (resp. *preinjective*) if it is of the form $\tau^{-n} P$ (resp. $\tau^n I$), where P is indecomposable projective, I is indecomposable injective, and $n \geqslant 0$. An indecomposable H-module is called regular if it is not preprojective nor preinjective. For the next theorem we refer to [45] and [28].

Theorem 8.3. *Let $\vec{\Delta}$ be a wild connected quiver. Then there is a trisection* mod-$H = \mathcal{P} \vee \mathcal{R} \vee \mathcal{I}$, *where $\mathcal{P}$, $\mathcal{I}$ and $\mathcal{R}$ is, respectively, the additive closure of all preprojective, preinjective or regular H-modules. Moreover,*

(i) *The preprojective (resp. the preinjective) indecomposable modules each form a single Auslander-Reiten component, the preprojective (resp. preinjective component).*

(ii) *The indecomposable regular modules decompose into components C_x, each of type $\mathbb{Z}\mathbb{A}_\infty$, where x belongs to some index set X.*

(iii) *If X and Y are indecomposable regular, then $\mathrm{Hom}(X, \tau^n Y) \neq 0$ for $n \gg 0$.*   □

Very little is known or conjectured about the, perhaps geometric, structure of the index set $X = X(H)$ parametrizing the regular components for a wild hereditary algebra H. It should perhaps be imagined to be a huge, exotic and complicated space. Some indication on the importance of $X(H)$ is given by the existence of so called *Kerner bijections*: If H and H' are connected, wild hereditary algebras there exist natural bijections $X(H) \to X(H')$, see [6].

9 Spectral properties of hereditary categories

Let $\mathcal{H}$ be a connected hereditary k-category with a tilting object. By Happel's theorem $\mathcal{H}$ is derived equivalent to a module category mod-H, H a hereditary k-algebra, or to a category coh $\mathbb{X}$ of coherent sheaves on a weighted projective line. The easiest way to decide which of the two cases happens, is to determine the *Coxeter polynomial $ch_\mathcal{H}$* of $\mathcal{H}$, defined as the characteristic polynomial of the Coxeter transformation, that is, the $\mathbb{Z}$-linear map $\mathrm{K}_0(\mathcal{H}) \to \mathrm{K}_0(\mathcal{H})$ induced by the Auslander-Reiten translation in $D^b\mathcal{H}$. By definition, the Coxeter polynomial is thus a derived invariant of $\mathcal{H}$.

Proposition 9.1. *Let $\mathcal{H}$ be a connected hereditary k-category with a tilting object. Then the following holds:*

(i) *If $\mathcal{H} = \mathrm{coh}\, \mathbb{X}$ is the category of coherent sheaves on a weighted projective line of weight type $(p_1, \ldots, p_t)$, then all roots of*

$$ch_\mathcal{H}(x) = (x - 1)^2 \prod_{i=1}^{t} \frac{x^{p_i} - 1}{x - 1}$$

lie on the unit circle and, moreover, 1 is a root of $ch_\mathcal{H}$.

(ii) *Assume $\mathcal{H} = \mathrm{mod}\text{-}H$ for a finite dimensional connected hereditary k-algebra. Then*

(a) *If H is representation-finite, then all roots of ch_H lie on the unit* *circle, and* 1 *is not a root.*

The Coxeter transformation Φ_H *is periodic, and its period* p *is the* Coxeter number *of* Δ, *given as* $p = n + 1$ *for* $\mathbb{A}_n$, $2(n-1)$ *for* $\mathbb{D}_n$ *and* $12, 18$ *and* 30 *for* $\mathbb{E}_6$, $\mathbb{E}_7$ *and* $\mathbb{E}_8$, *respectively.*

(b) *If* H *is tame with associated Dynkin type* $[p, q, r]$, *then*

$$ch_H(x) = (x-1)^2 \frac{x^p - 1}{x - 1} \cdot \frac{x^q - 1}{x - 1} \cdot \frac{x^r - 1}{x - 1}.$$

(c) *If* H *is wild hereditary, then the spectral radius* $\rho_H = \max\{|z| \,|\, z \in \mathbb{C}, \, ch_H(z) = 0\}$ *of the Coxeter transformation is* > 1 *and, moreover,* ρ_H *is a simple root of* ch_H.

Proof. For (i) we refer to [34] and [31]. By Corollary 10.15 the category of coherent sheaves of weight type (p, q, r), $1/p + 1/q + 1/r > 1$, and the category of modules over a connected tame hereditary algebra of type $[p, q, r]$ are derived equivalent, hence $(ii)(b)$ is a special case of (i). Assertion $(ii)(a)$ is a well known result from Lie theory, and besides easily established directly. Finally, $(ii)(c)$ is due to Ringel [47]. $\square$

For the path algebra $H = k[\vec{\Delta}]$ of a wild connected quiver with *bipartite orientation*, that is, each vertex of $\vec{\Delta}$ is either a sink or a source, there is additional information on the spectral behavior: In this case all roots of ch_H are either real or lie on the unit circle. For wild quivers in general this assertion is not correct as shows the following quiver

While it is easy to decide whether a polynomial appears as the Coxeter polynomial for a weighted projective line, no manageable criterion is known to decide the analogous question for hereditary algebras. Many questions in this field are still open.

10 Weighted projective lines

10.1 Hereditary categories with a tilting object

It is natural to aim for a classification of hereditary categories — if wanted with further properties — up to derived equivalence. In general this task is difficult and only partial solutions are known. If we assume the existence of a tilting object then, by Happel's theorem 6.3, there are only two cases to consider, the categories mod-Λ for a finite dimensional hereditary algebra and the categories $\operatorname{coh}\mathbb{X}$ of coherent sheaves on a weighted projective line $\mathbb{X}$.

Weighted projective lines were introduced by Geigle-Lenzing in [12] to analyze the interaction between preprojective and regular modules for tame hereditary algebras and to relate Ringel's classification of indecomposable modules over tubular algebras [46] to Atiyah's classification [1] of coherent sheaves over an elliptic curve. A weighted projective line $\mathbb{X}$ is defined through its attached category $\operatorname{coh}\mathbb{X}$ of coherent sheaves which is a natural generalization of the category of coherent sheaves on the projective line, but allowing a finite number of points x having more than one simple sheaf (always a finite number $p(x) > 1$) to be concentrated in x. We will see that the classification problem for indecomposables for $\mathcal{H} = \operatorname{coh}\mathbb{X}$ mainly depends on a homological invariant, the Euler characteristic $\chi_{\mathcal{H}}$ of $\mathcal{H}$ (or $\mathbb{X}$).

It may happen that a category $\operatorname{coh}\mathbb{X}$, $\mathbb{X}$ a weighted projective line, and a category mod-H, H a connected finite dimensional hereditary algebra, are derived equivalent. Actually this is going to happen exactly for the tame hereditary algebras Λ and the weighted projective lines of positive Euler characteristic. An overview of the situation is given by the following picture, which is up to derived equivalence:

	$\operatorname{coh}\mathbb{X}$ $\chi_{\mathbb{X}} < 0$ type: **wild**	
	$\operatorname{coh}\mathbb{X}$ $\chi_{\mathbb{X}} = 0$ type: **tame tubular**	
mod-H H **representation-finite**	mod-$H \sim_{der} \operatorname{coh}\mathbb{X}$ H tame hereditary $\chi_{\mathbb{X}} > 0$ type: **tame domestic**	mod-H H **wild hereditary**

Fig. 6.1. Hereditary categories with a tilting object

10.2 Generalities

Let k be an algebraically closed field. By a *category of coherent sheaves* $\operatorname{coh}\mathbb{X}$ *on a weighted projective line* $\mathbb{X}$ we mean in the sequel a small k-category $\mathcal{H}$ satisfying the axioms (H 1) to (H 7) below. Before stating the last three axioms, we prove some consequences of the first four ones.

(H 1) $\mathcal{H}$ is a connected abelian k-category, and each object in $\mathcal{H}$ is noetherian.

(H 2) $\mathcal{H}$ is (skeletally) small and Ext-finite, that is, all morphism and extension spaces in $\mathcal{H}$ are finite dimensional k-vector spaces.

(H 3) (Serre duality) We assume the existence of an equivalence $\tau : \mathcal{H} \to \mathcal{H}$ and of natural isomorphisms $\operatorname{D}\operatorname{Ext}^1(X, Y) = \operatorname{Hom}(Y, \tau X)$ for all objects X, Y of $\mathcal{H}$.

(H 4) $\mathcal{H}$ is noetherian, but not every object of $\mathcal{H}$ has finite length.

Assumptions (H 1) – (H 3) imply that the abelian category $\mathcal{H}$ is *hereditary*, that is, extension spaces $\operatorname{Ext}^d_{\mathcal{H}}(X, Y)$ vanish in degree $d \geqslant 2$. This holds since Serre duality in the form (H 3) implies that $\operatorname{Ext}^1(X, -)$ is a right exact functor for each X. As another consequence of Serre duality, for each indecomposable object X there is an almost split sequence $0 \to \tau X \to Z \to X \to 0$; in particular, τ serves as the Auslander-Reiten translation. Since τ is an equivalence, moreover, $\mathcal{H}$ does not have any nonzero projective or injective objects.

We denote by $\mathcal{H}_0$ the full subcategory of $\mathcal{H}$ consisting of all objects of finite length. Further $\mathcal{H}_+$ denotes the full subcategory of $\mathcal{H}$ consisting of all objects without a simple subobject. By definition there are no nonzero morphisms from (any object of) $\mathcal{H}_0$ to (any object of) $\mathcal{H}_+$.

Proposition 10.1. *Assume (H 1)–(H4). Then each indecomposable object from $\mathcal{H}$ either belongs to $\mathcal{H}_+$ or to $\mathcal{H}_0$. Moreover, for some index set C we have $\mathcal{H}_0 = \coprod_{x \in C} \mathcal{U}_x$, where each $\mathcal{U}_x$ is a connected uniserial length category of τ-period $p(x)$ which may be finite or infinite. If $\mathcal{H}$ has a tilting object, then $p(x)$ is always finite, and there are at most finitely many $x \in C$ with $p(x) > 1$.*

Proof. For an indecomposable object X let X_0 denote its maximal subobject of finite length. We claim that $\operatorname{Ext}^1(X/X_0, X_0) = 0$. Otherwise by Serre duality $\operatorname{Hom}(\tau^{-1}X_0, X/X_0)$ is non-zero, contradicting maximality of X_0. Hence the sequence $0 \to X_0 \to X \to X/X_0 \to 0$ splits, establishing the first claim. We now are going to prove the second assertion. The category $\mathcal{H}_0$ is a *length category*, that is, $\mathcal{H}_0$ is abelian and each of its objects has finite length. Moreover, as an exact extension-closed subcategory of $\mathcal{H}$, the category $\mathcal{H}_0$ is hereditary with τ acting as an equivalence; it follows that for each simple object S of $\mathcal{H}_0$, that is, of $\mathcal{H}$, there is — up to isomorphism — exactly one simple S' (S'') with $\mathrm{Ext}^1(S, S') \neq 0$ ($\mathrm{Ext}^1(S'', S) \neq 0$, respectively) and, moreover, the extension spaces in question are of dimension one over $\mathrm{End}(S)$. As in Proposition 10.1 it follows from [11, 8.3] that $\mathcal{H}_0$ is *uniserial*, meaning that each indecomposable object in $\mathcal{H}_0$ has a unique composition series. For the last assertion one uses that $K_0(\mathcal{H})$ has finite rank over $\mathbb{Z}$ if $\mathcal{H}$ has a tilting object. $\qquad\square$

The members of the index set C are called the *points* of $\mathcal{H}$, notation $C = C(\mathcal{H})$. The objects of $\mathcal{U}_x$ are said to be *concentrated in x*. The indecomposable objects of $\mathcal{U}_x$ form a tube as in Example 2.7. By the simples of the tube we mean the simple objects of the abelian category $\mathcal{U}_x$. Objects from $\mathcal{H}_+$ (respectively those of rank one) are called *bundles* (resp. *line bundles*).

(H 5) There is a linear form $\mathrm{rk} : K_0(\mathcal{H}) \to \mathbb{Z}$, called *rank*, that is τ-invariant, zero on objects of $\mathcal{H}_0$, and > 0 on nonzero objects of $\mathcal{H}_+$. Moreover, $\mathcal{H}_+$ contains a *line bundle*.

(H 6) Each tube in $\mathcal{H}_0$ has only finitely many simple objects. Moreover, if L is a line bundle and $\mathcal{U}_x$ is a tube in $\mathcal{H}_0$, then $\sum_S \dim_k \mathrm{Hom}(L, S) = 1$, where S runs through the simple objects from $\mathcal{U}_x$.

(H 7) $\mathcal{H}$ has a tilting object.

Let $x_1, \ldots, x_t$ be the finitely many points of $\mathcal{H}$ with $p_i := p(x_i) > 1$, then $(p_1, \ldots, p_t)$ is called the *weight type* of $\mathcal{H}$.

Remark 10.2. The axioms (H 1) to (H 7) focus on important properties of categories of coherent sheaves on weighted projective lines. Note however, that they form a redundant system:
(*i*) Note that $\mathcal{H}_0$ is a *Serre subcategory* of $\mathcal{H}$, that is, is closed under the formation of subobjects, quotients and extensions. Therefore the quotient category $\mathcal{H}/\mathcal{H}_0$, obtained from $\mathcal{H}$ by formally inverting all morphisms in $\mathcal{H}$ with kernel and cokernel in $\mathcal{H}_0$, is an abelian category. It can be shown [43, IV.1.4] or [36, Prop. 4.9] that each object in $\mathcal{H}/\mathcal{H}_0$

has finite length. (The proof, however, is not easy.) Defining rk$X$ as the length of X in $\mathcal{H}/\mathcal{H}_0$ then yields a rank function satisfying (H 5).
(*ii*) Assume (H 1) to (H 4) and (H 7). Then the remaining axioms follow, see [32].

Note further that properties (H 1) to (H 6) with the additional request that each tube $\mathcal{U}_x$ has exactly one simple object characterize the categories of coherent sheaves on smooth projective curves, as can be deduced from [43].

Lemma 10.3. *Any non-zero morphism from a line bundle L to a bundle E is a monomorphism. In particular, the endomorphism ring $\mathrm{End}(L)$ of a line bundle equals k.*

Proof. Let $u : L \to E$ be a non-zero morphism. By properties of the rank, the kernel of u has rank zero, therefore has finite length and consequently is zero. The second assertion follows from the first because $\mathrm{End}(L)$ has finite dimension over k and has no zero-divisors. $\square$

10.3 Shift action associated to a point

This section is an adaptation of a part of [35], dealing with module categories with separating tubular families, to the present context of hereditary categories satisfying (H 1)–(H 6). The request (H 7) of a tilting object is not of relevance, here. Let S_x be a simple object in $\mathcal{U}_x$. The additive closure $\mathcal{S}_x$ of the Auslander-Reiten orbit $\tau^j S_x$, $1 \leqslant j \leqslant p(x)$, of S_x consists of all semisimple objects from $\mathcal{U}_x$, hence is a semisimple abelian category. Therefore each k-linear functor $G : \mathcal{S}_x \to \mathrm{mod}\text{-}k$ is exact and hence representable (by an object from $\mathcal{S}_x$). This follows from [8, 51] or by a direct argument using that $\mathcal{S}_x$ is equivalent to $\mathrm{mod}\text{-}k^{p(x)}$. For instance, if G is contravariant, we get $G \cong \mathrm{Hom}(-, Z)$ with $Z = \bigoplus_{j=1}^{p(x)} G(\tau^j S_x) \otimes \tau^j S_x$; the covariant case is dual.

When applied to the restriction $\mathrm{Ext}^1(-, E)|_{\mathcal{S}_x}$ of $\mathrm{Ext}^1(-, E)$ to $\mathcal{S}_x$, the argument shows the existence of an object

$$E_x = \bigoplus_{j=1}^{p(x)} \mathrm{Ext}^1(\tau^j S_x, E) \otimes \tau^j S_x$$

from $\mathcal{S}_x$ and a natural isomorphism of functors $\eta_E : \mathrm{Hom}(-, E_x)|_{\mathcal{S}_x} \to \mathrm{Ext}^1(-, E)|_{\mathcal{S}_x}$. By means of the Yoneda lemma, we shall view η_E as a short exact sequence η_E : $\quad 0 \to E \to E(x) \to E_x \to 0$, $E_x \in \mathcal{S}_x$ with the property that Yoneda composition $\mathrm{Hom}(U, E_x) \to \mathrm{Ext}^1(U, E)$, $f \mapsto \eta_E.f$ with η_E induces an isomorphism for each U in $\mathcal{S}_x$. We call η_E the $\mathcal{S}_x$-*universal extension* of E. Notice that the identification $\mathrm{Hom}(-, E_x)|_{\mathcal{S}_x} = \mathrm{Ext}^1(-, E)|_{\mathcal{S}_x}$ turns the assignment $E \mapsto E_x$ into a functor such that, for each $u : E \to N$, we obtain $u.\eta_E = \eta_N.u_x$.

Similarly, the restriction $\mathrm{Hom}(E, -)|_{\mathcal{S}_x}$ is representable by an object E^x from $\mathcal{S}_x$, and the isomorphism $\mathrm{Hom}(E^x, -) \to \mathrm{Hom}(E, -)|_{\mathcal{S}_x}$ corresponds to a morphism $\pi^E : E \to E^x$, called the $\mathcal{S}_x$-*couniversal morphism* for E.

Proposition 10.4 (Shift by a universal extension). *For each $E \in \mathcal{H}_+$ we fix an $\mathcal{S}_x$-universal extension $\eta_E\colon 0 \to E \xrightarrow{\alpha_E} E(x) \xrightarrow{\beta_E} E_x \to 0$ for E. The following properties hold:*

(i) Also $E(x)$ belongs to $\mathcal{H}_+$ and β_E is the $\mathcal{S}_x$-couniversal morphism for $E(x)$.

(ii) For each morphism $u : E \to N$ in $\mathcal{H}_+$, there is a unique morphism $u(x) : E(x) \to N(x)$ yielding a commutative diagram

$$
\begin{array}{ccccccccc}
\eta_E : & 0 & \to & E & \xrightarrow{\alpha_E} & E(x) & \xrightarrow{\beta_E} & E_x & \to & 0 \\
 & & & u \downarrow & & u(x) \downarrow & & u_x \downarrow & & \\
\eta_N : & 0 & \to & N & \xrightarrow{\alpha_N} & N(x) & \xrightarrow{\beta_N} & N_x & \to & 0.
\end{array}
$$

(iii) The arising functor $\sigma'_x : \mathcal{H}_+ \to \mathcal{H}_+$, $E \mapsto E(x)$, is an equivalence which preserves the rank and additionally is exact on short exact sequences $0 \to E' \to E \to E'' \to 0$ with terms from $\mathcal{H}_+$.

We are going to see later, see Theorem 10.8, that σ'_x extends to a self-equivalence σ_x of $\mathcal{H}$.

Proof. *(i)*: The sequence $0 = \mathrm{Hom}(-, E)|_{\mathcal{S}_x} \to \mathrm{Hom}(-, E(x))|_{\mathcal{S}_x} \xrightarrow{-\circ\beta_E}$ $\mathrm{Hom}(-, E_x)|_{\mathcal{S}_x} \xrightarrow{\eta_E} \mathrm{Ext}^1(-, E)|_{\mathcal{S}_x} \xrightarrow{\mathrm{Ext}^1(-,\beta_E)} \mathrm{Ext}^1(-, E(x))|_{\mathcal{S}_x} \to \mathrm{Ext}^1(-, E_x)|_{\mathcal{S}_x} \to 0$ of functors on $\mathcal{S}_x$ is exact. By definition of $\mathcal{S}_x$-universal extensions, η_E is an isomorphism. This shows that $E(x)$ does not admit nonzero morphisms from $\mathcal{S}_x$, and moreover that $\mathrm{Ext}^1(-, \beta_E)$, hence by Serre duality $\mathrm{Hom}(\beta_E, -)$, is an isomorphism thus proving that $\beta_E : E(x) \to E_x$ is $\mathcal{S}_x$-couniversal. Since E belongs to $\mathcal{H}_+$, it is further obvious that $E(x)$ does not admit any non-zero morphism from $\mathcal{U}_y$, for $y \neq x$, so that $E(x)$ also is a member of $\mathcal{H}_+$.

(ii): Because $u.\eta_E = \eta_N.u_x$, we obtain a commutative diagram

$$
\begin{array}{ccccccccccc}
\eta_E : & & 0 & \to & E & \xrightarrow{\alpha_E} & E(x) & \xrightarrow{\beta_E} & E_x & \to & 0 \\
 & & & & u\downarrow & & \downarrow & & \| & & \\
u.\eta_E = \eta_N.u_x : & & 0 & \to & N & \to & X & \to & N_x & \to & 0 \\
 & & & & \| & & \downarrow & & u_x\downarrow & & \\
\eta_N : & & 0 & \to & N & \xrightarrow{\alpha_N} & N(x) & \xrightarrow{\beta_N} & N_x & \to & 0,
\end{array}
$$

allowing to define $u(x) : E(x) \to N(x)$ as the composition of the two vertical maps in the middle. Uniqueness of $u(x)$ follows from $\mathrm{Hom}(E_x, N(x)) = 0$.

(iii): For E from $\mathcal{H}_+$ let $\pi^E : E \to E^x$ be the surjective $\mathcal{S}_x$-couniversal homomorphism. We denote the corresponding kernel by $E(-x)$ which is a member of $\mathcal{H}_+$. Clearly, the assignment $E \mapsto E(-x)$ extends to a functor $\rho'_x : \mathcal{H}_+ \to \mathcal{H}_+$, which by assertion (i), is a left inverse to σ'_x. To see that ρ'_x also serves as a right inverse to σ'_x, we start with some E from $\mathcal{H}_+$, and form the sequence $\eta^E : 0 \to E(-x) \to E \xrightarrow{\pi^E} E^x \to 0$ which leads to an exact sequence $0 \to \mathrm{Hom}(-, E^x)|_{\mathcal{S}_x} \xrightarrow{\eta^E} \mathrm{Ext}^1(-, E(-x))|_{\mathcal{S}_x} \to \mathrm{Ext}^1(-, E)|_{\mathcal{S}_x} \xrightarrow{\mathrm{Ext}^1(-,\pi^E)} \mathrm{Ext}^1(-, E^x)|_{\mathcal{S}_x} \to 0$. By Serre duality, $\mathrm{Ext}^1(-, \pi^E)$ corresponds to $\mathrm{Hom}(\pi^E, -)$, so is an isomorphism, implying that η^E is an $\mathcal{S}_x$-universal extension.

Exactness of σ'_x on exact sequences $0 \to E' \to E \to E'' \to 0$ in $\mathcal{H}_x$ follows from inspection of the commutative diagram

$$
\begin{array}{ccccccccccc}
 & & & & 0 & & 0 & & 0 & & \\
 & & & & \downarrow & & \downarrow & & \downarrow & & \\
\eta_{E'} : & & 0 & \to & E' & \xrightarrow{\alpha_{E'}} & E'(x) & \xrightarrow{\beta_{E'}} & E_x & \to & 0 \\
 & & & & u\downarrow & & u(x)\downarrow & & u_x\downarrow & & \\
\eta_E : & & 0 & \to & E & \xrightarrow{\alpha_E} & E(x) & \xrightarrow{\beta_E} & E_x & \to & 0 \\
 & & & & v\downarrow & & v(x)\downarrow & & v_x\downarrow & & \\
\eta_{E''} : & & 0 & \to & E'' & \xrightarrow{\alpha_{E''}} & E''(x) & \xrightarrow{\beta_{E''}} & E''_x & \to & 0 \\
 & & & & \downarrow & & \downarrow & & \downarrow & & \\
 & & & & 0 & & 0 & & 0 & &
\end{array}
$$

whose rows and outer columns are exact: for the exactness of the right column notice that the functor $E \mapsto E_x$ is exact, and further $\mathrm{Ext}^1(U, -)$, with $U \in \mathcal{S}_x$, is exact on short exact sequences with terms in $\mathcal{H}_+$. The assertion on exactness follows. It is further obvious from the construction, that shift $E \mapsto E(x)$ preserves the rank.   □

The following proposition shows the importance of line bundles for a study of the category $\mathcal{H}$:

Proposition 10.5. *Each $E \in \mathcal{H}_+$ has a filtration $0 = E_0 \subseteq E_1 \subseteq \cdots \subseteq E_{r-1} \subseteq E_r = E$ with line bundle factors E_i/E_{i-1}.*

Proof. We argue by induction on the rank r of E. For $r = 1$ there is nothing to prove. For $r > 1$ we get an inclusion $L \hookrightarrow E$ in the quotient category $\tilde{\mathcal{H}} = \mathcal{H}/\mathcal{H}_0$ for some line bundle L. (Recall that $\mathcal{H}$ and $\tilde{\mathcal{H}}$ have the same objects.) By the definition of quotient categories this yields a nonzero morphism (necessarily monomorphic) $L' \hookrightarrow E$ for some line bundle $L' \subseteq L$. Consider the exact sequence $0 \to L' \to E \to E/L' \to 0$ and let $\bar{L}/L'$ denote the maximal subobject of E/L' having finite length. Then $\bar{L}$ is a line bundle contained in E such that $E/\bar{L}$ belongs to $\mathcal{H}_+$. The claim now follows by induction on the rank.   □

For $E \in \mathcal{H}_+$ we define $E(nx)$ to be $(\sigma'_x)^n(E)$ for each $n \geq 0$ and to be $(\rho'_x)^{-n}(E)$ for $n < 0$.

Corollary 10.6. *Let E and F be in $\mathcal{H}_+$ and x in C. Then $\mathrm{Hom}(E, F(nx)) \neq 0$ and $\mathrm{Ext}^1(E, F(nx)) = 0$ for $n \gg 0$.*

Proof. In view of the proposition, the two assertions reduce to the case where E and F are line bundles. The first assertion follows by observing that $\langle [E], [F(nx)] \rangle$ is unbounded as a function of n. For the second assertion we invoke Serre duality and use that any non-zero $F(nx) \to \tau E$ is a monomorphism with cokernel U_n from $\mathcal{H}_0$. Assume that $\mathrm{Ext}^1(E, F(nx)) \neq 0$ for infinitely many $n \geq 0$ then the formula $\langle [F], [F(nx)] \rangle + \langle [F], [U_n] \rangle = \langle [F], [\tau E] \rangle$ holding for such n's yields a contradiction since its left hand side is unbounded as a function of n.   □

Proposition 10.7 ([12, Cor. 1.8.3]). *There exists a countable generating family $\mathcal{L}$ of line bundles for $\mathcal{H}$, that is, for each object X of $\mathcal{H}$ there exists an epimorphism $L_1 \oplus \ldots \oplus L_n \to X$ with $L_1, \ldots, L_n$ from $\mathcal{L}$.*

Proof. We first collect a finite system $\mathcal{L}'$ of line bundles such that for each simple object S there is a non-zero morphism $L' \to S$ with $L' \in \mathcal{L}'$. Note that $\mathcal{L}'$ already generates $\mathcal{H}_0$. Denote by $\mathcal{L}'(nx)$ the system of all

$L'(nx)$ with L' in $\mathcal{L}$ and let $\mathcal{L}$ be the union of all $L'(-nx)$, $n \geqslant 0$. The preceding corollary implies that $\mathcal{L}$ generates $\mathcal{H}$. $\qquad\square$

For the next statement compare [33, 38, 35], where the context is however slightly different. Note that we use the notation from Section 4.

Theorem 10.8. *The equivalence $\sigma'_x : \mathcal{H}_+ \to \mathcal{H}_+$ extends to a self-equivalence σ_x of $\mathcal{H}$ and hence of $D^b\mathcal{H}$ such that for each Y from $D^b\mathcal{H}$ there is an exact triangle $\bigoplus_{j=1}^{p(x)} \mathrm{Hom}^\bullet(\tau^j S_x, Y) \otimes \tau^j S_x \overset{can}{\longrightarrow} Y \longrightarrow \sigma_x Y \longrightarrow \bigoplus_{j=1}^{p(x)} \mathrm{Hom}^\bullet(S_x, Y) \otimes \tau^j S_x[1]$.*

Proof. First we construct "functorial resolutions" of objects from $\mathcal{H}$ by objects of $\mathcal{H}_+$. Let $\mathcal{L}$ be a countable generating family of line bundles for $\mathcal{H}$. We represent $\mathcal{L}$ as the union $\mathcal{L}_0 \subset \mathcal{L}_1 \subset \cdots$ of finite subsets, each containing a generating system for $\mathcal{H}_0$. For each object Y from $\mathcal{H}$ and each $n \geqslant 0$ we form the canonical morphism $E_n(Y) := \bigoplus_{L \in \mathcal{L}_n} \mathrm{Hom}(L, Y) \otimes L \overset{\beta_{n,Y}}{\longrightarrow} Y$, yielding an exact sequence

$$0 \to F_n(Y) \overset{\alpha_{n,Y}}{\longrightarrow} E_n(Y) \overset{\beta_{n,Y}}{\longrightarrow} Y$$

which is functorial in Y. Note that $\beta_{n,Y}$ is an epimorphism for $n \gg 0$. By construction $E_n(Y)$ is a direct factor of $E_{n+1}(Y)$ and $F_n(Y)$ is a subobject of $F_{n+1}Y$; moreover the factors $E_{n+1}(Y)/E_n(Y)$ and $F_{n+1}(Y)/F_n(Y)$ belong to $\mathcal{H}_+$ for $n \gg 0$. Next we define $\sigma_n Y$ as the cokernel of $(F_n(Y))(x) \overset{\alpha_{n,Y}(x)}{\longrightarrow} (E_n(X))(x)$. By the preceding comments $\alpha_{n,Y}$ is a monomorphism and $\sigma_n(Y) \subseteq \sigma_{n+1}Y$ for $n \gg 0$. Notice moreover that for any $E \in \mathcal{H}_+$ we have $\sigma_n E \subseteq E(x)$ for all n and $\sigma_n E = E(x)$ for $n \gg 0$. It follows that the sequence $(\sigma_n Y)$ stabilizes for each $Y \in \mathcal{H}$, that is, we have $\sigma Y = \sigma_n Y$ for large n. The resulting functor $\sigma_x(Y) = \varinjlim \sigma_n Y$ is the wanted extension of σ'_x to $\mathcal{H}$. $\qquad\square$

10.4 Normal form of tilting object: canonical algebras

It is quite exceptional that a normal form for a tilting module and its endomorphism algebra exists. This is the case for the categories $\mathcal{H}$ of coherent sheaves on a weighted projective line. We fix a line bundle L of $\mathcal{H}$. Recall that for each point x of C there is a unique simple S_x in $\mathcal{H}$ with $\mathrm{Hom}(L, S_x) \neq 0$. By Serre duality $\tau^{-1}S_x$ is the unique simple concentrated in x extending non-trivially with L. Putting $L_0^{(x)} = L$ we obtain a nonsplit sequence $0 \to L_0^{(x)} \to L_1^{(x)} \to \tau^{-1}S_x \to 0$. Continuing we obtain short exact sequences $0 \to L_{i-1}^{(x)} \to L_i^{(x)} \to \tau^{-i}S_x \to 0$ for each integer $i = 1, \ldots, p(x)$. Since the class $\sum_{i=1}^{p(x)}[\tau^{-i}S_x]$ is easily seen to be independent of x it follows that the line bundle $\bar{L} := L_{p(x)}^{(x)}$ is independent of x, see Prop. 5.3.

Theorem 10.9. *Each category $\mathcal{H}$ of coherent sheaves on a weighted projective line over k of weight type $p = (p_1, \ldots, p_t)$ has a tilting object whose endomorphism ring is a canonical algebra of type $(p, \underline{\lambda})$ for a parameter sequence $(1 = \lambda_3, \ldots, \lambda_t)$ of pairwise distinct nonzero elements from k.*

Proof. We follow [32] and consider the line bundles L and $\bar{L}$ together with a representative system of all $L_j^{(x)}$, with $x \in C$ and $j = 1, \ldots, p(x) - 1$. Since only the exceptional points $x_1, \ldots, x_t$ matter, this yields the full subcategory consisting of the following objects

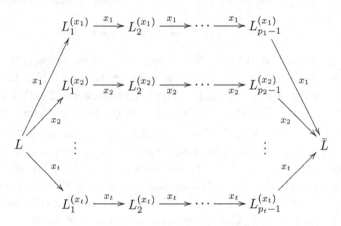

and actually generated by the above quiver. As is easily checked, the above line bundle configuration, called *canonical*, is a tilting object in $\mathcal{H}$. Note that $\mathrm{Hom}(L, \bar{L}) \cong k^2$ and that the cokernel of each $x_i^{p_i} : L \to \bar{L}$ belongs to the exceptional tube $\mathcal{U}_i = \mathcal{U}_{x_i}$ with the simple S_{x_i}, yielding $t-3$ relations $x_i^{p_i} = x_2^{p_2} - \lambda_i x_1^{p_1}$, $i = 3, \ldots, t$, for pairwise distinct nonzero elements $\lambda_3, \ldots, \lambda_t$ from k.   □

Denoting by $\Lambda(p, \underline{\lambda})$ the *canonical algebra* given by the above quiver with the above relations, we obtain the following consequence:

Corollary 10.10 ([12]). *Each category of weighted projective lines is* *derived equivalent to the category* mod-$\Lambda(p, \underline{\lambda})$ *over a canonical algebra.*

$\square$

10.5 Degree, slope and Euler characteristic

There are two important linear forms on $K_0(\mathcal{H})$, rank and degree. While the rank is zero on $\mathcal{H}_0$, positive on non-zero objects from $\mathcal{H}_+$ and further constant on τ-orbits, the degree is positive on non-zero objects from $\mathcal{H}_0$ and constant on τ-orbits in $\mathcal{H}_0$. While there is — up to scalars — only one choice for the rank, several choices are possible for the degree. It is customary to normalize the degree in such a way that it becomes zero on a preselected line bundle L_0.

Following [12] we fix a line bundle L_0, and denote by p the least common multiple of the τ-periods of exceptional simple objects from $\mathcal{H}$. Consider the following average of the Euler form

$$\langle\langle x, y \rangle\rangle = \frac{1}{p} \sum_{j=0}^{p-1} \langle \tau^j x, y \rangle$$

having values in $\frac{1}{p}\mathbb{Z}$. We then define the *degree* on $K_0(\mathcal{H})$ by the formula

$$\deg x = \langle\langle [L_0], x \rangle\rangle - \langle\langle [L_0], [L_0] \rangle\rangle \cdot \operatorname{rk}(x).$$

It follows from the definition that $\deg L_0 = 0$, and $\deg S_x = 1/p(x)$ for each simple S_x concentrated in x. If L is any further line bundle we have $[L] = [L_0] + u$ with $\operatorname{rk}(u) = 0$, hence $\deg \tau u = \deg u$. It follows that the difference $\deg \tau L - \deg L_0$ does not depend on the choice of the line bundle L. We thus call $\chi_{\mathcal{H}} = \deg L - \deg \tau L$ the *Euler characteristic* of $\mathcal{H}$. The Euler characteristic is an important invariant of $\mathcal{H}$, it has a significant influence on the representation type.

The *slope* of a non-zero object X of $\mathcal{H}$ is defined as the quotient $\mu X = \deg X / \operatorname{rk}(X) \in \mathbb{Q} \cup \{\infty\}$. (The zero object is allowed to have any slope). A non-zero object X is called *stable* (resp. *semistable*) if $\mu X' < \mu X$ (resp. $\mu X' \leqslant \mu X$) holds for any proper subobject X' of X. The zero object is also semistable. Let $f : X \to Y$ be a non-zero homomorphism between semistable objects, then $\mu X \leqslant \mu Y$. If moreover X and Y are stable and f is not an isomorphism, then $\mu X < \mu Y$. In particular, the endomorphism ring of a stable object is a division ring, hence in the present context isomorphic to k. For the next result, see [41] or [12, Prop. 5.2].

Proposition 10.11. *Let $q \in \mathbb{Q} \cup \{\infty\}$. Then the full subcategory $\mathcal{H}^{(q)}$ of $\mathcal{H}$ consisting of all semistable objects of slope q is a hereditary abelian category where each object has finite length. Moreover, the simple objects of $\mathcal{H}^{(q)}$ are exactly the stable objects of slope q.*

Proof. Clearly, the assertion holds for $q = \infty$, since $\mathcal{H}^{(\infty)} = \mathcal{H}_0$. We may thus assume q is a rational number. It is easy to check that $\mathcal{H}^{(q)}$ is closed under kernels, cokernels and extensions. As an exact subcategory of $\mathcal{H}$ the category $\mathcal{H}^{(q)}$ hence is itself an abelian category; moreover, extensions spaces $\mathrm{Ext}^1(X, Y)$, with X and Y from $\mathcal{H}^{(q)}$, taken in $\mathcal{H}^{(q)}$ agree with those taken in $\mathcal{H}$, hence $\mathcal{H}^q$ is also hereditary. Let X be a subobject of Y. If X and Y have the same slope and the same rank it follows from the properties of rank and degree that $X = Y$. As an immediate consequence the length of an object X in $\mathcal{H}^{(q)}$ is bounded by its rank. □

The importance of the Euler characteristic is underlined by the two following propositions.

Proposition 10.12. *Let $(p_1, \ldots, p_t)$ be the weight type of $\mathcal{H}$. Then $\chi_{\mathcal{H}} = 2 - \sum_{i=1}^t \left(1 - \frac{1}{p_i}\right)$. Moreover, for each non-zero bundle E we have $\mu(\tau E) = \mu(E) - \chi_{\mathcal{H}}$.*

Proof. Since $\mathcal{H}$ has a tilting object, the Euler form $\langle -, - \rangle$ is non-degenerate. Choose a line bundle L, a homogeneous simple S_0 and a representative system $S_1, \ldots, S_t$ of exceptional simples satisfying $\mathrm{Ext}^1(L, S_i) \neq 0$. We put $u = [L] + (t - 2)[S_0] - \sum_{i=1}^t [S_i])$. Using that $\langle u, - \rangle$ and $\langle [\tau L], - \rangle$ agree on a system of generators for $\mathrm{K}_0(\mathcal{H})$, it follows that $[\tau L] = u$. Passing to degrees and invoking Proposition 10.5 the two assertions follow. □

For the proof of the next proposition we refer to [12, prop. 5.5] or [36, prop. 4.1]. Note that $\chi_{\mathcal{H}} > 0$ if and only if the weight type of $\mathcal{H}$ is one of (), (n), (m, n), $(2, 2, n)$ $(n \geqslant 2)$, $(2, 3, 3)$, $(2, 3, 4)$ or $(2, 3, 5)$. Here () stands for the empty weight type. Further $\chi_{\mathcal{H}} = 0$ if and only if the weight type is one of $(2, 2, 2, 2)$, $(3, 3, 3)$, $(2, 4, 4)$ or $(2, 3, 6)$.

Proposition 10.13. *(i) If $\chi_{\mathcal{H}} \geqslant 0$ then each indecomposable bundle in*

$\mathcal{H}$ *is semistable.*

(ii) If $\chi_{\mathcal{H}} > 0$ *then each indecomposable bundle in* $\mathcal{H}$ *is stable.* □

10.6 Positive Euler characteristic: domestic case

The next theorem, due to Hübner [24], implies that a category $\mathcal{H}$ with positive Euler characteristic $\chi_{\mathcal{H}}$ is derived equivalent to the category of modules over a tame hereditary k-algebra H, that is, the path algebra of a quiver whose underlying graph is extended Dynkin. For the proof we refer to [36, prop. 6.5, prop. 4.3]. From property (ii) we conclude that for positive Euler characteristic there is an additional normal form for tilting bundles, called the *hereditary normal form.*

Theorem 10.14. *Assume that* $\mathcal{H}$ *has Euler characteristic* $\chi_{\mathcal{H}} > 0$. *Then the following properties hold:*

(i) Each indecomposable bundle is stable and exceptional.

(ii) The direct sum T *of a representative system (with respect to isomorphism) of indecomposable bundles* E *with* $0 \leqslant \mu E < \chi_{\mathcal{H}}$ *is a tilting object.*

(iii) The indecomposable bundles form a single Auslander-Reiten component.

(iv) The endomorphism ring of T *is isomorphic to the path algebra of an extended Dynkin quiver, which is either of type* $\widetilde{\mathbb{A}}_{p,q}$ *or of type* $\widetilde{\mathbb{D}}_n$ *($n \geqslant 4$),* $\widetilde{\mathbb{E}}_6$, $\widetilde{\mathbb{E}}_7$ *or* $\widetilde{\mathbb{E}}_8$ *with bipartite orientation.* □

Let T be a tilting object in $\mathcal{H}$. It is now easy to derive the properties of mod-H, $H = \text{End}(T)$, from the above description of $\mathcal{H}$. One obtains this way an alternative description of the module category for tame hereditary algebras, similarly for the larger class of *tame concealed algebras* defined as the endomorphism rings of tilting bundles of tilting objects in $\mathcal{H}$ with $\chi_{\mathcal{H}} > 0$.

Corollary 10.15. *Up to derived equivalence the following two classes of hereditary categories coincide:*

(i) the categories of coherent sheaves coh $\mathbb{X}$ *on a weighted projective line* $\mathbb{X}$ *with positive Euler characteristic.*

(ii) the module categories mod-H *over a connected finite dimensional tame hereditary algebra* H. □

10.7 Euler characteristic zero: tubular case

We assume that $\mathcal{H}$ is a hereditary noetherian category with Serre
duality and having a tilting object. We call the category $\mathcal{H}$ *tubular* if
its Euler characteristic is zero. It follows from Proposition 10.13 that
in this case each indecomposable bundle is semistable. The main result
of this section is that each Auslander-Reiten component of $\mathcal{H}$ is a tube.
More precisely we show that the indecomposables of a fixed slope, which
may be a rational number or infinity, form a one-parameter family of
tubes, naturally indexed by the points of the projective line. The proof
relies on the following two statements (compare [19] for the first one).

Lemma 10.16. *Let $\rho \geqslant 0$ be an integral-valued function on $\mathcal{H}$, additive
with respect to short exact sequences, such that the Serre subcategory
$\mathcal{H}' = \{C : \rho(C) = 0\}$ is noetherian. Assume further that $\mathrm{Hom}(\mathcal{H}', \mathcal{H}'') =
0$, where $\mathcal{H}''$ is the additive category generated by the indecomposable
objects not in $\mathcal{H}'$. Then $\mathcal{H}$ is also noetherian.*

Proof. Let $X_1 \subseteq X_2 \cdots \subseteq X_n \subseteq \cdots \subseteq X$ be an ascending chain of
subobjects of an object X from $\mathcal{H}$. Since $\rho \geqslant 0$ it follows that $\rho(X_1) \leqslant
\rho(X_2) \leqslant \cdots \leqslant \rho(X_n) \leqslant \cdots \leqslant \rho(X)$. Hence we can assume without
restriction that ρ is constant on (X_n). This yields an ascending chain
$X_2/X_1 \subseteq X_3/X_1 \subseteq \cdots \subseteq X_n/X_1 \subseteq \cdots \subseteq X/X_1$ of subobjects of X/X_1,
hence of the direct sum $(X/X_1)'$ of the indecomposable summands of
X/X_1 which are in $\mathcal{H}'$, since $\mathrm{Hom}(\mathcal{H}', \mathcal{H}'') = 0$. The claim follows since
$\mathcal{H}'$ is noetherian. $\square$

Assume now that $\chi_{\mathcal{H}} = 0$. Then each indecomposable bundle is
semistable and, moreover, the semistable bundles of slope q form a
hereditary abelian category $\mathcal{H}^{(q)}$ which is stable under the equivalence
τ. It follows [11] that $\mathcal{H}^{(q)}$ is a uniserial category decomposing into
connected uniserial categories.

Lemma 10.17. *Assume that $\chi_{\mathcal{H}} = 0$. For a rational number q we form the interval category $\mathcal{H}\langle q \rangle =
\mathrm{add}\left(\bigvee_{r \in \mathbb{Q} \cup \{\infty\}, q < r} \mathcal{H}^{(r)}[-1] \vee \bigvee_{s \in \mathbb{Q}, s \leqslant q} \mathcal{H}^{(s)}\right)$. Then $\mathcal{H}' = \mathcal{H}\langle q \rangle$
is a hereditary noetherian k-category with Serre duality, whose full
subcategory $\mathcal{H}'_0$ of finite length objects equals $\mathcal{H}^{(q)}$. Moreover, if $\mathcal{H}$ has
a tilting object, the same holds for $\mathcal{H}'$.*

Proof. In view of 7.1 $\mathcal{H}'$ is abelian. The construction yields that $\mathcal{H}'$ is hereditary and, with the equivalence τ induced by $\tau_{D^b\mathcal{H}}$, satisfies Serre duality. Writing $q = d/r$ with d, r coprime integers and $r > 0$, it follows in a straightforward manner that the linear form $\rho = d\,\mathrm{rk} - r\,\deg$ on $K_0(\mathcal{H}) = K_0(\mathcal{H}')$ is non-negative on all objects of $\mathcal{H}'$ and moreover vanishes exactly on objects from $\mathcal{H}'_0 := \mathcal{H}^{(q)}$, and hence acts as a rank function for $\mathcal{H}'$. It now follows from Lemma 10.16 that $\mathcal{H}'$ is noetherian. Let T be a tilting object in $\mathcal{H}$. We write $T = T_0 \oplus T_1$, where each indecomposable summand of T_0 (resp. T_1) has slope $\leqslant q$ (resp. $> q$). It follows that $T' = T_0 \oplus \tau T_1[-1]$ is a tilting object of $\mathcal{H}'$. $\square$

Note that the existence of a tilting object implies that all tubes of $\mathcal{H}^{(q)}$ have finite τ-period and only finitely many of them have τ-period > 1. A much stronger statement can be shown by invoking the shift functors from Section 10.3.

Proposition 10.18. *For any rational number q there is a selfequivalence of $D^b\mathcal{H}$ inducing equivalences from $\mathcal{H}$ to $\mathcal{H}\langle q \rangle$ and from $\mathcal{H}_0$ to $\mathcal{H}^{(q)}$.*

Proof. Let $x_1, \ldots, x_t$ denote the finite number of points x of $\mathcal{H}$ such that $\mathcal{U}_x$ has finite τ-period $p_1, \ldots, p_t$, respectively. Let p be the least common multiple of $p_1, \ldots, p_t$. By Theorem 10.8 the shift σ_x associated to the point x yields a selfequivalence of $D^b\mathcal{H}$. By construction σ_x satisfies $\mu(\sigma_x E) = \mu E + 1/p(x)$ for each non-zero E. Since $p/p_1, \ldots, p/p_t$ are collectively coprime, there exists an selfequivalence σ of $D^b\mathcal{H}$, belonging to the subgroup generated by the shifts associated to the $x_1, \ldots, x_t$, satisfying $\mu(\sigma E) = \mu E + 1/p$.

Applying the same argument to $\mathcal{H}\langle q \rangle$, we obtain a selfequivalence ρ of $D^b\mathcal{H}' = D^b\mathcal{H}$ satisfying $\mu(\rho E) = 1/(1/p + \mu E)$. It follows that the subgroup $\langle \sigma, \rho \rangle$ of selfequivalences generated by σ and ρ acts transitively on the set $\mathbb{Q} \cup \{\infty\}$ of all slopes, and the assertion follows. $\square$

Theorem 10.19. *Let $\mathcal{H}$ be hereditary noetherian category of Euler characteristic zero and having a tilting object. Then*

$$\mathcal{H} = \bigvee_{q \in \mathbb{Q} \cup \{\infty\}} \mathcal{H}^{(q)}, \quad \text{where for each } q \text{ we have } \mathcal{H}^{(q)} \cong \mathcal{H}_0.$$

Proof. Since $\mathcal{H}$ has Euler characteristic zero, each indecomposable object is semistable, hence belongs to some $\mathcal{H}^{(q)}$. It follows that $\mathcal{H} = \bigvee_{q \in \mathbb{Q} \cup \{\infty\}} \mathcal{H}^{(q)}$. The last claim now follows from Proposition 10.18. $\square$

The Auslander-Reiten quiver of $\mathcal{H}$ is a rational family — indexed by $\mathbb{Q} \cup \{\infty\}$ — of one-parameter families of tubes, each of them indexed by the set of points $k \cup \{\infty\}$ of the projective line over k. The theorem, with a different proof, is due to Happel and Ringel [22]. We further note that — with minor modifications — the proof also holds for a category of coherent sheaves on a smooth elliptic curve, thus covering Atiyah's theorem [1].

10.8 The module category over a tubular algebra

By definition, a finite dimensional k-algebra Λ is called *tubular* if there exists a hereditary noetherian category $\mathcal{H}$ which is tubular and a tilting object T in $\mathcal{H}$ such that Λ is isomorphic to the endomorphism algebra of T. Our prime examples of tubular algebras are the canonical algebras $\Lambda(2,2,2,2;\lambda)$, $\Lambda(3,3,3)$, $\Lambda(2,4,4)$ and $\Lambda(2,3,6)$. We are going to describe the shape of the module category mod-Λ if Λ is a tubular canonical algebra. The case of an arbitrary tubular algebra is similar, but technically more involved.

Let $\Lambda = \Lambda(p_1, \ldots, p_t)$ be a tubular canonical algebra. By Λ' we denote the algebra arising from Λ by deleting the sink vertex c. Obviously Λ' is the path algebra of the star $[p_1, \ldots, p_t]$ having t branches of length $p_1, \ldots, p_t$, respectively. (The length of a branch counts the number of its vertices.) If P_c denotes the indecomposable projective Λ-module corresponding to c, then mod-Λ' can be viewed as the full subcategory of mod-Λ consisting of all Λ-modules with $\operatorname{Hom}(P_c, M) = 0$, that is, taking value zero at the sink vertex c of the quiver of Λ. We describe the position of the indecomposable Λ-modules by means of the identification mod-$\Lambda \subseteq \mathcal{H} \vee \mathcal{H}[1]$.

Theorem 10.20. *Let* $\Lambda = \Lambda(p_1, \ldots, p_t)$ *be a tubular canonical algebra. We put* $p = \operatorname{lcm}(p_1, \ldots, p_t)$. *Then the following assertions hold:*
(i) The category mod-Λ *has a preprojective component* $\mathcal{P}(\Lambda)$ *containing all indecomposable projectives* L_v, *where the vertex* v *is different from the sink vertex of* Λ. *Moreover,* $\mathcal{P}(\Lambda)$ *is lying in* $\mathcal{H}$ *and agrees with the preprojective component* $\mathcal{P}(\Lambda')$ *of* Λ'.
(ii) $\mathcal{P}(\Lambda)$ *consists exactly of the indecomposable* Λ-*modules* P *with slope* $q = \mu(P)$ *in the range* $0 \leqslant q < p$.
(iii) The remaining indecomposable Λ-*modules in* $\mathcal{H}$ *have slope* $\geqslant p$. *In*

more detail,

(a) *An object M of* $\mathcal{H}^{(p)}$ *belongs to* mod-Λ *if and only if* $\mathrm{Ext}^1(L_c, M) = 0$.

(b) *The category* $L_c^{\perp} \cap \mathcal{H}^{(p)}$ *agrees with the category of regular* Λ'-*modules.*

(c) *For* $q > p$ *all objects of* $\mathcal{H}^{(q)}$ *are* Λ-*modules.*

The corresponding dual assertions (and proofs) concerning the Λ-modules belonging to $\mathcal{H}[1]$ are left to the reader. We thus arrive to the following visualization of mod-Λ in terms of its Auslander-Reiten quiver, taken from [26].

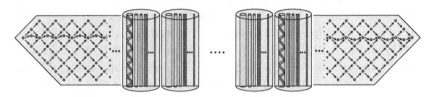

Proof. We first show that the right perpendicular category of L_c, formed in $\mathcal{H}$, agrees with mod-Λ'. Notice that this part of the proof is independent of the assumption on tubular weights. Assume first that M belongs to $\mathcal{H}$ and belongs to $L_c^{\perp}$. Since each indecomposable summand L_v of T admits an embedding $L_v \hookrightarrow L_c$, the condition $\mathrm{Ext}^1(L_c, M) = 0$ implies that $\mathrm{Ext}^1(L_v, M) = 0$ for each vertex v of Λ, and hence M belongs to mod-Λ with the additional property that its value $M_c = \mathrm{Hom}(L_c, M)$ at the vertex c vanishes. This proves that M — under the above identifications — is a Λ'-module. Conversely, assume M is indecomposable and belongs to mod-Λ', that is, is a Λ-module satisfying $\mathrm{Hom}(L_c, M) = 0$. We know that M belongs to $\mathcal{H} \vee \mathcal{H}[1]$. If M belongs to $\mathcal{H}$ the assertion is obvious. We thus need to exclude that M has the form $M = X[1]$ for some X in $\mathcal{H}$ satisfying $\mathrm{Hom}(T, X) = 0$. Since $M_c = 0$ we obtain $\mathrm{Ext}^1(L_c, X) = 0$, hence $\mathrm{Ext}^1(T, X) = 0$ by the previous argument. Since T is tilting, this implies $X = 0$, and we are done.

Next we have a closer look at mod-Λ'. Note that $\mathcal{H}^{(p)} = \coprod_{x \in X} \mathcal{V}_x$ for uniserial categories (tubes) $\mathcal{V}_x$. Moreover, L_c belongs to some $\mathcal{V}_{x_0}$. Since the $\mathcal{V}_x$ are pairwise orthogonal, clearly each $\mathcal{V}_x$, $x \neq x_0$, belongs to mod-Λ'. Since L_c is quasisimple in the tube $\mathcal{V}_c$ it follows that $L_c^{\perp} \cap \mathcal{V}_{x_0}$ is again a tube and that its tubular rank is one less than the tubular rank of $\mathcal{V}_{x_0}$. To conclude: $L_c^{\perp} \cap \mathcal{H}^{(p)} = \coprod_{x \in X} \mathcal{V}_x'$ with connected uniserial categories $\mathcal{V}_x' = \mathcal{V}_x$ for $x \neq x_0$ and $\mathcal{V}_{x_0}' = L_c^{\perp} \cap \mathcal{V}_{x_0}$.

We consider the additive function $\lambda = p\,\mathrm{rk} - \deg : \text{mod-}\Lambda' \to \mathbb{Z}$. On the system of indecomposable projective Λ'-modules this takes values as follows

$$
\lambda : p \begin{array}{l} \nearrow \\ \to \\ \searrow \end{array}
\begin{array}{ccccc}
p - p/p_1 & \to & p - 2p/p_1 & \to \cdots \to & p - (p_1 - 1)p/p_1 \\[2mm]
p - p/p_2 & \to & p - 2p/p_2 & \to \cdots \to & p - (p_2 - 1)p/p_2 \\[2mm]
p - p/p_t & \to & p - 2p/p_t & \to \cdots \to & P - (p_t - 1)p/p_t.
\end{array}
$$

For the tubular weights these stars $[p_1, \ldots, p_t]$ are the extended Dynkin quivers $[2,2,2,2]$, $[3,3,3]$, $[2,4,4]$ and $[2,3,6]$, and λ induces the unique normalized additive function on each of these quivers. Assume now that M is an indecomposable Λ'-module. It is well-known that $\lambda(M) > 0$, $\lambda(M) = 0$ or $\lambda(M) < 0$ if and only if M is respectively preprojective, regular or preinjective. This happens if and only if respectively $\mu(M) < p$, $\mu(M) = p$ or $\mu(M) > 0$, and proves the claims of the theorem. In particular the category $\mathcal{R}(\Lambda')$ of regular Λ'-modules agrees with $L_c^\perp \cap \mathcal{H}^{(p)}$. $\qquad\square$

10.9 Negative Euler characteristic: the wild case

For a detailed study of this case, and the relationship to the representation theory of path algebras of wild hereditary stars, we refer to [34]. Here, we just mention a fundamental fact:

Proposition 10.21. *Let $\mathcal{H}$ be a category of coherent sheaves on a weighted projective line of negative Euler characteristic. Then the following holds:*

(i) Each Auslander-Reiten component in $\mathcal{H}_+$ has shape $\mathbb{Z}\mathbb{A}_\infty$.

(ii) If E and F are nonzero bundles, then for large positive n we have $\mathrm{Hom}(E, \tau^n F) \neq 0$ *and* $\mathrm{Hom}(\tau^n E, F) = 0$. $\qquad\square$

11 Quasitilted algebras

We recall that a finite dimensional k-algebra Λ is called *quasitilted* if there exists a tilting object T in a hereditary abelian category $\mathcal{H}$ such that Λ is isomorphic to the endomorphism ring of T. Quasitilted algebras are an important class of algebras for many reasons: Many interesting algebras are quasitilted, tilted algebras of finite or tame type provide the central clue in quite a number of classification results. Canonical algebras and tubular algebras are quasitilted. Further, quasitilted algebras have a clear homological description: By [21] an algebra Λ is quasitilted if and only if it satisfies the two following two conditions:

(i) Λ has global dimension $\leqslant 2$.

(ii) Each indecomposable Λ-module M has projective or injective dimension $\leqslant 1$.

Restricting to connected algebras, Happel's theorem (Theorem 6.3) implies — with a little extra work classifying hereditary categories with a tilting object up to derived equivalence — that quasitilted algebras occur in two types:

(a) the tilted algebras [23] $\Lambda = \mathrm{End}(T)$, where T is a tilting object in a module category mod-H over a finite dimensional hereditary algebra H.

(b) the quasitilted algebras $\Lambda = \mathrm{End}(T)$ of canonical type [37], where T is a tilting object in a hereditary category $\mathcal{H}$, derived-equivalent to a category of coherent sheaves on a weighted projective line.

Due to Corollary 10.15 there is some overlap between the two types. Note, in particular, that any representation-finite quasitilted algebra is actually tilted [21, Cor. 2.3.6]. According to [37] a representation-infinite quasitilted algebra of canonical type can be characterized as a semiregular branch enlargement of a concealed canonical algebra, that is, a semiregular branch extension of the endomorphism ring of a tilting bundle on a weighted projective line.

References

[1] M. F. Atiyah. Vector bundles over an elliptic curve. *Proc. London Math. Soc.*, **7** (1957), 414–452.

[2] A.A. Beilinson. Coherent sheaves on P_n and problems of linear algebra. *Funct. Anal. Appl.* **12** (1979), 214–216.

[3] I.N. Bernstein, I.M. Gelfand and V.A. Ponomarev. Coxeter functors and Gabriel's theorem. *Russ. Math. Surv.* **28** (1973), 17–32.

[4] A.I. Bondal and M.M. Kapranov. Representable functors, Serre functors, and mutations. *Math. USSR, Izv.* **35** (1990), 519–541.

[5] K. Bongartz. Tilted algebras. In: *Representations of algebras, Springer Lect. Notes Math.* **903** (1981), 26–38.

[6] W. Crawley-Boevey and O. Kerner. A functor between categories of regular modules for wild hereditary algebras. *Math. Ann.* **298** (1994), 481–487.

[7] V. Dlab, C. M. Ringel, *Indecomposable representations of graphs and algebras*, Mem. Am. Math. Soc. **173** (1976).

[8] S. Eilenberg, Abstract description of some basic functors. *J. Indian Math. Soc.* **24** (1960), 231–234.

[9] P. Gabriel. Des catégories abéliennes. *Bull. Soc. Math. France* **90** (1962), 323–448.

[10] P. Gabriel. Unzerlegbare Darstellungen I. *Manuscripta Math.* **6** (1972) 71–103.

[11] P. Gabriel. Indecomposable representations II. *Symposia Mat.Inst. Naz. Alta Mat.* **11**, 1973, 81–104.

[12] W. Geigle and H. Lenzing. A class of weighted projective lines arising in representation theory of finite dimensional algebras. *Lect. Notes Math.* **1273**, (1987), 265–297.

[13] W. Geigle, H. Lenzing, Perpendicular categories with applications to representations and sheaves, *J. Algebra* **144** (1991), 273-343.

[14] S.I. Gelfand and Yu.I. Manin. *Methods of homological algebra.* Second edition. Springer-Verlag, 2003.

[15] D. Happel. *Triangulated categories and the representation theory of finite dimensional algebras.* Cambridge University Press, Cambridge, London Math. Soc. Lect. Notes Ser. **119**, 1988.

[16] D. Happel. Quasitilted algebras. *CMS Conf. Proc.* **23** (1998), 55–83.

[17] D. Happel. A characterization of hereditary categories with tilting object. *Invent. math.* **144** (2001), 381–398.

[18] D. Happel and I. Reiten. Directing objects in hereditary categories. In: Trends in the representation theory of finite-dimensional algebras (Seattle, WA, 1997) *Contemp. Math.* **229** (1998), 169–179.

[19] D. Happel and I. Reiten. A combinatorial characterization of hereditary categories containing simple objects, Representations of Algebras (Sao Paulo 1999), *Lect. Notes in Pure and Applied Math.* **224**, Dekker, New York 2002, 91-97.

[20] D. Happel and I. Reiten, Hereditary abelian categories with tilting object over arbitrary base fields. *J. Algebra* **256** (2002), 414–432.

[21] D. Happel, I. Reiten and S. Smalø. Tilting in abelian categories and quasitilted algebras. *Mem. Amer. Math. Soc.* **575**, 1996.

[22] D. Happel and C. M. Ringel. The derived category of a tubular algebra. Springer Lecture Notes in Mathematics **1177** (1986), 156–180.

[23] D. Happel and C. M. Ringel. Tilted Algebras. *Trans. Amer. Math. Soc.* **274** (1982), 399–443.

[24] T. Hübner. *Classification of indecomposable vector bundles on weighted projective curves.* Diplomarbeit Universität Paderborn, 1986.

[25] T. Hübner. *Exzeptionelle Vektorbündel und Reflektionen an Kippgarben über projektiven gewichteten Kurven.* Dissertation Universität Paderborn, 1996.

[26] M. Jesse. *Die Klassifikation unzerlegbarer Objekte erblicher Kategorien vom Garbentyp im tubularen Fall.* Diplomarbeit, Universität Paderborn, 2003.

[27] B. Keller. Derived categories and tilting. *This Handbook.*

[28] O. Kerner. Representations of wild quivers. *CMS Conf. Proc.* **19** (1996),

65–107.

[29] D. Kussin. *Aspects of hereditary representation theory over non-algebraically closed fields*, Habilitationsschrift, Universität Paderborn, 2004.

[30] H. Lenzing. Wild canonical algebras and rings of automorphic forms. In: *Finite dimensional algebras and related topics* (V. Dlab and L.L. Scott ets.), Kluwer 1994, 191-212.

[31] H. Lenzing. A K-theoretic study of canonical algebras. *Can. Math. Soc. Conf. Proc.* **18** (1996), 433–454 .

[32] H. Lenzing. Hereditary noetherian categories with a tilting complex, *Proc. Am. Math. Soc.* **125** (1997), 1893-1901.

[33] H. Lenzing and H. Meltzer. Sheaves on a weighted projective line of genus one, and representations of a tubular algebra. *Can. Math. Soc. Conf. Proc.* **14** (1993), 313-337.

[34] H. Lenzing and J.A. de la Peña. Wild canonical algebras. *Math. Z.* **224** (1997), 403–425.

[35] H. Lenzing, J. A. de la Peña, Concealed-canonical algebras and algebras with a separating tubular family, Proc. London Math. Soc. **78** (1999), 513–540.

[36] H. Lenzing and I. Reiten, Hereditary noetherian categories of positive Euler characteristic. Math. Z., to appear.

[37] H. Lenzing and A. Skowroński. Quasi-tilted algebras of canonical type. *Coll. Math.* **71** (1996), 161–181.

[38] H. Meltzer. Tubular mutations. *Colloq. Math.* **74** (1997), 267–274.

[39] H. Meltzer. *Exceptional vector bundles, tilting sheaves and tilting complexes for weighted projective lines.* Mem. Amer. Math. Soc. **171** (2004).

[40] B. Mitchell, Theory of Categories, Academic Press, New York-London (1965).

[41] C.S. Seshadri. Fibrés vectoriels sur les courbes algébriques. *Astérisque* **96** (1982), 1-209.

[42] N. Popescu. *Abelian Categories with Applications to Rings and Modules,* Academic Press, London, New York (1973).

[43] I. Reiten and M. Van den Bergh. Noetherian hereditary categories satisfying Serre duality, *J. Amer. Math. Soc.* **15** (2002), 295–366.

[44] J. Rickard. Morita theory for derived categories. *J. London Math. Soc.* **39** (1989), 436–456.

[45] C.M. Ringel. Finite dimensional hereditary algebras of wild representation type. *Math. Z.* **161** (1978), 235–255.

[46] C.M. Ringel. *Tame algebras and integral quadratic forms.* Lect. Notes Math. **1099** (1984).

[47] C. M. Ringel. The spectral radius of the Coxeter transformations for a generalized Cartan matrix. *Math. Ann.* **300** (1994), 331–339.

[48] A.N. Rudakov. *Helices and vector bundles: seminaire Rudakov.* London Math. Soc. Lecture Notes series **148**. Cambridge University Press, 1990.

[49] J.-P. Serre. Faisceaux algébriques cohérents. *Annals of Math.* **61** (1955), 197–278

[50] M. Takane. On the Coxeter transformation of a wild algebra. *Arch. Math.* **63** (1994), 128–135.

[51] C. E. Watts. Intrinsic characterisations of some additive functors. *Proc. Amer. Math. Soc.* **11** (1960), 5–8.

Helmut Lenzing
Institut für Mathematik-Informatik
Universität Paderborn
D-33095 Paderborn, Germany
Email: helmut@math.uni-paderborn.de

7

Fourier-Mukai transforms

Lutz Hille and Michel Van den Bergh

Dedicated to Claus Michael Ringel on the occasion of his 60th birthday.

Abstract

In this paper we discuss some of the recent developments on derived equivalences in algebraic geometry.

1 Some background

In this paper we discuss some of the recent developments on derived equivalences in algebraic geometry but we don't intend to give any kind of comprehensive survey. It is better to regard this paper as a set of pointers to some of the recent literature.

To put the subject in context we start with some historical background. Derived (and triangulated) categories were introduced by Verdier in his thesis (see [26, 79]) in order to simplify homological algebra. From this point of view the role of derived categories is purely technical.

The first non-trivial derived equivalence in the literature is between the derived categories of sheaves on a sphere bundle and its dual bundle [70]. The equivalence resembles Fourier-transform and is now known as a "Fourier-Sato" transform.

The first purely algebro-geometric derived equivalence seems to appear in [54] where is it is shown that an abelian variety A and its dual $\hat{A}$ have equivalent derived categories of coherent sheaves. Again the equivalence is similar to a Fourier-transform and is therefore called a "Fourier-Mukai" transform.

In [7] Beilinson showed that $\mathbb{P}^n$ is derived equivalent to a (noncommutative) finite dimensional algebra. This explained earlier results by Barth and Hulek on the relation between vector bundles and linear algebra. Beilinson's result has been generalized to other varieties and has evolved into the theory of exceptional sequences (see for example [9]). The observation that derived equivalences do not preserve commutativity is significant for non-commutative algebraic geometry (see for example [29]).

Most algebraists probably became aware of the existence non-trivial derived equivalences when Happel showed that "tilting" (as introduced by Brenner and Butler [15]) leads to a derived equivalence between finite dimensional algebras [32]. This was generalized by Rickard who worked out the Morita theory for derived categories of rings [62, 63]; see also [41].

Hugely influential was the so-called homological mirror symmetry conjecture by Kontsevich [46] which states (very roughly) that for two Calabi-Yau manifolds X, Y in a mirror pair, the bounded derived category of coherent sheaves on X is equivalent to a certain triangulated category (the Fukaya category) related to the symplectic geometry of Y. The homological mirror symmetry conjecture was recently proved by Seidel for quartic surfaces (which are the simplest Calabi-Yau manifolds after elliptic curves) [72].

Finally this introduction would be incomplete without at least mentioning the celebrated Riemann-Hilbert correspondence [14, 38, 50, 51] which gives a derived equivalence between sheaves of vector spaces and regular holonomic D-modules on a complex manifold or a smooth algebraic variety (depending on context). This is a far reaching generalization of the classical correspondence between local systems and vector bundles with flat connections.

Acknowledgment. The authors would like to thank Dan Abramovich, Paul Balmer, Alexei Bondal, Tom Bridgeland, Daniel Huybrechts, Pierre Schapira, Paul Smith and the anonymous referee for helpful comments on the first version of this paper.

There are many other survey papers dedicated to Fourier-Mukai transforms. We refer in particular to Raphael Rouquier's "Catégories dérivées et géometrie algebriques" [68]. Another good source of information is given by preliminary course notes by Daniel Huybrechts [35].

2 Notations and conventions

Throughout we work over the base field $\mathbb{C}$. The bounded derived category of coherent sheaves on a variety X is denoted by $\mathcal{D}^b(X)$. Similarly, the bounded derived category of finitely generated modules over an algebra A is denoted by $\mathcal{D}^b(A)$. The shift functor in the derived category is denoted by [1]. All functors between triangulated categories are additive and exact (i.e. they commute with shift and preserve distinguished triangles).

A sheaf is a coherent $\mathcal{O}_X$–module and a point in X is always a closed point. The structure sheaf of a point x will be denoted by $\mathcal{O}_x$. The canonical divisor of a smooth projective variety is denoted by K_X and the canonical sheaf is denoted by ω_X.

3 Basics on Fourier-Mukai transforms

Let X and Y be connected smooth projective varieties. We are interested in equivalences of the derived categories $\Phi\colon \mathcal{D}^b(Y) \longrightarrow \mathcal{D}^b(X)$. Such varieties X and Y are also called *Fourier-Mukai partners* and the equivalence Φ is called a *Fourier-Mukai transform*. In this section we will discuss some properties which remain invariant under Fourier-Mukai transforms. The main technical tool is Orlov's theorem (see below) which states that any derived equivalence $\Phi\colon \mathcal{D}^b(Y) \longrightarrow \mathcal{D}^b(X)$ is coming from a complex on the product $Y \times X$.

Given Fourier-Mukai X, Y it is also interesting to precisely classify the Fourier-Mukai transforms $\mathcal{D}^b(Y) \longrightarrow \mathcal{D}^b(X)$ (it is usually sufficient to consider $X = Y$). This is generally a much harder problem which has been solved in only a few special cases, notably abelian varieties [59] and varieties with ample canonical or anti-canonical divisor (see Theorem 4.4 below).

To start one has the following simple result.

Lemma 3.1 ([21, Lemma 2.1]). *If X and Y are Fourier-Mukai partners, then $\dim_(X) = \dim_(Y)$ and the canonical line bundles ω_X and ω_Y have the same order.*

Proof. The proof is an exercise in the use of *Serre functors* [13]. The Serre functor $S_X = -\otimes\omega_X[\dim_(X)]$ on X is uniquely characterized by the existence of natural isomorphisms

$$\operatorname{Hom}_{\mathcal{D}^b(X)}(\mathcal{E},\mathcal{F}) \cong \operatorname{Hom}_{\mathcal{D}^b(X)}(\mathcal{F}, S_X\mathcal{E})^*. \tag{3.0.1}$$

By uniqueness it is clear that any Fourier-Mukai transform commutes with Serre functors. Pick a point $y \in Y$ and put $\mathcal{E} = \Phi(\mathcal{O}_y)$. The fact that $S_Y[-\dim_Y](\mathcal{O}_y) \cong \mathcal{O}_y$ yields $S_X[-\dim_Y](\mathcal{E}) \cong \mathcal{E}$, or $\mathcal{E} \otimes_X \omega_X[\dim_X - \dim_Y] \cong \mathcal{E}$. Looking at the homology of $\mathcal{E}$ we see that this impossible if $\dim_Y \neq \dim_X$. The statement about the orders of ω_X and ω_Y follows by considering the orders of the functors $S_X[-\dim_X]$ and $S_Y[-\dim_Y]$. $\qquad\square$

The following important result tells that any derived equivalence between $\mathcal{D}^b(Y)$ and $\mathcal{D}^b(X)$ is obtained from an object on the product $Y \times X$.

Theorem 3.2 ([58]). *Let $\Phi\colon \mathcal{D}^b(Y) \longrightarrow \mathcal{D}^b(X)$ be a fully faithful functor. Then there exists an object $\mathcal{P}$ in $\mathcal{D}^b(Y \times X)$, unique up to isomorphism, such that Φ is isomorphic to the functor*

$$\Phi^{\mathcal{P}}_{Y\to X}(-) := \pi_{X*}(\mathcal{P} \otimes_{\mathcal{O}_{Y\times X}} \pi_Y^*(-)),$$

where π_X and π_Y are the projection maps and π_{X*}, $\otimes$, and π_Y^* are the appropriate derived functors.

In the original statement of this theorem Φ was required to have a right adjoint but this condition is automatically fulfilled by [12, 13].

The object $\mathcal{P}$ in the theorem above is also called the *kernel* of the Fourier-Mukai transform.

Remark 3.3. Theorem 3.2 is quite remarkable as for example its analogue for affine varieties or finite dimensional algebras is unknown (except for hereditary algebras [52]). Projectivity is used in the proof in the following way: let $\mathcal{L}$ be an ample line bundle on a projective variety X. Then for any coherent sheaf $\mathcal{F}$ on X one has $\operatorname{Hom}_{\operatorname{coh}(X)}(\mathcal{F}, \mathcal{L}^{-n}) = 0$ for large n. If X is for example affine then $\mathcal{O}_X$ is ample but this additional property does not hold.

It would seem useful to generalize Theorem 3.2 to singular varieties, in particular those occurring in the minimal model program (see below). A first result in this direction has been obtained by Kawamata [40] who proves the analogue of Theorem 3.2 for orbifolds.

The real significance of Theorem 3.2 is that it makes it possible to define Φ on objects functorially derived from X and Y. For example (see [22, 60]) let $\operatorname{ch}'_X(-) = \operatorname{ch}_X(-) . \operatorname{Td}(X)^{1/2}$ (where $\operatorname{ch}_X(-)$ is the Chern character and $\operatorname{Td}(X)$ is the Todd class of X). Using $\operatorname{ch}'_{Y \times X}(\mathcal{P})$ as kernel one finds a linear isomorphism of vector spaces

$$H^*(\Phi) \colon H^*(Y, \mathbb{Q}) \longrightarrow H^*(X, \mathbb{Q})$$

preserving parity of degree. Since the Chern character of $\mathcal{P}$ and the Todd class on $Y \times X$ may have denominators the same result is not a priory true for $H^*(X, \mathbb{Z})$. However it is true for elliptic curves (trivial) and for abelian and K3-surfaces [55].

Remark 3.4. In order to circumvent the non-preservation of integrality it may be convenient to replace $H^*(X, \mathbb{Z})$ by topological K-theory [37] $K^*(X)^{\text{top}} = K^0(X)^{\text{top}} \oplus K^1(X)^{\text{top}}$ which is the K-theory of complex vector bundles (not necessarily holomorphic) on the underlying real manifold of X. Topological K-theory is a cohomology theory satisfying the usual Eilenberg-Steenrod axioms except the dimension axiom (which fixes the cohomology of a point). Since $K^*(-)^{\text{top}}$ has the appropriate functoriality properties [37] one proves that Φ induces an isomorphism

$$K^*(\Phi)^{\text{top}} \colon K^*(Y)^{\text{top}} \to K^*(X)^{\text{top}}$$

It follows from the Atiyah-Hirzebruch spectral sequence that $K^*(X)^{\text{top}}$ is a finitely generated $\mathbb{Z}/2\mathbb{Z}$ graded abelian group such that the Chern-character

$$\operatorname{ch} \colon K^*(X)^{\text{top}} \to H^*(X, \mathbb{Q})$$

induces an isomorphism [34, Eq (3.21)]

$$K^*(X)^{\text{top}} \otimes_{\mathbb{Z}} \mathbb{Q} \cong H^*(X, \mathbb{Q})$$

In good cases the lattices given by $K^*(X)^{\text{top}}$ and $H^*(X, \mathbb{Z})$ are the same. This is for example the case for curves, K3 surfaces and abelian varieties.

By Riemann-Roch the following diagram is commutative

$$
\begin{array}{ccc}
K^0(Y) & \xrightarrow{\ K^0(\Phi)\ } & K^0(X) \\
\Big\downarrow \text{ch}'_Y(-) & & \Big\downarrow \text{ch}'_X(-) \\
H^*(Y,\mathbb{Q}) & \xrightarrow{\ H^*(\Phi)\ } & H^*(X,\mathbb{Q})
\end{array}
$$

$K^0(X)$ is equipped with the so-called Euler form

$$
e([E],[F]) = \sum_i (-)^i \dim_{\text{Hom } \mathcal{D}^b(X)}(E, F[i])
$$

which is of course preserved by $K^0(\Phi)$. The map $\text{ch}'_X(-)$ is compatible with the Euler form up to sign provided one twists the standard bilinear form on cohomology (obtained from Poincare duality) slightly [22]. More precisely put

$$
\check{v} = i^{\deg v} e^{-(1/2)K_X} v
$$

and

$$
\langle v, w \rangle = \deg(\check{v} \cup w)
$$

Then

$$
e([E],[F]) = -\langle \text{ch}'_X(E), \text{ch}'_X(F) \rangle
$$

The map $H^i(\Phi)$ is an isometry for $\langle -, - \rangle$.

The standard grading on $H^*(X,\mathbb{C})$ is of course not preserved by a Fourier-Mukai transform. However there is a different grading which is preserved. Define

$$
{}^n H^*(X,\mathbb{C}) = \bigoplus_{j-i=n} H^{i,j}(X)
$$

where $H^m(X,\mathbb{C}) = \oplus_{i+j=m} H^{i,j}(X,\mathbb{C}) = \oplus_{i+j=m} H^i(X, \Omega_X^j)$ is the Hodge decomposition [31, §0.6]. It is classical that algebraic cycles lie in ${}^0 H^*(X,\mathbb{C})$. From the fact that the kernel of $H^*(\Phi)$ is algebraic it follows that $H^*(\Phi)$ preserves the $*(-)$ grading.

As another application of functoriality note that if S is of finite type then there is an equivalence

$$
\Phi_S \colon \mathcal{D}^b(Y_S) \to \mathcal{D}^b(X_S)
$$

induced by $\mathcal{P}_S$ (i.e. a Fourier-Mukai transform extends to families).

Example 3.5. Here we give an example of a Fourier-Mukai transform which is very important for mirror-symmetry. Assume first that Z is a four dimensional symplectic manifold and let $i\colon S^2 \to Z$ be an embedding of a sphere as a Lagrangian submanifold. Then there exists a symplectic automorphism τ of Z which is trivial outside a tubular neighborhood of S^2 and which is the antipodal map on S^2 itself [73]. τ is called the *symplectic Dehn twist* of Z associated to i.

By the homological mirror symmetry conjecture there should be an analogous notion for derived categories of varieties. This was worked out in [74] (see also [42, 69]). It turns out that the analogue of a Lagrangian sphere is a so-called spherical object. To be more precise $\mathcal{E} \in \mathcal{D}^b(X)$ is *spherical* if $\mathrm{Hom}^i_{\mathcal{D}^b(X)}(\mathcal{E}, \mathcal{E})$ is equal to $\mathbb{C}$ for $i = 0, \dim_X$ and is zero in all other degrees and if in addition $\mathcal{E} \cong \mathcal{E} \otimes \omega_X$.

Associated to a spherical object $\mathcal{E} \in \mathcal{D}^b(X)$ there is an auto-equivalence $T_{\mathcal{E}}$ of $\mathcal{D}^b(X)$, informally defined by

$$T_{\mathcal{E}}(\mathcal{F}) = \mathrm{cone}\left(\mathrm{RHom}_{\mathcal{D}^b(X)}(\mathcal{E}, \mathcal{F}) \otimes_{\mathbb{C}} \mathcal{E} \xrightarrow{\text{evaluation}} \mathcal{F}\right)$$

The non-functoriality of cones leads to a slight technical problem with the naturality of this definition. This would be a problem for abstract triangulated categories but it can be rectified here using the fact that $\mathcal{D}^b(X)$ (being a derived category) is the H^0-category of an exact DG-category.

It is easy to show that the kernel of $T_{\mathcal{E}}$ is given by

$$\mathrm{cone}\left(\check{\mathcal{E}} \boxtimes \mathcal{E} \xrightarrow{\varphi} \mathcal{O}_{\Delta}\right)$$

where $\check{\mathcal{E}} = \mathrm{RHom}_{\mathcal{O}_X}(\mathcal{E}, \mathcal{O}_X)$, $\mathcal{O}_{\Delta}$ is the structure sheaf of the diagonal and φ is the obvious map.

If X is a K3-surface then $\mathcal{O}_X$ is spherical and the kernel of $T_{\mathcal{O}_X}$ is given by $\mathcal{O}_X(-\Delta)$. Other examples of spherical objects are structure sheaves of a rational curve on a smooth surface with self intersection -2 and restrictions of exceptional objects to anticanonical divisors. In particular this last construction yields spherical objects on hypersurfaces of degree $n + 1$ in $\mathbb{P}^n$.

It is convenient to have a criterion for a functor of the form $\Phi^{\mathcal{P}}_{Y \to X}(-) := \pi_{X,*}(\mathcal{P} \otimes \pi_Y^*(-))$ to be an equivalence. The following result originally due to Bondal and Orlov [9] and slightly amplified by Bridgeland [18, Theorem 1.1] shows that we can use the skyscraper sheaves as test objects.

Theorem 3.6. *Let* $\mathcal{P}$ *be an object in* $\mathcal{D}^b(Y \times X)$. *Then the functor* $\Phi := \Phi^{\mathcal{P}}_{Y \to X}(-) \colon \mathcal{D}^b(Y) \longrightarrow \mathcal{D}^b(X)$ *is fully faithful if and only if the following conditions hold*

    1. *for each point* y *in* Y

$$\operatorname{Hom}_{\mathcal{D}^b(X)}(\Phi(\mathcal{O}_y), \Phi(\mathcal{O}_y)) = \mathbb{C}$$

    2. *for each pair of points* y_1 *and* y_2 *and each integer* i

$$\operatorname{Hom}^i_{\mathcal{D}^b(X)}(\Phi(\mathcal{O}_{y_1}), \Phi(\mathcal{O}_{y_2})) = 0 \ \textit{unless } y_1 = y_2 \textit{ and } 0 \leqslant i \leqslant \dim_Y.$$

If these conditions hold then Φ *is an equivalence if and only if* $\Phi(\mathcal{O}_y) \otimes \omega_X \cong \Phi(\mathcal{O}_y)$ *for all* $y \in Y$.

Remark 3.7. Assume that $\mathcal{P}$ is an object in $\operatorname{coh}(Y \times X)$ flat over Y and write $\mathcal{P}_y = \Phi(\mathcal{O}_y)$. Then the previous theorem implies that Φ is fully faithful if and only if

    1. for each point y in Y

$$\operatorname{Hom}_{\mathcal{D}^b(X)}(\mathcal{P}_y, \mathcal{P}_y) = \mathbb{C}$$

    2. for each pair of points $y_1 \neq y_2$ and each integer i

$$\operatorname{Ext}^i_{\mathcal{O}_X}(\mathcal{P}_{y_1}, \mathcal{P}_{y_2}) = 0.$$

It is obvious that the conditions for Theorem 3.6 are necessary. Proving that they are also sufficient is much harder. Since the proof in [9] only works for derived categories of coherent sheaves, we make explicit some of the steps in Bridgeland's proof (see [18]) which are valid for more general triangulated categories.

Let $\mathcal{A}$ be a triangulated category. A subset Ω is called *spanning* if for each object a in $\mathcal{A}$ each of the following conditions implies $a = 0$:

    1. $\operatorname{Hom}^i(a, b) = 0$ for all $b \in \Omega$ and all $i \in \mathbb{Z}$,

    2. $\operatorname{Hom}^i(b, a) = 0$ for all $b \in \Omega$ and all $i \in \mathbb{Z}$.

It is easy to see that the set of all skyscraper sheaves on a smooth projective variety X is a spanning class for $\mathcal{D}^b(X)$. Note that a spanning class will not usually generate $\mathcal{A}$ in any reasonable sense.

Theorem 3.8 ([18, Theorem 2.3]). *Let* $F\colon \mathcal{A} \longrightarrow \mathcal{B}$ *be an exact functor between triangulated categories with left and right adjoint. Then F is fully faithful if and only if there exists a spanning class Ω for $\mathcal{A}$ such that for all elements a_1, a_2 in Ω, and all integers i, the homomorphism*

$$F\colon \operatorname{Hom}^i_{\mathcal{A}}(a_1, a_2) \longrightarrow \operatorname{Hom}^i_{\mathcal{B}}(Fa_1, Fa_2)$$

is an isomorphism.

Recall that a category is called indecomposable if it is not the direct sum of two non-trivial subcategories. The derived category $\mathcal{D}^b(X)$ is indecomposable for X connected. For a finite dimensional algebra A the derived category $\mathcal{D}^b(A)$ is connected precisely when A is connected.

Theorem 3.9 ([20, Theorem 2.3]). *Let* $F\colon \mathcal{A} \to \mathcal{B}$ *be a fully faithful functor between triangulated categories with Serre functors $S_{\mathcal{A}}$, $S_{\mathcal{B}}$ (see (3.0.1)) possessing a left adjoint. Suppose that $\mathcal{A}$ is non-trivial and $\mathcal{B}$ is indecomposable. Let Ω be a spanning class for $\mathcal{A}$ and assume that $FS_{\mathcal{A}}(\omega) \cong S_{\mathcal{B}}F(\omega)$ for all $\omega \in \Omega$. Then F is a equivalence of categories.*

It follows from [12, 13] that $\Phi^{\mathcal{P}}_{Y \to X}$ has both a right and a left adjoint. Explicit formulas for the left and the right adjoint are [18, Lemma 4.5]:

$$\Phi^{\check{\mathcal{P}} \otimes \pi_X^* \omega_X [\dim_X]}_{X \to Y}(-) \quad \text{and} \quad \Phi^{\check{\mathcal{P}} \otimes \pi_Y^* \omega_Y [\dim_Y]}_{X \to Y}(-)$$

Applying Theorems 3.8, 3.9 with $F = \Phi^{\mathcal{P}}_{Y \to X}$ and $\Omega = \{\mathcal{O}_y \mid y \in Y\}$ almost proves Theorem 3.6 except that we seem to need additional information on $\operatorname{Hom}^i_{\mathcal{D}^b(X)}(\Phi(\mathcal{O}_y), \Phi(\mathcal{O}_y))$ for $i > 0$. It is not at all obvious but it turns out that this extra information is unnecessary. Although it is not clear how to formalize it, it seems that this part of the proof may generalize whenever Y is the solution of some type of moduli problem in a triangulated category $\mathcal{B}$ (with $\mathcal{P}$ being the universal family). See [19, 20, 77] for other manifestations of this principle.

4 The reconstruction theorem

It is quite trivial to reconstruct X from the abelian category $\operatorname{coh}(X)$ [28, 66, 68]. For example the points of X are in one-one correspondence with the objects in $\operatorname{coh}(X)$ without proper subobjects. With a little more work one can also recover the Zariski topology on X as well as the structure sheaf.

It is similarly of interest to know to which extent one can reconstruct a variety from its derived category. The existence of non-isomorphic Fourier-Mukai partners shows that this cannot be done in general, but it is possible if the canonical sheaf or the anticanonical sheaf is ample. Later Balmer and Rouquier [3, 68] have shown independently that one can reconstruct the variety from the category of coherent sheaves viewed as a tensor category, the crucial point is that the tensor product allows to reconstruct the point objects for any variety.

Theorem 4.1 ([10, Theorem 2.5]). *Let X be a smooth connected projective variety with either ω_X ample or ω_X^{-1} ample. Assume $\mathcal{D}^b(X)$ is equivalent to $\mathcal{D}^b(Y)$. Then X is isomorphic to Y.*

Proof. We give a proof based on Orlov's theorem. For another proof see [68]. Note that Y is also connected since $\mathcal{D}^b(Y) \cong \mathcal{D}^b(X)$ is connected.

Let $\Phi \colon \mathcal{D}^b(Y) \to \mathcal{D}^b(X)$ be the derived equivalence and let S be the Serre functor $-\otimes \omega_X[\dim_X]$ on X. Recall that it is intrinsically defined by (3.0.1). We say that E in $\mathcal{D}^b(X)$ is a *point object* if

1. $E \cong S(E)[i]$ for some integer i,

2. $\mathrm{Hom}^i(E, E) = 0$ for all $i < 0$, and

3. $\mathrm{Hom}(E, E) = \mathbb{C}$.

It is easy to prove that the only point objects in $\mathcal{D}^b(X)$ (under the assumptions on ω_X) are the shifts of the skyscraper sheaves. The main point is 1., since this condition and the ampleness of $\omega_X^{\pm 1}$ easily implies that E has finite length cohomology.

It follows that Φ sends skyscraper sheaves to shifts of skyscraper sheaves. Then the proof may then be finished using Corollary 4.3 below. □

We need the following standard fact.

Proposition 4.2. *Let $\pi \colon Z \to S$ be a flat morphism of schemes of finite type with S connected. Let $\mathcal{P} \in \mathcal{D}^-(\mathrm{coh}(Z))$ and and assume that for all $s \in S$ we have that $\mathcal{P} \overset{L}{\otimes}_{\mathcal{O}_Z} \pi^* \mathcal{O}_s \cong \mathcal{O}_z[n]$ for some $n \in \mathbb{Z}$, $z \in Z$. Then $\mathcal{P} \cong i_* \mathcal{L}[m]$ where $i \colon S \to Z$ is a section of π, $\mathcal{L} \in \mathrm{Pic}(S)$ and $m \in \mathbb{Z}$.*

Proof. We claim first that the support of the cohomology $\mathcal{P}$ is finite over

S. Assume that this is false and let $H^i(\mathcal{P})$ be the highest cohomology group with non-finite support. Then, up to finite length sheaves we have $H^i(\mathcal{P}) \otimes_{\mathcal{O}_Z} \pi^* \mathcal{O}_s \cong H^i(\mathcal{P} \overset{L}{\otimes}_{\mathcal{O}_Z} \pi^* \mathcal{O}_s)$. Hence $H^i(\mathcal{P}) \otimes_{\mathcal{O}_Z} \pi^* \mathcal{O}_s$ has finite length for all s which is a contradiction.

It is now sufficient to prove that $\mathcal{P}_0 = \pi_*(\mathcal{P})$ is a shifted line bundle given that $\mathcal{P}_0 \overset{L}{\otimes}_{\mathcal{O}_S} \mathcal{O}_s$ has one-dimensional cohomology for all s.

Fix $s \in S$ and assume $\mathcal{P}_0 \overset{L}{\otimes}_{\mathcal{O}_S} \mathcal{O}_s \cong \mathcal{O}_s[n]$. Using Nakayama's lemma we deduce that there is a neighborhood U of s such that $H^i(\mathcal{P}_0 \mid U) = 0$ for $i > -n$. We temporarily replace S by U.

Applying $- \overset{L}{\otimes}_{\mathcal{O}_S} \mathcal{O}_s$ to the triangle

$$\tau_{\leqslant -n-1}\mathcal{P}_0 \to \mathcal{P}_0 \to H^{-n}(\mathcal{P}_0)[n] \to$$

we find $H^{-n}(\mathcal{P}_0) \otimes_{\mathcal{O}_S} \mathcal{O}_s \cong \mathcal{O}_s$ and $\mathrm{Tor}_1^{\mathcal{O}_S}(H^{-n}(\mathcal{P}_0), \mathcal{O}_s) = 0$. Hence $H^{-n}(\mathcal{P})$ is a line bundle on a neighborhood of s. Shrinking S further we may assume $\mathcal{P}_0 \cong \tau_{\leqslant -n-1}\mathcal{P}_0 \oplus H^{-n}(\mathcal{P}_0)[n]$ and hence $\tau_{\leqslant -n-1}\mathcal{P}_0 \overset{L}{\otimes}_{\mathcal{O}_S} \mathcal{O}_s = 0$. Shrinking S once again we have $\tau_{\leqslant -n-1}\mathcal{P}_0 = 0$ and thus $\mathcal{P}_0 \cong H^0(\mathcal{P}_0)[n]$ is a line bundle on a neighborhood of s.

Since this works for any s and S is connected we easily deduce that $\mathcal{P}_0$ is itself a shifted line bundle. $\qquad\square$

We deduce the following

Corollary 4.3. *Assume that* $\Phi \colon \mathcal{D}^b(Y) \to \mathcal{D}^b(X)$ *is a Fourier-Mukai transform between smooth connected projective varieties which sends skyscraper sheaves to shifted skyscraper sheaves. Then* Φ *is of the form* $\sigma_*(-\otimes_{\mathcal{O}_X} \mathcal{L})[n]$ *for an isomorphism* $\sigma \colon Y \to X$, $\mathcal{L} \in \mathrm{Pic}(Y)$ *and* $n \in \mathbb{Z}$.

Proof. By Proposition 4.2 the kernel of Φ must be of the form $\mathcal{P} = (1, \sigma_*)_* \mathcal{L}[n]$ for some map $\sigma \colon Y \to X$. The resulting $\Phi_{Y \to X}^{\mathcal{P}} = \sigma_*(-\otimes_{\mathcal{O}_X})[n]$ will be a derived equivalence if and only if σ is an isomorphism. $\qquad\square$

One also obtains as a corollary the following result.

Theorem 4.4 ([10, Theorem 3.1]). *Let* X *be a smooth connected projective variety with ample canonical or anticanonical sheaf. Then the* *group of isomorphism classes of auto-equivalences of* $\mathcal{D}^b(X)$ *is generated by the automorphisms of* X*, the twists by line bundles and the translations.*

Remark 4.5. It is clear that the notion of point object make sense for arbitrary triangulated categories with Serre functor.

Let $\mathcal{D}$ be the bounded derived category of modules over a connected finite dimensional hereditary $\mathbb{C}$–algebra A. Then point objects only exist for A tame (or in the trivial case $A \cong \mathbb{C}$). For Dynkin quivers or wild quivers the structure of the Auslander-Reiten components is well-known (Gabriels work on Dynkin-quivers and Ringels work on wild hereditary alegebras), consequently, point objects do not exist. In the tame case the point objects are the shifts of quasi-simple modules in homogeneous tubes (see [64]). Let A be not necessary hereditary and we assume $\mathcal{D}^b(A)$ is equivalent to $\mathcal{D}^b(X)$ for some smooth projective variety X. Then $\mathcal{D}^b(A)$ has point objects. The situation is similar if we replace X by a weighted projective variety. However, it is an open problem to construct algebras A having (sufficiently many) point objects without knowing such an equivalence between $\mathcal{D}^b(A)$ and $\mathcal{D}^b(X)$ for some (weighted) projective variety X.

Note that there is a subtle point in the statement of Theorem 4.1. One does not *apriori* require Y to have ample canonical or anti-canonical divisor. If we preimpose this condition then Theorem 4.1 also follows from Theorem 4.6 below which morally corresponds to the fact that derived equivalences commute with Serre functors.

Theorem 4.6 ([60]). *Let* X *be a smooth projective variety. Then the integers* $\dim_{\Gamma}(X, \omega_X^{\otimes m})$ *as well as the the canonical and anti-canonical rings are derived invariants.*

Assume that X is connected. For a Cartier divisor D denote by $R(X, D)$ the ring

$$R(X, D) = \bigoplus_{n \geqslant 0} \Gamma(X, \mathcal{O}_X(nD))$$

and by $K(X, D)$ the part of degree zero of the graded quotient field of $R(X, D)$. We have $K(X, D) \subset K(X)$ where $K(X)$ is the function field of X. By [76, Prop 5.7] $K(X, D)$ is algebraically closed in $K(X)$. If some positive multiple of D is effective then the D-Kodaira dimension $\kappa(X, D)$ of $K(X, D)$ is the transcendence degree of $K(X, D)$, otherwise we set $k(X, D) = -\infty$. It is clear that we have

$$\kappa(X, D) \leqslant \dim_X$$

and in case of equality we have $K(X) = K(X, D)$.

The Kodaira dimension $\kappa(X)$ of X is $\kappa(X, K_X)$. X is of general type if $\kappa(X, K_X) = \dim_X$.

Corollary 4.7 ([39, Theorem 2.3]). *The Kodaira dimension is invariant under Fourier-Mukai transforms. If X is of general type then any Fourier-Mukai partner of X is birational to X.*

Proof. This follows directly from Theorem 4.6 and the preceding discussion. $\qquad\Box$

5 Curves and surfaces

In this section we consider Fourier-Mukai transforms for smooth projective curves and smooth projective surfaces. For curves the situation is rather trivial: only elliptic curves admit non-trivial Fourier-Mukai transforms $\mathcal{D}^b(C) \cong \mathcal{D}^b(D)$, and in that case the curves C and D must be isomorphic. The group of auto-equivalences of $\mathcal{D}^b(C)$ is generated by the trivial ones and the classical Fourier-Mukai transform (which is almost the same as the auto-equivalence associated to the spherical object $\mathcal{O}_E$).

For surfaces the situation is more complicated and is worked out in detail in [21]. The classification of possible non-trivial Fourier-Mukai transforms is based on the classification of complex surfaces (see [4, page 188]). This classification is summarized in Table 1.

Let us start with the case of curves. Let C be a smooth projective curve and denote by g_C the genus of C. According to the degree of the canonical divisor K_C there are three distinct classes:

1. $K_C < 0$: C is the projective line $\mathbb{P}^1(\mathbb{C})$ and $g_C = 0$,

2. $K_C = 0$: C is an elliptic curve and $g_C = 1$,

3. $K_C > 0$: C is a curve of general type and $g_C > 1$.

Class of X	$\kappa(X)$	n_X	$b_1(X)$	c_1^2	c_2
1) minimal rational surfaces	$-\infty$		0	8,9	4,3
3) ruled surfaces of genus $g \geqslant 1$	$-\infty$		$2g$	$8(1-g)$	$4(1-g)$
4) Enriques surfaces	0	2	0	0	12
5) hyperelliptic surfaces	0	$2,3,4,6$	2	0	0
7) K3-surfaces	0	1	0	0	24
8) tori	0	1	4	0	0
9) minimal properly elliptic surfaces	1			0	$\geqslant 0$
10) minimal surfaces of general type	2		$\equiv 0 \bmod 2$	> 0	> 0

Table 1. Classification of algebraic smooth complex surfaces

Using the reconstruction theorems 4.1 and 4.4 it is obvious that nontrivial Fourier-Mukai transforms can only exist for elliptic curves since K_C^{-1} is ample in case 1. and K_C is ample in case 3..

We will now look in somewhat more detail at the interesting case of elliptic curves. Note that if C, D are abelian varieties then it is known precisely when C and D are derived equivalent and furthermore the group $\mathrm{Aut}(\mathcal{D}^b(C))$ consisting of auto-equivalences of $\mathcal{D}^b(C)$ (up to isomorphism) is also completely understood [59]. Here we give an elementary account of the one-dimensional case. This is well-known and was explained to us by Tom Bridgeland. First we have the following result.

Theorem 5.1. *If C, D are derived equivalent elliptic curves then $C \cong D$.*

Proof. By the discussion in §3 the Hodge structures on $H^1(C,\mathbb{C})$ and $H^1(D,\mathbb{C})$ are isomorphic. Since the isomorphism class of an elliptic curve is encoded in its Hodge structure on $H^1(-,\mathbb{C})$ we are done. $\square$

Determining the structure of $\mathrm{Aut}(\mathcal{D}^b(C))$ requires slightly more work. For an elliptic curve C let e_C be the Euler form on $K^0(C)$. By Serre duality e_C is skew symmetric. Put $\mathcal{N}(C) = K^0(C)/\mathrm{rad}\, e_C \cong \mathbb{Z}^2$. e_C defines a non-degenerate skew symmetric form (i.e. a symplectic form) on $\mathcal{N}(C)$ which we denote by the same symbol.

$\mathcal{N}(C)$ has a canonical basis given by $v_1 = [\mathcal{O}_C]$, $v_2 = [\mathcal{O}_x]$ ($x \in C$ arbitrary). The matrix of $e_C(v_i, v_j)_{ij}$ with respect to this basis is

$$\begin{pmatrix} 0 & 1 \\ -1 & 0 \end{pmatrix}$$

With respect to the standard basis the group of symplectic automorphisms of $\mathcal{N}(C)$ may be identified with $\mathrm{Sl}_2(\mathbb{Z})$.

Let T_1, T_2 be the auto-equivalences of C associated to the spherical objects $\mathcal{O}_C$ and $\mathcal{O}_x$. It is not hard to see that $T_2 = - \otimes_{\mathcal{O}_C} \mathcal{O}_C(x)$ so only T_1 is a non-trivial Fourier-Mukai transform.

One computes that with respect to the standard basis the action of T_1, T_2 on $\mathcal{N}(C)$ is given by matrices

$$T_1 = \begin{pmatrix} 1 & -1 \\ 0 & 1 \end{pmatrix}$$

$$T_2 = \begin{pmatrix} 1 & 0 \\ 1 & 1 \end{pmatrix}$$

These matrices are standard generators for $\mathrm{Sl}_2(\mathbb{Z})$ which satisfy the braid relation

$$T_1 T_2 T_1 = T_2 T_1 T_2. \tag{5.0.2}$$

Remark 5.2. Since the objects $\mathcal{O}_C$, $\mathcal{O}_x$ form a so-called A_2 configuration [74] the relation (5.0.2) actually holds in $\mathrm{Aut}(\mathcal{D}^b(C))$.

We have:

Theorem 5.3. *Let* $\mathrm{Aut}^0(\mathcal{D}^b(C))$ *be the subgroup of* $\mathrm{Aut}(\mathcal{D}^b(C))$ *consisting of auto-equivalences of the form* $\sigma_*(- \otimes_{\mathcal{O}_C} \mathcal{L})[n]$ *where* $\sigma \in \mathrm{Aut}(C)$, $\mathcal{L} \in \mathrm{Pic}^0(C)$ *and* $n \in 2\mathbb{Z}$. *Then the symplectic action of* $\mathrm{Aut}(\mathcal{D}^b(C))$ *on* $\mathcal{N}(C)$ *yields an exact sequence*

$$0 \to \mathrm{Aut}^0(\mathcal{D}^b(C)) \to \mathrm{Aut}(\mathcal{D}^b(C)) \to \mathrm{Sl}_2(\mathbb{Z}) \to 0.$$

Proof. The existence of T_1, T_2 implies that the map $\mathrm{Aut}(\mathcal{D}^b(C)) \to \mathrm{Sl}_2(\mathbb{Z})$ is onto.

Assume that $\Phi \in \mathrm{Aut}(\mathcal{D}(C))$ act trivially on $\mathcal{N}(C)$. It is easy to see that for an object $\mathcal{E} \in \mathcal{D}^b(C)$ this implies

$$\begin{aligned} \deg \Phi(\mathcal{E}) &= \deg \mathcal{E} \\ \mathrm{rk}\, \Phi(\mathcal{E}) &= \mathrm{rk}\, \mathcal{E} \end{aligned} \tag{5.0.3}$$

The abelian category coh(D) is hereditary and hence every object in $\mathcal{D}^b(D)$ is the direct sum of its cohomology. Since $\Phi(\mathcal{O}_y)$ must be indecomposable we deduce from (5.0.3) that $\Phi(\mathcal{O}_y)$ is a twisted skyscraper sheaf.

We find by Corollary 4.3 that $\Phi = \sigma_*(- \otimes_{\mathcal{O}_C} \mathcal{L})[n]$. The fact that Φ acts trivially on $\mathcal{N}(C)$ implies $\deg \mathcal{L} = 0$ and n is even.   □

Remark 5.4. Using similar arguments as above it is easy to see that the orbits of the action $\mathrm{Aut}(\mathcal{D}^b(C))$ on the indecomposable objects in $\mathcal{D}^b(C)$ are indexed by $\mathbb{N}\backslash\{0\}$. The quotient map is given by

$$E \mapsto \gcd(\mathrm{rk}(E), \deg(E))$$

In particular any indecomposable vector bundle is in the orbit of an indecomposable finite length sheaf.

Remark 5.5. The situation for elliptic curves is very similar to the situation for tubular algebras [64, 33], tubular canonical algebras, or tubular weighted projective curves (weighted projective curves of genus one) [48]. We quickly explain how these three categories $\mathcal{D}^b(C)$ (C an elliptic curve), $\mathcal{D}^b(\mathbb{X})$ ($\mathbb{X}$ a tubular weighted projective curve) and $\mathcal{D}^b(\Lambda)$ (Λ a tubular canonical algebra or a tubular algebra) are related to each other. Any elliptic curve C admits a non-trivial automorphism $\varphi \colon C \longrightarrow C$, $x \mapsto -x$. Let $G \cong \mathbb{Z}/2\mathbb{Z}$, generated by φ. The category of G-equivariant sheaves on C is isomorphic to the category of coherent sheaves on a weighted projective line of type $\tilde{\mathbb{D}}_4$. For the remaining types $\mathbb{E}_{6,7,8}$ we consider elliptic curves with complex multiplication of order 3, 4 or 6, respectively. Then an analogous result holds for those curves (see also [71]).

Now we discuss the case of surfaces. In the rest of this section a surface will be a smooth projective surface.

Remember that a surface X is called minimal if it does not contain an exceptional curve C (i.e. a smooth rational curve with self intersection -1). The possible non-trivial Fourier-Mukai partners for minimal surfaces were classified by Bridgeland and Maciocia in [21]. This classification is based on the classification of surfaces (see [4, page 188]) as summarized in Table 1 (we have only listed the algebraic surfaces as these are the only ones of interest to us).

Table 1 is in terms of some standard invariants which we first describe. We have already mentioned the Kodaira dimension $\kappa(X)$. It is either

$-\infty, 0, 1$ or 2 and divides the minimal surfaces into four classes. For an arbitrary surface X there is always a map $X \to X_0$ to a minimal surface. If $k(X) \geqslant 0$ then X_0 depends only on the birational equivalence class of X [4, Proposition (4.6)].

Further invariants are the first Betti number $b_1(X) = \dim_H {}^1(X, \mathbb{C})$, the square of the first Chern class $c_1^2(X) = K_X^2$ and the second Chern class $c_2(X)$ (where $c_i = c_i(T_X)$). Finally, for surfaces of Kodaira dimension zero one also needs the smallest natural number n_X with $n_X K_X = 0$.

The invariants $b_1(X)$, $c_1(X)^2$, $c_2(X)$ contain exactly the same information as the (numeric) Hodge diamond of X:

$$
\begin{array}{ccccc}
 & & 1 & & \\
 & q(X) & & q(X) & \\
p_g(X) & & h^{1,1}(X) & & p_g(X) \\
 & q(X) & & q(X) & \\
 & & 1 & & \\
\end{array}
$$

where $p_g(X)$ is the geometric genus of X, $q(X)$ is the Noether number of X and $h^{ij}(X) = \dim_H {}^{ij}(X, \mathbb{C})$. One has

$$b_1(X) = 2q(x)$$
$$c_2(X) = 2 + 2p_g(X) - 4q(X) + h^{1,1}(X)$$
$$\frac{1}{12}(c_1(X)^2 + c_2(X)) = 1 - q(x) + p_g(X))$$

The second line is the Gauss-Bonnet formula [31, §3.3] which says that $c_2(X)$ is equal to the Euler number $\sum_i \dim_(-1)^i \dim_H {}^i(X, \mathbb{C})$ of X. The third formula is Noether's formula. It follows from applying the Riemann-Roch theorem [4, Thm I.(5.3)] to the structure sheaf.

For abelian and K3-surfaces the so-called transcendental lattice is of interest. First note that $H^2(X, \mathbb{Z})$ is free. For abelian surfaces this is clear since they are tori and for K3 surfaces it is [4, Prop VIII(3.2)]. The Neron-Severi lattice is $N_X = H^2(X, \mathbb{Z}) \cap H^{1,1}(X)$ and the transcendental lattice T_X is the sublattice of $H^2(X, \mathbb{Z})$ orthogonal to N_X.

Theorem 5.6 ([21, Theorem 1.1]). *Let X and Y be a non-isomorphic smooth connected complex projective surfaces with equivalent derived categories $\mathcal{D}^b(X)$ and $\mathcal{D}^b(Y)$ such that X is minimal. Then either*

1. *X is a torus (an abelian surface, in class 8)) and Y is also a torus with Hodge-isometric transcendental lattice,*

2. *X is a K3-surface (a surface in class 7)) and Y is also a K3-surface with Hodge isometric transcendental lattice, or*

3. *X is an elliptic surface and Y is another elliptic surface obtained by taking a relative Picard scheme of the elliptic fibration on X.*

A Hodge isometry between transcendental lattices is an isometry under which the one dimensional subspaces $H^0(X, \omega_X)$ and $H^0(Y, \omega_Y)$ of $T_X \otimes_{\mathbb{R}} \mathbb{C}$ and $T_Y \otimes_{\mathbb{R}} \mathbb{C}$ correspond.

The proof of Theorem 5.6 is quite involved and uses case by case analysis quite essentially. As a very rough indication of some of the methods one might use, let us show that if X is minimal then so is Y and X and Y are in the same class. Along the way we will settle the easy case $\kappa(X) = 2$.

Step 1:   By Corollary 4.7 and the discussion in §3 X and Y have the same Kodaira dimension and the same Hodge diamond. In particular they have the same $b_1(-)$, $c_1(-)^2$ and $c_2(-)$. Hence if they are both minimal then they are in the same class.

Step 2:   Assume now that X is minimal and let $Y \to Y_0$ be a minimal model of Y. We have $b_1(Y) = b_1(Y_0)$ [4, Theorem I.(9.1)]. If $\kappa(X) = -\infty, 1, 2$ then the class of X is recognizable from $b_1(X)$ and hence Y_0 must be in the same class as X. If Y_0 is not in class 1,10) then it follows from the classification that $c_1(Y_0)^2 = c(X)^2$ and hence $c_1(Y_0)^2 = c_1(Y)^2$. If Y_0 is in class 10) then by Corollary 4.7 we have $X = Y_0$ and hence we also have $c_1(Y_0)^2 = c_1(Y)^2$. Since $c_1(-)^2$ changes by one under a blowup [4, Theorem I.(9.1)(vii)] it follows in these cases that $Y = Y_0$.

If Y_0 is is in class 1) then in principle we could have $c_1(Y_0)^2 = 9$, $c_1(Y)^2 = c_1(X)^2 = 8$. But then in Y is the blowup of $\mathbb{P}^2$ in a point and hence is Del-Pezzo. We conclude by the reconstruction theorem 4.1 that $X = Y$ which is a contradiction.

Step 3:   If $\kappa(X) = 0$ then ω_X has finite order and hence the same is true for Y by Lemma 3.1. This is impossible if Y is not minimal.

Let us also say a bit more on the K3 and abelian case. Assume that X is a a K3 or abelian surface. Then according [55] the Chern character $K^0(X) \to H^*(X, \mathbb{Q})$ takes it values in $H^*(X, \mathbb{Z})$. As before let $\mathcal{N}(X)$

be $K^0(X)$ modulo the radical of the Euler form. Since the intersection form on $H^*(X, \mathbb{Z})$ is non-degenerate it follows that $\mathcal{N}(X)$ is the image of $K^0(X)$ in $H^*(X, \mathbb{Z})$. It is easy to see that the orthogonal to $\mathcal{N}(X)$ is T_X.

Now assume that X and Y are derived equivalent K3 or abelian surfaces. Again by [55] the induced isometry between $H^*(X, \mathbb{Q})$ and $H^*(Y, \mathbb{Q})$ yields an isometry between $H^*(X, \mathbb{Z})$ and $H^*(X, \mathbb{Z})$. By the above discussion there is an isometry between T_X and T_Y. This is a Hodge isometry since $H^0(X, \omega_X) = {}^2H^*(X, \mathbb{C})$.

The complete result for K3 or abelian surfaces is as follows.

Theorem 5.7 ([58], see also [21]). *Let X and Y be a pair of either K3-surfaces or abelian surfaces (tori) then the following statements are equivalent.*

1. *There exists a Fourier-Mukai transform $\Phi \colon \mathcal{D}^b(Y) \longrightarrow \mathcal{D}^b(X)$.*

2. *There is an Hodge isometry $\varphi^t \colon T(Y) \longrightarrow T(X)$.*

3. *There is an Hodge isometry $\varphi \colon H^{2*}(Y, \mathbb{Z}) \longrightarrow H^{2*}(X, \mathbb{Z})$.*

4. *Y is isomorphic to a fine, two-dimensional moduli space of stable sheaves on X.*

The non minimal case is covered by the following result of Kawamata.

Theorem 5.8 ([39, Theorem 1.6]). *Assume that X, Y are Fourier-Mukai partners but with X not minimal. Then there are only a finite number of possibilities for Y (as in the minimal case). If X is not isomorphic to a relatively minimal elliptic rational surface then X and Y are isomorphic.*

It remains to classify the auto-equivalences of the derived category $\mathcal{D}^b(X)$ for a surface X. Orlov solved this problem for an abelian surface [59] (and more generally for abelian varieties). Ishii and Uehara [36] solve the problem for the minimal resolutions of A_n-singularities on a surface (so this is a local result). The most interesting open case is given by K3-surfaces although here important progress has recently been made by Bridgeland [16, 17]. For any X Bridgeland constructs a finite dimensional complex manifold $\mathrm{Stab}(X)$ on which $\mathrm{Aut}(D^b(X))$ acts naturally. Roughly speaking the points of $\mathrm{Stab}(X)$ correspond to t-structures on $D^b(X)$ together with extra data defining Harder-Narasimhan filtrations on objects in the heart. The definition of $\text{Stab}(X)$ was directly inspired by work of Michael Douglas on stability in string theory [27]. It seems very important to obtain a better understanding of the space $\text{Stab}(X)$.

6 Threefolds and higher dimensional varieties

If X is a projective smooth threefolds then just as in the surface case one would like to find a unique smooth minimal X_0 birationally equivalent to X. Unfortunately it is well known that this is not possible so some modifications have to be made. In particular one has to allow X_0 to have some mild singularities, and furthermore X_0 will in general be far from unique.

Throughout all our varieties are projective. We say that X is *minimal* if X is $\mathbb{Q}$-Gorenstein and K_X is numerically effective. I.e. for any curve $C \subset X$ we have $K_X \cdot C \geqslant 0$.

A natural category to work in are varieties with *terminal* singularities. Recall that a projective variety X has terminal singularities if it is $\mathbb{Q}$-Gorenstein and for a (any) resolution $f\colon Z \to X$ the discrepancy $(\mathbb{Q}\text{-})$divisor $K_Z - f^*K_X$ contains every exceptional divisor with strictly positive coefficients. If $\dim_X \leqslant 2$ and X has terminal singularities then X is smooth. So terminal singularities are indeed very mild.

If X is a threefold with terminal singularities then there exists a map $f\colon Z \to X$ which is an isomorphism in codimension one such that Z is terminal, and $\mathbb{Q}$-factorial [45, Theorem 6.25]. Minimal threefolds with $\mathbb{Q}$-factorial terminal singularities are the "end products" of the three dimensional minimal program. Such minimal models are however not unique. One has the following classical result by Kollar [44].

Theorem 6.1. *Any birational map between minimal threefolds with $\mathbb{Q}$-factorial terminal singularities can be decomposed as a sequence of flops.*

Recall that a flop is a birational map which factors as $(f^+)^{-1}f$

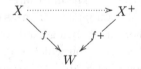

where f, f^+ are isomorphisms in codimension one such that K_X and

K_{X^+} are $\mathbb{Q}$-trivial on the fibers of f and f^+ respectively and such that there is a $\mathbb{Q}$-Cartier divisor D on X with the property that D is relatively ample for f and $-D$ is relatively ample for f^+.

Example 6.2. The easiest (local) example of a flop is the Atiyah flop [61]: Let $W = \mathrm{Spec}\,(\mathbb{C}[x, y, z, u]/(xu - yz)$ be the affine cone over $\mathbb{P}^1 \times \mathbb{P}^1$ associated to the line bundle $\mathcal{O}_{\mathbb{P}^1 \times \mathbb{P}^1}(1, 1)$. W has an isolated singularity in the origin which may be resolved in two different ways $X \longrightarrow W \longleftarrow X^+$ by blowing up the ideals (x, y) and (x, z). The varieties X and X^+ are related by a flop.

How does one construct a minimal model? Assume that X has $\mathbb{Q}$-factorial terminal singularities such that K_X is not numerically effective. The celebrated cone theorem [24, 45] allows one to construct a map $f : X \to W$ with relatively ample $-K_X$ such that one of the following properties holds [24, Thm (5.9)]

1. $\dim_X > \dim_W$ and f is a $\mathbb{Q}$-Fano fibration.

2. f is birational and contracts a divisor.

3. f is birational and contracts a subvariety of codimension $\geqslant 2$.

Case 1. is what one would get by applying the cone theorem to $\mathbb{P}^2$. The result would be the contraction $\mathbb{P}^2 \to \mathrm{pt}$. In the case of surfaces 2. corresponds to blowing down exceptional curves. In general the result is again a variety with terminal singularities and smaller Neron-Severi group. Case 3. represents an new phenomenon which only occurs in dimension three and higher. In this case W may be not be $\mathbb{Q}$-Gorenstein so one is out of the category one wants to work in. In order to continue at this point one introduces a new operation called a *flip*. A flip is a birational map which factors as $(f^+)^{-1} f$

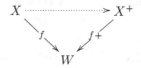

where f, f^+ are isomorphisms in codimension one such that $-K_X$ is relatively ample for f, K_X is relatively ample for f^+ and X^+ again has $\mathbb{Q}$-factorial terminal singularities. The existence of three dimensional flips was settled by Mori in [53]. In higher dimension it is still open.

Example 6.3. Let us give an easy example of a (higher dimensional) flip generalizing Example 6.2. Let W be the affine cone over $\mathbb{P}^m \times \mathbb{P}^n$ ($m \leqslant n$) associated to the line bundle $\mathcal{O}_{\mathbb{P}^m \times \mathbb{P}^n}(1,1)$. W has two canonical resolutions, the first one X being given as the total space of the vector bundle $\mathcal{O}(1)^{\oplus n}$ over $\mathbb{P}^m$ and the second one X^+ as the total space of the vector bundle $\mathcal{O}(1)^{\oplus m}$ over $\mathbb{P}^n$. The birational map $X \dashrightarrow X^+$ is a flip.

Following (and slightly generalizing) [11] (see also [39]) let us say that a birational map $X \dashrightarrow X^+$ between $\mathbb{Q}$-Gorenstein varieties is a *generalized flip* if there is a commutative diagram with $\widetilde{X}$ smooth

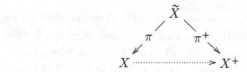

such that $D = \pi^* K_X - \pi^{+*}(K_{X^+})$ is effective. If $D = 0$ then $X \dashrightarrow X^+$ is a generalized flop.

Bondal and Orlov [11] state the following conjecture (see also [39]).

Conjecture 6.4. *For any generalized flip $X \dashrightarrow X^+$ between smooth projective varieties there is a full faithful functor $D^b(X^+) \to D^b(X)$. This functor is an equivalence for generalized flops.*

One could think of this conjecture as the foundation for a "derived minimal model" program.

As evidence of the fact that smooth projective varieties related by a generalized flop are expected to have many properties in common we recall the following very general result by Batyrev and Kontsevich.

Theorem 6.5. *If X and X^+ smooth varieties related by a generalized flop then they have the same Hodge numbers.*

Proof. (see [6, 25, 49]) If X and X^+ are related by a generalized flop then they have the same "stringy E-function". Since X and X^+ are smooth the stringy E-function is equal to usual E-function which encodes the Hodge numbers. $\square$

Remark 6.6. The relation between Conjecture 6.4 and Theorem 6.5 seems rather subtle. Indeed a non-trivial Fourier-Mukai transform does not usually preserve cohomological degree and hence certainly does not preserve the Hodge decomposition.

For non-smooth varieties $D^b(X)$ is probably not the correct object to consider. If X is $\mathbb{Q}$-Gorenstein then every point $x \in X$ has some neighborhood U_x such that on U_x there is some positive number m_x with the property $m_x K_x = 0$. Then K_x generates a cover $\tilde{U}_x$ of U_x on which $\mathbb{Z}/m\mathbb{Z}$ is acting naturally. Gluing the local quotient stacks $\tilde{U}_x/(\mathbb{Z}/m\mathbb{Z})$ defines a Deligne-Mumford stack [47] $\mathcal{X}$ birationally equivalent to X. As usual we write $D^b(\mathcal{X})$ for $D^b(\mathrm{coh}(\mathcal{X}))$. The following result summarizes what is currently known in dimension three concering the categories $D^b(\mathcal{X})$.

Theorem 6.7. *Let* $\alpha\colon X \dashrightarrow X^+$ *be a generalized flop between threefolds with $\mathbb{Q}$-factorial terminal singularities.*

1. α *is a composition of flops.*

2. *There is a corresponding equivalence* $D^b(\mathcal{X}) \to D^b(\mathcal{X}^+)$.

In this generality this result was proved by Kawamata in [39]. The corresponding result in the smooth case was first proved by Bridgeland in [19]. By 1) it is sufficient to consider the case of flops. While trying to understand Bridgeland's proof the second author produced a mildly different proof of the result [78]. Some of the ingredients in this new proof were adapted to the case of stacks by Kawamata. We should also mention [23] which uses a different method to extend Bridgeland's result to singular spaces.

Let us give some more comments on flips and flops. Flips and flops occur very naturally in invariant theory [75] and toric geometry and, as a particular case, for moduli spaces of thin sincere representations of quivers.

Batyrev's construction of Calabi-Yau varieties [5] uses toric geometry, in particular toric Fano varieties. Those varieties correspond to reflexive polytopes. Reflexive polytopes can also be constructed directly from quivers, however, this class of reflexive polytopes is very small. For moduli spaces of thin sincere quiver representations of dimension three all flips are actually flops.

Remark 6.8. The results above should have consequences for derived categories of modules over finite dimensional algebras. However, no example is known of a derived equivalence between a bounded derived category $\mathcal{D}^b(A)$ of modules over finite dimensional algebra A and $\mathcal{D}^b(X)$, where X admits a flop. The "closest" examples to such an equivalence are the fully faithful functors constructed in [1]. If one allows flips (instead of flops) such equivalences exist, one may find toric varieties Y with a full strong exceptional sequence of line bundles. However, for its counterpart W under the flip such sequences are not known. Strongly related to this problem is a conjecture of A. King, that each smooth toric variety admits a full strong exceptional sequence of line bundles, however, even the existence of a full exceptional sequence of line bundles is an open problem (see [43] and [2]).

7 Non-commutative rings in algebraic geometry

In the previous section we considered mainly Fourier-Mukai transforms between algebraic varieties. There are also species of Fourier-Mukai transforms where one of the partners is non-commutative. In this section we discuss some examples. In contrast to the previous sections our algebraic varieties will not always be projective.

Let $f\colon X \to W$ be a projective birational map between Gorenstein varieties. f is said to be a crepant resolution if X is smooth and if $f^*\omega_W = \omega_X$. A variant of Conjecture 6.4 is the following:

Conjecture 7.1. *Assume that W has Gorenstein singularities and that we have two crepant resolutions.*

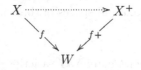

Then X and X^+ are derived equivalent.

This conjecture is known in a number of special cases. See the previous section and [9, 19, 56, 39]. There was some initial hope that the derived equivalence between X and X^+ would always be induced by $\mathcal{O}_{X \times_W X^+}$ but this turned out to be false for certain so-called "stratified Mukai-flops". See [57].

We will now consider a mild non-commutative situation to which a similar conjecture applies. Let $G \subset \mathrm{Sl}_n(\mathbb{C})$ be a finite group and put $W = \mathbb{C}^n/G$. Write $D_G^b(\mathbb{C}^n)$ for the category of G equivariant coherent sheaves on $\mathbb{C}^n$ and let $X \to W$ be a crepant resolution W.

Conjecture 7.2. $D^b(X)$ and $D_G^b(\mathbb{C}^n)$ *are derived equivalent.*

If A is the skew group ring $\mathcal{O}(\mathbb{C}^n) * G$ then one may view A as a *non-commutative crepant resolution* of $\mathbb{C}^n/G$. Conjecture 7.2 may be reinterpreted as saying that all commutative crepant resolutions are derived equivalent to a non-commutative one. So in that sense it is an obvious generalization of Conjecture 7.1. A proper definition of a non-commutative crepant resolution together with a suitably generalized version of Conjecture 7.2 was given in [77]. An example where this generalized conjecture applies is [30]. A similar but slightly different conjecture is [11, Conjecture 5.1].

Conjecture 7.2 has now been proved in two cases. First let X be the irreducible component of the G-Hilbert scheme of $\mathbb{C}^n$ containing the regular representation. Then we have the celebrated BKR-theorem [20].

Theorem 7.3. *Assume that* $\dim_X \leqslant n+1$ *(this holds in particular if* $n \leqslant 3$*). Then* X *is a crepant resolution of* W *and* $D^b(X)$ *is equivalent to* $D_G^b(\mathbb{C}^n)$*.*

Note that this theorem, besides establishing the expected derived equivalence, also produces a specific crepant resolution of W. For $n = 3$ this was done earlier by a case by case analysis (see [65] and the references therein).

Very recently the following result was proved.

Theorem 7.4 ([8]). *Assume that* G *acts symplectically on* $\mathbb{C}^n$ *(for some arbitrary linear symplectic form). Then Conjecture 7.2 is true.*

Somewhat surprisingly this result is proved by reduction to characteristic p.

Let us now discuss a similar but related problem. For a given scheme X one may want to find algebras A derived equivalent to X. One has the following very general result.

Theorem 7.5 ([12], see also [67]). *Assume that X is separated. Then there exists a perfect complex E such that $D(\mathrm{Qcoh}(X))$ is equivalent to $D(A)$ where A is the DG-algebra $\mathrm{RHom}_{\mathcal{O}_X}(E, E)$.*

Recall that a perfect complex is one which is locally quasi-isomorphic to a finite complex of finite rank vector bundles.

In order to replace DG-algebras by real algebras let us say that a perfect complex $E \in D(\mathrm{Qcoh}(X))$ is *classical tilting* if it generates $D(\mathrm{Qcoh}(X))$ (in the sense that $\mathrm{RHom}_{\mathcal{O}_X}(E, U) = 0$ implies $U = 0$) and $\mathrm{Hom}^i_{\mathcal{O}_X}(E, E) = 0$ for $i \neq 0$. One has the following result.

Theorem 7.6. *Assume that X is projective over a noetherian affine scheme of finite type and assume $E \in D(\mathrm{Qcoh}(X))$ is a classical tilting object. Put $A = \mathrm{End}_{\mathcal{O}_X}(E)$. Then*

1. *$\mathrm{RHom}_{\mathcal{O}_X}(E, -)$ induces an equivalence between $D(\mathrm{Qcoh}(X))$ and $D(A)$.*

2. *This equivalence restricts to an equivalence between $D^b(\mathrm{coh}(X))$ and $D^b(\mathrm{mod}(A))$.*

3. *If X is smooth then A has finite global dimension.*

Proof. 1) is just a variant on Theorem 7.5. The inverse functor is $-\overset{L}{\otimes}_A E$. To prove 2) note that the perfect complexes are precisely the compact objects (see [12, Theorem 3.1.1] for a very general version of this statement). Hence perfect complexes are preserved under $-\overset{L}{\otimes}_A E$. An object U has bounded cohomology if and only for any perfect complex C one has $\mathrm{Hom}(C, U[n]) = 0$ for $|n| \gg 0$. Hence objects with bounded cohomology are preserved as well. Now let Z be an object in $D^b(\mathrm{mod}(A))$. Then it easy to see that $\tau_{\geq n}(Z \overset{L}{\otimes}_A E)$ is in $D^b(\mathrm{coh}(X))$ for any n. Since $Z \overset{L}{\otimes}_A E$ has bounded cohomology we are done. To prove 3) note that for any $U, V \in \mathrm{mod}(A)$ we have $\mathrm{Ext}^i_A(U, V)$ for $i \gg 0$. Since A has finite type this implies that A has finite global dimension. $\square$

Classical tilting objects (and somewhat more generally: "exceptional collections") exist for many classical types of varieties [9]. The following somewhat abstract result was proved in [78].

Theorem 7.7. *Assume that* $f : Y \to X$ *is a projective map between varieties, with* X *affine such that* $Rf_* \mathcal{O}_Y = \mathcal{O}_X$ *and such that* $\dim_f {}^{-1}(x) \leqslant 1$ *for all* $x \in X$. *Then* Y *has a classical tilting object.*

This result was inspired by Bridgeland's methods in [19]. It applies in particular to resolutions of three-dimensional Gorenstein terminal singularities. It also has a globalization if X is quasi-projective instead of affine.

References

[1] K. Altmann and L. Hille, *Strong exceptional sequences provided by quivers*, Algebr. Represent. Theory **2** (1999), no. 1, 1–17.

[2] D. Auroux, L. Katzarkov, and D. Orlov, *Mirror symmetry for weighted projective planes and their noncommutative deformations*, math.AG/0404281.

[3] P. Balmer, *Presheaves of triangulated categories and reconstruction of schemes*, Math. Ann. **324** (2002), no. 3, 557–580.

[4] W. Barth, C. Peters, and A. Van de Ven, *Compact complex surfaces*, Ergebnisse der Mathematik und ihrer Grenzgebiete (3) [Results in Mathematics and Related Areas (3)], vol. 4, Springer-Verlag, Berlin, 1984.

[5] V. V. Batyrev, *Dual polyhedra and mirror symmetry for Calabi-Yau hypersurfaces in toric varieties*, J. Algebraic Geom. **3** (1994), no. 3, 493–535.

[6] ———, *Stringy Hodge numbers of varieties with Gorenstein canonical singularities*, Integrable systems and algebraic geometry (Kobe/Kyoto, 1997), World Sci. Publishing, River Edge, NJ, 1998, pp. 1–32.

[7] A. A. Beilinson, *The derived category of coherent sheaves on* $\mathbf{P}^n$, Selecta Math. Soviet. **3** (1983/84), no. 3, 233–237, Selected translations.

[8] R. Bezrukavnikov and D. Kaledin, *McKay equivalence for symplectic resolutions of singularities*, math.AG/0401002.

[9] A. Bondal and D. O. Orlov, *Semi-orthogonal decompositions for algebraic varieties*, math.AG 9506012.

[10] ———, *Reconstruction of a variety from the derived category and groups of autoequivalences*, Compositio Math. **125** (2001), no. 3, 327–344.

[11] ———, *Derived categories of coherent sheaves*, Proceedings of the International Congress of Mathematicians, Vol. II (Beijing, 2002) (Beijing), Higher Ed. Press, 2002, pp. 47–56.

[12] A. Bondal and M. Van den Bergh, *Generators and representability of functors in commutative and noncommutative geometry*, Mosc. Math. J. **3** (2003), no. 1, 1–36, 258.

[13] A. I. Bondal and M. M. Kapranov, *Representable functors, Serre functors, and reconstructions*, Izv. Akad. Nauk SSSR Ser. Mat. **53** (1989), no. 6, 1183–1205, 1337.

[14] A. Borel, P.-P. Grivel, B. Kaup, A. Haefliger, B. Malgrange, and F. Ehlers, *Algebraic D-modules*, Perspectives in Mathematics, vol. 2, Academic Press Inc., Boston, MA, 1987.

[15] S. Brenner and M. C. R. Butler, *Generalizations of the Bernstein-Gel'fand-Ponomarev reflection functors*, Representation theory, II (Proc. Second Internat. Conf., Carleton Univ., Ottawa, Ont., 1979), Lecture Notes in Math., vol. 832, Springer, Berlin, 1980, pp. 103–169.

[16] T. Bridgeland, *Stability conditions on K3 surfaces*, math.AG/0307164.

[17] _____, *Stability conditions on triangulated categories*, math.AG/0212237.

[18] _____, *Equivalences of triangulated categories and Fourier-Mukai transforms*, Bull. London Math. Soc. **31** (1999), no. 1, 25–34.

[19] _____, *Flops and derived categories*, Invent. Math. **147** (2002), no. 3, 613–632.

[20] T. Bridgeland, A. King, and M. Reid, *The McKay correspondence as an equivalence of derived categories*, J. Amer. Math. Soc. **14** (2001), no. 3, 535–554 (electronic).

[21] T. Bridgeland and A. Maciocia, *Complex surfaces with equivalent derived categories*, Math. Z. **236** (2001), no. 4, 677–697.

[22] A. Caldararu, *The Mukai pairing, II: the Hochschild-Kostant-Rosenberg isomorphism*, math.AG/0308080.

[23] J.-C. Chen, *Flops and equivalences of derived categories for threefolds with only terminal Gorenstein singularities*, J. Differential Geom. **61** (2002), no. 2, 227–261.

[24] H. Clemens, J. Kollár, and S. Mori, *Higher-dimensional complex geometry*, Astérisque (1988), no. 166, 144 pp. (1989).

[25] A. Craw, *An introduction to motivic integration*, math.AG/9911179.

[26] P. Deligne, *Cohomologie étale*, Springer-Verlag, Berlin, 1977, Séminaire de Géométrie Algébrique du Bois-Marie SGA 4½, Avec la collaboration de J. F. Boutot, A. Grothendieck, L. Illusie et J. L. Verdier, Lecture Notes in Mathematics, Vol. 569.

[27] M. R. Douglas, *Dirichlet branes, homological mirror symmetry, and stability*, Proceedings of the International Congress of Mathematicians, Vol. III (Beijing, 2002) (Beijing), Higher Ed. Press, 2002, pp. 395–408.

[28] P. Gabriel, *Des catégories abéliennes*, Bull. Soc. Math. France **90** (1962), 323–448.

[29] W. Geigle and H. Lenzing, *A class of weighted projective curves arising in representation theory of finite-dimensional algebras*, Singularities, representation of algebras, and vector bundles (Lambrecht, 1985), Lecture Notes in Math., vol. 1273, Springer, Berlin, 1987, pp. 265–297.

[30] I. Gordon and S. P. Smith, *Representations of symplectic reflection algebras and resolutions of deformations of symplectic quotient singularities*, math.RT/0310187.

[31] P. Griffiths and J. Harris, *Principles of algebraic geometry*, Wiley Classics Library, John Wiley & Sons Inc., New York, 1994, Reprint of the 1978 original.

[32] D. Happel, *Triangulated categories in the representation theory of finite-dimensional algebras*, London Mathematical Society Lecture Note Series, vol. 119, Cambridge University Press, Cambridge, 1988.

[33] D. Happel and C. M. Ringel, *The derived category of a tubular algebra*, Representation theory, I (Ottawa, Ont., 1984), Lecture Notes in Math., vol. 1177, Springer, Berlin, 1986, pp. 156–180.

[34] P. Hilton, *General cohomology theory and K-theory*, Course given at the University of São Paulo in the summer of 1968 under the auspices of the

Instituto de Pesquisas Matemáticas, Universidade de São Paulo. London Mathematical Society Lecture Note Series, vol. 1, Cambridge University Press, London, 1971.

[35] D. Huybrechts, *Fourier-Mukai transforms in algebraic geometry*, http://www.institut.math.jussieu.fr/~huybrech/FM.ps, preliminary lecture notes.

[36] A. Ishii and H. Uehara, *Autoequivalences of derived categories on the minimal resolutions of A_n-singularities on surfaces*, math.AG/0409151.

[37] M. Karoubi, *K-theory*, Springer-Verlag, Berlin, 1978, An introduction, Grundlehren der Mathematischen Wissenschaften, Band 226.

[38] M. Kashiwara, *The Riemann-Hilbert problem for holonomic systems*, Publ. Res. Inst. Math. Sci. **20** (1984), no. 2, 319–365.

[39] Y. Kawamata, *D-equivalence and K-equivalence*, J. Differential Geom. **61** (2002), no. 1, 147–171.

[40] _____, *Equivalences of derived categories of sheaves on smooth stacks*, Amer. J. Math. **126** (2004), no. 5, 1057–1083.

[41] B. Keller, *Derived categories and tilting*, this volume.

[42] M. Khovanov and P. Seidel, *Quivers, Floer cohomology, and braid group actions*, J. Amer. Math. Soc. **15** (2002), no. 1, 203–271 (electronic).

[43] A. King, *Tilting bundles on some rational surfaces*, preprint http://www.maths.bath.ac.uk/~masadk/papers/.

[44] J. Kollár, *Flops*, Nagoya Math. J. **113** (1989), 15–36.

[45] J. Kollár and S. Mori, *Birational geometry of algebraic varieties*, Cambridge Tracts in Mathematics, vol. 134, Cambridge University Press, Cambridge, 1998, With the collaboration of C. H. Clemens and A. Corti, Translated from the 1998 Japanese original.

[46] M. Kontsevich, *Homological algebra of mirror symmetry*, Proceedings of the International Congress of Mathematicians, Vol. 1, 2 (Zürich, 1994) (Basel), Birkhäuser, 1995, pp. 120–139.

[47] G. Laumon and L. Moret-Bailly, *Champs algébriques*, Ergebnisse der Mathematik und ihrer Grenzgebiete. 3. Folge. A Series of Modern Surveys in Mathematics [Results in Mathematics and Related Areas. 3rd Series. A Series of Modern Surveys in Mathematics], vol. 39, Springer-Verlag, Berlin, 2000.

[48] H. Lenzing and H. Meltzer, *Sheaves on a weighted projective line of genus one and representations of a tubular algebra*, Proceedings of the Sixth International Conference on Representations of Algebras (Ottawa, ON, 1992) (Ottawa, ON), Carleton-Ottawa Math. Lecture Note Ser., vol. 14, Carleton Univ., 1992, p. 25.

[49] E. Looijenga, *Motivic measures*, Astérisque (2002), no. 276, 267–297, Séminaire Bourbaki, Vol. 1999/2000.

[50] Z. Mebkhout, *Une autre équivalence de catégories*, Compositio Math. **51** (1984), no. 1, 63–88.

[51] _____, *Une équivalence de catégories*, Compositio Math. **51** (1984), no. 1, 51–62.

[52] J.-i. Miyachi and A. Yekutieli, *Derived Picard groups of finite-dimensional hereditary algebras*, Compositio Math. **129** (2001), no. 3, 341–368.

[53] S. Mori, *Flip theorem and the existence of minimal models for 3-folds*, J. Amer. Math. Soc. **1** (1988), no. 1, 117–253.

[54] S. Mukai, *Duality between $D(X)$ and $D(\hat{X})$ with its application to Picard*

sheaves, Nagoya Math. J. **81** (1981), 153–175.

[55] _____, *On the moduli space of bundles on K3 surfaces. I*, Vector bundles on algebraic varieties (Bombay, 1984), Tata Inst. Fund. Res. Stud. Math., vol. 11, Tata Inst. Fund. Res., Bombay, 1987, pp. 341–413.

[56] Y. Namikawa, *Mukai flops and derived categories*, J. Reine Angew. Math. **560** (2003), 65–76.

[57] _____, *Mukai flops and derived categories. II*, Algebraic structures and moduli spaces, CRM Proc. Lecture Notes, vol. 38, Amer. Math. Soc., Providence, RI, 2004, pp. 149–175.

[58] D. O. Orlov, *Equivalences of derived categories and K3 surfaces*, J. Math. Sci. (New York) **84** (1997), no. 5, 1361–1381, Algebraic geometry, 7.

[59] _____, *Derived categories of coherent sheaves on abelian varieties and equivalences between them*, Izv. Ross. Akad. Nauk Ser. Mat. **66** (2002), no. 3, 131–158.

[60] _____, *Derived categories of coherent sheaves and equivalences between them*, Uspekhi Mat. Nauk **58** (2003), no. 3, 89–172.

[61] M. Reid, *What is a flip?*, colloquium talk at Utah 1992.

[62] J. Rickard, *Morita theory for derived categories*, J. London Math. Soc. (2) **39** (1989), no. 3, 436–456.

[63] _____, *Derived equivalences as derived functors*, J. London Math. Soc. (2) **43** (1991), no. 1, 37–48.

[64] C. M. Ringel, *Tame algebras and integral quadratic forms*, Lecture Notes in Mathematics, vol. 1099, Springer-Verlag, Berlin, 1984.

[65] S. Roan, *Minimal resolutions of Gorenstein orbifolds in dimension three*, Topology **35** (1996), no. 2, 489–508.

[66] A. L. Rosenberg, *The spectrum of abelian categories and reconstruction of schemes*, Rings, Hopf algebras, and Brauer groups (Antwerp/Brussels, 1996), Lecture Notes in Pure and Appl. Math., vol. 197, Dekker, New York, 1998, pp. 257–274.

[67] R. Rouquier, *Dimensions of triangulated categories*, math.CT/0310134.

[68] _____, *Catégories dérivées et géométrie algébrique*, http://www.math.jussieu.fr/~rouquier/preprints/luminy.pdf, 2004, preprint.

[69] R. Rouquier and A. Zimmermann, *Picard groups for derived module categories*, Proc. London Math. Soc. (3) **87** (2003), no. 1, 197–225.

[70] M. Sato, T. Kawai, and M. Kashiwara, *Microfunctions and pseudo-differential equations*, Hyperfunctions and pseudo-differential equations (Proc. Conf., Katata, 1971; dedicated to the memory of André Martineau), Springer, Berlin, 1973, pp. 265–529. Lecture Notes in Math., Vol. 287.

[71] O. Schiffmann, *Noncommutative projective curves and quantum loop algebras.*, math.QA 0205267.

[72] P. Seidel, *Homological mirror symmetry for the quartic surface*, math.SG/0310414.

[73] _____, *Lectures on four-dimensional Dehn twists*, math.SG/0309012.

[74] P. Seidel and R. Thomas, *Braid group actions on derived categories of coherent sheaves*, Duke Math. J. **108** (2001), no. 1, 37–108.

[75] M. Thaddeus, *Geometric invariant theory and flips*, J. Amer. Math. Soc. **9** (1996), no. 3, 691–723.

[76] K. Ueno, *Classification theory of algebraic varieties and compact com-*

plex spaces, Springer-Verlag, Berlin, 1975, Notes written in collaboration with P. Cherenack, Lecture Notes in Mathematics, Vol. 439.

[77] M. Van den Bergh, *Non-commutative crepant resolutions*, The legacy of Niels Henrik Abel, Springer, Berlin, 2004, pp. 749–770.

[78] ———, *Three-dimensional flops and noncommutative rings*, Duke Math. J. **122** (2004), no. 3, 423–455.

[79] J.-L. Verdier, *Des catégories dérivées des catégories abéliennes*, Astérisque (1996), no. 239, xii+253 pp. (1997), With a preface by Luc Illusie, Edited and with a note by Georges Maltsiniotis.

Lutz Hille
Mathematisches Seminar
Universität Hamburg
20146 Hamburg
Germany
E-mail: hille@math.uni-hamburg.de
http://www.math.uni-hamburg.de/home/hille/

Michel Van den Bergh
Departement WNI
Limburgs Universitair Centrum
Universitaire Campus
3590 Diepenbeek
Belgium
E-mail: vdbergh@luc.ac.be
http://alpha.luc.ac.be/Research/Algebra/Members/michel_id.html

8

Tilting theory and homologically finite subcategories with applications to quasihereditary algebras

Idun Reiten

Introduction

Tilting theory is a central topic in the representation theory of algebras and related areas, as illustrated by the diverse contributions to this volume. One of the important aspects has been connections with homologically finite subcategories, which is a common term for contravariantly, covariantly and functorially finite subcategories. One main result along these lines is the following correspondence theorem.

For an artin algebra Λ of finite global dimension there is a one-one correspondence between isomorphism classes of basic cotilting Λ-modules and contravariantly finite resolving subcategories of the category of finitely generated Λ-modules.

This result has interesting applications to the quasihereditary algebras introduced in [18], with the category of modules having standard filtration as the relevant subcategory [44]. Through the above result, there is an associated tilting module, which since the early nineties has had great influence in the theory of algebraic groups and Lie algebras (see [22] and the references there).

This paper is centered around the above correspondence theorem. Actually, we discuss a more general version valid beyond finite global dimension, which has applications to generalizations of quasihereditary algebras.

We start with the basic setup in Chapter 1. First we discuss some generalities on correspondences between modules and subcategories, or pairs of subcategories. Then we recall results from tilting theory which are relevant for our discussions in this paper, and we give a brief introduction to the theory of contravariantly and covariantly finite subcategories [11] [12]. We also discuss some connections between the topics.

In Chapter 2 we give the correspondence theorem for any artin algebra from [7]. We first explain the special case which was the starting point for the connection along these lines from [12], and we discuss the close interplay with, and influence by, the theory of maximal Cohen-Macaulay modules over Cohen-Macaulay rings. We also include some old results on abelian groups from [45] related to these ideas.

In Chapter 3 we discuss the application to quasihereditary algebras. Here the categories of modules with standard filtrations play a central role. They are shown to be contravariantly finite resolving, so that there is a naturally associated (co)tilting module [44]. This module is called the characteristic tilting module, and plays a central role in the theory, in addition to providing a beautiful illustration of the correspondence theorem. We also discuss the generalizations to properly stratified and standardly stratified algebras, giving interesting illustrations beyond finite global dimension.

There are generalizations of the correspondence theorem in various directions. One direction is to consider more general modules than (co)tilting modules, and try to describe corresponding subcategories. Some work is done in this direction for Wakamatsu tilting modules [39]. For generalizations to arbitrary modules, we refer to other papers in this volume [47, 48]. Also we discuss some results of this nature for the bounded derived category [17][13].

All our rings, unless otherwise stated, will be artin algebras. The modules will be finitely generated, and we denote by mod Λ the category of finitely generated (left) Λ-modules. By a subcategory of mod Λ we mean a full subcategory closed under isomorphisms and finite direct sums and, if not otherwise stated, also closed under direct summands. We shall in an informal way refer to the number of indecomposable modules, or to the direct sum of indecomposable modules with a certain property, and let it be understood that we choose one module from each isomorphism class.

I would like to thank A. Beligiannis, C.F. Berg, A. Buan, V. Mazorchuk, S.O. Smalø, Ø. Solberg, and especially the referee, for helpful comments and suggestions.

1 The Basic Ingredients

In this chapter we provide relevant background material from tilting theory and homologically finite subcategories, in order to discuss connections and interplay between the topics in later chapters. The main theme of the paper is correspondences between modules and subcategories, so we start with a general discussion of this type of problem.

1.1 General correspondences

In this section we discuss some natural types of correspondences between modules and subcategories for artin algebras.

Let Λ be an artin algebra and T a module in $\operatorname{mod}\Lambda$ which is selforthogonal, that is, $\operatorname{Ext}^i_\Lambda(T,T) = 0$ for $i > 0$. We say that a module C in a subcategory $\mathcal{Y}$ of $\operatorname{mod}\Lambda$ is Ext-*projective* in $\mathcal{Y}$ if $\operatorname{Ext}^1_\Lambda(C,\mathcal{Y}) = 0$, and *strong* Ext-*projective* if $\operatorname{Ext}^i_\Lambda(C,\mathcal{Y}) = 0$ for all $i > 0$. Denote by $T^\perp$ the subcategory of $\operatorname{mod}\Lambda$ whose objects are the X with $\operatorname{Ext}^i_\Lambda(T,X) = 0$ for $i > 0$. Then $T^\perp$ is the largest subcategory $\mathcal{Y}$ of $\operatorname{mod}\Lambda$ such that T is strong Ext-projective in $\mathcal{Y}$. We have assumed that T is selforthogonal to make sure that T is in the subcategory $T^\perp$. Denote by $\mathcal{Y}_T$ the subcategory of $\operatorname{mod}\Lambda$ whose objects are the Y in $T^\perp$ for which there is an exact sequence $\cdots \to T_n \xrightarrow{f_n} T_{n-1} \to \cdots \to T_1 \xrightarrow{f_1} T_0 \xrightarrow{f_0} Y \to 0$, with T_i in $\operatorname{add} T$ and $\operatorname{Ker} f_i$ in $T^\perp$. Recall that the objects in $\operatorname{add} T$ are summands of finite direct sums of copies of T. Then $\mathcal{Y}_T$ is the largest subcategory of $\operatorname{mod}\Lambda$ such that T, in addition to being strong Ext-projective, is a *generator* of $\mathcal{Y}_T$ in the sense that if X is in $\mathcal{Y}_T$ there is an exact sequence $0 \to Y \to T_0 \to X \to 0$ with T_0 in $\operatorname{add} T$ and Y in $\mathcal{Y}_T$.

Conversely, starting with a subcategory $\mathcal{Y}$ of $\operatorname{mod}\Lambda$, a natural way of associating a module (or a subcategory which often will turn out to be given by a module), is to consider the Ext-projective or the strong Ext-projective objects. This is of course closely connected with finding an inverse for associating $T^\perp$ or $\mathcal{Y}_T$ with T.

Natural general questions are then the following. Let T be a selforthogonal module. Then we can associate with T a subcategory $\mathcal{Y}$ which is $T^\perp$ or $\mathcal{Y}_T$, and ask when the Ext-projectives or strong Ext-projectives in $\mathcal{Y}$ are given by $\operatorname{add} T$. Similarly, given a subcategory $\mathcal{Y}$ of $\operatorname{mod}\Lambda$, when are the Ext-projectives or strong Ext-projectives in $\mathcal{Y}$ given by $\operatorname{add} T$ for some selforthogonal module T, such that $\mathcal{Y}$ is $T^{\perp}$ or $\mathcal{Y}_T$?

Assume that T belongs to a class of selforthogonal modules such that we have a correspondence $T \longleftrightarrow \mathcal{Y}$ of one of the above types. Then it is natural to investigate which properties of T correspond to which properties of $\mathcal{Y}$. Such a type of result can be used in both directions, according to whether there is a subcategory $\mathcal{Y}$ or a module T that appears naturally.

Sometimes it is interesting to investigate correspondences between modules and certain pairs of subcategories, rather than just one subcategory. This is closely related to the above discussion, but provides a different point of view which may be useful to keep in mind.

For a subcategory $\mathcal{Y}$ of $\operatorname{mod} \Lambda$, denote by $\operatorname{Ker} \operatorname{Ext}^1(\ ,\mathcal{Y})$ the subcategory of $\operatorname{mod} \Lambda$ whose objects are the C with $\operatorname{Ext}^1_{\Lambda}(C,Y) = 0$ for all $Y \in \mathcal{Y}$, and let $^{\perp}\mathcal{Y} = \{C; \operatorname{Ext}^i_{\Lambda}(C,Y) = 0 \text{ for } Y \in \mathcal{Y} \text{ and } i > 0\}$. Then clearly $\mathcal{Y} \cap {}^{\perp}\mathcal{Y}$ consists of the strong Ext-projective objects in $\mathcal{Y}$ and $\mathcal{Y} \cap \operatorname{Ker}(\operatorname{Ext}^1(\ ,\mathcal{Y}))$ consists of the Ext-projective ones. So we can describe one direction as the correspondence $\mathcal{Y} \rightsquigarrow \mathcal{Y} \cap {}^{\perp}\mathcal{Y}$ or $\mathcal{Y} \rightsquigarrow \mathcal{Y} \cap \operatorname{Ker}(\operatorname{Ext}^1(\ ,\mathcal{Y}))$. Hence it is also natural to consider correspondences between selforthogonal modules T and pairs of subcategories $(^{\perp}\mathcal{Y},\mathcal{Y})$, or $(\operatorname{Ker}(\operatorname{Ext}^1(\ ,\mathcal{Y})),\mathcal{Y})$.

We also have the dual considerations. Then Ext-injective and strong Ext-injective objects in a subcategory $\mathcal{X}$ of $\operatorname{mod} \Lambda$ are defined in the obvious way, along with the concept of a cogenerator. We associate with a selforthogonal module T the subcategory $^{\perp}T = {}^{\perp}(\operatorname{add} T)$ of $\operatorname{mod} \Lambda$. Then $^{\perp}T$ is the largest subcategory $\mathcal{X}$ of $\operatorname{mod} \Lambda$ where T is strong Ext-injective, and we denote by $\mathcal{X}_T$ the largest subcategory of $\operatorname{mod} \Lambda$ where T is also a cogenerator.

1.2 Background on tilting theory

In this section we recall some basic material on tilting theory relevant for this paper, from [15] [30] [14] [29] [36].

Denote by $\operatorname{pd}_{\Lambda} C$ the projective dimension of a Λ-module C, and by $\operatorname{id}_{\Lambda} C$ the injective dimension. Let first T be a tilting Λ-module of projective dimension at most one, that is:

(i) $\operatorname{pd}_{\Lambda} T \leqslant 1$

(ii) $\operatorname{Ext}^1_\Lambda(T, T) = 0$

(iii) there exists an exact sequence $0 \to \Lambda \to T_0 \to T_1 \to 0$ with T_0 and T_1 in $\operatorname{add} T$.

Denote by $(\mathcal{T}, \mathcal{F})$ the associated torsion pair, where $\mathcal{T}$ is the torsion class $\operatorname{Fac} T$ whose objects are the factors of finite direct sums of copies of T, and $\mathcal{F}$ the torsionfree class $\{C; \operatorname{Hom}_\Lambda(T, C) = 0\}$. Recall that in our setting a class of modules $\mathcal{T}$ is a torsion class if and only if it is closed under extensions and quotients, and a class of modules $\mathcal{F}$ is a torsionfree class if and only if it is closed under extensions and submodules. Alternative descriptions are $\mathcal{F} = \operatorname{Sub} D \operatorname{Tr} T$ and $\mathcal{T} = \{C; \operatorname{Ext}^1_\Lambda(T, C) = 0\} = \{C; \operatorname{Ext}^i_\Lambda(T, C) = 0 \text{ for } i > 0\} = T^\perp$. Here the objects of $\operatorname{Sub} D \operatorname{Tr} T$ are the submodules of finite direct sums of copies of $D \operatorname{Tr} T$, and D denotes the ordinary duality and Tr the transpose. Also, for each $X \in \mathcal{T}$ there is an exact sequence $0 \to Y \to T_0 \to X \to 0$ with $T_0 \in \operatorname{add} T$ and $Y \in \mathcal{T}$.

Dually, a module U in $\operatorname{mod} \Lambda$ is a cotilting module of injective dimension at most one if $D(U)$ is a tilting $\Lambda^{\operatorname{op}}$-module of projective dimension at most one. Associated with U is a torsion pair $(\mathcal{U}, \mathcal{V})$, where $\mathcal{V}$ is the torsionfree class $\operatorname{Sub} U$ and $\mathcal{U}$ is the torsion class $\{Y; \operatorname{Hom}_\Lambda(Y, U) = 0\}$. Alternative descriptions are $\mathcal{V} = \{C; \operatorname{Ext}^1_\Lambda(C, U) = 0\} = \{C; \operatorname{Ext}^i_\Lambda(C, U) = 0 \text{ for } i > 0\} = {}^\perp U$ and $\mathcal{U} = \operatorname{Fac} \operatorname{Tr} DU$.

Denote by $\Gamma = \operatorname{End}_\Lambda(T)^{\operatorname{op}}$ the endomorphism algebra of the tilting module T with $\operatorname{pd}_\Lambda T \leqslant 1$. Then $U = D(T)$ is a cotilting Γ-module with $\operatorname{id}_\Lambda U \leqslant 1$, $\operatorname{End}_\Gamma(U)^{\operatorname{op}} \simeq \Lambda$, and we have the following basic Brenner-Butler theorem [15].

Theorem 1.1. *Let T be a tilting module with $\operatorname{pd}_\Lambda T \leqslant 1$ over an artin algebra Λ. Then we have the following, where $(\mathcal{T}, \mathcal{F})$ denotes the torsion pair in $\operatorname{mod} \Lambda$ associated with the tilting Λ-module T, and $(\mathcal{U}, \mathcal{V})$ the torsion pair in $\operatorname{mod} \Gamma$ associated with the cotilting Γ-module $D(T)$.*

(a) *There are inverse equivalences $\operatorname{Hom}_\Lambda(T, \ ): \mathcal{T} \to \mathcal{V}$ and $D \operatorname{Hom}_\Gamma(\ , D(T)): \mathcal{V} \to \mathcal{T}$.*

(b) *There are inverse equivalences $\operatorname{Ext}^1_\Lambda(T, \ ): \mathcal{F} \to \mathcal{U}$ and $D \operatorname{Ext}^1_\Gamma(\ , D(T)): \mathcal{U} \to \mathcal{F}$.*

This result is the basis for comparing $\operatorname{mod} \Lambda$ and $\operatorname{mod} \Gamma$, and is especially interesting when Λ is hereditary, in which case the algebras Γ obtained this way are the *tilted* algebras, and the torsion pair $(\mathcal{U}, \mathcal{V})$ is split, that is, each indecomposable Γ-module is in $\mathcal{U}$ or $\mathcal{V}$ [30].

Let now T be an arbitrary tilting Λ-module, that is

(i) $\mathrm{pd}_\Lambda T < \infty$

(ii) $\mathrm{Ext}_\Lambda^i(T,T) = 0$ for $i > 0$

(iii) there is an exact sequence $0 \to \Lambda \to T_0 \to \cdots \to T_n \to 0$ with the T_i in $\mathrm{add}\, T$.

Whereas it is no longer true in general that $\mathrm{Fac}\, T$ is a torsion class and we may have $T^\perp \neq \mathrm{Fac}\, T$, we do have $T^\perp = \mathcal{Y}_T$, in the terminology of section 1.1 (see [7]). This means that in the largest subcategory where T is strong Ext-projective, T is also a generator. It will be the topic of the next chapter to investigate the correspondence $T \mapsto T^\perp$ more closely, in view of the general considerations from the previous section.

For an arbitrary tilting module T there are, like in the case when $\mathrm{pd}_\Lambda T \leqslant 1$, induced equivalences between certain subcategories of modules over Λ and $\Gamma = \mathrm{End}_\Lambda(T)^{\mathrm{op}}$, but not all of these subcategories have especially nice properties. Nevertheless, some of these subcategories have been important. Let T be a tilting module with $\mathrm{pd}_\Lambda T = n$. While for $n = 1$ we had the associated subcategories $\mathcal{T} = \{C; \mathrm{Ext}^1(T,C) = 0\}$ and $\mathcal{F} = \{C; \mathrm{Hom}(T,C) = 0\}$, we have in general for any i with $0 \leqslant i \leqslant n$ an associated subcategory $\mathcal{Y}_i = \{C; \mathrm{Ext}_\Lambda^t(T,C) = 0$ for $t \neq i, t \geqslant 0\}$, which clearly generalizes the case $n = 1$. Then for a cotilting module U with $\mathrm{id}_\Lambda U = n$ we have for any i with $0 \leqslant i \leqslant n$ an associated subcategory $\mathcal{X}_i = \{C; \mathrm{Ext}^t(C,U) = 0$ for $t \neq i, t \geqslant 0\}$. Also in this generality, we have the following, where gl. dim. Λ denotes the global dimension of Λ.

Proposition 1.2. *Let T be a tilting Λ-module with $\mathrm{pd}_\Lambda T = n$ and $\Gamma = \mathrm{End}_\Lambda(T)^{\mathrm{op}}$.*

(a) *Then the Γ-module $U = D(T)$ is a cotilting module with $\mathrm{id}_\Gamma D(T) = n$ and $\mathrm{End}_\Gamma(D(T))^{\mathrm{op}} \simeq \Lambda$.*

(b) *If gl. dim. $\Lambda = t < \infty$, then gl. dim. $\Gamma \leqslant t + n$.*

There is the following relationship between the subcategories introduced above [36] [29].

Theorem 1.3. *Let T be a tilting Λ-module with $\operatorname{pd}_\Lambda T = n$, and let $\mathcal{Y}_i = \{C; \operatorname{Ext}_\Lambda^t(T, C) = 0 \text{ for } t \neq i, t \geqslant 0\}$ be the associated sub-categories of $\operatorname{mod}\Lambda$ for $i = 0, 1, \dots n$. Let $\Gamma = \operatorname{End}_\Lambda(T)^{\mathrm{op}}$, and let $\mathcal{X}_i = \{C; \operatorname{Ext}_\Gamma^t(C, U) = 0 \text{ for } t \neq i, t \geqslant 0\}$ be the subcategories of $\operatorname{mod}\Gamma$ associated to the cotilting Γ-module $U = D(T)$.*

Then we have inverse equivalences of categories $\operatorname{Ext}_\Lambda^i(T, \ ): \mathcal{Y}_i \to \mathcal{X}_i$ and $D\operatorname{Ext}_\Gamma^i(\ , U): \mathcal{X}_i \to \mathcal{Y}_i$.

Note that when $n = 1$, this specializes to Theorem 1.1, where $\mathcal{F} = \mathcal{Y}_0$, $\mathcal{T} = \mathcal{Y}_1$, $\mathcal{U} = \mathcal{X}_0$ and $\mathcal{V} = \mathcal{X}_1$.

These equivalences are naturally explained by the existence of a derived equivalence $R\operatorname{Hom}_\Lambda(T, \ ): D^b(\Lambda) \to D^b(\Gamma)$ [29], which restricts to give these equivalences. This result has had enormous influence, via the tilting complexes of Rickard [42], in the representation theory of finite groups.

We shall especially be dealing with the case $i = 0$, when we have $\mathcal{Y} = \mathcal{Y}_0 = T^\perp$ and $\mathcal{X} = \mathcal{X}_0 = {}^\perp U$, so that we get inverse equivalences $\operatorname{Hom}_\Lambda(T, \ ): T^\perp \to {}^\perp(D(T))$ and $D\operatorname{Hom}_\Gamma(\ , D(T)): {}^\perp(D(T)) \to T^\perp$. We shall in addition deal with the subcategories ${}^\perp\mathcal{Y}_0 = \{C; \operatorname{Ext}_\Lambda^i(C, \mathcal{Y}_0) = 0 \text{ for } i > 0\}$ and $\mathcal{X}_0^\perp = \{C; \operatorname{Ext}_\Lambda^i(\mathcal{X}_0, C) = 0 \text{ for } i > 0\}$ naturally associated with $\mathcal{Y}_0$ and $\mathcal{X}_0$, and we shall later see that there are sometimes useful induced equivalences involving such subcategories.

An important aspect of tilting modules of special interest for the applications is the following [30].

Proposition 1.4. *Let T be a tilting module over an artin algebra Λ. Then the number of indecomposable summands of T equals the number of simple Λ-modules.*

For tilting modules of projective dimension at most one, the following, which does not hold for tilting modules in general [43], is also useful [14].

Proposition 1.5. *Let T' be a selforthogonal module with $\operatorname{pd}_\Lambda T' \leqslant 1$ over an artin algebra Λ. Then there is some Λ-module T'' with $\operatorname{pd}_\Lambda T'' \leqslant 1$ such that $T = T' \coprod T''$ is a tilting module.*

1.3 Homologically finite subcategories

In this section we provide some background material for the theory of contravariantly finite, covariantly finite and functorially finite subcategories, as introduced in [11]. This includes motivation and basic examples, with special emphasis on results relevant to the focus of this paper.

Let C be a subcategory of $\operatorname{mod}\Lambda$. Then C is *contravariantly finite* in $\operatorname{mod}\Lambda$ if for all $X \in \operatorname{mod}\Lambda$ there is a map $f: C \to X$ which is a *right C-approximation*, that is, $C \in C$ and given any map $t: C' \to X$ with $C' \in C$, there is some $s: C' \to C$ such that $fs = t$. The notions of *covariantly finite* subcategories and left C-approximations are defined dually. Finally, C is by definition *functorially finite* in $\operatorname{mod}\Lambda$ if it is both covariantly and contravariantly finite. A common name for these three kinds of subcategories is *homologically finite*. More generally, if C is a subcategory of an arbitrary category $\mathcal{D}$, we define C to be contravariantly, covariantly or functorially finite in $\mathcal{D}$ in the obvious way.

The importance of these subcategories is that many properties of $\operatorname{mod}\Lambda$ are inherited by them. The motivation for introducing this type of subcategories comes more specifically from the theory of preprojective and preinjective partitions on one hand and the development of a general theory for almost split sequences in subcategories on the other hand [11] [12] (see also [8]). Central for the first topic is that a subcategory C of $\mathcal{D}$ in $\operatorname{mod}\Lambda$ is functorially finite in $\mathcal{D}$ if C is obtained from $\mathcal{D}$ by removing only a finite number of indecomposable modules.

For the second topic, it was proved that a functorially finite subcategory C, which is in addition closed under direct summands and extensions, has almost split sequences [12]. This means in particular that for any indecomposable module C in C which is not Ext-projective in C there is some almost split sequence $0 \to A \to B \to C \to 0$, and the same for each indecomposable module A in C which is not Ext-injective. So a natural problem investigated in [12] was to try to identify the Ext-projective and Ext-injective objects in C, in particular to decide whether (1) the number of indecomposable Ext-projectives and the number of indecomposable Ext-injectives objects are finite and coincide, and, (2) if the numbers coincide, whether they are the same as the number of simple Λ-modules.

These questions were investigated under appropriate additional assumptions on the subcategories. Central examples of contravariantly finite subcategories C are those closed under factors. And there is the following nice criterion for when they are covariantly finite [11].

Proposition 1.6.   (a) *Let C be a subcategory of* mod Λ *closed under factors. Then C is covariantly finite if and only if it is of the form* Fac X *for some X in C.*

   (b) *Dually, let C be a subcategory of* mod Λ *closed under submodules. Then C is contravariantly finite if and only if it is of the form* Sub X *for some X in C.*

The following result was proved about Ext-projectives and Ext-injectives [12].

Proposition 1.7. *For the subcategories of the form* Sub X *or* Fac X, *there is only a finite number of indecomposable* Ext-*projectives and* Ext-*injectives.*

The subcategories closed under factors which are in addition closed under extensions are exactly the torsion classes, and the extension closed ones closed under submodules are the torsionfree classes. When $(\mathcal{T}, \mathcal{F})$ is a torsion pair, we have the following result on covariantly and contravariantly finite [4] [46].

Theorem 1.8. *Let $(\mathcal{T}, \mathcal{F})$ be a torsion pair in* mod Λ. *Then $\mathcal{T}$ is covariantly (functorially) finite in* mod Λ *if and only if $\mathcal{F}$ is contravariantly (functorially) finite in* mod Λ.*

While torsion pairs $(\mathcal{T}, \mathcal{F})$ are characterized via $\mathcal{T}$ having as objects the C such that $\operatorname{Hom}(C, \mathcal{F}) = 0$ and $\mathcal{F}$ having the objects B such that $\operatorname{Hom}(\mathcal{T}, B) = 0$, we can instead consider pairs of subcategories $(\mathcal{X}, \mathcal{Y})$ such that $\mathcal{Y} = \operatorname{Ker} \operatorname{Ext}^1(\mathcal{X},)$ and $\mathcal{X} = \operatorname{Ker} \operatorname{Ext}^1(, \mathcal{Y})$. We say that such a pair of subcategories $(\mathcal{X}, \mathcal{Y})$ forms a *cotorsion pair*, with $\mathcal{X}$ being the cotorsionfree class and $\mathcal{Y}$ the cotorsion class [45] (see also [48]). Like torsion pairs, the cotorsion pairs also play an important role in connection with tilting modules, as we shall see later. It is immediate that for a cotorsion pair $(\mathcal{X}, \mathcal{Y})$, $\mathcal{X}$ is closed under extensions (and summands) and contains the projective modules, and $\mathcal{Y}$ is closed under extensions (and summands) and contains the injective modules. Recall that $\mathcal{X}$ is *resolving* if in addition it is closed under kernels of epimorphisms. Dually $\mathcal{Y}$ is *coresolving* if in addition it is closed under cokernels of monomorphisms. When $\mathcal{X}$ is resolving, it is easy to see that

$\operatorname{Ker} \operatorname{Ext}^1(\mathcal{X}, \ ) = \mathcal{X}^\perp$, and if $\mathcal{Y}$ is coresolving, then $^\perp \mathcal{Y} = \operatorname{Ker} \operatorname{Ext}^1(\ , \mathcal{Y})$. The following result, analogous to Theorem 1.8, will be of interest for us [7].

Theorem 1.9. *Let $(\mathcal{X}, \mathcal{Y})$ be a cotorsion pair in* $\operatorname{mod} \Lambda$. *Then we have the following.*

    (a) *$\mathcal{X}$ is contravariantly finite in* $\operatorname{mod} \Lambda$ *if and only if $\mathcal{Y}$ is covariantly finite.*

    (b) *$\mathcal{X}$ is resolving if and only if $\mathcal{Y}$ is coresolving.*

A cotorsion pair $(\mathcal{X}, \mathcal{Y})$ with $\mathcal{X}$ contravariantly finite is often called a *complete cotorsion* pair.

There are interesting exact sequences associated with complete cotorsion pairs. The following result of Wakamatsu is important in this connection [49] [8]. Recall that a map $g \colon X \to C$ is *right minimal* if for any map $h \colon X \to X$ with $gh = g$, the map h is an isomorphism. For a subcategory $\mathcal{X}$ of $\operatorname{mod} \Lambda$, a right $\mathcal{X}$-approximation $g \colon X \to C$ is said to be a minimal right $\mathcal{X}$-approximation if $g \colon X \to C$ is also right minimal. The notion of a left minimal map is dual.

Lemma 1.10. *Let $\mathcal{X}$ be a contravariantly finite extension closed subcategory of* $\operatorname{mod} \Lambda$ *and let $g \colon X \to C$ a minimal right $\mathcal{X}$-approximation. Then $\operatorname{Ext}_\Lambda^1(\mathcal{X}, \operatorname{Ker} g) = 0$.*

We can now describe the associated exact sequences [6].

Theorem 1.11. *Let $(\mathcal{X}, \mathcal{Y})$ be a complete cotorsion pair in* $\operatorname{mod} \Lambda$.

    (a) *For any C in* $\operatorname{mod} \Lambda$ *there is an exact sequence $0 \to Y_C \to X_C \xrightarrow{g} C \to 0$ with $Y_C \in \mathcal{Y}$, $X_C \in \mathcal{X}$ and $g \colon X_C \to C$ a minimal right $\mathcal{X}$-approximation.*

    (b) *For any A in* $\operatorname{mod} \Lambda$ *there is an exact sequence $0 \to A \xrightarrow{f} Y^A \to X^A \to 0$, with $Y^A \in \mathcal{Y}$, $X^A \in \mathcal{X}$ and $f \colon A \to Y^A$ a minimal left $\mathcal{Y}$-approximation.*

So, similar to the situation for torsion pairs, $\operatorname{mod} \Lambda$ is built up from a complete cotorsion pair in the sense that each Λ-module is a quotient of a module from the cotorsionfree class by a module from the cotorsion class.

There are some useful general criteria for a subcategory of $\operatorname{mod}\Lambda$ to be contravariantly or covariantly finite. We state an important result of this type, which will be useful later [44]. For a finite set $X = \{X_1, \ldots X_n\}$ of Λ-modules we shall denote by $\mathcal{F}(X)$ the subcategory of $\operatorname{mod}\Lambda$ consisting of the modules which have filtrations with factors amongst $X_1, \ldots X_n$, and by $\bar{\mathcal{F}}(X)$ the subcategory of $\operatorname{mod}\Lambda$ whose objects are direct summands of the objects in $\mathcal{F}(X)$.

Theorem 1.12. *Let* $X = \{X_1, \ldots X_n\}$ *be a set of* Λ-*modules satisfying* $\operatorname{Ext}^1_\Lambda(X_i, X_j) = 0$ *for all* $i \geqslant j$. *Then the subcategories* $\mathcal{F}(X)$ *and* $\bar{\mathcal{F}}(X)$ *of* $\operatorname{mod}\Lambda$ *are functorially finite in* $\operatorname{mod}\Lambda$.

This has the following interesting consequence [44].

Corollary 1.13. *For a set* $X = \{X_1, \ldots X_n\}$ *of* Λ-*modules with* $\operatorname{Ext}^i_\Lambda(X_i, X_j) = 0$ *for all* $i \geqslant j$, *the subcategory* $\bar{\mathcal{F}}(X)$ *has almost split sequences.*

An important general property of a covariantly finite subcategory is the following, as the substitute for not being closed under factors ([11], Lemma 3.11).

Proposition 1.14. *A covariantly finite subcategory* $\mathcal{Y}$ *of* $\operatorname{mod}\Lambda$ *has a cover, that is, there is some* X *in* $\operatorname{add}\mathcal{Y}$ *such that* $\mathcal{Y} \subset \operatorname{Fac} X$.

Amongst frequently investigated subcategories of $\operatorname{mod}\Lambda$ it is of interest to know if they are contravariantly/covariantly finite, and if they do not have these properties in general, what are some sufficient conditions for these properties to hold?

For example the category $\Omega^d(\operatorname{mod}\Lambda)$ whose objects are the dth syzygy modules, including the projectives, is always functorially finite (see [9], [10]). On the other hand the category $\{C; \operatorname{pd}_\Lambda C < \infty\}$ is in general not contravariantly finite, even though there are interesting sufficient conditions [33], and it is also not always covariantly finite [31].

1.4 Basic interplay

In this section we discuss some of the basic interplay between tilting theory and functorially finite subcategories [12] [6] [7].

The first main connection was given in [12], where tilting theory was used to show the following (in a dual formulation). Here we denote by ann X the annihilator of a module X.

Theorem 1.15. *Assume that the subcategory* Fac X *of* mod Λ *is closed under extensions. Then there is the same number of indecomposable* Ext-*projective and indecomposable* Ext-*injective objects in* Fac X, *and the number coincides with the number of simple* $\Lambda/(\text{ann }X)$-*modules.*

The idea of the proof is the following. By Proposition 1.7 there is only a finite number of indecomposable Ext-projectives and Ext-injectives. Let T be the direct sum of the indecomposable Ext-projective modules in Fac X. Then it is shown that T is a tilting $\Lambda/(\text{ann }T)$-module, and the Ext-injective $\Lambda/(\text{ann }T)$-modules in Fac X coincide with the injective $\Lambda/(\text{ann }T)$-modules. Then we use that the number of indecomposable summands of T equals the number of simple $\Lambda/(\text{ann }T)$-modules, which is again the number of indecomposable injective $\Lambda/(\text{ann }T)$-modules.

On the other hand there are interesting contravariantly and co-variantly finite subcategories associated with tilting and cotilting modules. We have already mentioned that the category $\mathcal{T} = \text{Fac } T = \{C; \text{Ext}^1_\Lambda(T, C) = 0\} = T^\perp$ associated with a tilting module T with $\text{pd}_\Lambda T \leqslant 1$ is functorially finite. And dually, the category $\mathcal{F} = \text{Sub } U$ associated with a cotilting module with $\text{id}_\Lambda U \leqslant 1$ is functorially finite. More generally we have the following, part of which will be discussed in the next chapter [7].

Proposition 1.16.   (a) *If* T *is a tilting* Λ-*module, then the subcategory* $T^\perp$ *is functorially finite in* mod Λ.

   (b) *If* U *is cotilting* Λ-*module, then the subcategory* $^\perp U$ *is functorially finite in* mod Λ.

In particular, since $T^\perp$ and $^\perp U$ are clearly closed under extensions (and direct summands), they have almost split sequences.

Via Proposition 1.16, the tilting and cotilting modules also give rise to interesting complete cotorsion pairs [7]. For a subcategory $\mathcal{C}$ of mod Λ we denote by $\check{\mathcal{C}}$ the subcategory of mod Λ consisting of objects X for which there is an exact sequence $0 \to X \to C_0 \to C_1 \to \cdots \to C_n \to 0$, with the C_i in $\mathcal{C}$. Dually, the objects of the subcategory $\hat{\mathcal{C}}$ are the Y in mod Λ for which there is an exact sequence $0 \to C_n \to \cdots C_1 \to C_0 \to Y \to 0$

with the C_i in $\mathcal{C}$. We then have the following, which will be discussed more in the next chapter [7].

Proposition 1.17.   (a) *For a tilting module T we have the complete cotorsion pair* $(\mathrm{add}T, T^\perp)$.

(b) *For a cotilting module U we have the complete cotorsion pair* $({}^\perp U, \widehat{\mathrm{add}}U)$.

In view of previous remarks, we have a complete cotorsion pair $({}^\perp(T^\perp), T^\perp)$, so we only need to show ${}^\perp(T^\perp) = \widehat{\mathrm{add}}T$. Similarly we have a complete cotorsion pair $({}^\perp U, ({}^\perp U)^\perp)$, so we only need to show $({}^\perp U)^\perp = \widehat{\mathrm{add}}U$. We will comment more on this in the next chapter.

As a consequence of the general results in the previous section, we then have the following.

Corollary 1.18. *Let T be a tilting Λ-module. For any C in $\mathrm{mod}\,\Lambda$ there are exact sequences $0 \to Y_C \to X_C \xrightarrow{g} C \to 0$ and $0 \to C \xrightarrow{h} Y^C \to X^C \to 0$, with X_C, X^C in $\widehat{\mathrm{add}}T$ and Y_C, Y^C in $T^\perp$, and $g \colon X_C \to C$ is a minimal right $\widehat{\mathrm{add}}T$-approximation and $h \colon C \to Y^C$ a minimal left $T^\perp$-approximation.*

If C is in $\widehat{\mathrm{add}}T$, then since X^C is in $\widehat{\mathrm{add}}T$, which is closed under extensions, it follows that Y^C is in $\mathrm{add}\,T$. So there is associated with any indecomposable object in $\widehat{\mathrm{add}}T$ a unique module in $\mathrm{add}\,T$. We shall later consider a situation where $\widehat{\mathrm{add}}T$ has n distinguished indecomposable modules, where n is the number of simple modules, and the associated modules $Y_1, \ldots Y_n$ in $\mathrm{add}\,T$ are indecomposable. Then $Y_1, \ldots Y_n$ are exactly the indecomposable summands of T.

2 The Correspondence Theorem

In this chapter we discuss a correspondence theorem between tilting modules on one hand and covariantly or contravariantly finite subcategories with additional properties on the other hand. We discuss interplay with commutative ring theory and related work from abelian group theory.

2.1 The main result

This section is devoted to discussing the correspondence theorem. Such a theorem was first discovered for tilting modules of projective dimension at most one, at the time when more general tilting modules had not yet been introduced [12], and we consider this case first.

Let T be a tilting Λ-module with $\text{pd}_\Lambda T \leqslant 1$, and $\mathcal{T} = \text{Fac}\,T = \{C; \text{Ext}^1_\Lambda(T, C) = 0\} = T^\perp$. Then T is Ext-projective in $\mathcal{T}$, and $\text{add}\,T$ is given by the Ext-projectives in $\mathcal{T}$. For assume that T_1 is indecomposable Ext-projective in $\mathcal{T}$ and not in $\text{add}\,T$. Then $\text{Ext}^1_\Lambda(T \oplus T_1, T \oplus T_1) = 0$, so that $T \oplus T_1$ can be extended to a tilting module by Proposition 1.5, which gives a contradiction to Proposition 1.4.

Start conversely with a torsion class $\mathcal{T} = \text{Fac}\,X$ for some X, with $\text{ann}\,\mathcal{T} = (0)$, where $\text{ann}\,\mathcal{T} = \bigcap_{C \in \mathcal{T}} \text{ann}\,C$. Let T be a direct sum of the indecomposable Ext-projective modules in $\mathcal{T}$. Then we have seen that T is a tilting module. Recall that T is *basic* if $T = \coprod_{s=1}^n T_i$ with T_i indecomposable and $T_i \not\cong T_j$ for $i \neq j$. So we have the following (see [12]).

Theorem 2.1. *Let Λ be an artin algebra and T a Λ-module. Then $T \mapsto \text{Fac}\,T$ and $\mathcal{T} \mapsto \{T; \text{Ext}^1_\Lambda(T, \mathcal{T}) = 0\}$ give one-one inverse correspondences between basic tilting modules of projective dimension at most one and torsion classes $\mathcal{T} = \text{Fac}\,X$ with $\text{ann}\,\mathcal{T} = (0)$.*

Now consider arbitrary tilting modules T. Associated with T there is the subcategory $\mathcal{Y} = T^\perp = \{C; \text{Ext}^i_\Lambda(T, C) = 0 \text{ for } i > 0\}$, which is not closed under factors in general. But for a torsion class $\mathcal{T}$, being of the form $\text{Fac}\,X$ is equivalent to $\mathcal{T}$ being covariantly finite by Proposition 1.6, and by Proposition 1.16 we see that $\mathcal{Y}$ has this latter property in general. It is obvious that $\mathcal{Y}$ is closed under extensions and contains the injective modules. While $\mathcal{Y}$ is not closed under arbitrary quotients, it is easily seen to be closed under cokernels of monomorphisms and under direct summands. Hence $\mathcal{Y}$ is coresolving.

Let conversely $\mathcal{Y}$ be a covariantly finite coresolving subcategory of $\text{mod}\,\Lambda$. If Λ is of finite global dimension, then it turns out that the direct sum of the indecomposable Ext-projective modules is a tilting module. In the general case we need additional assumptions on $\mathcal{Y}$ to make the same conclusion. This way we get a correspondence between basic tilting modules and certain covariantly finite subcategories $\mathcal{Y}$.

By considering $\mathcal{X}$ such that $(\mathcal{X}, \mathcal{Y})$ is a cotorsion pair, we also obtain a related correspondence between basic tilting modules and certain contravariantly finite subcategories. Of course there are dual statements for cotilting modules, which we also include in the following main result [7].

Theorem 2.2. *Let Λ be an artin algebra.*

(a) *There is a one-one correspondence between basic tilting modules T and covariantly finite coresolving subcategories $\mathcal{Y}$ of $\operatorname{mod} \Lambda$ with $\check{\mathcal{Y}} = \operatorname{mod} \Lambda$, given by $T \mapsto T^{\perp}$ and $\mathcal{Y} \mapsto T = $ direct sum of the indecomposable Ext-projectives in $\mathcal{Y}$.*

(b) *There is a one-one correspondence between basic tilting modules T and contravariantly finite resolving subcategories $\mathcal{X}$ of $\operatorname{mod} \Lambda$ with $\mathcal{X} \subset \{C; \operatorname{pd}_{\Lambda} C < \infty\}$, given by $T \mapsto \widetilde{\operatorname{add}T}$ and $\mathcal{X} \mapsto T = $ direct sum of the indecomposable Ext-injectives in $\mathcal{X}$.*

(c) *There is a one-one correspondence between basic cotilting modules U and contravariantly finite resolving subcategories $\mathcal{X}$ of $\operatorname{mod} \Lambda$ with $\hat{\mathcal{X}} = \operatorname{mod} \Lambda$, given by $U \mapsto {}^{\perp}U$ and $\mathcal{X} \mapsto U = $ direct sum of the indecomposable Ext-injectives in $\mathcal{X}$.*

(d) *There is a one-one correspondence between basic cotilting modules U and covariantly finite coresolving subcategories $\mathcal{Y}$ of $\operatorname{mod} \Lambda$ with $\mathcal{Y} \subset \{C; \operatorname{id}_{\Lambda} C < \infty\}$, given by $U \mapsto \widehat{\operatorname{add}U}$ and $\mathcal{Y} \mapsto U = $ direct sum of the indecomposable Ext-projectives in $\mathcal{Y}$.*

Using that the torsion class $\mathcal{T}$ is of the form $\operatorname{Fac} X$ if and only if it is covariantly finite by Proposition 1.6, and that $\operatorname{ann} \mathcal{T} = (0)$ if and only if $\mathcal{T}$ contains all injectives, it is not hard to see that this specializes to Theorem 2.1.

We give a sketch of the main steps in the proof of Theorem 2.2 (c)(d). So let U be a cotilting Λ-module. Then ${}^{\perp}U$ is clearly resolving, and one proves ${}^{\perp}U = \mathcal{X}_U$, in the notation of section 1. Then we show $\widehat{\mathcal{X}_U} = \operatorname{mod} \Lambda$, and use a result from [6] saying that since $\mathcal{X}_U$ is resolving with an Ext-injective cogenerator U, then $\mathcal{X}_U$ is contravariantly finite and $\mathcal{X}_U^{\perp} = \widehat{\operatorname{add}U}$.

Assume conversely that $\mathcal{X}$ is contravariantly finite resolving in $\operatorname{mod} \Lambda$ with $\hat{\mathcal{X}} = \operatorname{mod} \Lambda$. Then one can use the first part to find a cotilting module U such that $\mathcal{X} \subset \mathcal{X}_U$, and U is an Ext-injective cogenerator also for $\mathcal{X}$. Since both $\mathcal{X}$ and $\mathcal{X}_U$ are resolving with U as Ext-injective cogenerator and $\widehat{\mathcal{X}} = \operatorname{mod}\Lambda = \widehat{\mathcal{X}_U}$, we have $\mathcal{X}_U^{\perp} = \widehat{\operatorname{add}U} = \mathcal{X}^{\perp}$. Using that $(\mathcal{X}_U, \widehat{\operatorname{add}U})$ and $(\mathcal{X}, \widehat{\operatorname{add}U})$ are then both cotorsion pairs, it follows that $\mathcal{X} = \mathcal{X}_U$.

We have pointed out in Proposition 1.16 that when T is a tilting module, then $T^{\perp}$ is also functorially finite. In view of Theorem 2.2 this can be restated by saying that a covariantly finite coresolving subcategory $\mathcal{Y}$ with the additional property that $\check{\mathcal{Y}} = \operatorname{mod}\Lambda$ must be functorially finite. It has been shown in [35] that the additional condition can be dropped, so that any covariantly finite coresolving subcategory of $\operatorname{mod}\Lambda$ is in fact functorially finite.

We can clearly also talk about a correspondence theorem between tilting or cotilting modules and certain cotorsion pairs.

Corollary 2.3. *We have the following for an artin algebra Λ.*

(a) *There is a one-one correspondence between basic tilting modules T and complete cotorsion pairs $(\mathcal{X}, \mathcal{Y})$ with $\mathcal{Y}$ coresolving and $\check{\mathcal{Y}} = \operatorname{mod}\Lambda$, given by $T \mapsto (\operatorname{add}T, T^{\perp})$ and $(\mathcal{X}, \mathcal{Y}) \mapsto$ direct sum of the indecomposable modules in $\mathcal{X} \cap \mathcal{Y}$.*

(b) *There is a one-one correspondence between basic cotilting modules U and complete cotorsion pairs $(\mathcal{X}, \mathcal{Y})$ with $\mathcal{X}$ resolving and $\widehat{\mathcal{X}} = \operatorname{mod}\Lambda$, given by $U \to ({}^{\perp}U, \widehat{\operatorname{add}U})$ and $(\mathcal{X}, \mathcal{Y}) \mapsto$ direct sum of the indecomposable modules in $\mathcal{X} \cap Y$.*

Note that when $\operatorname{pd}_{\Lambda} T \leqslant 1$, then the corresponding cotorsion pair is $(\operatorname{Sub}T, \operatorname{Fac}T)$.

Also note that the conditions on the corresponding subcategories simplify for algebras of finite global dimension. For when $\mathcal{X}$ contains the projectives, we have automatically $\widehat{\mathcal{X}} = \operatorname{mod}\Lambda$, and when $\mathcal{Y}$ contains the injectives, we have $\check{\mathcal{Y}} = \operatorname{mod}\Lambda$. Similarly, all subcategories are contained in $\{C; \operatorname{pd}_{\Lambda} C < \infty\}$ and in $\{C; \operatorname{id}_{\Lambda} C < \infty\}$. So in this case we have the following.

Corollary 2.4. *For an algebra of finite global dimension the tilting modules coincide with the cotilting modules.*

Corollary 2.5. *Let Λ be an artin algebra of finite global dimension.*

(a) *There is a one-one corespondence between basic (co)tilting mod-*

ules T and covariantly finite coresolving subcategories $\mathcal{Y}$ of $\operatorname{mod}\Lambda$ given by $T \mapsto T^{\perp} = \widehat{\operatorname{add}}T$.

(b) *There is a one-one correspondence between basic (co)tilting modules T and contravariantly finite resolving subcategories $\mathcal{Y}$ of $\operatorname{mod}\Lambda$ given by $T \mapsto {}^{\perp}T = \widehat{\operatorname{add}}T$.*

Let T be a tilting module over an artin algebra Λ, and let $(\mathcal{X}, \mathcal{Y}) = (\widehat{\operatorname{add}}T, T^{\perp})$ be the associated cotorsion pair. For $\Gamma = \operatorname{End}_\Lambda(T)^{\operatorname{op}}$, let $(\mathcal{X}', \mathcal{Y}') = ({}^{\perp}(D(T)), \widehat{\operatorname{add}}D(T))$ be the cotorsion pair in $\operatorname{mod}\Gamma$ associated with the cotilting Γ-module $D(T)$. We have seen that we have an equivalence $F = \operatorname{Hom}_\Lambda(T, \ ) \colon \mathcal{Y} \to \mathcal{X}'$, but we do not know whether there is a (canonical) equivalence between $\mathcal{X}$ and $\mathcal{Y}'$ in general. This is however the case when Λ, and consequently also Γ by Proposition 1.2, has finite global dimension. Since then $\widehat{\operatorname{add}}T = {}^{\perp}T$ and $\widehat{\operatorname{add}}D(T) = (D(T))^{\perp}$ by Corollary 2.5, it follows from section 1.2 that $G = D\operatorname{Hom}_\Lambda(\ , T) \colon \mathcal{X} \to \mathcal{Y}'$ is an equivalence.

We do not know whether there is an equivalence $G = D\operatorname{Hom}_\Lambda(\ , T) \colon \widehat{\operatorname{add}}T \to \widehat{\operatorname{add}}D(T)$ in general. Similarly, let U be a cotilting Λ-module and $({}^{\perp}U, \widehat{\operatorname{add}}U)$ the associated cotorsion pair. For $\Gamma = \operatorname{End}_\Lambda(U)^{\operatorname{op}}$, let $(\widehat{\operatorname{add}}D(U), D(U)^{\perp})$ be the cotorsion pair in $\operatorname{mod}\Gamma$ associated with the tilting Γ-module $D(U)$. Then we have seen that we have an equivalence $G = D\operatorname{Hom}(\ , U) \colon {}^{\perp}U \to D(U)^{\perp}$, and for Λ of finite global dimension an equivalence $F = \operatorname{Hom}_\Lambda(U, \ ) \colon \widehat{\operatorname{add}}U \to \widehat{\operatorname{add}}D(U)$. But we do not know if there is such an equivalence F in general.

There are however cases beyond finite global dimension where we have such equivalences, and we shall see such a class in Chapter 3. Another case is provided by the following. We say that a cotilting module U is a *strong cotilting module* if $\widehat{\operatorname{add}}U = \{C; \operatorname{id}_\Lambda C < \infty\}$ and that a tilting module T is a *strong tilting module* if $\widehat{\operatorname{add}}U = \{C; \operatorname{pd}_\Lambda C < \infty\}$ [7]. This concept is motivated by the case of a commutative complete local noetherian Cohen-Macaulay ring and its dualizing module. If U is a strong cotilting Λ-module, and $D(U)$ is a strong tilting Γ-module, where $\Gamma = \operatorname{End}_\Lambda(U)^{\operatorname{op}}$, then there is an equivalence $F = \operatorname{Hom}_\Lambda(U, \ ) \colon \widehat{\operatorname{add}}U \to \widehat{\operatorname{add}}D(U)$ [7]. A special case is that Λ is a Gorenstein algebra, that is, Λ has finite injective dimension both as a left and as a right module over Λ. Then Λ is itself a cotilting module, and $\Gamma = \operatorname{End}_\Lambda(\Lambda)^{\operatorname{op}} \simeq \Lambda$. Further $\widehat{\operatorname{add}}\Lambda = \{C; \operatorname{id}_\Lambda C < \infty\} = \{C; \operatorname{pd}_\Lambda C < \infty\} = \widehat{\operatorname{add}}D(\Lambda)$, and $F = \operatorname{Hom}_\Lambda(\Lambda, \ ) \colon \widehat{\operatorname{add}}\Lambda \to \widehat{\operatorname{add}}D(\Lambda)$ is the identity functor.

When Λ is an algebra of finite global dimension, we have seen that a tilting module T is automatically a cotilting module and that the two associated covariantly finite subcategories $T^{\perp}$ and $\widetilde{\mathrm{add}T}$ coincide and the two associated contravariantly finite subcategories ${}^{\perp}T$ and $\widehat{\mathrm{add}T}$ coincide. When Λ has infinite global dimension, we may still have tilting modules T which are also cotilting modules, for example $T = \Lambda$ for a Gorenstein algebra Λ. It is easy to see that $\widehat{\mathrm{add}T}$ and $T^{\perp}$ never coincide for algebras of infinite global dimension, and similarly $\widehat{\mathrm{add}T}$ and ${}^{\perp}T$ do not coincide.

2.2 *Some applications*

When we encounter a contravariantly finite resolving subcategory, or a covariantly finite coresolving one, there is always a natural question whether we have any of the additional conditions satisfied which ensure the existence of an associated tilting or cotilting module.

In the next chapter we discuss the main and most influential application of the correspondence theorem, to quasihereditary algebras and their generalizations. In this section we indicate two other applications of a quite different nature.

There is an interesting illustration of the theory for the closed model structures of Quillen [41] (see [13] for details). We do not recall the relevant definitions here, but just say that in a closed model category in the sense of Quillen [41] there are associated some important subcategories in a natural way, namely on one hand the subcategory Cof, respectively T Cof, of cofibrant, respectively trivially cofibrant objects, and on the other hand the subcategory Fib, respectively T Fib, of fibrant, respectively trivially fibrant, objects. Then the subcategories Cof and T Cof are contravariantly finite and the subcategories Fib and T Fib are covariantly finite. In addition we have nice illustrations of cotorsion pairs, since (Cof, T Fib) and (T Cof, Fib) are both examples of such pairs. In particular, there are associated tilting modules, and any tilting module gives rise to a cotorsion pair of this kind. Via showing that sometimes such pairs (i.e. when $\mathcal{X} \cap \mathcal{Y}$ is covariantly finite) give rise to closed model structures, it follows that tilting modules determine closed model structures.

Another illustration is given by the following. Let Λ be an artin algebra satisfying the following condition:

(*) *If* $0 \to \Lambda \to I_0 \to I_1 \to \cdots \to I_j \to \cdots$ *is a minimal injective resolution of* Λ *as right module, then* $\operatorname{pd}_{\Lambda^{\mathrm{op}}} I_j \leqslant j + 1$ *for all* $j \geqslant 0$.

This holds in particular for the Auslander rings, that is, the rings Λ which are k-Gorenstein for all k, that is $\operatorname{pd}_\Lambda I_j \leqslant j$ for all $j \geqslant 0$. For each d denote by $\mathcal{X}_d$ the subcategory of $\operatorname{mod}\Lambda$ whose objects are the direct summands of objects in the syzygy category $\Omega^d(\operatorname{mod}\Lambda)$. Then $\mathcal{X}_d$ is closed under extensions [10], and hence also under kernels of epimorphisms [8]. We have mentioned that $\Omega^d(\operatorname{mod}\Lambda)$ is functorially finite, and hence $\mathcal{X}_d$ also has the same property [11]. Then we have the following.

Proposition 2.6. *For an artin algebra* Λ *satisfying (*) the categories* $\mathcal{X}_d$ *are functorially finite resolving with* $\widehat{\mathcal{X}_d} = \operatorname{mod}\Lambda$.

As a consequence there is a cotilting module U such that $\mathcal{X}_d = {}^\perp U$, and we have a cotorsion pair $(\mathcal{X}_d, \mathcal{Y}_d)$, with $\mathcal{Y}_d = \widehat{\operatorname{add}U}$. It is easy to see that $\mathcal{Y}_d = \{C; \operatorname{id}_\Lambda C \leqslant d\}$, which is hence covariantly finite coresolving. The indecomposable summands of U are the indecomposable projective modules P with $\operatorname{id}_\Lambda P \leqslant d$ and the $\Omega^d I$ for I indecomposable injective with $\operatorname{pd}_\Lambda I > d$ [9]. It also follows from Proposition 2.6 that $\mathcal{X}_d$ has almost split sequences for Λ satisfying (*).

2.3 Interplay with the commutative case

In this section we discuss modules analogous to cotilting modules in commutative ring theory, and how the interplay between artin algebras and commutative rings has been fruitful in both directions.

The cotilting modules T have many analogous properties with the dualizing module ω for a complete local commutative Cohen-Macaulay ring R. The module ω has finite injective dimension, and $\operatorname{Ext}^i_R(\omega, \omega) = 0$ for all $i > 0$. The condition for a cotilting module that there is an exact sequence $0 \to T_n \to \cdots \to T_1 \to T_0 \to D\Lambda \to 0$, says that the category $\widehat{\operatorname{add}T}$ contains the injective Λ-modules, or in other words, is a cogenerator for $\operatorname{mod}\Lambda$. This last property also holds for ω (see [6]). The category ${}^\perp\omega$ is the category of maximal Cohen-Macaulay modules $CM(R)$, which plays a central role in commutative ring theory. In this setting the duality $\mathrm{Hom}_R(\ ,\omega)\colon CM(R) \to CM(R)$ is important. This corresponds to the fact that $\mathrm{Hom}_\Lambda(T,\ )$ for a tilting module T induces an equivalence of categories between $T^\perp$ and the subcategory $^\perp(D(T))$ of $\mathrm{mod}\,\mathrm{End}_\Lambda(T)^{\mathrm{op}}$, or dually, that a cotilting module U induces an equivalence $D\,\mathrm{Hom}_\Lambda(\ ,U)\colon {}^\perp U \to D(U)^\perp$, or rather a duality $\mathrm{Hom}_\Lambda(\ ,U)\colon {}^\perp U \to {}^\perp U_{\mathrm{End}(U)}$. Note that in the commutative case we have $\mathrm{End}_R(\omega) \simeq R$. But it is rare for cotilting modules over artin algebras that we get the same algebra back again.

Here we see two different lines of developments leading to similar concepts. The dualizing modules have been around for a long time in commutative ring theory, and the roots of tilting theory, starting much later, is a different kind of story. Whereas the theory of tilting and of cotilting modules are dual for finitely generated modules over artin algebras, and hence equally interesting, the concept of tilting module is not interesting in the commutative case. For if R is a local complete noetherian commutative Cohen-Macaulay ring, then R is the only basic tilting module.

The Memoir by Auslander-Bridger [5] has served as an important foundation for many ideas in commutative and noncommutative ring theory. There are also various examples of the behavior of left and right approximations in [5], and hence some interesting predecessors of the theory of contravariantly and covariantly finite subcategories. For example, in this new language there is the following result from [5] (p. 64, p. 85).

Proposition 2.7. *For a k-Gorenstein ring Λ the category $\{C; \mathrm{pd}_\Lambda\, C \leqslant k\}$ is covariantly finite in $\mathrm{mod}\,\Lambda$, and the category $\{\mathrm{Tr}\,\Omega^k C; C \in \mathrm{mod}\,\Lambda\}$ is contravariantly finite.*

After the work of Auslander-Smalø on the theory of contravariantly and covariantly finite subcategories, there was further work for Cohen-Macaulay rings, actually formulated more generally in the context of abelian categories [6]. This specialized to the concept of maximal Cohen-Macaulay approximation, and the construction $\breve{\mathcal{Y}}$ (and $\widehat{\mathcal{X}}$) first appeared here, in particular the category $\widehat{\omega}$ and the fact that $\widehat{CM}(R) = \mathrm{mod}\,R$ [6]. This work served again as inspiration for the work on the connection between (co)tilting modules and covariantly/contravariantly finite subcategories, discussed in the previous section. The pair $(CM(R), \widehat{\omega})$ was a model example for a pair of contravariantly finite/covariantly finite subcategories.

The notions of contravariantly and covariantly finite subcategories were also independently explored by Enochs, in the language of (pre)covers and (pre)envelopes [25].

2.4 Predecessors for abelian groups

We have seen that associated to tilting or cotilting modules there are natural cotorsion pairs $(\mathcal{X}, \mathcal{Y})$ where $\mathcal{X}$ is contravariantly finite (resolving) and $\mathcal{Y}$ is covariantly finite (coresolving). The same thing holds for the pair $(CM(R), \hat{\omega})$ when R is a commutative complete local noetherian Cohen-Macaulay ring and ω the dualizing module.

The idea of cotorsion pairs/theories goes back to the work of Salce on abelian groups [45], where it is defined the way we give it here. Note that since $\mathrm{Ext}^i(\ ,\ ) = 0$ for $i > 1$ for abelian groups, it does not make any difference whether we deal with $\mathrm{Ker}(\mathrm{Ext}^1(\ ,\mathcal{Y}))$ or $^\perp\mathcal{Y}$. The classical example was $(\mathcal{X}, \mathcal{Y})$, where $\mathcal{X}$ is the class of torsionfree groups, and $\mathcal{Y}$ the class of cotorsion groups, which by definition is $\mathcal{X}^\perp$. Salce is concerned with cotorsion pairs cogenerated by a class $\mathcal{A}$, in the sense that $\mathcal{X} = \mathrm{Ker}(\mathrm{Ext}^1(\ ,\mathcal{A}))$ and $\mathcal{Y} = \mathrm{Ker}(\mathrm{Ext}^1(\mathcal{X},\ ))$. He investigates especially those which are cogenerated by subgroups of $\mathbb{Q}$ containing $\mathbb{Z}$.

It was already known that for every group G there is an exact sequence $0 \to G \to Y \to X \to 0$ with Y cotorsion and X torsionfree. For a general cotorsion theory Salce refers to this property as the cotorsion pair having enough injectives, and the dual property as having enough projectives. He shows that in this setting a cotorsion theory $(\mathcal{X}, \mathcal{Y})$ has enough injectives if and only if it has enough projectives. Note that these conditions are closely related to $\mathcal{X}$ being contravariantly finite and $\mathcal{Y}$ being covariantly finite. These are clearly a consequence, and the statements are equivalent for mod Λ with Λ an artin algebra, because of the validity of the Wakamatsu lemma in this case.

Some problems are posed at the end of the paper of Salce, which have been considered recently [28] [24]: Find all cotorsion pairs $(\mathcal{X}, \mathcal{Y})$ cogenerated by torsionfree groups. Does a cotorsion pair $(\mathcal{X}, \mathcal{Y})$ of abelian groups always have enough projectives?

3 Quasihereditary algebras and their generalizations

The main application of the correspondence theorem between tilt-
ing/cotilting modules and contravariantly/covariantly finite subcategories is Ringel's application to quasihereditary algebras. There are
follow-up results to more general classes of algebras, including properly
stratified and standardly stratified algebras, which we also discuss.

3.1 Preliminaries

We start with the basic definitions and background material for quasi-
hereditary algebras and their generalizations. There are various equivalent definitions in the literature, and we choose the one most convenient
for our discussion.

Let Λ be an artin algebra, and $P_1, \ldots P_n$ a fixed ordering of the indecomposable projective Λ-modules, with corresponding simple tops $S_1, \ldots S_n$.
Let Δ_i be the largest factor of P_i with no composition factor S_j for $j > i$.
Denote by $\bar{\Delta}_i$ the largest factor of Δ_i where S_i occurs only once as a
composition factor. The Δ_i are called the *standard* modules, and the $\bar{\Delta}_i$
the *proper standard* modules. Then for $\Delta = \{\Delta_1, \ldots \Delta_n\}$ the objects of
$\mathcal{F}(\Delta)$ are those with *standard filtrations*, and for $\bar{\Delta} = \{\bar{\Delta}_1, \ldots \bar{\Delta}_n\}$ the
objects of $\mathcal{F}(\bar{\Delta})$ are those with *proper standard filtrations*.

Dually, let $I_1, \ldots I_n$ be the indecomposable injective modules, with
$\operatorname{soc} I_j \simeq S_j$. Let ∇_i be the largest submodule of I_i with no composition factor S_j for $j > i$, and $\bar{\nabla}_i$ the largest submodule of ∇_i with S_i
occurring only once as composition factor, and let $\nabla = \{\nabla_1, \ldots \nabla_n\}$ and
$\bar{\nabla} = \{\bar{\nabla}_1, \ldots \bar{\nabla}_n\}$. Then the objects of $\mathcal{F}(\nabla)$ are those with *costandard
filtrations*, and the objects of $\mathcal{F}(\bar{\nabla})$ are those with *proper costandard
filtrations*.

Having possible applications of the correspondence theorem in mind, it
is of interest to investigate the relevant properties of the above subcategories. We then have the following, where part (b) is a consequence of
(a) and Theorem 1.12 [44].

Theorem 3.1.   (a) *With   the   previous   notation,   we   have*
$\operatorname{Ext}^1_\Lambda(\Delta_i, \Delta_j) = 0$, *for* $i \geqslant j$ *and* $\operatorname{Ext}^1_\Lambda(\nabla_i, \nabla_j) = 0$ *for*
$i \leqslant j$.

(b) *The subcategories $\mathcal{F}(\Delta)$ and $\mathcal{F}(\nabla)$ are functorially finite in* mod Λ.

It is obvious that our subcategories are closed under extensions, and one can also show that $\mathcal{F}(\Delta)$ and $\mathcal{F}(\bar{\Delta})$ are closed under summands and kernels of epimorphisms and that $\mathcal{F}(\nabla)$ and $\mathcal{F}(\bar{\nabla})$ are closed under summands and cokernels of monomorphisms [44] [23] [1]. So the only thing missing for the first two to be resolving or the second two to be coresolving is that they contain the projectives, respectively the injectives. From this point of view, the following definitions are natural.

The algebra Λ is said to be *standardly stratified* if $\mathcal{F}(\Delta)$ contains the projectives, and Λ is *quasihereditary* if in addition $\Delta_i = \bar{\Delta}_i$ for all $i = 1, \ldots n$. If Λ is standardly stratified and the Δ_i have filtrations using the $\bar{\Delta}_j$, that is, $\mathcal{F}(\Delta) \subset \mathcal{F}(\bar{\Delta})$, then Λ is said to be *properly stratified* [20]. The quasihereditary algebras are characterized amongst the standardly stratified ones as follows (see [2]).

Proposition 3.2. *Let Λ be a standardly stratified artin algebra. Then Λ is quasihereditary if and only if* gl. dim. $\Lambda < \infty$.

The quasihereditary algebras were introduced by Cline-Parshall-Scott [18], motivated by the theory of Lie algebras and algebraic groups. Also the standardly stratified algebras appear naturally in Lie theory (see [26] [37]), and were introduced in [19] with the weaker condition that there is only a preorder on the indecomposable projective modules. The categories $\mathcal{F}(\Delta)$ and $\mathcal{F}(\nabla)$ play a central role in the theory of these algebras. In particular the subcategory $\mathcal{F}(\Delta) \cap \mathcal{F}(\nabla)$ of modules having both standard and costandard filtrations is important.

The central role of quasihereditary algebras in representation theory is also stressed by Iyama's work on the representation dimension [32], where he shows that for any artin algebra there is a module M which is a projective generator and injective cogenerator, such that $\mathrm{End}_\Lambda(M)^{\mathrm{op}}$ is quasihereditary. This answered Auslander's conjecture that the representation dimension is always finite.

3.2 Quasihereditary algebras

In this section we apply the correspondence theorem to quasihereditary algebras. General references for this section are [44] [23] [34].

Let $\Lambda = (\Lambda, \leqslant)$ be a quasihereditary algebra with respect to a fixed ordering $P_1, \ldots P_n$ of the indecomposable projective modules. As we have seen, gl. dim. $\Lambda < \infty$, so the tilting and cotilting modules coincide, and to apply the correspondence theorem we only need to check that our subcategories are contravariantly finite resolving/covariantly finite coresolving. We have the following [44].

Theorem 3.3. *Let* $(\Lambda, \leqslant)$ *be a quasihereditary algebra. Then there is a basic (co)tilting Λ-module T such that* $^\perp T = \mathcal{F}(\Delta)$ *and* $T^\perp = \mathcal{F}(\nabla)$, *and* $(\mathcal{F}(\Delta), \mathcal{F}(\nabla))$ *is a cotorsion pair.*

We make some remarks about the proof. As we have already mentioned, $\mathcal{F}(\Delta)$ is contravariantly resolving. So, since gl. dim. $\Lambda < \infty$, we have our desired (basic) (co)tilting module T such that $^\perp T = \mathcal{F}(\Delta)$. It is then shown that $\mathcal{F}(\nabla) = \mathcal{F}(\Delta)^\perp$, so that $(\mathcal{F}(\Delta), \mathcal{F}(\nabla))$ is the associated cotorsion pair, and consequently $\mathcal{F}(\nabla) = T^\perp$, since $(^\perp T, T^\perp)$ is a cotorsion pair by section 2.

This particular (basic) tilting module T has some extra nice properties. While in general we have $T = T_1 \coprod \cdots \coprod T_n$, where n is the number of simple modules S_i, we normally do not have a natural correspondence between the T_i and the S_i. For quasihereditary algebras we do however have such a correspondence. This can be seen as follows. Let T' be one of the T_i. Since T' is in $\mathcal{F}(\Delta)$, we have an exact sequence $0 \to \Delta_i \xrightarrow{f} T' \to X \to 0$, with X in $\mathcal{F}(\Delta)$, for some i. Then $f \colon \Delta_i \to T'$ is a left $\mathcal{F}(\nabla)$-approximation since $\mathrm{Ext}^1_\Lambda(X, \mathcal{F}(\nabla)) = 0$, necessarily minimal since T' is indecomposable. By uniqueness of left minimal $\mathcal{F}(\nabla)$-approximations, Δ_i uniquely determines T', and since the number of T_i is the same as the number of Δ_i, then T' also uniquely determines Δ_i. Since $\mathrm{Ext}^1_\Lambda(\Delta_j, \Delta_i) = 0$ for $j \geqslant i$, X must be filtered by Δ_j's for $j < i$. So S_i occurs as a composition factor of T', and no S_j with $j > i$ occurs. Hence T' is naturally associated with S_i, and we write $T' = T_i$. Dually we have, since T' is in $\mathcal{F}(\nabla)$, an exact sequence $0 \to Y \to T_i \xrightarrow{g} \nabla_i \to 0$ with Y in $\mathcal{F}(\nabla)$. Then $g \colon T_i \to \nabla_i$ is a minimal right $\mathcal{F}(\Delta)$-approximation.

The above (basic) (co)tilting module T is canonically associated with a quasihereditary algebra, and is called the *characteristic tilting module*. Hence there is canonically associated with T the algebra $\Gamma = \mathrm{End}_\Lambda(T)^{\mathrm{op}}$ called the *Ringel dual*. The equivalences between subcategories associated with cotorsion pairs are nicely illustrated here.

Using the notation $T = T_1 \coprod \cdots \coprod T_n$ as explained above, we consider Γ with the opposite of the ordering for Λ [44].

Theorem 3.4. *Let $(\Lambda, \leqslant)$ be a quasihereditary algebra, with characteristic tilting module T.*

    (a) *Then $\Gamma = \mathrm{End}_\Lambda(T)^{\mathrm{op}}$ is a quasihereditary algebra, with the induced opposite ordering for Γ, and $\Lambda \simeq \mathrm{End}_\Gamma(D(T))^{\mathrm{op}}$.*

    (b) *Let Δ_i' and ∇_i' be the standard and costandard modules for Γ. Then we have equivalences of categories $F = \mathrm{Hom}_\Lambda(T, \ ) \colon \mathcal{F}(\nabla) \to \mathcal{F}(\Delta')$ and $G = D\,\mathrm{Hom}_\Lambda(\ , T) \colon \mathcal{F}(\Delta) \to \mathcal{F}(\nabla')$.*

For (b), we have seen in section 2.1 that we have equivalences $F \colon \mathcal{F}(\nabla) \to \mathcal{X}'$ and $G \colon \mathcal{F}(\Delta) \to \mathcal{Y}'$, where $(\mathcal{X}', \mathcal{Y}')$ is the cotorsion pair associated with the cotilting Γ-module $D(T)$. One can consider the set of modules $F(\nabla_i)$ and show that they have to coincide with the standard modules $\Delta_1', \ldots \Delta_n'$ for Γ, in opposite order. Similarly the $G(\Delta_i)$ are the costandard modules $\nabla_1', \ldots \nabla_n'$ for Γ, so that $\mathcal{X}' = \mathcal{F}_\Gamma(\Delta')$ and $\mathcal{Y}' = \mathcal{F}_\Gamma(\nabla')$. Hence $D(T)$ is the characteristic tilting module for Γ, which is quasihereditary since $F(T) = \Gamma \in \mathcal{F}_\Gamma(\Delta')$.

In addition to giving a beautiful illustration of the general theory, this result also has the following important application, taking advantage of basic properties of tilting modules.

Corollary 3.5. *Let Λ be a quasihereditary algebra with n simple modules. Then there are exactly n indecomposable Λ-modules which have both a standard and costandard filtration.*

For note that $\mathcal{F}(\Delta) \cap \mathcal{F}(\nabla) = \mathrm{add}\,T$, where T is the characteristic tilting module, and T has exactly n indecomposable summands by Proposition 1.4.

The part of the theory of quasihereditary algebras related to tilting theory has had a large influence in the theory of algebraic groups and related topics (see [22]). It has also inspired further developments within the theory of quasihereditary algebras. For natural questions are to investigate more closely the relationship between a quasihereditary algebra and its Ringel dual (see [50]), as well as using the relationship to investigate when $\mathcal{F}(\Delta)$ is of finite (representation) type. For example $\mathcal{F}_\Lambda(\Delta)$ is clearly of finite type if the Ringel dual Γ of Λ is of finite type. And results for quasihereditary algebras involving tilting suggest problems within tilting theory more generally. For example, how generally do we have a natural correspondence between the indecomposable summands of a tilting module and the simple modules. Interesting are also the generalizations of quasihereditary algebras to be discussed in the next section, which are still closely related to tilting theory.

3.3 Standardly stratified algebras

In this section we discuss a correspondence theorem for standardly stratified algebras, and the smaller class of properly stratified algebras [1] [2] [20] [21] [27] [37] [40].

When a standardly stratified algebra is not quasihereditary, we have seen in Proposition 3.2 that it must have infinite global dimension, so we get a nice illustration of the theory also beyond finite global dimension.

Let Λ be a standardly stratified algebra, that is, Λ is in $\mathcal{F}(\Delta)$, or equivalently, $D(\Lambda)$ is in $\mathcal{F}(\bar{\nabla})$ [1]. Then $\mathcal{F}(\Delta)$ is functorially finite in $\mathrm{mod}\,\Lambda$ by Theorem 1.12. It is clearly closed under extensions, and under summands and kernels of epimorphisms by [44] [23]. It is easy to see by induction that the modules in $\mathcal{F}(\Delta)$ have finite projective dimension, starting with $\Delta_n = P_n$ being projective [2] [40]. Hence there is by Theorem 2.2 (b) a tilting module T such that $\mathrm{add}T = \mathcal{F}(\Delta)$. Since $\mathcal{F}(\Delta)^{\perp} = \mathcal{F}(\bar{\nabla})$ by [1], we have $\mathcal{F}(\bar{\nabla}) = T^{\perp}$, and $(\mathcal{F}(\Delta), \mathcal{F}(\bar{\nabla}))$ is a complete torsion pair. So we have the following result, where the second part follows by duality.

Theorem 3.6. (a) *Let Λ be standardly stratified. Then $\mathcal{F}(\Delta) \subset \{C; \mathrm{pd}_{\Lambda} C < \infty\}$, and there is a tilting module T such that $\mathcal{F}(\Delta) = \mathrm{add}T$ and $\mathcal{F}(\bar{\nabla}) = T^{\perp}$, and $(\mathcal{F}(\Delta), \mathcal{F}(\bar{\nabla}))$ is a complete cotorsion pair.*

(b) *Assume that Λ^{op} is standardly stratified. Then $\mathcal{F}(\nabla) \subset \{C; \mathrm{id}_{\Lambda} C < \infty\}$, and there is a cotilting module U such that $\mathcal{F}(\nabla) = \mathrm{add}U$ and $\mathcal{F}(\bar{\Delta}) = {}^{\perp}U$, and $(\mathcal{F}(\bar{\Delta}), \mathcal{F}(\nabla))$ is a complete cotorsion pair.*

When Λ is standardly stratified, the associated tilting module is also called the *characteristic tilting module*, and when Λ^{op} is standardly stratified, then the associated cotilting Λ-module U is called the *characteristic cotilting module.* When Λ is standardly stratified, Λ^{op} does not necessarily have the same property (see [27]). For we can have that $\Lambda \in \mathcal{F}(\Delta)$, but $\Lambda \notin \mathcal{F}(\bar{\Delta})$. Actually, if Λ is standardly stratified, then it is properly stratified if and only if Λ^{op} is standardly stratified, and so Λ is properly stratified if and only if Λ^{op} is [21].

Using the complete cotorsion pair $(\mathcal{F}(\Delta), \mathcal{F}(\bar{\nabla}))$ for a standardly stratified algebra Λ we get in the same way as for quasihereditary algebras a one-one correspondence between indecomposable summands T_i of T and simple modules S_i, such that S_i is a composition factor of T_i, but no S_j with $j > i$ is. In particular, there is induced an ordering of the T_i, so that we again can use the opposite ordering for the Ringel dual $\Gamma = \mathrm{End}_{\Lambda}(T)^{\mathrm{op}}$. Denote by Δ_i' and $\bar{\Delta}_i'$ the standard and proper standard modules for Γ, and by ∇_i' and $\bar{\nabla}_i'$ the costandard and proper costandard modules. It turns out that Γ is not necessarily standardly stratified, but the opposite ring has this property. And we have a similar relationship between subcategories as before [2].

Theorem 3.7. *Let Λ be a standardly stratified algebra with characteristic tilting module T and let $\Gamma = \mathrm{End}_{\Lambda}(T)^{\mathrm{op}}$. Then we have the following.*

(a) *Γ^{op} is standardly stratified, and Γ has a characteristic cotilting module $D(T)$.*

(b) *$\mathcal{F}(\bar{\Delta}') = {}^{\perp}(D(T))$ and $\mathcal{F}(\nabla') = \widehat{\mathrm{add}}D(T)$, and we have equivalences $F = \mathrm{Hom}_{\Lambda}(T,)\colon \mathcal{F}(\bar{\nabla}) \to \mathcal{F}(\bar{\Delta}')$ and $G = D\,\mathrm{Hom}_{\Lambda}(\,,T)\colon \mathcal{F}(\Delta) \to \mathcal{F}(\nabla')$.*

We have the equivalence $F = \mathrm{Hom}_{\Lambda}(T,\ )\colon T^{\perp} \to {}^{\perp}(D(T))$ from general tilting theory, and $T^{\perp} = \mathcal{F}(\bar{\nabla})$. By considering the Γ-modules $F(\bar{\nabla}_i)$ one shows that $F(\mathcal{F}(\bar{\nabla})) = \mathcal{F}(\bar{\Delta})$. From general theory one has that $G\colon \mathcal{F}(\Delta) \to \mathrm{mod}\,\Gamma$ is full and faithful, using that $\mathcal{F}(\Delta) = \widehat{\mathrm{add}}T$, but to identify the image as $\mathcal{F}(\nabla')$ one considers the $G(\Delta_i)$. In particular, this gives examples of $D\,\mathrm{Hom}_{\Lambda}(\,,T)\colon \widehat{\mathrm{add}}T \to \widehat{\mathrm{add}}D(T)$ being an equivalence for a tilting module T, for algebras of infinite global dimension.

Let now Λ be properly stratified. Then we have both a characteristic tilting module T and a characteristic cotilting module U. Then $\Gamma = \mathrm{End}_{\Lambda}(T)^{\mathrm{op}}$ and Γ^{op} are not properly stratified in general. But if $T = U$, then Γ is again properly stratified, with the characteristic tilting and cotilting module coinciding [27].

Note that when the characteristic tilting module T is also a characteristic cotilting module for a properly stratified algebra Λ, which can happen also for infinite global dimension, the four subcategories $T^\perp$, $^\perp T$, $\widetilde{\mathrm{add}T}$ and $\widetilde{\mathrm{add}T}$ associated with T in Theorem 2.2 have a nice illustration as the four subcategories with filtrations which we are dealing with.

Assume more generally that Λ is a standardly stratified algebra and that Γ is properly stratified. Let T be the characteristic tilting module for Λ, and let T' be the characteristic tilting module for Γ. Consider the equivalence $F = \mathrm{Hom}_\Lambda(T, \ ) \colon T^\perp \to \ _\Gamma^\perp D(T) = \mathcal{F}_\Gamma(\bar{\Delta})$. Then $T' \in \mathcal{F}_\Gamma(\Delta) \subset \mathcal{F}_\Gamma(\bar{\Delta})$, so we can consider $H = F^{-1}(T')$ [27]. This module is shown to have some interesting properties [27]. If $T' = \ _\Gamma D(T)$, then H is just $_\Lambda D(\Lambda)$. Recall that the finitistic dimension of Λ, denoted fin. dim. Λ, is $\sup\{\mathrm{pd}_\Lambda C; \mathrm{pd}_\Lambda C < \infty, C \in \mathrm{mod}\,\Lambda\}$.

Theorem 3.8. *For a standardly stratified algebra Λ such that the Ringel dual Γ is properly stratified, we have the following, with the above notation.*

(a) *H is a tilting Λ-module.*

(b) *$\mathcal{X} = \{C; \mathrm{pd}_\Lambda C < \infty\}$ is contravariantly finite in $\mathrm{mod}\,\Lambda$ and $\mathcal{X} = \widetilde{\mathrm{add}H}$.*

(c) *fin. dim. $\Lambda = \mathrm{pd}_\Lambda H$.*

Note that this motivates interesting questions for tilting theory more generally. Let Λ be an artin algebra with a pair of modules (T, U), where T is a tilting module and $U \not\simeq T$ is a cotilting module, such that $\widetilde{\mathrm{add}U} \subset T^\perp$, that is, $U \in T^\perp$, or equivalently $T \in {}^\perp U$. Let $\Gamma = \mathrm{End}_\Lambda(T)^{\mathrm{op}}$ and $F = \mathrm{Hom}_\Lambda(T, \ ) \colon T^\perp \to {}^\perp(D(T))$. Then $\widetilde{\mathrm{add}D(U)} \subset {}^\perp(D(T))$, and we let $H = F^{-1}(D(U))$. When does H have similar properties to those stated in the above theorem? Also it would be interesting to find natural sources of such pairs (T, U).

For more information on the finitistic dimension for standardly stratified or properly stratified algebras, see [38]. In particular, for a properly stratified algebra Λ fin. dim. $\Lambda \leqslant \mathrm{pd}_\Lambda T + \mathrm{id}_\Lambda U$, where T is the characteristic tilting module and U is the characteristic cotilting module.

4 Generalizations

In this chapter we consider on one hand generalizations of the correspondence theorem for (co)tilting modules for $\operatorname{mod}\Lambda$ using the more general class of Wakamatsu tilting modules. There are also generalizations to categories of arbitrary modules over a ring, with direct extensions of the definitions of tilting/cotilting modules. For this we refer to the survey [47]. On the other hand there are similar correspondence theorems for triangulated categories, in particular bounded derived categories of finitely generated modules, and we discuss some of the work which has been done in this direction.

4.1 Wakamatsu tilting modules

Generalizations of tilting modules (of finite projective dimension) have been considered, amongst others by Wakamatsu [49]. A module T in $\operatorname{mod}\Lambda$ for an artin algebra Λ is said to be a *Wakamatsu tilting module* if $\operatorname{Ext}^i_\Lambda(T,T) = 0$ for $i > 0$ and there is an exact sequence $0 \to \Lambda \xrightarrow{f_0} T_0 \xrightarrow{f_1} T_1 \to \cdots T_i \to \cdots$ with the T_i in $\operatorname{add} T$ and $\operatorname{Coker} f_i \in {}^{\perp}T$ for all $i \geqslant 0$.

A Wakamatsu cotilting module is defined dually, and it is known that T is a Wakamatsu tilting module if and only if it is a Wakamatsu cotilting module, so we do not really need the latter concept [49] [16]. It is an interesting open problem, closely related to the homological conjectures for artin algebras, whether a Wakamatsu tilting module of finite projective dimension must be a tilting module. This is called the *Wakamatsu tilting conjecture* in [13].

Recall that for a tilting module T we had two associated subcategories, on one hand $T^{\perp}$, where $\widetilde{T^{\perp}} = \operatorname{mod}\Lambda$, and on the other hand $\widetilde{\operatorname{add}T}$ whose modules have finite projective dimension. We had for a general selforthogonal module T considered the subcategory $\mathcal{Y}_T$ of $T^{\perp}$, whose modules Y admit an exact sequence $\cdots \to T_i \xrightarrow{f_i} T_{i-1} \to \cdots \xrightarrow{f_2} T_1 \xrightarrow{f_1} T_0 \to Y \to 0$, with T_i in $\operatorname{add}T$ and $\operatorname{Ker} f_i$ in $T^{\perp}$. For a tilting module T we had $\mathcal{Y}_T = T^{\perp}$, but it is not known whether this equality holds for Wakamatsu tilting modules. So it is not so clear to start with which of the categories $\mathcal{Y}_T$ and $T^{\perp}$ would be the best candidate for the subcategory in a correspondence theorem generalizing $T \mapsto T^{\perp}$ for T a tilting module. Such questions are discussed and investigated in [39], where the following is proved.

Theorem 4.1 (Correspondence theorem). *Let T be a selforthogonal module over an artin algebra Λ.*

(a) *Then $T \mapsto \mathcal{Y}_T$ and $\mathcal{Y} \mapsto \mathcal{Y} \cap {}^\perp\mathcal{Y}$ give a one-one correspondence between Wakamatsu tilting modules and coresolving subcategories with Ext-projective generator, maximal amongst those with the same Ext-projective generator.*

(b) *Then $T \mapsto \mathcal{X}_T$ and $\mathcal{X} \mapsto \mathcal{X} \cap \mathcal{X}^\perp$ give a one-one correspondence between Wakamatsu tilting modules and resolving subcategories with Ext-injective cogenerator, maximal amongst those with the same Ext-injective cogenerator.*

We point out that if $\mathcal{Y}$ is a coresolving subcategory of $\operatorname{mod}\Lambda$, then $\mathcal{Y}$ being covariantly finite and $\check{\mathcal{Y}} = \operatorname{mod}\Lambda$ is equivalent to $\mathcal{Y}$ having an Ext-projective generator and $\check{\mathcal{Y}} = \operatorname{mod}\Lambda$ [6]. This makes it clear that the class of subcategories appearing in the above theorem is an extension of the class corresponding to tilting modules.

An example is given in [39] to show that the condition on maximality in the above theorem can not be dropped.

Also here one can alternatively formulate the relationship between Wakamatsu tilting modules and the associated cotorsion pair $({}^\perp\mathcal{Y}, \mathcal{Y})$ for case (a) and $(\mathcal{X}, \mathcal{X}^\perp)$ for case (b).

4.2 Correspondences for derived categories

There is an analogue of the main correspondence theorem in the setting of bounded derived categories $D^b(\Lambda)$ of an artin algebra Λ [17], which we discuss in this section. We also point out connections with torsion pairs associated with tilting objects (see [13] [3]).

The first problem is to extend the essential notions involved in the statement of the correspondence theorem for $\operatorname{mod}\Lambda$ in a natural way to the setting of derived categories. The notion of tilting object, called tilting complex, in $D^b(\Lambda)$, is central and has been introduced in [42]. We say that a subcategory $\mathcal{Y}$ of $D^b(\Lambda)$ is *coresolving* if $\mathcal{Y}$ is closed under extensions and the shift [1] [17]. This is a natural analogue of a subcategory of $\operatorname{mod}\Lambda$ being closed under extensions and cokernels of monomorphisms.

There does not seem to be a natural analogue of containing all injective modules. But this is not so serious since in the correspondence theorem for mod Λ this condition could have been dropped, because it is a consequence of $\breve{\mathcal{Y}} = \text{mod}\,\Lambda$.

Denote by $K^b(\mathcal{P}(\Lambda))$ the bounded complexes of finitely generated projective modules. Recall that a complex $T^\bullet$ is a *tilting complex* if

(i) $T^\bullet \in K^b(\mathcal{P}(\Lambda))$

(ii) $\text{Hom}(T, T[i]) = 0$ for $i \neq 0$

(iii) $T^\bullet$ generates $K^b(\mathcal{P}(\Lambda))$, that is , $K^b(\mathcal{P}(\Lambda))$ is the smallest triangulated subcategory of $D^b(\Lambda)$ containing $T^\bullet$.

Also define $\breve{\mathcal{Y}}$ to be the full subcategory of $D^b(\Lambda)$ with objects C such that there is an integer n, objects $Y_0, \ldots Y_n$ in $\mathcal{Y}$ and objects $K_1, \ldots K_{n-1}$ in $D^b(\Lambda)$, and a sequence of triangles

$$C \to Y_0 \to K_1 \to$$
$$\vdots$$
$$K_{n-2} \to Y_{n-2} \to K_{n-1} \to$$
$$K_{n-1} \to Y_{n-1} \to Y_n \to$$

Then the following is proved [17]. Here a subcategory $\mathcal{C}$ of $D^b(\Lambda)$ is *selforthogonal* if $\text{Hom}(A, B[i]) = 0$ for $A, B \in \mathcal{C}$ and $i \neq 0$, and $T^{\bullet\perp}$ is defined to be $\{C; \text{Hom}(T^\bullet, C[i] = 0 \text{ for } i > 0\}$.

Theorem 4.2. *There is a one-one correspondence between basic tilting complexes $T^\bullet$ in $D^b(\Lambda)$ and full subcategories $\mathcal{Y}$ of $D^b(\Lambda)$ having the properties*

(i) *$\mathcal{Y}$ is covariantly finite*

(ii) *$\mathcal{Y}$ is coresolving*

(iii) *$\breve{\mathcal{Y}} = D^b(\Lambda)$*

(iv) *$^\perp\mathcal{Y}$ generates $K^b(\mathcal{P}(\Lambda))$*

(v) *$\mathcal{Y} \cap {}^\perp\mathcal{Y}$ is selforthogonal.*

The correspondence is given by $T^\bullet \mapsto T^{\bullet\perp}$ and $\mathcal{Y} \mapsto \mathcal{Y} \cap {}^\perp\mathcal{Y}$.

Note that compared to the corresponding result for mod Λ, two additional conditions have been added. There might have been a different possible choice for the definition of the basic notions, although this one seems very natural. But given this choice, the last two conditions can not be dropped, and it is shown in [17] that neither condition is a consequence of the others.

We also point out that an analogue of the Wakamatsu lemma is proved in [17], and from this it follows that when $T^{\bullet}$ is a tilting object in $D^b(\Lambda)$, then $({}^{\perp}\mathcal{Y}, \mathcal{Y})$ is a complete cotorsion pair, with the obvious definitions analogous to those in mod Λ.

For artin algebras Λ there is a nice correspondence between tilting modules T with $\mathrm{pd}_{\Lambda} T \leqslant 1$ and certain torsion pairs $(\mathcal{T}, \mathcal{F})$, where the associated torsion class coincides with the associated cotorsion class. But there is not such a correspondence for arbitrary tilting modules. We remark that we do however have an induced torsion pair $(\mathcal{X}/\operatorname{add} T, \mathcal{Y}/\operatorname{add} T)$ in the (stable) category mod $\Lambda/\operatorname{add} T$, with the appropriate definitions, where mod $\Lambda/\operatorname{add} T$ is a pretriangulated category, which is triangulated when Λ is selfinjective and $T = \Lambda$ [13] [6]. We now explain that for $D^b(\Lambda)$ the situation is analogous to the case of projective dimension at most one for mod Λ. Before giving the definition of a torsion pair in $D^b(\Lambda)$, it is useful to recall the definitions in an abelian category.

For an abelian category $\mathcal{A}$, the subcategory $\mathcal{T}$ is a torsion class if it is closed under extensions and factors and the inclusion $i\colon \mathcal{T} \to \mathcal{A}$ has a right adjoint (which in particular implies that $\mathcal{T}$ is contravariantly finite in $\mathcal{A}$). The last condition is automatic in mod Λ. There is a natural generalization of torsion class and torsion pair to triangulated categories, in particular to $D^b(\Lambda)$, which is closely related to the better known notion of t-structure (see [13]).

We say that a pair $(\mathcal{T}, \mathcal{F})$ of full subcategories of $D^b(\Lambda)$ is a *torsion pair* if

(i) $\mathrm{Hom}(\mathcal{T}, \mathcal{F}) = 0$

(ii) $\mathcal{T}[1] \subset \mathcal{T}$ or $\mathcal{F}[-1] \subset \mathcal{F}$

(iii) For each $C \in D^b(\Lambda)$ there is a triangle $X \to C \to Y \to$ with $X \in \mathcal{T}$ and $Y \in \mathcal{F}$. Then $\mathcal{T}$ is the *torsion class* and $\mathcal{F}$ the *torsionfree class*.

An equivalent characterization is the following, using that we have a

Krull-Schmidt category: A full subcategory $\mathcal{T}$ of $D^b(\Lambda)$ is a torsion class if and only if $\mathcal{T}$ is coresolving and contravariantly finite in $D^b(\Lambda)$ [13].

For a tilting object $T^\bullet$ in $D^b(\Lambda)$ (that is, a tilting complex), we associate a torsion pair $(\mathcal{T}, \mathcal{F})$ with $T^\bullet$ as follows [13]. We first define the torsionfree class $\mathcal{F}$. We clearly want $\mathrm{Hom}(T^\bullet, \mathcal{F}) = 0$, but with this condition alone the subcategory $\mathcal{F}$ will not have the required properties. The idea is then to take the largest subcategory of $\{C; \mathrm{Hom}(T^\bullet, C) = 0\}$ which is closed under extensions and [-1]. So we let $\mathcal{F} = \{C; \mathrm{Hom}(T^\bullet, C[i]) = 0 \ \forall \ i \leqslant 0\}$, and then define $\mathcal{T} = \{X; \mathrm{Hom}(X, \mathcal{F}) = 0\}$. We then have the following ([13] III.2, III.4).

Theorem 4.3. *The pair $(\mathcal{T}, \mathcal{F})$ as defined above is a torsion pair in* $D^b(\Lambda)$ *and* $\mathcal{T} = \{C; \mathrm{Hom}(T^\bullet, C[i]) = 0 \text{ for } i > 0\} = T^{\bullet\perp}$.

The proof of this result goes via the unbounded derived category $D(\mathrm{Mod}\,\Lambda)$ of all Λ-modules.

Note that it follows from Theorem 4.2 and Theorem 4.3 that the torsion class and the cotorsion class associated with a tilting complex in $D^b(\Lambda)$ coincide. In particular, a subcategory $\mathcal{Y}$ with the properties (i)-(v) from Theorem 4.2 must be contravariantly finite. It would be interesting to see if, like for $\mathrm{mod}\,\Lambda$ [35], the last three conditions can be dropped, so that covariantly finite coresolving implies contravariantly finite also in this context.

Also note that by Theorem 4.2 we get a one-one correspondence between (basic) tilting complexes in $D^b(\Lambda)$ and torsion pairs $(\mathcal{T}, \mathcal{F})$ where $\mathcal{T}$ satisfies (i) (ii) (iv) (v) in Theorem 4.2. The tilting complex $T^\bullet$ associated with $(\mathcal{T}, \mathcal{F})$ is then given by $\mathcal{T} \cap {}^\perp\mathcal{T}$. There is also an alternative way of getting back to $T^\bullet$. For the projective objects of the heart $\mathcal{T} \cap \mathcal{F}[1]$ of the associated t-structure is equivalent to $\mathrm{add}\,\Gamma$ (see [13]).

References

[1] I. Ágoston, V. Dlab, E. Lukacs, Stratified algebras, C.R. Math. Acad. Sci. Soc. R. Can. 20 (1998) No. 1, 22 - 28.

[2] I. Ágoston, D. Happel, E. Lukacs, L. Unger, Standardly stratified algebras and tilting. J. Algebra 226 (2000), no. 1, 144-160.

212   *I. Reiten*

[3] L. Alonso Tarrío, A. Jeremías López, M. Souto Solario, Construction of t-structures and equivalences of derived categories, Trans. Amer. Math. Soc. 355 (2003) no. 6, 2523-2543.

[4] I. Assem, Torsion theories induced by tilting modules, Canada. J. Math. 36 (5) (1984) 899 - 913.

[5] M. Auslander, M. Bridger, Stable module theory, Mem. Amer. Math. Soc. 94 (1969).

[6] M. Auslander, R.O. Buchweitz, Maximal Cohen-Macaulay approximations, Soc. Math. France 38 (1989) 5 - 37.

[7] M. Auslander, I. Reiten, Applications to contravariantly finite subcategories, Adv. Math. 86 (1991) 111 - 152.

[8] M. Auslander, I. Reiten, Homologically finite subcategories of Algebras, London Math. Soc., Lecture Notes in Math. 168, Cambridge Univ. Press, Cambridge (1992) 1 - 42.

[9] M. Auslander, I. Reiten, k-Gorenstein algebras and syzygy modules, J. Pure Applied Algebra 92 (1994) 1 - 27.

[10] M. Auslander, I. Reiten, Syzygy modules for noetherian rings, J. Algebra 183, No. 1 (1996) 167 - 185.

[11] M. Auslander, S.O. Smalø , Preprojective modules over artin algebras, J. Algebra 66 (1980) 61 - 122.

[12] M. Auslander, S.O. Smalø , Almost split sequences in subcategories, J. Algebra 69 (1981) 426 - 454. Addendum, J. Algebra 69 (1981).

[13] A. Beligiannis, I. Reiten, Homological aspects of torsion theories, Mem. Amer. Math. Soc. (to appear).

[14] K. Bongartz, Tilted algebras, in: "Proc. ICRAIII, Puebla (1980)" 26 - 38, Lecture Notes in Mathematics, Vol 903 (1981) 26-38, Springer-Verlag, New York/Berlin.

[15] S. Brenner, M.C.R. Butler, Generalizations of the Bernstein-Gelfand-Ponomarev reflection functors, Lecture Notes in Mathematics 832, Springer-Verlag, Berlin (1980) 103 - 169.

[16] A. Buan, Ø. Solberg, Relative cotilting theory and almost complete cotilting modules, Algebras and Modules II, CMS Conf. Proc. 24 (1998), Amer. Math. Soc., 77 - 92.

[17] A. Buan, Subcategories of the derived category and cotilting comoplexes, Colloq. Math. Vol. 88 No. 1 (2001) 1 - 11.

[18] E. Cline, B. Parshall, L. Scott, Finite dimensional algebras and highest weight categories. J. Reine Angew. Math. 391, 85 - 99 (1988).

[19] E. Cline, B. Parshall, L. Scott, Stratifying endomorphism algebras. Mem. Amer. Math. Soc. 124 (1996).

[20] V. Dlab, Quasi-hereditary algebras revisited, An. St. Univ. Ovidius Constanta 4 (1996) 43 - 54.

[21] V. Dlab, Properly stratified algebras, C.R. Acad. Sci. Paris Ser. I Math 331 (2000) No. 3, 191 - 196.

[22] S. Donkin, Tilting modules for algebraic groups and finite dimensional algebras, this volume.

[23] V. Dlab, C.M. Ringel, The module theoretic approach to quasihereditary algebras, in "Representations of Algebras and Related Topics" (H. Tachikava and S. Brenner, Eds.) 200 - 224. London Mathematical Society Lecture Note Series, Vol. 168 (1992).

[24] P. Eklof, S. Shelah, On the existence of precovers, I.U.J. Math. 47 (2003) 173 - 188.

[25] E. *Enochs*, Injective and flat covers, envelopes and resolvents, J. Algebra 39 (1981) 189 - 209.

[26] V. *Futorny*, S. *König*, V. *Mazorchuk*, Categories of induced modules and standardly stratified algebras, Algebras and Representation Theory 5 (2002), 259 - 276.

[27] A. *Frisk*, V. *Mazorchuk*, Properly stratified algebras, Proc. London Math. Soc., Vol. 92, Part 1, 29-61 (2004).

[28] R. *Göbel*, S. *Shelah*, L. *Wallutis*, On the lattices of cotorsion theories, J. Alg. 238 (2001), 292 - 313.

[29] D. *Happel*, Triangulated categories in the representation theory of finite dimensional algebras, London Math. Soc. Lecture Notes Series, Vol. 119, Cambridge University Press, Cambridge, 1988.

[30] D. *Happel*, C.M. *Ringel*, Tilted algebras, Trans. Amer. Math. Soc. 274 (1982) 399-443.

[31] B. *Huisgen-Zimmermann*, S. O. *Smalø* , Co-versus contravariant finiteness of categories of representations, Advances in Ring Theory, Trends in mathematics, Birkhäuser 1997, ed. S.K. Jain and S. Tariq Rizvi.

[32] O. *Iyama*, Finitenes of representation dimension, Proc. Amer. Math. Soc. (3) (2003), No. 4, 1011 - 1014.

[33] K. *Igusa*, S.O. *Smalø* , G. *Todorov*, Finite projectivity and contravariant finitness, Proc. Amer. Math. Soc. 109 no. 4 (1990), 937 - 941.

[34] M. *Klucznik*, S. *König*, Characteristic tilting modules over quasihereditary algebras, preprint, http://www.mathematik.uni-bielefeld.de/~koenig/tilting.ps.

[35] H. *Krause*, Ø. *Solberg*, Applications of cotorsion pairs, J. London Math. Soc., Vol. 68, No. 3 (2003) 631 - 650.

[36] Y. *Miyashita*, Tilting modules of finite projective dimension, Math. Z. 193, 113 - 146 (1986).

[37] V. *Mazorchuk*, Stratified algebras arising in Lie theory, in: Representations of finite dimensional algebras and related topics in Lie theory and geometry, Fields Institute Communications, Vol. 40 (2004) 245-260, V. Dlab, C.M. Ringel ed.

[38] V. *Mazorchuk*, On finitistic dimension of stratified algebras, Algebra Discrete Math. (2004) no.3, 77-88.

[39] F. *Mantese*, I. *Reiten*, Wakamatsu tilting modules, J. Alg. 278 (2004), no.2, 532-552.

[40] M.I. *Platzeck*, I. *Reiten*, Modules of finite projective dimension over standardly stratified algebras, Comm. in Alg. 29 3 (2001) 973 - 986.

[41] D. *Quillen*, Homotopical Algebra, Springer Lecture Notes in Math. 43 (1967).

[42] J. *Rickard*, Morita theory for derived categories, J. London Math. Soc. (2) 39 no. 3 (1989) 436 - 456.

[43] J. *Rickard*, A. *Schofield*, Cocovers and tilting modules, Math. Proc. Camb. Phil. Soc. 106 (1989) 1-5.

[44] C.M. *Ringel*, The category of modules with good filtrations over a quasi-hereditary algebra has almost split sequences, Math. Z. 208 (1991) 209 - 233.

[45] L. *Salce*, Cotorsion theories for abelian groups, Symposia Math. 23 (1979) 11 - 32.

[46] S.O. *Smalø* , Torsion theories and tilting modules, Bull. London Math. Soc 16 (1984) 518 - 522.

[47] Ø. *Solberg*, Infinite dimensional tilting modules over finite dimensional algebras, this volume.

[48] J. *Trlifaj*, Infinite dimensional tilting modules and cotorsion pairs, this volume.

[49] T. *Wakamatsu*, On modules with trivial self-extensions, J. Algebra 114 (1988) 106 - 114.

[50] C. *Xi*, Characteristic tilting modules and Ringel duals, Science in China A., Vol. 43, No. 11 (2000) 1121 - 1130.

Idun Reiten

Department of Mathematical Sciences

Norwegian University of Science and Technology

N-7491 Trondheim

Norway

idunr@math.ntnu.no

9

Tilting modules for algebraic groups and finite dimensional algebras

Stephen Donkin

Introduction

This is a survey article describing some of the ways in which the theory of rational representations of algebraic groups interacts with the representation theory of finite dimension algebras, with particular emphasis on tilting modules.

In its simplest form the connection between the two areas is the following. Let G be a linear algebraic group over an algebraically closed field k. Then the coordinate algebra $k[G]$ is naturally a commutative Hopf algebra, in particular a coalgebra. A coalgebra is the union of its finite dimensional subcoalgebras and, for a finite dimensional coalgebra C, say, there is a natural equivalence of categories between the category of C-comodules and the category of modules for the dual algebra C^*. A (rational) G-module is, more or less by definition, a $k[G]$-comodule. If V is a finite dimensional right comodule, with structure map $\tau\colon V \to V \otimes k[G]$, then the image of τ lies in $V \otimes C$, for some finite dimensional subcoalgebra C of $k[G]$ and V is naturally a right C-comodule and hence a left C^*-module. Thus the (finite dimensional) representation theory of G is simply the union of the (finite dimensional) representation theories of the finite dimensional algebras C^*, as C ranges over finite dimensional subcoalgebras of $k[G]$.

Of course all of this is so far too general to be of any particular use. However, when G is reductive, $k[G]$ may be written as the ascending union of finite dimensional subcoalgebras C, defined by the weight theory of G, whose dual algebra C^* is quasi-hereditary, in the sense of Cline, Parshall and Scott, [7]. These algebras (or rather their module categories) have

a very tight structure, in particular they have finite global dimension
and have decomposition numbers which obey the "Brauer-Humphreys
reciprocity formulas (analogues of the well known Bernstein-Gelfand-
Gelfand reciprocity formula in the category $\mathcal{O}$ of modules for a semisimple, complex Lie algebra).

There are already several survey articles on tilting modules for finite
dimensional algebras and for algebraic groups, see [22], [11],[29], [2], [45]
(see also the appendix of [23]): [22] deals with the relations with invariant theory, [11] and [23, Appendix] deal only with finite dimensional
algebra aspects, the article [29] is particularly concerned with applications of Schur algebras - as quasi-hereditary algebras - to the representation theory of symmetric groups, whereas [2] (dealing in particular with
sum formulas for tilting modules for algebraic and quantum groups) and
[45] (which includes an extensive account of fundamental theorems on
induced modules for algebraic groups) are written from a thoroughly
algebraic group theory perspective. Our aim here (partly as in order to
offer something different and partly because of our limited expertise) is
to describe, in some detail, the relationship between the representation
theory of algebraic groups and quasi-hereditary algebras which arise in
this context (the so-called generalized Schur algebras). We consider in
some detail the usual Schur algebras, and give some applications of our
point of view.

We make no attempt to be encyclopedic (in fact we are highly selective)
and instead refer the reader to the excellent survey articles mentioned
above for details of the many important applications described therein.

There are a few new proofs and results and a new conjecture scattered
throughout. We give a somewhat new treatment of some recent work of
Hemmer and Nakano, which shows, in particular, that Specht module
multiplicities are well defined if the base field does not have characteristic
2 or 3, with an improvement on degree in certain extension groups for
Hecke algebras. We give a homological treatment of a recent result
of Fayers and Lyle on homomorphisms between symmetric groups; our
version is valid also for Hecke algebras and is generalized to extension
groups in small degree. This can be found in Section 10. We give,
in Section 5, a character formula for certain tilting modules which are
projective as modules for the restricted enveloping algebra. We make a
conjecture describing the support variety of tilting modules for general
linear groups in characteristic 2 (also in Section 5).

Finally we mention a couple of further applications which would have fitted well here but have been omitted to keep the size of the paper down. The first is the application of tilting modules to the calculation of the character of the cohomology of line bundles on the generalized flag variety. We refer the reader to the papers [25] and [27]. The second application is a tilting modules treatment of Adams operations on group representations, with applications to Lie representations of general linear groups, due to R. Bryant, and for this we refer to [4].

1 Quasi-hereditary algebras

We quickly review the theory of quasi-hereditary algebras, due to Cline, Parshall and Scott. A fuller account, from the point of view relevant here, can be found in [23, Appendix]. We fix a field k which, for convenience, we take to be algebraically closed. For a k-algebra S we denote by $\mathrm{Mod}(S)$ the category of left S-modules and by $\mathrm{mod}(S)$ the category of finite dimensional left S-modules. We write $V \in \mathrm{mod}(S)$ to indicate that V is a finite dimensional left S-module.

Let S be a finite dimensional k-algebra. Let $\{L(\lambda) \mid \lambda \in \Lambda\}$ be a complete set of pairwise non-isomorphic irreducible S-modules. We introduce some notation that will also be used in the strongly related contexts of the representation theory of coalgebras and of reductive algebraic groups. Let π be a subset of Λ. We say that $V \in \mathrm{mod}(S)$ *belongs to* π if all composition factors of V belong to $\{L(\lambda) \mid \lambda \in \pi\}$. For an arbitrary $V \in \mathrm{mod}(S)$ the set of submodules belonging to π has a unique maximal element, and we denote it $O_\pi(V)$. Similarly, among all submodules U of V such that V/U belongs to π there is a unique minimal one, and we denote this $O^\pi(V)$. (The notation comes from an analogy with finite group theory where, for a set of primes π, the standard notation for the largest normal subgroup of a finite group G whose order is divisible only by primes in π is $O_\pi(G)$, with $O^\pi(G)$ defined similarly.)

Let $P(\lambda)$ be a minimal projective cover and $I(\lambda)$ a minimal injective envelope of $L(\lambda)$, for $\lambda \in \Lambda$. Assume now that Λ is given a partial order. We put $\pi(\lambda) = \{\mu \in \Lambda \mid \mu < \lambda\}$. Let $M(\lambda)$ denote the maximal submodule of $P(\lambda)$.

The module $\Delta(\lambda)$ is defined by $\Delta(\lambda) = P(\lambda)/O^{\pi(\lambda)}(M(\lambda))$ and the module $\nabla(\lambda)$ is the submodule of $I(\lambda)$ containing $L(\lambda)$ defined by

$\nabla(\lambda)/L(\lambda) = O_{\pi(\lambda)}(I(\lambda)/L(\lambda))$. A module isomorphic to $\Delta(\lambda)$ (resp. $\nabla(\lambda)$) for some $\lambda \in \Lambda$ will be called a standard module (resp. a costandard module).

By construction $L(\lambda)$ occurs precisely once as a composition factor of $\Delta(\lambda)$ and of $\nabla(\lambda)$ and other composition factors have the form $L(\mu)$, with $\mu < \lambda$. Thus we have the following.

Lemma 1.1. *The Grothendieck group* $\mathrm{Grot}(S)$ *of* $\mathrm{mod}(S)$ *has the following* $\mathbb{Z}$-*bases:*

(i) $\{[L(\lambda)] \mid \lambda \in \Lambda\}$;

(ii) $\{[\Delta(\lambda)] \mid \lambda \in \Lambda\}$;

(iii) $\{[\nabla(\lambda)] \mid \lambda \in \Lambda\}$.

Here $[X]$ denotes the class in $\mathrm{Grot}(S)$ of $X \in \mathrm{mod}(S)$.

For $X \in \mathrm{mod}(S)$ and $\lambda \in \Lambda$ we write, as usual, $[X : L(\lambda)]$ for the multiplicity of $L(\lambda)$ as a composition factor of X. In addition we define the "multiplicities" $(X : \Delta(\lambda)), (X : \nabla(\lambda)) \in \mathbb{Z}$ by the formulas

$$[X] = \sum_{\lambda \in \Lambda}(X : \Delta(\lambda))[\Delta(\lambda)] \text{ and } [X] = \sum_{\lambda \in \Lambda}(X : \nabla(\lambda))[\nabla(\lambda)].$$

For $X \in \mathrm{mod}(S)$ a filtration $0 = X_0 < X_1 < \cdots < X_n$ will be called a standard filtration (resp. costandard filtration) if each X_i/X_{i-1} is a standard module (resp. costandard module) for $1 \leqslant i \leqslant n$. We write $X \in \mathcal{F}(\Delta)$ (resp. $X \in \mathcal{F}(\nabla)$) to indicate that X is a finite dimensional module which admits a standard filtration (resp. costandard filtration). Note that, for $X \in \mathcal{F}(\Delta)$ (resp. $X \in \mathcal{F}(\nabla)$), the integer $(X : \Delta(\lambda))$ (resp. $(X : \nabla(\lambda))$) is the multiplicity of $\Delta(\lambda)$ (resp. $\nabla(\lambda)$) as a section in *every* standard (resp. costandard) filtration.

Definition 1.2. The category $\mathrm{mod}(S)$ is said to be a highest weight category (with respect to the labelling of a full set of irreducible modules by the poset Λ) if $P(\lambda) \in \mathcal{F}(\Delta)$, $(P(\lambda) : \Delta(\lambda)) = 1$ and $(P(\lambda) : \Delta(\mu)) = 0$ whenever $\mu \in \Lambda$ is not greater than or equal to λ.

There is also the equivalent notion that a finite dimensional algebra S over k is quasi-hereditary (see e.g. [23, Appendix]. However, from our point of view the important definition is the one just given and we shall say (by a somewhat casual use of terminology) that S is a quasi-hereditary algebra if its module category is a high weight category.

We now assume that S is quasi-hereditary. An easy dimension shifting argument gives the following crucial homological property.

Theorem 1.3. *For* $\lambda, \mu \in \Lambda$, *we have*

$$\operatorname{Ext}^i_S(\Delta(\lambda), \nabla(\mu)) = \begin{cases} k, & \text{if } i = 0 \text{ and } \lambda = \mu; \\ 0, & \text{otherwise.} \end{cases}$$

It is easy to deduce a symmetry property between projectives and injectives; namely that an injective module has a costandard filtration, that $(I(\lambda) : \nabla(\lambda)) = 1$ and that $(I(\lambda) : \nabla(\mu)) = 0$ unless μ is greater than or equal to λ (for $\lambda, \mu \in \Lambda$). Indeed one can use this property to formulate the definition of high weight category as a condition on injective modules instead of projective modules.

Note that Theorem 1.3 implies that for $X \in \mathcal{F}(\Delta)$ (resp. $X \in \mathcal{F}(\nabla)$) the filtration multiplicity $(X : \Delta(\mu))$ (resp. $(X : \nabla(\mu))$) is the dimension of $\operatorname{Hom}_S(X, \nabla(\mu))$ (resp. $\operatorname{Hom}_S(\Delta(\mu), X)$), for $\mu \in \Lambda$. Applying this to $P(\lambda)$ and $I(\lambda)$ we obtain the following, which is known as "Brauer-Humphreys reciprocity".

Theorem 1.4. *For* $\lambda, \mu \in \Lambda$ *we have*

$$(P(\lambda) : \Delta(\mu)) = [\nabla(\mu) : L(\lambda)] \text{ and } (I(\lambda) : \nabla(\mu)) = [\Delta(\mu) : L(\lambda)].$$

We shall write $X \in \mathcal{F}(\Delta) \cap \mathcal{F}(\nabla)$ to indicate that X is a finite dimensional S-module which has both a standard filtration and a costandard filtration. Such a module will be called a *partial tilting module* or just a *tilting module*. There is a very nice parametrization of the indecomposable partial tilting modules, due to Ringel, [49] (see also [48]).

Theorem 1.5.   (i) *For* $\lambda \in \Lambda$ *there exists an indecomposable partial tilting module* $T(\lambda)$ *which has the property that* $(T(\lambda) : \Delta(\lambda)) = 1$ *and* $(T(\lambda) : \Delta(\mu)) = 0$, *for* μ *not less than or equal to* λ *(for* $\mu \in \Lambda$).

(ii) $\{T(\lambda) \mid \lambda \in \Lambda\}$ *is a complete set of pairwise non-isomorphic indecomposable partial tilting modules.*

(iii) *We have $(T(\lambda) : \nabla(\lambda)) = 1$ and $(T(\lambda) : \nabla(\mu)) = 0$ for μ not less than or equal to λ (for $\lambda, \mu \in \Lambda$).*

The existence of tilting modules leads to the construction of a related algebra S' called the *Ringel dual* of S. By a *full tilting module* we mean a tilting module T such that $T(\lambda)$ occurs as a component of T for each $\lambda \in \Lambda$. Let T be a full tilting module. We define $S' = \mathrm{End}_S(T)^{\mathrm{op}}$, the opposite algebra of $\mathrm{End}_S(T)$. Note that S' is determined up to Morita equivalence. We have the natural left exact functor $F = \mathrm{Hom}_S(T, -) \colon \mathrm{mod}(S) \to \mathrm{mod}(S')$. It follows immediately from Theorem 1.3 that F is exact on short exact sequences of modules in $\mathcal{F}(\nabla)$. We define $P'(\lambda) = FT(\lambda)$, $\lambda \in \Lambda$.

Theorem 1.6. (i) $\{P'(\lambda) \mid \lambda \in \Lambda\}$ *is a complete set of pairwise non-isomorphic projective indecomposable left S'-modules.*

(ii) *Defining $L'(\lambda)$ to be the head of $P'(\lambda)$ $(\lambda \in \Lambda)$ and letting Λ' be the poset with underlying set that of Λ and partial order the reverse of that of the poset Λ, we have that $\mathrm{mod}(S')$ is a high weight category, with respect to the labelling $L'(\lambda)$, $\lambda \in \Lambda'$, of a complete set of pairwise non-isomorphic irreducible left S'-modules.*

(iii) *The module $\Delta'(\lambda) = F\nabla(\lambda)$ is the standard S'-module with head $L'(\lambda)$ and moreover, we have,*

$$(T(\lambda) : \nabla(\mu)) = (P'(\lambda) : \Delta'(\mu)) = [\nabla'(\mu) : L'(\lambda)]$$

(where $\nabla'(\mu)$ denotes the costandard S'-module with socle $L'(\mu)$), for $\lambda, \mu \in \Lambda$.

The sense in which S' is dual to S is explained by the following.

Theorem 1.7. *For $\lambda \in \Lambda$ the module $T'(\lambda) = FI(\lambda)$ is the indecomposable partial tilting module for S' labelled by λ. Furthermore, a suitable choice of full tilting modules for S and S' gives rise to an isomorphism $S \to S''$ of quasi-hereditary algebras.*

2 Coalgebras and Comodules

The connection between representations of reductive algebraic groups and quasi-hereditary algebras is made (as we shall see) via coalgebras. On the one hand the linear dual of a coalgebra is naturally an associative algebra and on the other hand a rational module for an algebraic group is (more or less by definition) a comodule for the coordinate algebra.

We refer the reader to J.A. Green's paper [33] for a thorough treatment of the representation theory of coalgebras. Let (A, δ, ϵ) be a k-coalgebra. Thus A is a k-vector space and $\delta\colon A \to A \otimes A$, $\epsilon\colon A \to k$ are linear maps such that

$$(\delta \otimes 1) \circ \delta = (1 \otimes \delta) \circ \delta \text{ and } (\epsilon \otimes 1) \circ \delta = (1 \otimes \epsilon) \circ \delta = 1$$

(where 1 here denotes the identity map on V and the identity map on A).

The linear dual A^* of A is an associative algebra with product $\alpha\beta = (\alpha \otimes \beta)\delta$ and identity ϵ. By a right comodule we mean a k-vector space V together with a linear map (called the structure map) $\tau\colon V \to V \otimes A$ such that $(1 \otimes \delta) \circ \tau = (\tau \otimes 1) \circ \tau\colon V \to V \otimes A \otimes A$ (where 1 here denotes the identity map on V). Left comodules are defined similarly. The right A-comodules (resp. finite dimensional right comodules) naturally form the objects of a category which we denote $\mathrm{Comod}(A)$ (resp. $\mathrm{comod}(A)$).

A right A-comodule V, with structure map $\tau\colon V \to V \otimes A$, is naturally a left A^*-module with action $\alpha v = (1 \otimes \alpha)\tau(v)$, $\alpha \in A^*$, $v \in V$. Moreover, if A is finite dimensional then we obtain in this way an equivalence of categories from $\mathrm{Comod}(A)$ to $\mathrm{Mod}(A^*)$. Given a right A-comodule V we shall in fact regard V as a left A^*-module (as above) and if A is finite dimensional and W is a left A^*-module we shall also regard W as a right A-comodule without further comment.

For $V \in \mathrm{Comod}(A)$ there is an injective comodule I, determined up to isomorphism, such that V embeds in I in such a way that the embedding induces an isomorphism between the socle of V and the socle of I. The injective comodule I is called the injective envelope of V. Suppose that $\{L(\lambda) \mid \lambda \in \Lambda\}$ is a complete set of pairwise non-isomorphic irreducible right A-comodules. For each $\lambda \in \Lambda$ we choose an injective indecomposable comodule $I(\lambda)$ containing $L(\lambda)$. Then $\{I(\lambda) \mid \lambda \in \Lambda\}$ is a complete set of pairwise non-isomorphic injective indecomposable A-comodules and every injective right A-comodule is a direct sum of copies of the comodules $I(\lambda)$, $\lambda \in \Lambda$ (see [33]).

Let π be a subset of Λ. We say that a left comodule V belongs to π if all composition factors of V come from $\{L(\lambda) \mid \lambda \in \pi\}$. Let V be an arbitrary left A-comodule. The set of subcomodules of V which belong to π has a unique maximal element and we denote this $O_\pi(V)$.

Regarding A itself as a right A-comodule via the structure map $\delta: A \to A \otimes A$ we obtain a subspace $A(\pi) = O_\pi(A)$ and it is easy to check that in fact $A(\pi)$ is a subcoalgebra of A. Moreover, if V is a right A-comodule then the structure map $V \to V \otimes A$ has image in $V \otimes A(\pi)$ if and only if V belongs to π. A right A-comodule V belonging to π is thus an $A(\pi)$-comodule and we identify, in this way, the category of right A-comodules belonging to π with the category of all right $A(\pi)$-comodules. Thus, in particular, $\{L(\lambda) \,|\, \lambda \in \pi\}$ is a complete set of pairwise non-isomorphic right $A(\pi)$-comodules.

If A is finite dimensional then we shall call it quasi-hereditary if its dual algebra A^* is quasi-hereditary, in other words A is quasi-hereditary if the category of finite dimensional comodules $\mathrm{comod}(A)$ is a high weight category.

This definition may be given intrinsically in the comodule category as follows. We assume the notation above and that Λ is a poset. For $\lambda \in \Lambda$ define $\pi(\lambda) = \{\mu \in \Lambda \,|\, \mu < \lambda\}$. We define the submodule $\nabla(\lambda)$ of $I(\lambda)$ containing $L(\lambda)$ by the equation

$$\nabla(\lambda)/L(\lambda) = O_{\pi(\lambda)}(I(\lambda)/L(\lambda)).$$

We call the comodules $\nabla(\lambda)$, $\lambda \in \Lambda$, the costandard comodules. A filtration $0 = X_0 < X_1 < \cdots < X_n = X$ of a left A-comodule X is called costandard if each section X_i/X_{i-1} is a costandard module $(1 \leqslant i \leqslant n)$. For an A-comodule X we write $X \in \mathcal{F}(\nabla)$ to indicate that X admits a costandard filtration. As in the finite dimensional algebra case, the multiplicity of $\nabla(\lambda)$ as a section in a costandard filtration of X is independent of the choice of filtration and is denoted $(X : \nabla(\lambda))$.

Definition 2.1. The finite dimensional coalgebra A is quasi-hereditary (with respect to the labelling of the irreducible modules by the poset Λ) if $I(\lambda) \in \mathcal{F}(\nabla)$ if $(I(\lambda) : \nabla(\lambda)) = 1$ and if $(I(\lambda) : \nabla(\mu)) = 0$ unless μ is greater than or equal to λ.

However, we do not just want to deal with finite dimensional coalgebras (if G is a linear algebraic group over k then the coordinate algebra $k[G]$ is in general an infinite dimensional coalgebra which plays a central role in the sequel). So it is convenient to have a definition of quasi-hereditary valid in the more general context. In fact, for many purposes, it would be enough to work with coalgebras which locally are finite dimensional and quasi-hereditary, i.e. coalgebras with the property that each finite dimensional subspace is contained in a finite dimensional quasi-hereditary coalgebra. The coalgebras arising as coordinate algebras of reductive algebraic groups have a stronger property, however, and we shall take this as our definition.

By a *saturated* subset π of a poset Λ we mean one which is downwardly closed, i.e has the property that whenever $\lambda \in \pi$ and $\mu \leqslant \lambda$ then $\mu \in \pi$.

Definition 2.2. Let Λ be a poset such that for each $\lambda \in \Lambda$ the set $\{\mu \in \Lambda \,|\, \mu \leqslant \lambda\}$ is finite. Suppose that A is a coalgebra over k and that $\{L(\lambda) \,|\, \lambda \in \Lambda\}$ is a complete set of pairwise non-isomorphic irreducible right A-comodules. We say that A is a quasi-hereditary coalgebra (with respect to the labelling of irreducible comodules by the poset Λ) if, for every finite saturated subset π of Λ, the coalgebra $A(\pi) = O_\pi(A)$ is finite dimensional and quasi-hereditary (with respect to the labelling of the irreducible right $A(\pi)$-comodules by the poset π).

Suppose that A is a quasi-hereditary k-coalgebra, as above. We shall define, a standard comodule $\Delta(\lambda)$, a costandard comodule $\nabla(\lambda)$, and a (partial) tilting module $T(\lambda)$, by a local finiteness argument. So let $\lambda \in \Lambda$ and let $\pi = \{\mu \in \Lambda \,|\, \mu \leqslant \lambda\}$. We say that a finite dimensional right A-comodule V is "λ-admissible" if it is finite dimensional, belongs to π, has simple head $L(\lambda)$ and $L(\lambda)$ occurs exactly once as a composition factor of V. Any such V is a homomorphic image of the projective cover of $L(\lambda)$ as a left $S(\pi)$-module, where $S(\pi) = A(\pi)^*$ is the dual algebra. Hence there is a bound on the dimension of λ-admissible comodules and it follows that there is a unique (up to isomorphism) universal one, i.e. a λ-admissible V with the property that every λ-admissible comodule is a homomorphic image of V. We denote such a comodule by $\Delta(\lambda)$ (and call it the standard comodule labelled by λ). We say that a right comodule V is "λ-coadmissible" if it is finite dimensional right comodule which has socle $L(\lambda)$, belongs to π and $L(\lambda)$ occurs exactly once as a composition factor. We deduce in a similar way that all λ-coadmissible comodules embed in a common one, which we denote $\nabla(\lambda)$. (Note that in fact $\nabla(\lambda)$ is isomorphic to the subcomodule $\nabla'(\lambda)$ of $I(\lambda)$ containing $L(\lambda)$ defined by $\nabla'(\lambda)/L(\lambda) = O_{\pi'}(I(\lambda)L(\lambda))$, where $\pi' = \{\mu \in \Lambda \,|\, \mu < \lambda\}$.) As usual we shall write $X \in \mathcal{F}(\Delta)$ (resp. $X \in \mathcal{F}(\nabla)$) to indicate that $X \in \mathrm{comod}(A)$ admits a standard (resp. costandard) filtration, and for such X and $\lambda \in \Lambda$ write the corresponding filtration multiplicity as

$(X : \Delta(\lambda))$ (resp. $(X : \nabla(\lambda))$). By a (right) tilting comodule for A we mean a finite dimensional right A comodule which admits both a standard filtration and a costandard filtration.

From the construction we see that if σ is any finite saturated subset of Λ containing π then $\Delta(\lambda)$ (resp. $\nabla(\lambda)$) is the standard (resp. costandard) $A(\sigma)$-comodule corresponding to λ. Let $T(\lambda)$ be the indecomposable tilting comodule for $A(\pi)$ labelled by λ. Then $T(\lambda)$ is an indecomposable right comodule for $A(\sigma)$ which admits both a standard filtration and a costandard filtration and in which $\nabla(\lambda)$ occurs as a composition factor with multiplicity 1. Hence $T(\lambda)$ is also the tilting comodule, labelled by λ, for $A(\sigma)$.

Thus we have, for each $\lambda \in \Lambda$, an indecomposable finite dimensional right A-comodule $T(\lambda) \in \mathcal{F}(\Delta) \cap \mathcal{F}(\nabla)$ with $(X : \Delta(\lambda)) = 1$ and $(X : \Delta(\mu)) = 0$ for μ not less than or equal to λ. If $T'(\lambda)$ is also such a right A-comodule and σ is a finite saturated subset of Λ such that both $T(\lambda)$ and $T'(\lambda)$ belong to σ then $T(\lambda)$ and $T'(\lambda)$ are indecomposable tilting comodules, labelled by λ, for $A(\sigma)$ and hence are isomorphic as $A(\sigma)$-comodules, and hence as A-comodules. Similarly one gets that a finite dimensional right A-comodule in $\mathcal{F}(\Delta) \cap \mathcal{F}(\nabla)$ is a direct sum of copies of $T(\lambda)$, $\lambda \in \Lambda$. To summarize, we have the following extension of Theorem 1.5.

Theorem 2.3.   (i) *For $\lambda \in \Lambda$ there exists an indecomposable finite dimensional right A-comodule $T(\lambda)$ which is a tilting module and has the property that $(T(\lambda) : \Delta(\lambda)) = 1$ and $(T(\lambda) : \Delta(\mu)) = 0$, for μ not less than or equal to λ (for $\mu \in \Lambda$).*

   (ii) *$\{T(\lambda) \mid \lambda \in \Lambda\}$ is a complete set of pairwise non-isomorphic indecomposable tilting modules.*

   (iii) *We have $(T(\lambda) : \nabla(\lambda)) = 1$ and $(T(\lambda) : \nabla(\mu)) = 0$ for μ not less than or equal to λ (for $\lambda, \mu \in \Lambda$).*

We now give a criterion for a k-coalgebra A to be quasi-hereditary.

Theorem 2.4. *Let $\{L(\lambda) \mid \lambda \in \Lambda\}$ be a complete set of pairwise non-isomorphic irreducible right A-comodules, labelled by the poset Λ. Suppose that:*

   (i) *for each $\lambda \in \Lambda$ there exists a universal (finite dimensional)*

λ-admissible right comodule $\Delta(\lambda)$ and universal (finite dimensional) λ-coadmissible right comodule $\nabla(\lambda)$; and

(ii) $\mathrm{Ext}^i_A(\Delta(\lambda), \nabla(\mu)) = 0$ for all $\lambda, \mu \in \Lambda$ and $i = 1, 2$.

Then A is quasi-hereditary (with respect to the labelling of irreducible comodules by the poset Λ).

Here $\mathrm{Ext}^i_A(X, Y)$, for right comodules X, Y, is computed via an injective comodule resolution of Y. The argument of proof is a straightforward adaptation to the coalgebra context of arguments which appear in the algebraic group context in [14].

In view of the following result, the homological algebra of finite dimensional comodules for a quasi-hereditary coalgebra may be understood in terms of that of its finite dimensional quasi-hereditary subcoalgebras. For the proof we refer to the argument of proof of the corresponding result for modules for algebraic groups, see [16, (2.1f) Theorem].

Theorem 2.5. *Let A be a quasi-hereditary coalgebra as above. Let π be a saturated subset of of Λ. Then for right $A(\pi)$-comodules (equivalently right A-comodules belonging to π) X, Y, we have $\mathrm{Ext}^i_{A(\pi)}(X, Y) \cong \mathrm{Ext}^i_A(X, Y)$, for all $i \geqslant 0$.*

3 Linear Algebraic Groups

We now see how all of this theory comes to life in the context of reductive groups. By an algebraic group we mean a linear algebraic group over the algebraically closed field k. Let G be an algebraic group, with coordinate algebra $k[G]$. The canonical example is $G = \mathrm{GL}(n)$, the group of invertible $n \times n$ matrices with entries in k. In this case we have $k[G] = k[c_{11}, \ldots, c_{nn}, 1/d]$, where $c_{ij}(g)$ is the (i, j)-entry, and $d(g)$ is the determinant, of $g \in G$.

A left kG-module V is called locally finite (dimensional) if every cyclic submodule (and hence every finitely generated submodule) of V is finite dimensional. A finite dimensional left kG-module V is called *rational* if for some (and hence every) basis $v_1, \ldots, v_n$ of V the corresponding coefficient functions f_{ij}, defined by the equations

$$gv_i = \sum_{j=1}^{n} f_{ji}(g)v_j$$

(for $g \in G$, $1 \leqslant i \leqslant n$) belong to $k[G]$. A rational (left) kG-module (of arbitrary dimension) is a kG-module V such that:

(i) V is locally finite; and

(ii) every finite dimensional submodule of V is rational.

Given a rational kG-module V with basis $\{v_i \mid i \in I\}$ and "coefficient functions" $f_{ij} \in k[G]$ defined by

$$gv_i = \sum_{j \in I} f_{ji}(g)v_j$$

($g \in G$, $i \in I$) we have on V the structure of a right $k[G]$-comodule via the structure map $\tau: V \to V \otimes k[G]$, given by $\tau(v_i) = \sum_{j \in I} v_j \otimes f_{ji}$ (for $i \in I$). The definition of τ is independent of the choice of basis and one obtains in this way an equivalence of categories between rational kG-module and right $k[G]$-comodules. We identify a rational (left) G-module with a (right) $k[G]$-comodule via this equivalence.

The coefficient space $\mathrm{cf}(V)$ is, by definition, the subspace of $k[G]$ spanned by all coefficient functions f_{ij}. It is independent of the choice of basis. Note also that the image of the structure map τ lies in $V \otimes \mathrm{cf}(V)$.

From now on a G-module will mean a rational kG-module. We note that the category of G-modules is closed under the operations of taking submodules, quotient modules and tensor products. Moreover if V is a finite dimensional rational G-module then so is the dual module V^*. We write $\mathrm{Mod}(G)$ (resp. $\mathrm{mod}(G)$) for the category of G-modules (resp. finite dimensional G-modules).

Examples of rational modules for linear algebraic groups

We give some example of rational modules for an arbitrary linear algebraic group G.

(i) If V is a k-vector space viewed as a kG-module with trivial action ($gv = v$ for all $g \in G$, $v \in V$) then V is a rational G-module. In particular we have the one dimensional trivial kG-module, denoted k.

The condition that the trivial module k is (up to isomorphism) the only simple G-module is equivalent to the condition that G is a unipotent group, i.e. the image of G in some (and hence every) faithful matrix representation (affording a rational module) should be conjugate to a subgroup of the group of upper unitriangular matrices.

(ii) Suppose G has an algebraic right action on an affine algebraic variety V. Thus the action $\mu\colon V \times G \to V$ is required to be a morphism of affine varieties. The comorphism $\mu^*\colon k[V] \to k[V \times G] = k[V] \otimes k[G]$ makes the coordinate algebra $k[V]$ into a rational G-module (on which G acts as k-algebra automorphisms).

In particular we can take $V = G$ with the action by right multiplication: $\mu(x,g) = xg$, for $x, g \in G$. The coordinate algebra $k[G]$, considered as a rational G-module in this way, is called the left regular G-module.

Alternatively, we get a G-module structure on $k[G]$ via the conjugation action $(\kappa(x,g) = g^{-1}xg)$ of G on itself. Explicitly, the module action is given by $(g \cdot f)(x) = f(g^{-1}xg)$, for $g, x \in G$, $f \in k[G]$. Note that the augmentation ideal $\mathcal{M} = \{f \in k[G] \mid f(1) = 0\}$ is a G-submodule. Hence $\mathcal{M}^2$ is also a submodule and we have the finite dimensional rational module $\mathcal{M}/\mathcal{M}^2$. The action of G on $k[G]$ gives rise to a kG-module structure on the dual $k[G]^*$, in the usual way, i.e. in such a way that $(g\gamma)(f) = \gamma(g^{-1} \cdot f)$, for $g \in G$, $\gamma \in k[G]^*$ and $f \in k[G]$. This action is not, in general, rational. However, the Lie algebra

$$\mathrm{Lie}(G) = \{\gamma \in k[G]^* \mid \gamma(ab) = \gamma(a)b(1) = a(1)\gamma(b), \text{ for all } a, b \in k[G]\}$$

is a G-submodule of $k[G]^*$ and naturally dual to $\mathcal{M}/\mathcal{M}^2$. Thus the Lie algebra $\mathrm{Lie}(G)$ is naturally a rational G-module.

Examples of rational modules for general linear groups

We now take $G = \mathrm{GL}(n)$ and provide some additional examples of rational modules in this case. Let E be the natural module of column vectors of length n and let $e_1, \ldots, e_n$ be the standard basis (so e_i has 1 in the ith position and 0s elsewhere). The (i,j)-matrix coefficient function with respect to this basis is the function c_{ij} which picks out the (i,j)-entry of an invertible matrix. Thus E is a rational module. Hence the rth tensor power $E^{\otimes r} = E \otimes \cdots \otimes E$ is also rational as are the rth symmetric power $S^r E$ and the rth exterior power $\bigwedge^r E$ (quotient modules of $E^{\otimes r}$). Thus we have, for any finite sequence $\alpha = (\alpha_1, \alpha_2, \ldots, \alpha_m)$ a rational module $S^\alpha E = S^{\alpha_1} E \otimes \cdots \otimes S^{\alpha_m} E$ and a rational module $\bigwedge^\alpha E = \bigwedge^{\alpha_1} E \otimes \cdots \otimes \bigwedge^{\alpha_m} E$.

4 Reductive Groups

The rational representation theory of reductive groups is highly developed and we need to recall now some of its features. A full account is given in the book by Jantzen, [39]. We begin with the simplest kind of reductive algebraic group. An algebraic group T is called a torus if it is isomorphic to a product of copies of the multiplicative group of the field k, equivalently if T is isomorphic to the group of invertible diagonal $n \times n$-matrices, for some n. The character group $X(G)$, of an algebraic group G, is the (abelian) group of all algebraic group homomorphisms from G to $k^\times$, the multiplicative group of the field. The group operation on $X(G)$ is given by $(\lambda + \mu)(g) = \lambda(g)\mu(g)$, for $g \in G$, $\lambda, \mu \in X(G)$.

To start with, for definiteness, we take T to be the group of invertible diagonal $n \times n$-matrices. Let $\epsilon_i : T \to k$ be the group homomorphism which takes $t \in T$ to its (i, i)-entry, for $1 \leqslant i \leqslant n$. Then $X(T)$ is the free abelian group on $\epsilon_1, \ldots, \epsilon_n$, and $X(T)$ may be identified with $\mathbb{Z}^n$ in such a way that $\lambda = (\lambda_1, \ldots, \lambda_n) \in \mathbb{Z}^n$ is the homomorphism taking $t \in T$ to $t_1^{\lambda_1} t_2^{\lambda_2} \ldots t_n^{\lambda_n}$, where t_i is the (i, i)-entry of t, for $1 \leqslant i \leqslant n$.

Thus, if T is any n-dimensional torus, then the character group $X(T)$ is free abelian of rank n. For $V \in \mathrm{Mod}(T)$ and $\lambda \in X(T)$ we have the λ *weight space* $V^\lambda = \{v \in V \mid tv = \lambda(t)v \text{ for all } t \in T\}$. We have $V = \bigoplus_{\lambda \in X(T)} V^\lambda$. In particular all T-modules are semisimple. We say that $\lambda \in X(T)$ is a *weight of* V if $V^\lambda \neq 0$.

To keep track of weight space multiplicities, one assigns a character to a finite dimensional T-module. We form the integral group ring $\mathbb{Z}X(T)$. Thus $X(T)$ has a $\mathbb{Z}$-basis of "formal exponentials" e^λ, $\lambda \in X(T)$, which multiply according to the rule $e^\lambda e^\mu = e^{\lambda+\mu}$. The formal character of a finite dimensional module V is defined by

$$\mathrm{ch}\, V = \sum_{\lambda \in X(T)} \dim V^\lambda e^\lambda.$$

We need to recall now how a root system may be attached to a reductive algebraic group. We fix a maximal torus T in G and a Borel subgroup (

i.e. a maximal closed solvable connected subgroup of G) containing T. We have the normalizer N of T in G and the Weyl group $W = N/T$. The action of N on T by conjugation gives rise to a $\mathbb{Z}W$-module structure on $X(T)$ and hence a $\mathbb{R}W$-module structure on the real vector space $\mathbb{R} \otimes_\mathbb{Z} X(T)$. Let $(\, , \,)$ be a real positive definite, symmetric, W-invariant bilinear form on $\mathbb{R} \otimes_\mathbb{Z} X(T)$. Let Φ denote the set of roots, i.e. the non-zero weights for the action of T on $\mathrm{Lie}(G)$. Then $\mathrm{Lie}(G)$ decomposes as $\mathrm{Lie}(G) = \mathrm{Lie}(T) \bigoplus (\bigoplus_{\alpha \in \Phi} \mathrm{Lie}(G)^\alpha)$, and $\mathrm{Lie}(G)^\alpha$ is one dimensional, for $\alpha \in \Phi$. We identify $X(T)$ with a subgroup of $\mathbb{R} \otimes_\mathbb{Z} X(T)$ and let $\mathbb{E}$ be the $\mathbb{R}$-span of Φ in $\mathbb{R} \otimes_\mathbb{Z} X(T)$. Then the induced bilinear form makes $(\mathbb{E}, \Phi)$ into a root system with Weyl group W. We choose the system of positive roots which makes B the *negative* Borel subgroup. In other words we choose the system of positive roots $\Phi^+ = \{-\alpha \,|\, \alpha \in \Phi^-\}$, where Φ^- is the set of non-zero weights for the action of T on $\mathrm{Lie}(B)$.

We have a natural partial order on $X(T)$. We declare $\lambda \leq \mu$ if the difference $\mu - \lambda$ has the form $\sum_{\alpha \in \Phi^+} n_\alpha \alpha$, for non-negative integers n_α. For $\alpha \in \Phi$ we put $\check{\alpha} = 2\alpha/(\alpha, \alpha)$. Then $X^+(T) = \{\lambda \in X(T) \,|\, (\lambda, \check{\alpha}) \geq 0 \text{ for all } \alpha \in \Phi^+\}$ is the set of dominant weights.

We consider the representation theory of B. Now B is the semidirect product of T with U, the unipotent radical of B, i.e. the largest normal unipotent subgroup of B. Each $\lambda \in X(T)$ can be uniquely extended to a multiplicative character of B which, by abuse of notation, we also write λ. Explicitly, we have $\lambda(tu) = \lambda(t)$, for $t \in T$, $u \in U$. We identify $X(B)$ with $X(T)$ in this way.

The irreducible G-modules are classified as follows.

Theorem 4.1. (i) *For $\lambda \in X^+(T)$ there exists an irreducible G-module $L(\lambda)$ such that $\dim L(\lambda)^\lambda = 1$ and all weights of $L(\lambda)$ are less than or equal to λ.*

 (ii) *The modules $L(\lambda)$, $\lambda \in X^+(T)$, form a complete set of pairwise non-isomorphic irreducible G-modules.*

Example 4.2. We take $G = \mathrm{SL}_2(k)$. We take T to be the group of diagonal matrices in G and take B to the group of lower triangular matrices in G. Then $N = T \cup sT$, where $s = \begin{pmatrix} 0 & 1 \\ -1 & 0 \end{pmatrix}$. Then $X(T) = \mathbb{Z}\rho$ and $X^+(T) = \mathbb{N}_0\rho$, where $\rho \begin{pmatrix} a & 0 \\ 0 & a^{-1} \end{pmatrix} = a$, $a \in k^\times$ (and where

$\mathbb{N}_0$ denotes the set of non-negative integers). Let E denote the natural G-module of column vectors. If k has characteristic 0 then, as one may readily check, the rth symmetric power $S^r E$ is irreducible and has highest weight $r\rho$. Hence we have $L(r\rho) \cong S^r E$, for $r \geqslant 0$. This is not the case for all r if k has characteristic p (one may easily check that $S^p E$, for example, is reducible). However we do have $L(r\rho) \cong S^r E$ for $0 \leqslant r \leqslant p - 1$.

In general, the irreducible modules may be constructed via certain naturally occurring induced modules. We shall not develop the theory of induction for algebraic groups in general, but only give what we need for the sequel. Let V be a B-module. By definition the induced module $\operatorname{Ind}_B^G V$ is the space of maps $f : G \to V$ such that:

(i) f is regular, i.e. $f : G \to V$ has image in a finite dimensional subspace V_0, say, of V and the restriction $f : G \to V_0$ is a morphism of affine varieties (where V_0 is viewed as affine n-space, $n = \dim V_0$); and

(ii) f is B-equivariant, i.e. $f(bx) = bf(x)$, for all $b \in B$, $x \in G$.

The action of G on $\operatorname{Ind}_B^G V$ is given by $(gf)(x) = f(xg)$, $f \in \operatorname{Ind}_B^G V$, $g, x \in G$. To check that this module is rational one may first restrict to the case in which V is finite dimensional, with basis $v_1, \ldots, v_n$, say. For $f \in \operatorname{Ind}_B^G V$ we have $f_1, \ldots, f_n \in k[G]$, defined by $f(g) = f_1(g)v_1 + \cdots + f_n(g)v_n$. Then $\operatorname{Ind}_B^G V$ embeds (as a G-module) in $k[G] \oplus \cdots \oplus k[G]$ (the direct sum of n copies of the left regular module) via the map sending $f \in \operatorname{Ind}_B^G V$ to $(f_1, \ldots, f_n)$.

One may regard $\operatorname{Ind}_B^G V$ as a left exact functor, from $\operatorname{Mod}(B)$ to $\operatorname{Mod}(G)$. The natural evaluation map $\epsilon : \operatorname{Ind}_B^G V \to V$, $\epsilon(f) = f(1)$, gives rise to Frobenius Reciprocity, i.e. the isomorphism $\operatorname{Hom}_G(M, \operatorname{Ind}_B^G V) \to \operatorname{Hom}_B(M, V)$, $\theta \mapsto \epsilon \circ \theta$, for $M \in \operatorname{Mod}(G)$.

In addition to Frobenius reciprocity one also has the tensor identity. There is a natural isomorphism $\operatorname{Ind}_B^G(M \otimes V) \to M \otimes \operatorname{Ind}_B^G V$, for M a G-module and V a B-module. Furthermore this gives rise to an isomorphism in each degree: $R^i \operatorname{Ind}_B^G(M \otimes V) \to M \otimes R^i \operatorname{Ind}_B^G V$, where $R^i \operatorname{Ind}_B^G$ denotes the ith derived functor of the induction functor.

The right derived functors $R^i \operatorname{Ind}_B^G$ take finite dimensional B-modules to finite dimensional G-modules. This may be seen by associating to a finite dimensional B-module V a vector bundle $\mathcal{L}(V)$ on the projective variety G/B and identifying $R^i\mathrm{Ind}_B^G V$ with the ith (coherent sheaf) cohomology space $H^i(G/B, \mathcal{L}(V))$. In the case $i = 0$ there is a direct elementary argument to see that $R^0\mathrm{Ind}_B^G V = \mathrm{Ind}_B^G V$ is finite dimensional (see [15, Section 1.5]). The case in which $V = k_\lambda$, a one dimensional B-module on which T acts according to λ (and $\mathcal{L}(V)$ is a line bundle) is especially important and one has the following results.

Theorem 4.3.   (i) $\mathrm{Ind}_B^G k_\lambda$ *is non-zero if and only if* $\lambda \in X^+(T)$;

(ii) $R^i\mathrm{Ind}_B^G k_\lambda = 0$ *for all* $i \geqslant 0$ *if* $(\lambda, \check{\alpha}) = -1$, *for some simple root* α;

(iii) $R^i\mathrm{Ind}_B^G k_\lambda = 0$ *if* $\lambda \in X^+(T)$ *and* $i > 0$.

The third property is known as Kempf's vanishing theorem and it is this which is behind the connection between reductive groups and quasi-hereditary algebras.

We put $\nabla(\lambda) = \mathrm{Ind}_B^G k_\lambda$, for $\lambda \in X^+(T)$. We now describe the character of $\nabla(\lambda)$. It is convenient to enlarge $X(T)$ slightly and work in the group ring $\mathbb{Z}X'(T)$, where $X'(T)$ is the $\mathbb{Q}$ vector subspace of $\mathbb{R}\otimes_\mathbb{Z} X(T)$ spanned by $X(T)$. Thus we have $X(T) \leqslant X'(T) \leqslant \mathbb{R}\otimes_\mathbb{Z} X(T)$ (and $X'(T) \cong \mathbb{Q}\otimes_\mathbb{Z} X(T)$). Then $\mathbb{Z}X'(T)$ has $\mathbb{Z}$ basis e^λ, $\lambda \in X'(T)$, and we have $e^\lambda e^\mu = e^{\lambda+\mu}$, $\lambda, \mu \in X'(T)$.

For $\mu \in X'(T)$ we put $A(\mu) = \sum_{w\in W} \mathrm{sgn}(w)e^{w\mu}$, where $\mathrm{sgn}(w)$ denotes the sign of a Weyl group element w. Let $\rho \in X'(T)$ be half the sum of the positive roots. Then for $\lambda \in X(T)$ the alternating sum $A(\lambda + \rho)$ is divisible by $A(\rho)$ and the Weyl character $\chi(\lambda)$ is defined to be the quotient $A(\lambda + \rho)/A(\rho)$. The "dot action" of the Weyl group on $X'(T)$ is given by $w \cdot \nu = w(\nu + \rho) - \rho$. The Weyl character satisfies $\chi(w \cdot \nu) = \mathrm{sgn}(w)\chi(\nu)$, and moreover we have $\chi(\nu) = 0$ if $(\nu, \check{\alpha}) = -1$ for any simple root α. It follows that, for any $\nu \in X(T)$, the Weyl character is either 0 or $\pm\chi(\lambda)$ for some dominant weight λ.

It is not difficult to show that the Euler character $\sum_{i\geqslant 0}(-1)^i\mathrm{ch}\, R^i\mathrm{Ind}_B^G k_\lambda$ is equal to the Weyl character $\chi(\lambda)$, for $\lambda \in X(T)$, (see e.g. [15, Section 2.2]) and so from (iii) above we get the following explicit description of the character of an induced module.

Theorem 4.4. $\mathrm{ch}\,\nabla(\lambda) = \chi(\lambda)$, *for* $\lambda \in X^+(T)$.

It follows from Frobenius Reciprocity that there is an embedding of $L(\lambda)$

into $\nabla(\lambda)$. Moreover, its follows from the density of the "big cell" U^+B in G (where U^+ is the group generated by all roots subgroups U_α, α a positive root) that $\nabla(\lambda)$ has a simple socle as a B-module. Hence $\nabla(\lambda)$ has G-module socle $L(\lambda)$.

We write w_0 for the longest element of the Weyl group W. For $\lambda \in X^+(T)$ we put $\lambda^* = -w_0\lambda$ and $\Delta(\lambda) = \nabla(\lambda^*)^*$. The following consequences of Kempf's vanishing theorem were obtained by Cline, Parshall, Scott and van der Kallen [6].

Theorem 4.5.   (i) $\mathrm{Ext}^i_G(\Delta(\lambda), \nabla(\mu)) = \begin{cases} k, & \textit{if } i = 0 \textit{ and } \lambda = \mu; \\ 0, & \textit{otherwise.} \end{cases}$

(ii) *If* $\mathrm{Ext}^1_G(\nabla(\lambda), \nabla(\mu)) \neq 0$ *then* $\lambda > \mu$ *(for* $\lambda, \mu \in X^+(T)$*).*

For $\lambda \in X^+(T)$ let $I(\lambda)$ be the injective envelope $L(\lambda)$ (as a rational G-module, or $k[G]$-comodule). Then $\nabla(\lambda)$ embeds in $I(\lambda)$. The main result of [14] restated in the language of Section 2 is now the following.

Theorem 4.6. *We regard* $X^+(T)$ *as a poset with order induced from the natural order on* $X(T)$*. Then* $k[G]$ *is a quasi-hereditary coalgebra with respect to the labelling* $L(\lambda)$, $\lambda \in X^+(T)$*, of irreducible modules and moreover, for* $\lambda \in X^+(T)$*, the corresponding costandard* $k[G]$ *comodule (i.e. G-module) is* $\nabla(\lambda) = \mathrm{Ind}_B^G k_\lambda$ *and the standard module is* $\Delta(\lambda) = \nabla(\lambda^*)^*$*.*

Remark 4.7. In fact $k[B]$ is a quasi-hereditary coalgebra. This follows from van der Kallen's paper [41]. This is a deep and sophisticated work which generalizes the above (restriction gives a full embedding of the category of G-modules into the category of B-modules) and whose consequences, to the best of my knowledge, have so far not been investigated or exploited.

Definition 4.8. Let $A = k[G]$ and let π be a finite saturated subset of $X^+(T)$. The dual algebra $S(\pi)$ of the coalgebra $A(\pi)$ is called a generalized Schur algebra.

We will see in Section 8 that the Schur algebras $S(n, r)$, described by Green in [34], are generalized Schur algebras. The importance of these finite dimensional algebras algebras lies in that fact that the homological algebra of finite dimensional G-modules may be understood in terms of them, in view of the following result (which follows immediately from

Theorem 2.5).

Theorem 4.9. *If M, N are finite dimensional G-modules and π is a saturated subset of of $X^+(T)$ such that M, N belong to π then we have* $\mathrm{Ext}^i_G(M, N) = \mathrm{Ext}^i_{S(\pi)}(M, N)$*, for all $i \geqslant 0$.*

We conclude this section by stating a result which, though not part of the standard quasi-hereditary set-up, is nevertheless very useful in applications. It was proved by J-P Wang for large characteristics, by the author except for small characteristics and by O. Mathieu in general (for a full account of this and related issues see [42]).

Theorem 4.10. $\nabla(\lambda) \otimes \nabla(\mu) \in \mathcal{F}(\nabla)$ *for all $\lambda, \mu \in X^+(T)$.*

From this we get that if $X, Y \in \mathcal{F}(\nabla)$ then $X \otimes Y \in \mathcal{F}(\nabla)$. Moreover, since $X \in \mathcal{F}(\Delta)$ if and only if $X^* \in \mathcal{F}(\nabla)$ we get that if $X, Y \in \mathcal{F}(\Delta)$ then $X \otimes Y \in \mathcal{F}(\Delta)$. Hence if X, Y are tilting modules then $X \otimes Y$ is a tilting module.

5 Infinitesimal Methods

We consider some ways in which the tilting theory for a reductive group G interacts with the representation theory of the Lie algebra of G. In due course we shall (for convenience) take G to be semisimple and simply connected. However, to begin we let G be any linear algebraic group over an algebraically closed field k of characteristic $p > 0$. A left G-module V is naturally a right $k[G]$-comodule, and hence a left $k[G]^*$-module with action given by $X \cdot v = \sum_i X(f_i)v_i$, for $X \in k[G]^*$, $v \in V$ with $\tau(v) = \sum_i v_i \otimes f_i$ (where $\tau \colon V \to V \otimes k[G]$ is the structure map). This applies in particular to the left regular G-module, i.e the right regular $k[G]$-comodule $(k[G], \delta)$. Then we have $X(a) = (X \cdot a)(1)$, $X \in k[G]^*$, $a \in k[G]$. For $X \in \mathrm{Lie}(G)$, $a, b \in k[G]$ we have

$$
\begin{aligned}
X \cdot (ab) &= \textstyle\sum_{i,j} X(a'_i b'_j) a_i b_j \\
&= \textstyle\sum_{i,j} X(a'_i) b'_j(1) a_i b_j + \sum_{i,j} a'_i(1) X(b'_j) a_i b_j \\
&= (X \cdot a)b + a(X \cdot b)
\end{aligned}
$$

(where $\delta(a) = \sum_i a_i \otimes a'_i$, $\delta(b) = \sum_j b_j \otimes b'_j$). It follows that

$$
X^r \cdot (ab) = \sum_{r=i+j} \binom{r}{i} (X^i \cdot a)(X^j \cdot b)
$$

for $r \geqslant 0$, $a, b \in k[G]$. In particular $X^p \cdot (ab) = (X^p \cdot a)b + a(X^p \cdot b)$, for $a, b \in k[G]$, and hence $X^p \in \mathrm{Lie}(G)$ showing that $\mathrm{Lie}(G)$ is a p-Lie algebra with operation $X^{[p]} = X^p$, where X^p is computed in the associative algebra $k[G]^*$. (For generalities on p-Lie algebras see e.g. [3, Section 3.1].) One may check that in the case $G = \mathrm{GL}_n(k)$, identifying $\mathfrak{g} = \mathrm{Lie}(G)$ with the Lie algebra of $n \times n$-matrices $\mathrm{gl}_n(k)$ in the usual way, the p-operation is given by $X^{[p]} = X^p$ (the pth power of the matrix X).

If $\mathfrak{g}$ is a p-Lie algebra then we have the restricted enveloping algebra $u(\mathfrak{g})$ of $\mathfrak{g}$. By definition $u(\mathfrak{g})$ is $U(\mathfrak{g})/I$, where $U(\mathfrak{g})$ is the universal enveloping of $\mathfrak{g}$ and I is the ideal generated by $\{X^{[p]} - X^p \mid X \in \mathfrak{g}\}$. If $\mathfrak{g}$ has finite dimension n then $u(\mathfrak{g})$ has finite dimension p^n. Thus if $\mathfrak{g} = \mathrm{Lie}(G)$ then $u(\mathfrak{g})$ is a finite dimensional associative algebra of dimension $p^{\dim G}$. Moreover, the inclusion $\mathfrak{g} \to k[G]^*$ induces a monomorphism of algebras $u(\mathfrak{g}) \to k[G]^*$. A left G-module is naturally a $k[G]^*$-module and hence a $u(\mathfrak{g})$-module.

Now suppose that G is semisimple and simply connected. Then, in the notation of Section 4, the character group $X(T)$ is free abelian on the fundamental dominant weights $\omega_1, \ldots, \omega_l$ defined by $(\omega_i, \check{\alpha}_j) = \delta_{ij}$, $1 \leqslant i, j \leqslant l$, where $\alpha_1, \ldots, \alpha_l$ are the simple roots. We define $X_1(T)$ to be the set of dominant weights of the form $a_1\omega_1 + \cdots + a_l\omega_l$, with $0 \leqslant a_1, \ldots, a_l \leqslant p - 1$, i.e. $X_1(T) = \{\lambda \in X(T) \mid 0 \leqslant (\lambda, \check{\alpha}_j) \leqslant p - 1 \text{ for } 1 \leqslant j \leqslant l\}$.

We have the following fundamental theorem of Curtis.

Theorem 5.1. $\{L(\lambda) \mid \lambda \in X_1(T)\}$ *is a complete set of pairwise non-isomorphic simple $u(\mathfrak{g})$-modules.*

By analogy with Curtis's Theorem one would like to know that the projective indecomposable $u(\mathfrak{g})$-modules come from G-modules. The following conjecture (taken from [19]) states that in fact they are the restrictions of certain tilting modules.

Conjecture 5.2. *For $\lambda \in X_1(T)$ the restriction of $T(2(p-1)\rho + w_0\lambda)$ to $u(\mathfrak{g})$ is the projective cover of the $u(\mathfrak{g})$-module $L(\lambda)$.*

It is known that $T(2(p-1)\rho + w_0\lambda)$ is projective as a $u(\mathfrak{g})$-module and that the projective cover of $L(\lambda)$ occurs as a summand with multiplicity one. Thus the key question is whether $T(2(p-1)\rho + w_0\lambda)$ is indecomposable, as a $u(\ )$-module. The conjecture is known to be true for $p \geqslant 2h - 2$, where h is the Coxeter number of G, if the root system Φ is indecomposable, thanks to earlier work of Jantzen. This is discussed in [22]. (The Coxeter number is defined by $h - 1 = (\rho, \check{\beta}_0)$, where β_0 is the highest short root - we have $h = n$ for $G = \mathrm{SL}_n(k)$.)

In case $\lambda = (p-1)\rho$ we have $L(\lambda) = \nabla(\lambda) = \Delta(\lambda) = T(\lambda)$. This module is known as the Steinberg module and denoted St. It exerts an enormous moderating influence over the representation theory of G. In particular, the conjecture holds for $\lambda = (p-1)\rho$.

We have the Frobenius endomorphism of a general linear group $F \colon \mathrm{GL}_n(k) \to \mathrm{GL}_n(k)$ taking a matrix (a_{ij}) to the matrix (a_{ij}^p). One can realize G as a closed subgroup of some general linear group $\mathrm{GL}_n(k)$ in such a way that G, B and T are F-stable, and T is realized as a group of diagonal matrices (e.g. via the Chevalley construction). The restriction of F to G will also be denoted F, and called the Frobenius morphism of G (defined by this embedding of G in $\mathrm{GL}_n(k)$). For $M \in \mathrm{mod}(G)$ affording a representation $\psi \colon G \to \mathrm{GL}(M)$ we write M^F for the G-module with underlying vector space M on which G acts via $\psi \circ F$. For $\varphi = \sum_\lambda a_\lambda e^\lambda \in \mathbb{Z}X(T)$ we define $\varphi^F = \sum_\lambda a_\lambda e^{p\lambda}$. Then (from the definitions) we have $\mathrm{ch}\, M^F = (\mathrm{ch}\, M)^F$, for $M \in \mathrm{mod}(T)$.

For $\lambda = (p-1)\rho + p\tau$ with $\mu \in X_1(T)$, $\tau \in X^+(T)$, we have $\nabla(\lambda) = \mathrm{St} \otimes \nabla(\mu)^F$ (see [39, II,3.19 Proposition]). Moreover, if $M \in \mathrm{mod}(G)$ is indecomposable as a -module and $N \in \mathrm{mod}(G)$ is indecomposable then $M \otimes N^F$ is indecomposable (see [13, Section 2, Lemma]). We get the following.

Theorem 5.3. *Let* $\mu \in X_1(T)$ *and* $\tau \in X^+(T)$. *If* $T(2(p-1)\rho + w_0\mu)$ *is indecomposable as a* $u(\ )$-module *then* $T(2(p-1)\rho + w_0\mu + p\tau) \cong T(2(p-1)\rho + w_0\mu) \otimes T(\tau)^F$. *In particular we have* $\mathrm{St} \otimes T(\tau)^F \cong T((p-1)\rho + p\tau)$.

This leads, via Brauer's Formula (see e.g. [15, (2.2.3)]) to the following character formula.

Theorem 5.4. *Suppose* $\mu \in X_1(T)$, $\tau \in X^+(T)$ *and* $T(2(p-1)\rho + w_0\mu)$ *is indecomposable, as a* $u(\ )$-module. *We write the character* $\mathrm{ch}\, T(2(p-1)\rho + w_0\mu) = \chi((p-1)\rho)\psi_\mu$, *where* $\psi_\mu = \sum_{\nu \in X(T)} a_\nu e^\nu$.

Then we have

$$\operatorname{ch} T(2(p-1)\rho + w_0\mu + p\tau)$$
$$= \sum_{\nu \in X(T), \zeta \in X^+(T)} a_\nu (T(\tau) : \nabla(\zeta)) \chi((p-1)\rho + \nu + p\zeta).$$

A Weyl character $\chi(\sigma)$ is either 0 or $\pm\chi(\xi)$, for some $\xi \in X^+(T)$, (as mentioned in Section 4) so that any multiplicity $(T(2(p-1)\rho + w_0\mu + p\tau) : \nabla(\lambda))$ may be derived, at least in principle, from Theorem 5.4, provided that the multiplicities for $T(2(p-1)\rho+w_0\mu)$ and $T(\tau)$ are known.

We see how this works out in a special case. We assume, for convenience, that the root system Φ is connected. Let λ be a dominant weight such that $(\lambda, \check{\beta}_0) \leqslant p$, where β_0 is the highest short root. If μ is any dominant weight less than λ then we also have $(\mu, \check{\beta}_0) \leqslant p$.

Let V be module with character $\chi(\lambda)$ (e.g. $V = \nabla(\lambda)$). Then we have a B-module filtration $0 = V_0 < V_1 < \ldots < V_n = V$ with $V_i/V_{i-1} \cong k_{\nu_i}$ where $\chi(\lambda) = \sum_{i=1}^n e^{\nu_i}$. Hence $k_{(p-1)\rho} \otimes V$ has a B-module filtration with sections $k_{(p-1)\rho+\nu_i}$. Let ν be a weight of V and let α be a simple root. Then $w\nu = \tau$, say, is dominant for some $w \in W$ and $(\nu, \check{\alpha}) = (w\nu, \check{\gamma}) \geqslant (\tau, -\check{\beta}_0) \geqslant -p$, where $\gamma = w\alpha$. Hence we have $((p-1)\rho + \nu, \check{\alpha}) \geqslant p - 1 - p = -1$. Hence, by Theorem 4.3, $R\operatorname{Ind}_B^G(V_i/V_{i-1}) = 0$ and $\operatorname{St} \otimes V = \operatorname{Ind}_B^G(k_{(p-1)\rho} \otimes V)$ has a filtration with sections $\nabla((p-1)\rho+\nu_j)$, where j runs over the set

$$Z = \{1 \leqslant i \leqslant n \mid (p-1)\rho + \nu_i \in X^+\}.$$

We write $s(\lambda)$ for the orbit sum, i.e. $s(\lambda) = \sum_{\nu \in W\lambda} e^\nu$. From Brauer's formula we get $\chi((p-1)\rho)s(\lambda) = \sum_{\nu \in Z_\lambda} \chi((p-1)\rho + \nu)$, where

$$Z_\lambda = \{\nu \in W\lambda \mid (p-1)\rho + \nu \in X^+\}.$$

We identify W with a group of permutations of $X(T)$ via its natural (faithful) action. The affine Weyl group W_p is the group of permutations of $X(T)$ generated by W and all translations by $p\theta$, $\theta \in X(T)$. Then W_p is the semidirect product of W and the translation subgroup. The dot action of W on $X(T)$ extends to an action of W_p (given by $w \cdot \lambda = w(\lambda+\rho)-\rho$). If simple G-modules $L(\lambda)$, $L(\mu)$ lie in the same block then λ and μ lie in the same orbit of $X(T)$ for the dot action of W_p (see e.g. [39, II,7.1]).

We now consider the block component M, say, of $\operatorname{St} \otimes V$ for the block containing $L((p-1)\rho+\lambda)$. Note that, for $w \in W$, the weight $(p-1)\rho+w\lambda$ belongs to the orbit of $(p-1)\rho+\lambda$ under the dot action of the affine Weyl group W_p. On the other hand, if $(M : \nabla((p-1)\rho + \nu_j)) \neq 0$, for some $j \in Z$, then $(p-1)\rho+\nu_j \in W_p \cdot ((p-1)\rho+\lambda)$. Now ν_j is conjugate to a dominant weight τ, say, under W and $(p-1)\rho+\tau \in W_p \cdot ((p-1)\rho+\nu_j))$. Hence $(p-1)\rho+\lambda$ and $(p-1)\rho+\tau$ belong to same W_p orbit (for the dot action). But then $-\rho+\lambda$ and $-\rho+\tau$ belong to the same orbit. However, $-\rho + \lambda$ and $-\rho + \tau$ belong to the well known fundamental domain (see [39, II. Section 6.1]) for the action of W_p. So $-\rho + \lambda = -\rho + \tau$, $\lambda = \tau$ and $\nu_j \in Z_\lambda$. Thus, for $j \in Z$, we get

$$(M : \nabla((p-1)\rho + \nu_j)) = \begin{cases} 1, & \text{if } \nu_j \in Z_\lambda; \\ 0, & \text{otherwise.} \end{cases}$$

This proves that $\operatorname{ch} M = \chi((p-1)\rho)s(\lambda)$. Moreover, M has a component $T((p-1)\rho+\lambda)$ the character of $T((p-1)\rho+\lambda)$ is divisible by $\chi((p-1)\rho)$ and invariant under W. It follows that $\operatorname{ch} M = \operatorname{ch} T((p-1)\rho + \lambda)$, $M = T((p-1)\rho + \lambda)$ and therefore

$$\operatorname{ch} T((p-1)\rho + \lambda) = \chi((p-1)\rho)s(\lambda).$$

In [38] Jantzen considers a category of finite dimensional modules for $u(\ )$ and T together, the category of $(u(\ ), T)$-modules (equivalently, in [39], the category of G_1T-modules). It is shown in particular that for $\mu \in X(T)$ there is a unique simple $(u(\ ), T)$-module of high weight μ, this module has an injective envelope, denoted $\hat{Q}_1(\mu)$, and for $\mu \in X_1(T)$, the restriction of $\hat{Q}_1(\mu)$ to $u(\ )$ is the injective (and projective) indecomposable module corresponding to the simple $u(\ )$-module $L(\mu)$. Now, by character considerations, $\hat{Q}((p-1)\rho+w_0\lambda)$ is a direct summand of $T((p-1)\rho+\lambda)$ (as a $(u(\ ), T)$-module). Moreover, $\hat{Q}((p-1)\rho + w_0\lambda)$ has character which is divisible by $\chi((p-1)\rho)$ and W-invariant. Hence we must have

$$T((p-1)\rho + \lambda) \cong \hat{Q}((p-1)\rho + w_0\lambda)$$

as a $(u(\ ), T)$-module, and hence $T((p-1)\rho + \lambda)|_{u(\ )}$ is the projective indecomposable $u(\ )$-module corresponding to the simple module $L((p-1)\rho + w_0\lambda)$. This is a special case of the above Conjecture. It has a consequence for filtration multiplicities of some tilting modules and hence (see Section 9,(3)) for some decomposition numbers for symmetric groups.

Let $\mu \in X^+(T)$. Then, by [19, (2.1) Proposition], we have

$$T((p-1)\rho + \lambda + p\mu) \cong T((p-1)\rho + \lambda) \otimes T(\mu)^F.$$

Hence we have

$$
\begin{aligned}
\operatorname{ch} T&((p-1)\rho + \lambda + p\mu) \\
&= \textstyle\sum_{\xi \in X^+(T)} (T(\mu) : \nabla(\xi)) \chi((p-1)\rho) s(\lambda) \chi(\xi)^F \\
&= \textstyle\sum_{\xi \in X^+(T)} (T(\mu) : \nabla(\xi)) \chi((p-1)\rho + p\xi) s(\lambda) \\
&= \textstyle\sum_{\xi \in X^+(T), \nu \in W\lambda} (T(\mu) : \nabla(\xi)) \chi((p-1)\rho + \nu + p\xi).
\end{aligned}
$$

We summarize our findings.

Proposition 5.5. *Let* $\lambda, \mu \in X^+(T)$ *and assume* $(\lambda, \check{\beta}_0) \leqslant p$. *Then:*

(i) $\operatorname{ch} T((p-1)\rho + \lambda) = \chi((p-1)\rho) s(\lambda)$;

(ii) $T((p-1)\rho + \lambda)$, *as a* $u(\ )$-*module, is the projective cover of* $L((p-1)\rho + w_0\lambda)$;

(iii) $(T((p-1)\rho + \lambda + p\mu) : \nabla(\tau)) = \sum_{\xi \in N(\tau)} (T(\mu) : \nabla(\xi))$, *for* $\tau \in X^+(T)$, *where* $N(\tau) = \{\xi \in X^+(T) \mid \tau + \rho - p(\xi + \rho) \in W\lambda\}$.

We refer the reader to the papers of Cox, [8], and Erdmann, [30], for related applications to decomposition numbers of the tensor product theorem for tilting modules, [19, (2.1) Proposition].

6 Some support for tilting modules

Let be a p-Lie algebra. Attached to each restricted -module (equivalently $u(\ )$-module) M there is an affine algebraic variety $V(M)$, called its *support variety*. This is defined via the cohomology of M. However, we shall not need to deal with the cohomology directly: there is a more concrete realization of $V(M)$ due to Friedlander-Parshall, [32], and we shall use this instead. The support variety $V(M)$ may be identified with the subvariety of consisting of 0 together with all elements X such that $X^{[p]} = 0$ and M is not projective as a module for the subalgebra of $u(\ )$ generated by X (i.e for the restricted enveloping algebra of kX).

The problem of describing the support variety of a tilting module $T(\lambda)$ has been raised by J. E. Humphreys, in a talk at the meeting on Representations of Algebraic Groups at the Isaac Newton Institute in 1997. We make a modest contribution to this problem by offering a conjecture in the very special case in which $G = \mathrm{GL}_n(k)$ and k has characteristic 2.

Let $\lambda = (\lambda_1, \lambda_2, \ldots)$ be a partition of n. We write $J(\lambda)$ for the nilpotent $n \times n$ matrix with Jordan blocks of size $\lambda_1, \lambda_2, \ldots$. We say that a nilpotent matrix X has Jordan type λ if it is conjugate to $J(\lambda)$. Thus $V(k)$ is the variety of all nilpotent matrices whose Jordan type μ has all parts $\mu_1, \mu_2, \ldots$ of size at most p. Moreover, if M is a restricted -module then an explicit knowledge of $V(M)$ is equivalent to a knowledge of which $J(\mu)$ belong to $V(M)$, i.e. to knowing whether for such μ the module M is injective as a module for the subalgebra of $u(\ )$-generated by $J(\mu)$.

Let λ, μ be partitions of n. We write $\lambda \leqslant_R \mu$ if λ is a refinement of μ (as in [44]). The condition is there is some partition of sets $[1, n] = B_1 \cup \cdots \cup B_m$ with sizes $\mu_1, \mu_2, \ldots$ and partitions $B_r = B_{r1} \cup \cdots B_{rj_r}$ such that the subsets $B_{11}, \ldots, B_{mj_m}$ have sizes $\lambda_1, \lambda_2, \ldots$, taken in some order. For example we have $(5, 4, 2, 1, 1) \leqslant_R (7, 6)$ (since $6 = 1 + 1 + 4$ and $7 = 2 + 5$).

Let $\lambda \in \Lambda^+(n, r)$. Let a_0 be the number of parts equal to 0, let a_1 be the number of parts equal to 1 and so on. We write $\bar{\lambda}$ for the partition of n whose parts are $a_0, a_1, \ldots$ (arranged in descending order). We can now state our conjecture on the support variety of tilting modules.

Conjecture 6.1. *Suppose k has characteristic 2. Let $\lambda = (\lambda_1, \ldots, \lambda_n) \in \Lambda^+(n, r)$. For a partition μ of n we have $J(\mu) \in V(T(\lambda))$ if and only if $\mu \leqslant_R \bar{\lambda}$.*

By using [19, (1.5) Proposition] one gets one direction, namely that if $\mu \leqslant_R \bar{\lambda}$ then $J(\mu) \in V(T(\lambda))$. Moreover, we have checked the conjecture for $\mathrm{GL}(n)$ for all $\lambda = (\lambda_1, \ldots, \lambda_n)$ up to $n = 6$.

7 Invariant theory

Our interest in tilting modules was motivated originally by the possibility of using them in invariant theory. Nevertheless, we shall be brief in this section. There are two reasons for brevity. The first is that we have already had a chance to give a survey of this topic in [22]. The second reason is that the original application using tilting modules in a very explicit way has been superseded by the paper of Domokos and Zubkov, see [12] and related papers, and by further progress in the theory of semi-invariants by Domokos and Zubkov and also by Derksen and Weyman, see [9] and references given there. (Though in fact ∇-filtrations play a crucial role in these developments and modules which have both a ∇-filtration and a Δ-filtration play a key role in [12], even though the general theory of tilting modules is not used.) We shall content ourselves with stating the main general result from [20] and the application to invariant theory given there.

Let G be a reductive group over k. Let H be a closed subgroup of G. We consider the algebra of H class functions $C(G,H) = \{f \in A \mid f(hgh^{-1}) = f(g)$ for all $h \in H, g \in G\}$ (where $A = k[G]$ is the coordinate algebra). More generally, for a saturated subset π of $X^+(T)$ we put $C(G,H,\pi) = A(\pi) \cap C(G,H)$.

If $V \in \mathrm{mod}(G)$, affording the representation $\varphi \colon G \to \mathrm{GL}(V)$, and θ is an H-module endomorphism of V, we define $\chi_\theta \in C(G,H)$ by $\chi_\theta(g) = \mathrm{Trace}(\varphi(g) \circ \theta)$.

We say that a closed subgroup H is *saturated* if the induced module $\mathrm{Ind}_H^G k = k[H\backslash G] = \{f \in k[G] \mid f(hg) = f(g)$ for all $h \in H, g \in G\}$ admits an exhaustive, ascending filtration $0 = V_0 < V_1 < V_2 \ldots$ such that each V_i/V_{i-1} is isomorphic to $\nabla(\lambda_i)$, for some dominant weight λ_i.

The main general result of [20] is the following.

Theorem 7.1. *Suppose that H is a saturated subgroup of G. Then for any saturated subset π of $X^+(T)$ we have $C(G,H,\pi) = \{\chi_\theta \mid \theta \in \mathrm{End}_H(T(\lambda)), \lambda \in \pi\}$.*

This is applied to the special case in which G is a direct product of copies of $\mathrm{GL}(n)$, H is the diagonally embedded copy of $\mathrm{GL}(n)$ and π is the set of polynomial weights to prove the following result. We write $M(n)$ for the space of $n \times n$ matrices with entries in k and $\chi_s(y)$ for the sth coefficient of the characteristic polynomial of an $n \times n$-matrix y.

Theorem 7.2. *The algebra of invariants $k[M(n)^m]^{\mathrm{GL}(n)}$ (for the action of $\mathrm{GL}(n)$ on the variety $M(n)^m$, of m-tuples of matrices, by simultaneous conjugation) is generated by the functions $(x_1, \ldots, x_m) \mapsto \chi_s(x_{i_1}x_{i_2}\ldots x_{i_r})$, for $r, s \geqslant 1$ and $1 \leqslant i_1, \ldots, i_r \leqslant m$.*

For the corresponding result in characteristic zero see, e.g., [47]. The proof of the above given in [22] relies on a substantial amount of algebraic combinatorics, which is skillfully avoided in [12].

We mention that even the case $H = 1$ of Theorem 7.1 is not without interest. One may readily check that for $V \in \mathrm{mod}(G)$, the k-span of all $\chi_\theta, \theta \in \mathrm{End}_k(V)$, is the coefficient space $\mathrm{cf}(V)$. Hence we get the following description of $A(\pi)$.

Corollary 7.3. *(of Theorem 7.1) For a saturated subset π of $X^+(T)$, the subspace $A(\pi)$ is generated by the subspaces $\mathrm{cf}(V)$, as V ranges over all tilting modules $T(\lambda)$ with $\lambda \in \pi$.*

8 General Linear Groups

We consider the case of $G = \mathrm{GL}(n)$. Let $A(n) = k[c_{11}, \ldots, c_{nn}] \leqslant k[G]$. Then $A(n)$ is a subbialgebra of $k[G]$. A finite dimensional representation $\rho \colon G \to \mathrm{GL}_N(k)$ is said to be *polynomial* if it is given by $\rho(g) = (f_{ij}(g))$, $g \in G$, with all $f_{ij} \in A(n)$. A (finite dimensional) G-module V is said to be polynomial if it affords a polynomial matrix representation. The polynomial representation theory of G is studied, in great detail, by Green in the monograph [34], to which we refer the reader for background and the results not explicitly covered here. Note that if V is any finite dimensional G-module then $V \otimes D^{\otimes s}$ is polynomial, for some $s \geqslant 0$, where D is the (one dimensional) determinant module. Hence a knowledge of the representation theory of finite dimensional polynomial modules is equivalent to a knowledge of all finite dimensional G-modules.

Giving each c_{ij} degree one we get a grading $A(n) = \bigoplus_{r=0}^{\infty} A(n, r)$ and moreover, this is actually a coalgebra decomposition. The Schur algebra $S(n, r)$ is the dual algebra $A(n, r)^*$. A matrix representation $\rho \colon G \to \mathrm{GL}_N(k)$ is called *polynomial of degree r* if it given by $\rho(g) = (f_{ij}(g))$, $g \in G$, with all $f_{ij} \in A(n, r)$. A G-module V is called *polynomial of degree r* if it affords a matrix representation which is polynomial of degree r. An arbitrary polynomial G-module V has a unique decomposition $V = \bigoplus_{r=0}^{\infty} V(r)$, where $V(r)$ is polynomial of degree r. Thus the polynomial representation theory of G is determined by the representation theory of the finite dimensional coalgebras $A(n, r)$ and hence by the representation theory of the Schur algebras $S(n, r)$.

We may also see this set-up via the representation theory of reductive groups. We take for T the group of diagonal matrices and for B the group of lower-triangular matrices. Then $N = N_G(T)$ is the group of monomial matrices in G (matrices with precisely one non-zero entry in each row and column). Thus N is the semidirect product of T and the group of permutation matrices in G. We identify the Weyl group with the group of permutation matrices and hence with the symmetric group of degree n. We identify $X(T)$ with $\mathbb{Z}^n$, as in Section 4. Then we have

$$X^+(T) = \{\lambda = (\lambda_1, \ldots, \lambda_n) \in \mathbb{Z}^n \mid \lambda_1 \geqslant \ldots \geqslant \lambda_n\}.$$

We write $\Lambda(n)$ for $\{\lambda = (\lambda_1, \ldots, \lambda_n) \in \mathbb{Z}^n \mid \lambda_1, \ldots, \lambda_n \geqslant 0\}$ and write $\Lambda^+(n)$ for $X^+(n) \cap \Lambda(n)$, the set of partitions with at most n parts. We write $\Lambda^+(n, r)$ for the set of all partitions of r into at most n parts.

One may easily check that, for $r \leqslant n$, the exterior power $\bigwedge^r E$ is irreducible and (by Weyl's character formula) has character $\chi(1, \ldots, 1, 0, \ldots, 0)$. It follows that $\bigwedge^r E = \nabla(1, \ldots, 1, 0, \ldots, 0) = \Delta(1, \ldots, 1, 0, \ldots, 0)$ and hence is also the tilting module $T(1, \ldots, 1, 0, \ldots, 0)$. Hence (since the tensor product of tilting modules is a tilting module) $\bigwedge^\alpha E = \bigwedge^{\alpha_1} E \otimes \cdots \otimes \bigwedge^{\alpha_m} E$ is a tilting module for any m-tuple $\alpha = (\alpha_1, \ldots, \alpha_m)$ of non-negative integers.

Let $\lambda \in \Lambda^+(n, r)$ and let λ' denote the dual (or conjugate) partition, as in for example [44, Chapter I]. Then it is easy to check that the module $\bigwedge^{\lambda'} E$ has highest weight λ. Hence we have

$$\bigwedge^{\lambda'} E = T(\lambda) \oplus R$$

where R is a direct sum of tilting modules $T(\mu)$ with $\mu < \lambda$. Thus we get the following description of polynomial tilting modules for $\mathrm{GL}(n)$.

Theorem 8.1. *For $\lambda \in \Lambda^+(n, r)$ we have*

$$\bigwedge^{\lambda'} E = T(\lambda) \oplus R$$

where $R = \bigoplus_{\mu \in \Lambda^+(n,r), \mu < \lambda} T(\mu)^{(m_{\lambda\mu})}$ for certain non-negative integers $m_{\lambda\mu}$.

Let $\mathcal{P}$ denote the set of all partitions. Thus $\lambda \in \mathcal{P}$ is a string $\lambda = (\lambda_1, \lambda_2, \ldots)$ of non-negative integers, almost all zero with $\lambda_1 \geqslant \lambda_2 \geqslant \ldots$. For $\lambda \in \mathcal{P}$ let $|\lambda| = \sum_{i \geqslant 1} \lambda_i$. If $|\lambda| = r$ and $\lambda_{n+1} = 0$ we identify λ with an element of $\Lambda^+(n, r)$ in the obvious way. For $\lambda \in \Lambda^+(n, r)$ we write now $\nabla_n(\lambda)$ and $L_n(\lambda)$ for the induced $\mathrm{GL}_n(k)$-module and simple $\mathrm{GL}_n(k)$-module labelled by λ, to emphasize the role of n. Suppose now

$\lambda, \mu \in \mathcal{P}$, that $|\lambda| = |\mu|$ and that $\lambda_{n+1} = \mu_{n+1} = 0$. Then, by [34, Chapter 6], the decomposition number $[\nabla_n(\lambda) : L_n(\mu)]$ is equal to the decomposition number $[\nabla_N(\lambda) : L_N(\mu)]$, for all $N \geqslant n$. We denote this number simply $[\lambda : \mu]$.

Using the above theorem one may prove that $S(n,r)$ is its own Ringel dual for $r \leqslant n$ (by considering the exterior algebra $\bigwedge^* M(n)$ as a (G, G)-bimodule). The argument is given in [21, Section 5] (and an earlier proof, using the Schur functor, can be found in [19, Section 3]). This has the interesting consequence (see [19, Section 3]):

Theorem 8.2. *For all $\lambda, \mu \in \Lambda^+(n,r)$ we have $[T(\lambda) : \nabla(\mu)] = [\mu' : \lambda']$.*

A similar result for the other classical groups has been obtained by Adamovich and Rybnikov, see [1].

We end this section by noting that in fact the Schur algebra $S(n,r)$ is a generalized Schur algebra. We take $\pi = \Lambda^+(n,r)$. Let $A = k[G]$. Then by the Corollary of Section 8, $A(\pi)$ is spanned by $\mathrm{cf}(T(\lambda))$, as λ varies over $\Lambda^+(n,r)$. However, by Theorem 8.1, this is the span of $\mathrm{cf}(\bigwedge^\lambda E)$, as λ varies over all partitions of r whose parts have size at most n. Moreover, for such λ, the module $\bigwedge^\lambda E$ is a homomorphic image of $E^{\otimes r}$ so that $\mathrm{cf}(\bigwedge^\lambda E) \leqslant \mathrm{cf}(E^{\otimes r}) = A(n,r)$. Thus we have $A(\pi) \leqslant A(n,r)$ and indeed, since we may take $\lambda = (1, 1, \ldots, 1)$, we must have equality. Thus we get:

Theorem 8.3. *$S(n,r)$ is the generalized Schur algebra $S(\pi)$, for $\pi = \Lambda^+(n,r)$.*

Remark 8.4. This proof is different from the one given originally in [17]. The case of the general symplectic group is discussed in [18] where the two notions of Schur algebra are again identified. Running through the argument above again in this case gives a much shorter proof of this identification than in [18]. Furthermore, the notion of Schur algebra as the dual algebra of a graded component of a subalgebra of the coordinate algebra of reductive a group is discussed for other classical groups by Doty in [28]. Using the description of tilting modules for classical group which on finds in [1] it is easy to check that here too (in odd characteristic) these Schur algebras are generalized Schur algebras.

9 Connections with symmetric groups and Hecke algebras

We follow Green, [34], to make a connection between representations of general linear groups and representations of symmetric groups via the Schur functor. For this we need to set up some additional notation.

We write $\mathrm{Sym}(X)$ for the group of permutations of a set X. For $r \geqslant 1$ we write $\mathrm{Sym}(r)$ for $\mathrm{Sym}\{1, 2, \ldots, r\}$. We write $\Lambda(n, r)$ for the set of n-tuples $\alpha = (\alpha_1, \ldots, \alpha_n)$ of non-negative integers whose sum is r. For $\alpha \in \Lambda(r, r)$ we write $\mathrm{Sym}(\alpha)$ for the Young subgroup $\mathrm{Sym}\{1, \ldots, \alpha_1\} \times \mathrm{Sym}\{\alpha_1 + 1, \ldots, \alpha_1 + \alpha_2\} \times \cdots$ of $\mathrm{Sym}(r)$. We write $I(n, r)$ for the set of maps from $\{1, 2, \ldots, r\}$ to $\{1, 2, \ldots, n\}$. An element $i \in I(n, r)$ may be written as an r-tuple $i = (i_1, \ldots, i_r)$ of elements of $\{1, 2, \ldots, n\}$. Then $i \in I(n, r)$ has content $\alpha = (\alpha_1, \ldots, \alpha_r)$ where α_h is the number of $1 \leqslant a \leqslant r$ with $i_a = h$. Note that $i, j \in I(n, r)$ have the same content if and only if $i = j \circ \pi$ for some $\pi \in \mathrm{Sym}(r)$.

For $i, j \in I(n, r)$ we write c_{ij} for $c_{i_1 j_1} \ldots c_{i_r j_r} \in A(n, r)$. Then the elements c_{ij}, $i, j \in I(n, r)$, form a k-basis and we have $c_{ij} = c_{uv}$ if and only if there exists a permutation π of $\{1, 2, \ldots, r\}$ such that $i = i \circ u$, $v = j \circ \pi$. For $i, j \in I(n, r)$ the corresponding dual basis element of $S(n, r)$ is denoted ξ_{ij}. In particular for $\alpha \in \Lambda(n, r)$ we have the element ξ_α, defined by $\xi_\alpha = \xi_{ii}$, where i is any element of $I(n, r)$ which has content α. The elements ξ_α are idempotent and indeed

$$1 = \sum_{\alpha \in \Lambda(n,r)} \xi_\alpha$$

is an orthogonal decomposition of 1. One may check that if V is a module for $\mathrm{GL}_n(k)$ which is polynomial of degree r then we have $V^\alpha = \xi_\alpha V$, $\alpha \in \Lambda(n, r)$.

We now suppose that $r \leqslant n$ we let $u = (1, 2, \ldots, r) \in I(n, r)$ and let $\omega = (1, 1, \ldots, 1, 0, \ldots, 0)$, the content of u. We write $S = S(n, r)$ and $e = \xi_{uu}$ for short. Then eSe is naturally a k-algebra with identity e. Moreover, there is an algebra isomorphism from the group algebra $k\mathrm{Sym}(r)$ to eSe taking a permutation π to $\xi_{u, u\pi}$. In this way we identify eSe with the group algebra of $\mathrm{Sym}(r)$. We thus have a left exact functor, called the Schur functor, $f \colon \mathrm{mod}(S) \to \mathrm{mod}(\mathrm{Sym}(r))$ given on objects by $fV = eV$ (and where $\mathrm{mod}(\mathrm{Sym}(r))$ is short for $\mathrm{mod}(k\mathrm{Sym}(r))$).

Applying the Schur functor to familiar modules in the representation theory of general linear groups yields familiar modules in the representation theory of symmetric groups.

(1) (i) For $\alpha \in \Lambda(n,r)$ we have $fS^\alpha E = M(\alpha)$, where $M(\alpha)$ denotes the permutation module $k\mathrm{Sym}(r) \otimes_{k\mathrm{Sym}(\alpha)} k$.

  Now let $\lambda \in \Lambda^+(n,r)$.

  (ii) We have $f\bigwedge^{\lambda'} E = k_s \otimes M(\lambda)$, where k_s denotes the one dimensional $\mathrm{Sym}(r)$-module affording the sign representation.

  (iii) We have $f\nabla(\lambda) = \mathrm{Sp}(\lambda)$, where $\mathrm{Sp}(\lambda)$ denotes the Specht module corresponding to λ.

  (iv) Let $I_{n,r}(\lambda)$ denote the injective envelope of the S-module $L(\lambda)$. Then $fI_{n,r}(\lambda)$ is the Young module $Y(\lambda)$ corresponding to λ.

  (v) $fT(\lambda) = k_s \otimes Y(\lambda')$.

For constructions of Specht modules and Young modules entirely within the context of the symmetric groups see, for example, [36] and [37].

We write $\Lambda^+(n,r)_{\mathrm{row}}$ for the set of row regular partitions, i.e. the set of partitions $\lambda = (\lambda_1, \ldots, \lambda_n) \in \Lambda^+(n,r)$ which contain no sequence of p equal non-zero parts. We write $\Lambda^+(n,r)_{\mathrm{col}}$ for the set of column regular partitions, i.e. the set of $\lambda = (\lambda_1, \ldots, \lambda_n) \in \Lambda^+(n,r)$ such that $\lambda_i - \lambda_{i+1} < p$, for $1 \leq i < n$. Then we have the following results (see e.g. [34, Chapter 6]):

(2) (i) $\{fL(\lambda) \mid \lambda \in \Lambda^+(n,r)_{\mathrm{col}}\}$ is a complete set of pairwise non-isomorphic simple $k\mathrm{Sym}(r)$-modules.

  (ii) For $\lambda \in \Lambda^+(n,r)_{\mathrm{row}}$ the Specht module $\mathrm{Sp}(\lambda) = f\nabla(\lambda)$ has a simple head D^λ and $\{D^\lambda \mid \lambda \in \Lambda^+(n,r)_{\mathrm{row}}\}$ is a complete set of pairwise non-isomorphic $k\mathrm{Sym}(r)$-modules.

  (iii) The relationship between these two labellings of simple modules is: $k_s \otimes D^\lambda \cong fL(\lambda')$, $\lambda \in \Lambda^+(n,r)_{\mathrm{row}}$.

  (iv) For $\lambda \in \Lambda^+(n,r)$ and $\mu \in \Lambda^+(n,r)_{\mathrm{col}}$ we have an equality of decomposition numbers $[\nabla(\lambda) : L(\mu)] = [\mathrm{Sp}(\lambda) : fL(\mu)]$.

The last result comes by applying the Schur functor f to a composition series of $\nabla(\lambda)$. Now for $\lambda \in \Lambda^+(n,r)$ and $\mu \in \Lambda^+(n,r)_{\mathrm{row}}$ we have that $(T(\lambda) : \nabla(\mu))$ is $[\nabla(\mu'); L(\lambda')]$, by Section 8, and hence also $[\mathrm{Sp}(\mu') : fL(\lambda')]$ (applying the Schur functor) and hence also

$[k_s \otimes \mathrm{Sp}(\mu') : k_s \otimes fL(\lambda')] = [\mathrm{Sp}(\mu) : D^\lambda]$ (since $k_s \otimes \mathrm{Sp}(\mu')$ is isomorphic to the dual of $\mathrm{Sp}(\mu)$, see [36, (8.15)], and all irreducible $k\mathrm{Sym}(r)$-modules are self dual). Hence we have (as in [29, 4.5 Lemma]):

(3) $(T(\lambda) : \nabla(\mu)) = [\mathrm{Sp}(\mu) : D^\lambda]$, for $\lambda \in \Lambda^+(n,r)_{\mathrm{row}}, \mu \in \Lambda^+(n,r)$.

We should like now to point out that essentially all of this development extends to the case of quantum general linear groups and Hecke algebras. Full details can be found in [23]. We simply briefly describe the framework and leave it to the reader to formulate precise generalizations of the results for general linear groups and symmetric groups given above (or look them up in [23]).

So now let q be a non-zero element of k. We write $A_q(n)$ for the k-algebra given by generators c_{ij}, $1 \leqslant i,j \leqslant n$, subject to the relations:

$$\begin{aligned} c_{ir}c_{is} &= c_{is}c_{ir} & \text{for all } 1 \leqslant i,r,s \leqslant n \\ c_{jr}c_{is} &= qc_{is}c_{jr} & \text{for all } 1 \leqslant i < j \leqslant n, 1 \leqslant r \leqslant s \leqslant n \\ c_{js}c_{ir} &= c_{ir}c_{js} + (q-1)c_{is}c_{jr} & \text{for all } 1 \leqslant i < j \leqslant n, 1 \leqslant r < s \leqslant n. \end{aligned}$$

Then $A_q(n)$ is a bialgebra with comultiplication $\delta \colon A_q(n) \to A_q(n) \otimes A_q(n)$ and augmentation $\epsilon \colon A_q(n) \to k$ satisfying $\delta(c_{ij}) = \sum_{r=1}^n c_{ir} \otimes c_{rj}$ and $\delta(c_{ij}) = \delta_{ij}$, for $1 \leqslant i,j \leqslant n$. The determinant $d \in A_q(n)$ is defined by $d = \sum_\pi \mathrm{sgn}(\pi)c_{1,1\pi}c_{2,2\pi}\cdots c_{n,n\pi}$ where π runs over all permutations of $\{1,2,\ldots,n\}$ and where $\mathrm{sgn}(\pi)$ denotes the sign of a permutation π.

The bialgebra structure on $A_q(n)$ extends to the localization $A_q(n)_d$ of $A_q(n)$ at d (with $\delta(d^{-1}) = d^{-1} \otimes d^{-1}$ and $\epsilon(d^{-1}) = 1$). Furthermore, $A_q(n)_d$ is a Hopf algebra. We write $k[G(n)]$ for $A_q(n)_d$ and call $G(n)$ the quantum general linear group of degree n. We denote the antipode of $k[G(n)]$ by S. We have $S^2(f) = dfd^{-1}$, for $f \in k[G(n)]$. Explicitly, we have $S^2(c_{ij}) = dc_{ij}d^{-1} = q^{j-i}c_{ij}$ for $1 \leqslant i,j \leqslant n$.

By the expression "V is a left $G(n)$-module" we mean that V is a right $k[G(n)]$-comodule. We have, in particular, the natural left $G(n)$-module E with basis $e_1,\ldots,e_n$ and structure map $\tau \colon E \to E \otimes k[G(n)]$ given by $\tau(e_i) = \sum_{j=1}^n e_j \otimes c_{ji}$. More generally, for $r \geqslant 0$, we have the rth symmetric power $S^r E$ and rth exterior power $\bigwedge^r E$ (defined in [23]).

Let I be the Hopf ideal of $k[G(n)]$ generated by all c_{ij} with $i \neq j$. Then $k[G]/I$ is the algebra of Laurent series in $c_{11}+I,\ldots,c_{nn}+I$. Let $T(n)$ be the subgroup of $\mathrm{GL}_n(k)$ consisting of the diagonal matrices. We identify $k[G(n)]/I$ with $k[T(n)]$ in such a way that $c_{ii}+I$ is identified with the function taking $t \in T(n)$ to its (i,i)-entry. A $G(n)$-module is naturally a $k[G(n)]/I$-comodule and hence a $T(n)$-module. Hence we have a theory of weights for $G(n)$-modules. We write $X^+(n)$ for $\{\lambda = (\lambda_1, \ldots, \lambda_n) \in \mathbb{Z}^n \mid \lambda_1 \geqslant \ldots \geqslant \lambda_n\}$. For each $\lambda \in X^+(n)$ there is an irreducible G-module $L(\lambda)$ with unique highest weight λ (which occurs with multiplicity one). Moreover, $\{L(\lambda) \mid \lambda \in X^+(n)\}$ is a complete set of pairwise non-isomorphic irreducible $G(n)$-modules. Among all $G(n)$-modules M with socle $L(\lambda)$ and $M/L(\lambda)$ having only composition factors from $\{L(\mu) \mid \mu < \lambda\}$ there is one of largest dimension and we denote this module by $\nabla(\lambda)$. Then $k[G(n)]$ is a quasi-hereditary coalgebra (with respect to the usual dominance order on $X^+(n)$). There is a theory of polynomial $G(n)$-modules and the indecomposable polynomial tilting modules are exactly the components of the exterior powers $\bigwedge^\alpha E = \bigwedge^{\alpha_1} E \otimes \cdots \otimes \bigwedge^{\alpha_m} E$, as before.

For $i, j \in I(n,r)$ we write c_{ij} for $c_{i_1 j_1} \ldots c_{i_r j_r}$. We write $A(n,r)$ for the subcoalgebra of $k[G(n)]$ spanned (as a vector space) by the c_{ij} with $i, j \in I(n,r)$. The q-Schur algebra $S = S(n,r)$ is the dual algebra $A(n,r)^*$. For $\alpha \in \Lambda(n,r)$ we write ${}^\alpha A(n,r)$ for the subspace of $A(n,r)$ spanned by all c_{ij} with $i, j \in I(n,r)$ with i having content α. Then we have a $G(n)$-module decomposition $A(n,r) = \bigoplus_{\alpha \in \Lambda(n,r)} {}^\alpha A(n,r)$. Let $e \in S(n,r)$ be given by $e(f) = \epsilon(\pi(f))$, where $\pi \colon A(n,r) \to {}^\omega A(n,r)$ is the vector space projection (and $\omega = (1,1,\ldots,1,0,\ldots,0)$).

The Hecke algebra $\mathrm{Hec}(r)$ is the associative k-algebra defined by generators $T_1, \ldots, T_{r-1}$ subject to the relations:

$$
\begin{aligned}
(T_i - q)(T_i + 1) &= 0; \\
T_i T_{i+1} T_i &= T_{i+1} T_i T_{i+1} && \text{for } 1 \leqslant i \leqslant r-1; \\
T_i T_j &= T_j T_i, && \text{for } 1 \leqslant i,j \leqslant n-1, |i-j| > 1.
\end{aligned}
$$

There is a natural isomorphism $H(r) \to eSe$ (see [23, Section 2.1]) and we use this to identify $H(r)$ with the algebra eSe. We thus have the Schur functor $f \colon \mathrm{mod}(S) \to \mathrm{mod}(H(r))$ as in the case $q = 1$ discussed above.

10 Some recent applications to Hecke algebras

We give a couple of new applications. The first is related to recent work of Hemmer and Nakano and the second to recent work by Fayers and Lyle.

We use tilting modules to give a variant on the recent work of Hemmer and Nakano, [35] showing in particular that for a finite dimensional module X over a Hecke algebra $H(r)$ with parameter q such that $1 + q$ and $1 + q + q^2$ are non-zero the number of occurrences of $\mathrm{Sp}(\lambda)$, as a section is independent of the choice of the filtration. We deduce this from a special case of Proposition 10.5 (ii) which is in itself a generalization to arbitrary q of a result proved in the case $q = 1$ by Kleshschev and Nakano,[43]. (The second application is also derived from this result and general results for quantum general linear groups.)

As usual we denote by $H(r)$ the Hecke algebra of type A on generators $T_1, \ldots, T_{r-1}$, for $r \geqslant 2$, at a parameter $0 \neq q \in k$. We set $H(1) = k$.

Then $H(n)$ has the trivial module k (afforded by the representation sending T_i to q) and the sign module k_s (afforded by the representation sending T_i to -1). For a composition $\alpha = (\alpha_1, \ldots, \alpha_m)$ of r we write $H(\alpha)$ for the Young subalgebra $H(\alpha) \cong H(\alpha_1) \otimes \cdots \otimes H(\alpha_m)$.

We write $M(\lambda) = H(r) \otimes_{H(\alpha)} k$ for the "permutation module" and $M_s(\lambda) = H(r) \otimes_{H(\alpha)} k_s$ for the "signed permutation module", where α is the composition obtained by taking the parts of λ in reverse order. Then we have (cf Section 9,(1)) : $f M(\lambda) \cong S^\lambda E$ and $M_s(\lambda) = f \bigwedge^{\lambda'} E$.

Lemma 10.1. *Suppose $1 + q \neq 0$. Then we have:*

(i) $\mathrm{Hom}_{H(\alpha)}(k_s, k) = 0$ *for any composition $\alpha \neq (1^r)$ of r;*

(ii) $\mathrm{Hom}_{H(r)}(\mathrm{Sp}(\lambda), k) = 0$ *for any partition $\lambda \neq (r)$ of r.*

Proof. (i) Clear. (ii) There is an epimorphism $\bigwedge^{\lambda'} E \to \nabla(\lambda)$ giving rise to an epimorphism $M_s(\lambda) \to \mathrm{Sp}(\lambda)$. Hence $\mathrm{Hom}_{H(r)}(\mathrm{Sp}(\lambda), k)$ embeds in $\mathrm{Hom}_{H(r)}(M_s(\lambda), k) = \mathrm{Hom}_{H(\lambda)}(k_s, k)$, and this is 0 by (i). $\square$

Suppose $r \geqslant 1$ and $r = a + b$. We write $k_s \otimes k$ for the one dimensional $H(a, b) = H(a) \otimes H(b)$-module on which the first tensor factor acts via the sign representation and on which the second factor acts trivially.

If q does not satisfy an equation $q^m = 1$ for some $m \leqslant r$ then $\mathrm{Hec}(r)$ is semisimple (see e.g. [26, 1.4]) and what follows is trivially true. So we assume now that q is a root of unity and let l be the smallest positive integer such that $1 + q + \cdots + q^{l-1} = 0$.

Lemma 10.2. *Suppose* $r \geqslant 2$. *Then we have* $\mathrm{Ext}^d_{H(r)}(k_s, k) = \mathrm{Ext}^d_{H(r)}(k, k_s) = 0$ *in all degrees* $0 \leqslant d < l - 2$.

Proof. We have the involutory algebra anti-automorphism $\sigma : H(r) \to H(r)$, defined by $\sigma(T_i) = T_i$, $1 \leqslant i < r$. If V is a left $H(r)$-module then the dual space $V^* = \mathrm{Hom}_k(V, k)$ is naturally a left $H(r)$-module, with action given by $(h\alpha)(v) = \alpha(\sigma(h)v)$, $h \in H(r)$, $\alpha \in V^*$, $v \in V$. Moreover, the natural isomorphism $\mathrm{Hom}_{H(r)}(V, W)^* = \mathrm{Hom}_{H(r)}(W^*, V^*)$ (for $V, W \in \mathrm{mod}(H(r))$) extends in all degrees to give an isomorphism of vector spaces $\mathrm{Ext}^i_{H(r)}(V, W) \cong \mathrm{Ext}^i_{H(r)}(W^*, V^*)$, $i \geqslant 0$. We have $k_s^* \cong k_s$ and $k^* \cong k$ so that $\mathrm{Ext}^i_{H(r)}(k_s, k) \cong \mathrm{Ext}^i_{H(r)}(k, k_s)$ and it is enough to prove $\mathrm{Ext}^d_{H(r)}(k, k_s) = 0$ for all $0 \leqslant d < l - 2$.

Suppose for a contradiction that the Lemma is false and let r be as small as possible for which it fails. Suppose further that $d < l - 2$ is as small as possible such that $\mathrm{Ext}^d_{H(r)}(k, k_s) \neq 0$. Then $\mathrm{Hec}(r)$ is not semisimple so we have $r \geqslant l > d + 2$, in particular $r > 2$.

We choose $n \geqslant r$ and let $S = S(n, r)$. We identify $H(r)$ with eSe for the idempotent $e \in S$ considered in Section 9. We shall make use of the Schur functor $f : \mathrm{mod}(S) \to \mathrm{mod}(H(r))$.

We have the Koszul resolution

$$0 \to \bigwedge{}^r E \to \bigwedge{}^{r-1} E \otimes E \to \bigwedge{}^{r-1} E \otimes S^2 E \to \cdots$$

$$\to E \otimes S^{r-1} E \to S^r E \to 0$$

(cf [23, Section 4.8]). Applying the Schur functor we get an exact sequence

$$0 \to f \bigwedge{}^r E \to f(\bigwedge{}^{r-1} E \otimes E) \to f(\bigwedge{}^{r-2} E \otimes S^2 E) \to \cdots$$

$$\to f(E \otimes S^{r-1} E) \to f S^r E \to 0$$

of $H(r)$-modules. We leave it to the reader to check that

$f(\bigwedge^a E \otimes S^b E) \cong H(r) \otimes_{H(a,b)} (k_s \otimes k)$ for $r = a + b$ (cf [24, Lemma 3.1a]). In particular we have $f S^r E = k$.

We recall a fact from elementary homological algebra. Let R be a ring, V, W left R-modules and $t \geqslant 0$. If

$$\cdots \to X_1 \to X_0 \to V \to 0$$

is an exact sequence of R-modules then $\operatorname{Ext}_R^t(V, W) = 0$ provided that $\operatorname{Ext}_R^i(X_j, W) = 0$ whenever $i + j \leqslant t$.

Thus we must have $\operatorname{Ext}_{H(r)}^i(f(\bigwedge^{j+1} E \otimes S^{r-1-j} E), k_s) \neq 0$ for some $i + j \leqslant d$, equivalently $\operatorname{Ext}_{H(r)}^i(f(\bigwedge^j E \otimes S^{r-j} E), k_s) \neq 0$ for some $i + j \leqslant d + 1$, $j > 0$. Note that $i + j \leqslant d + 1 < d + 2 < l \leqslant r$ gives $j \leqslant r - 2$.

Now we have $f(E \otimes S^{r-1} E) = H(r) \otimes_{H(1,r-1)} k$ and so if $j = 1$ we get, by "Shapiro's Lemma"

$$0 \neq \operatorname{Ext}_{H(r)}^i(H(r) \otimes_{H(1,r-1)} k, k_s) = \operatorname{Ext}_{H(r-1)}^i(k, k_s)$$

contradicting the minimality of r. Hence we have $j > 1$. Now we have

$$\begin{aligned}
&\operatorname{Ext}_{H(r)}^i(f(\bigwedge^j E \otimes S^{r-j} E), k_s) \\
&= \operatorname{Ext}_{H(r)}^i(H(r) \otimes_{H(j,r-j)} (k_s \otimes k), k_s) \\
&= \operatorname{Ext}_{H(j) \otimes H(r-j)}^i(k_s \otimes k, k_s \otimes k_s)
\end{aligned}$$

which, by the Künneth formula, is

$$\bigoplus_{i = i_1 + i_2} \operatorname{Ext}_{H(j)}^{i_1}(k_s, k_s) \otimes \operatorname{Ext}_{H(r-j)}^{i_2}(k, k_s)$$

so that some $\operatorname{Ext}_{H(j)}^{i_1}(k_s, k_s) \otimes \operatorname{Ext}_{H(r-j)}^{i_2}(k, k_s)$ is non-zero. But, for $i + j \leqslant d + 1$ and $i = i_1 + i_2$, we get $i_2 \leqslant i \leqslant d + 1 - j < d$. We also have $r - j \geqslant 2$ and so get $\operatorname{Ext}_{H(r-j)}^{i_2}(k_s, k) = 0$ by minimality. This contradiction completes the proof. $\qquad\square$

Remark 10.3. To get that Specht module filtrations are well defined one only needs $\operatorname{Ext}_{H(r)}^1(k_s, k) = 0$ for $1 + q, 1 + q + q^2 \neq 0$ and this can be done by a more elementary argument showing that any extension of k by k_s splits (by considering the action of each generator T_i on such an extension).

Proposition 10.4. *If $X \in \operatorname{mod}(H(r))$ admits a Specht filtration then $\operatorname{Ext}_{H(r)}^d(X, M(\mu)) = 0$ for all $0 < d < l - 2$ and partitions μ of r.*

Proof. It is enough to consider $X = \operatorname{Sp}(\lambda)$ for λ a partition of r. Assume, for a contradiction, that the result is false and d is as small as possible such that some $\operatorname{Ext}_{H(r)}^d(\operatorname{Sp}(\lambda), M(\mu)) \neq 0$. By Shapiro's Lemma we have $\operatorname{Ext}_{H(\mu)}^d(\operatorname{Sp}(\lambda), k) \neq 0$. Moreover, the restriction of $\operatorname{Sp}(\lambda)$ to $H(\mu)$ has a filtration with sections $\operatorname{Sp}(\tau_1) \otimes \cdots \otimes \operatorname{Sp}(\tau_m)$, where τ_i is a partition of μ_i (see [23, Section 4.6]). Thus, by the Künneth formula once more, we are reduced to proving that $\operatorname{Ext}_{H(r)}^d(\operatorname{Sp}(\lambda), k) = 0$. We do this by induction on λ. For $\lambda = (1^r)$ this is true by Lemma 10.2. Now assume $\lambda \neq (1^r)$.

We have an exact sequence $0 \to J(\lambda) \to M_s(\lambda) \to \mathrm{Sp}(\lambda) \to 0$, where $J(\lambda)$ is filtered by Specht modules $\mathrm{Sp}(\mu)$ with $\mu < \lambda$. (The module $\bigwedge^{\lambda'} E$ has a good filtration with top term $\nabla(\lambda)$ and there is an exact sequence $0 \to K(\lambda) \to \bigwedge^{\lambda'} E \to \nabla(\lambda) \to 0$, where $K(\lambda)$ has a good filtration. Applying the Schur functor to this gives the required short exact sequence of $H(r)$-modules.) Thus we get an exact sequence

$$\mathrm{Ext}_{H(r)}^{d-1}(J(\lambda), k) \to \mathrm{Ext}_{H(r)}^d(\mathrm{Sp}(\lambda), k) \to \mathrm{Ext}_{H(r)}^d(M_s(\lambda), k).$$

We have $\mathrm{Ext}_{H(r)}^d(M_s(\lambda), k) = \mathrm{Ext}_{H(\lambda)}^d(k_s, k) = 0$, by Lemma 10.2. If $d = 1$ then $\mathrm{Ext}_{H(r)}^{d-1}(J(\lambda), k) = \mathrm{Hom}_{H(r)}(J(\lambda), k) = 0$ by Lemma 10.1(ii). If $d > 1$ then $\mathrm{Ext}_{H(r)}^{d-1}(J(\lambda), k) = 0$ by the inductive assumption, and hence $\mathrm{Ext}_{H(r)}^d(\mathrm{Sp}(\lambda), M(\mu)) = 0$. $\qquad\square$

We also have the "inverse Schur functor" $g: \mathrm{mod}(eSe) \to \mathrm{mod}(S)$ (see e.g. [34, Chapter 6]) given on objects by $gX = S \otimes_{eSe} X$. This functor is right exact and we consider its left derived functors $L^d g$.

Proposition 10.5. *If $X \in \mathrm{mod}(eSe)$ admits a Specht series then we have:*

(i) *$L^d g X = 0$ for $0 < d < l - 2$; and*

(ii) *$\mathrm{Ext}_S^d(gX, Y) = \mathrm{Ext}_{eSe}^d(X, fY)$, for $Y \in \mathrm{mod}(S)$ and $0 \leqslant d < l - 2$.*

Proof. (i) For $X \in \mathrm{mod}(eSe)$, $Y \in \mathrm{mod}(S)$ we have $\mathrm{Hom}_S(gX, Y) \cong \mathrm{Hom}_{eSe}(X, eY) = \mathrm{Hom}_{eSe}(X, fY)$ and hence a factorization of functors $\mathrm{Hom}_{eSe}(-, fY) = \mathrm{Hom}_S(-, Y) \circ g$. Moreover, g takes projective modules to projective modules so we get a Grothendieck spectral sequence, with second page $\mathrm{Ext}_S^i(L^j gX, Y)$, converging to $\mathrm{Ext}_{eSe}^*(X, fY)$. If Y is injective then this degenerates and we get

$$\mathrm{Hom}_S(L^i gX, Y) \cong \mathrm{Ext}_{eSe}^i(X, fY).$$

Thus we have $L^d gX = 0$ provided that $\mathrm{Ext}_{eSe}^d(X, fY) = 0$ for every injective $Y \in \mathrm{mod}(S)$. The injective indecomposable S-modules are summands of modules of the form $S^\mu E$ and $f S^\mu E = M(\mu)$. Hence it suffices to prove that $\operatorname{Ext}^d_{H(r)}(X, M(\mu)) = 0$ when X has a Specht series. This is true by Proposition 10.4.

(ii) This follows from (i) and the degeneration of the spectral sequence considered in the proof of (i).   □

Proposition 10.6. *Suppose* $1 + q \neq 0$ *and* $1 + q + q^2 \neq 0$.

(i) *We have* $g\operatorname{Sp}(\lambda) = \nabla(\lambda)$, *for* λ *a partition of* r.

(ii) *If* $X \in \operatorname{mod}(eSe)$ *has Specht series* $0 = X_0 < X_1 < \ldots < X_m = X$ *and* $0 = X'_0 < X'_1 < \ldots < X'_n = X$ *with* $X_i/X_{i-1} \cong \operatorname{Sp}(\lambda_i)$, $X'_j/X'_{j-1} \cong \operatorname{Sp}(\mu_j)$, *for partitions* λ_i, μ_j, *for* $1 \leqslant i \leqslant m$, $1 \leqslant j \leqslant n$, *then* $m = n$ *and for each partition* τ *we have* $|\{1 \leqslant i \leqslant m \mid \lambda_i = \tau\}| = |\{1 \leqslant j \leqslant n \mid \mu_j = \tau\}|$.

Proof. (i) The natural map $Se \otimes_{eSe} e\nabla(\lambda) \rightarrow \nabla(\lambda)$ is onto, by [23, 4.5.4], so that $\dim g\operatorname{Sp}(\lambda) \geqslant \nabla(\lambda)$, for all partitions λ of r. Now $Se \cong E^{\otimes r}$ has a ∇-filtration and $(Se : \nabla(\lambda)) = \dim \operatorname{Sp}(\lambda) \neq 0$, for λ a partition of r. Let $0 = X_0 < X_1 < \ldots < X_m = Se$ be such a filtration, with $X_i/X_{i-1} \cong \nabla(\lambda_i)$, for $1 \leqslant i \leqslant m$. Thus eSe has a filtration $0 = eX_0 < eX_1 < \ldots < eX_m = eSe$, with $eX_i/eX_{i-1} \cong \operatorname{Sp}(\lambda_i)$, for $1 \leqslant i \leqslant m$. Then, by Proposition 10.5(i), we have a filtration $0 = Y_0 < Y_1 < \cdots < Y_m = Se$ with $Y_i/Y_{i-1} \cong g\operatorname{Sp}(\lambda_i)$, $1 \leqslant i \leqslant m$. Thus $\dim Se = \sum_{i=1}^m \dim g\operatorname{Sp}(\lambda_i) = \sum_{i=1}^m \dim \nabla(\lambda_i)$ and it follows that $\dim g\operatorname{Sp}(\lambda_i) = \dim \nabla(\lambda_i)$ for all i and hence $g\operatorname{Sp}(\lambda) \cong \nabla(\lambda)$, for all λ.

(ii) This follows from (i) by applying g to X and computing ∇-filtration multiplicities.   □

We now turn our attention to a generalization of recent results of Fayers and Lyle, [31], on homomorphisms between certain Specht modules for symmetric groups. Similar results for decomposition numbers were proved in various degrees of generality by Gordon James and the author (see [23, Section 4.2]) for the result and history).

We fix a positive integer h. For a partition $\lambda = (\lambda_1, \lambda_2, \ldots)$ we define $\lambda^t = (\lambda_1, \ldots, \lambda_h)$ and $\lambda^b = (\lambda_{h+1}, \lambda_{h+2}, \ldots)$, the top and bottom parts of λ. We shall say that a pair of partitions (λ, μ) admits a *horizontal h-cut* if we have $|\lambda| = |\mu|$ and $\lambda_1 + \cdots + \lambda_h = \mu_1 + \cdots + \mu_h$. One similarly defines λ^l, λ^r, the left and right parts of the partition λ and the notion of a pair of partitions (λ, μ) admitting a *vertical h-cut* (for more details see e.g. [23, Section 4.2]).

For $\lambda \in X^+(n)$ we write $\nabla_n(\lambda)$ (instead of $\nabla(\lambda)$) for the induced module defined by λ if we wish to emphasize the role of n (as in Section 8). Recall (see e.g.[23]) that associated to $G(n)$ we have a root system $\Phi \subset X(T(n))$ and set of simple roots $\Pi = \{\alpha_1, \ldots, \alpha_n\}$, where $\alpha_i = \epsilon_i - \epsilon_{i+1}$, for $1 \leqslant i < n$, and $\epsilon_i = (0, \ldots, 0, 1, 0, \ldots, 0) \in X(T(n))$ (with 1 in the ith position). Now the assumption that (λ, μ) admits a horizontal h-cut is that $\lambda - \mu \in \mathbb{Z}\Sigma$, where $\Sigma = \Pi\backslash\{\alpha_h\}$. By [23, 4.2(17)] we therefore have that $\mathrm{Ext}^i_{G(n)}(\nabla(\lambda), \nabla(\mu)) = \mathrm{Ext}^i_{G_\Sigma}(\nabla_\Sigma(\lambda), \nabla_\Sigma(\mu))$, where G_Σ is the Levi subgroup corresponding to G_Σ, which we identify with $G(h) \times G(n-h)$ in the usual way, and $\nabla_\Sigma(\lambda), \nabla_\Sigma(\mu)$ are the corresponding induced modules for $G(h) \times G(n-h)$, which we identify with $\nabla_h(\lambda^t) \otimes \nabla_{n-h}(\lambda^b)$ and $\nabla_h(\mu^t) \otimes \nabla_{n-h}(\mu^b)$. Thus we have

$$\mathrm{Ext}^d_{G(n)}(\nabla(\lambda), \nabla(\mu))$$
$$= \mathrm{Ext}^d_{G(h)\times G(n-d)}(\nabla_h(\lambda^t) \otimes \nabla_{n-h}(\lambda^b), \nabla_h(\mu^t) \otimes \nabla_{n-h}(\mu^b))$$

and so, by the Künneth formula, obtain

$$\mathrm{Ext}^d_{G(n)}(\nabla(\lambda), \nabla(\mu))$$
$$= \bigoplus_{d=i+j} \mathrm{Ext}^i_{G(h)}(\nabla(\lambda^t), \nabla(\mu^t)) \otimes \mathrm{Ext}^j_{G(n-h)}(\nabla(\lambda^b), \nabla(\mu^b)).$$

Note, in particular, that if λ and μ are partitions into at most m parts then (λ, μ) admits a horizontal m-cut and we get $\mathrm{Ext}^d_{G(m)}(\nabla_m(\lambda), \nabla_m(\mu)) = \mathrm{Ext}^d_{G(n)}(\nabla_n(\lambda), \nabla_n(\mu))$ so we just write $\mathrm{Ext}^d(\nabla(\lambda), \nabla(\mu))$ for $\mathrm{Ext}^d_{G(r)}(\nabla_r(\lambda), \nabla_r(\mu))$, $r \geqslant m$. Returning to the case of a horizontal h-cut, we thus have

$$\mathrm{Ext}^d(\nabla(\lambda), \nabla(\mu)) = \bigoplus_{d=i+j} \mathrm{Ext}^i(\nabla(\lambda^t), \nabla(\mu^t)) \otimes \mathrm{Ext}^j(\nabla(\lambda^b), \nabla(\mu^b))$$

$$(*).$$

We now get the generalization of the results of Fayers and Lyle, [31].

Proposition 10.7. *Suppose that (λ, μ) is a pair of partitions of r, admitting a horizontal h-cut or a vertical h-cut for some $1 \leqslant h < r$. Then, for all $0 \leqslant d < l - 2$, we have*

$$\mathrm{Ext}^d_{H(r)}(\mathrm{Sp}(\lambda), \mathrm{Sp}(\mu))$$
$$= \bigoplus_{d=i+j} \mathrm{Ext}^i_{H(h)}(\mathrm{Sp}(\lambda^t), \mathrm{Sp}(\mu^t)) \otimes \mathrm{Ext}^j_{H(r-h)}(\mathrm{Sp}(\lambda^b), \mathrm{Sp}(\mu^b))$$

in the case of a horizontal h-cut and

$$\mathrm{Ext}^d_{H(r)}(\mathrm{Sp}(\lambda), \mathrm{Sp}(\mu))$$
$$= \bigoplus_{d=i+j} \mathrm{Ext}^i_{H(h)}(\mathrm{Sp}(\lambda^l), \mathrm{Sp}(\mu^l)) \otimes \mathrm{Ext}^j_{H(r-h)}(\mathrm{Sp}(\lambda^r), \mathrm{Sp}(\mu^r))$$

in the case of a vertical h-cut.

The statement for horizontal h-cuts follows from (*) and Proposition 10.5(ii) with $X = \mathrm{Sp}(\lambda)$, $Y = \nabla(\mu)$. The statement for vertical h-cuts may be deduced from that for horizontal h-cuts by dualizing and using the q-analogue of the usual isomorphism $\mathrm{Sp}(\lambda) \otimes k_s \cong \mathrm{Sp}(\lambda')^*$ (cf. [23, Proposition 4.5.9]) to interchange vertical and horizontal cuts (see also [19, (3.9) Corollary]).

The conscientious reader will have noticed that the case in which k has characteristic 3 and $q = 1$, of Fayers and Lyle's results, are not covered by the Proposition. However, by a result of Dipper and James. [10, 8.6 Corollary], (generalizing a result of Carter and Lusztig, [5, Theorem 3.7]) we have, $\mathrm{Hom}(\Delta(\mu), \Delta(\lambda)) = \mathrm{Hom}_{H(r)}(\mathrm{Sp}(\lambda), \mathrm{Sp}(\mu))$ and hence, by duality, $\mathrm{Hom}(\nabla(\lambda), \Delta(\mu)) = \mathrm{Hom}_{H(r)}(\mathrm{Sp}(\lambda), \mathrm{Sp}(\mu))$, for $l > 2$. Thus (*) also gives the Proposition in the case $l = 3$. I am grateful to Alison Parker for this remark.

References

[1] A. M. ADAMOVICH AND G.L. RYBNIKOV, Tilting modules for classical groups and Howe duality in positive characteristic, Transform. Groups **1** (1996), 1-34

[2] H. H. ANDERSEN, Tilting modules for algebraic groups, in: *Algebraic Groups and their Representations*, Proceedings of the NATO Advanced Study Institute on Molecular Representations and Subgroup Structure of Algebraic Groups and Related Finite Groups, Editors R. W. Carter and J. Saxl, pp 25-42, Kluwer, 1998

[3] A. BOREL, Linear Algebraic Groups, Second Edition, Graduate Texts in Mathematics 126, New York/Berlin/Heidelberg/London/Paris/Tokyo/Hong Kong/Barcelona, Springer 1991

[4] R. BRYANT, Free Lie algebras and Adams operations, Journal of the London Math. Society, **68** (2003), 355-370

[5] R. W. CARTER AND G. LUSZTIG, On the modular representations of the general linear and symmetric groups, Math. Zeit. **136** (1974), 193–242

[6] E. CLINE, B. PARSHALL, L. L. SCOTT AND W. VAN DER KALLEN, Rational and generic cohomology, Invent. Math. **39**, (1977), 143–163

[7] E. CLINE, B. PARSHALL AND L. L. SCOTT, Finite dimensional algebras and highest weight categories, J. reine angew. Math. **391** (1988), 85–99

[8] A. G. COX, The tilting tensor product theorem and decomposition numbers for symmetric groups, to appear in Algebras and Representation Theory

[9] H. DERKSEN AND J. WEYMAN, Semi-invariants for quivers with relations, *J. Algebra* **258**, (2002), 216-227

[10] R. DIPPER AND G. D. JAMES, q-tensor space and q-Weyl modules, Trans. Amer. Math. Soc. **327** (1991), 251-282

[11] V. DLAB AND C. M. RINGEL, The module theoretic approach to quasi-hereditary algebras, in H. Tachiwara and S. Brenner (eds), *Representations of Algebras and Related Topics*, pp. 200-224, London Math. Soc. Lecture Note Series **168**, Cambridge University Press 1992

[12] M. DOMOKOS AND A. N. ZUBKOV, Semi-invariants of quivers as determinants, Transform Groups. **6**, (2000), 9-24

[13] S. DONKIN, On a question of Verma, J. Lond. Math. Soc (2) **21** (1980), 445–455

[14] S. DONKIN, A filtration for rational modules, *Math. Zeit.* **177** (1981), 1–8

[15] S. DONKIN, Rational Representations of Algebraic Groups : Tensor Products and Filtrations *Lecture Notes in Mathematics* **1140**, Springer 1985, Berlin/Heidelberg/New York

[16] S. DONKIN, On Schur algebras and related algebras I, *J. Algebra* **104** (1986), 310–328

[17] S. DONKIN, On Schur algebras and related algebras II, *J. Algebra* **111** (1987), 354–364

[18] S. DONKIN, Good filtrations of rational modules for reductive groups. *The Arcata Conference on Representations of Finite Groups*, Part 1, 69–80, Proceedings of Symposia in Pure Mathematics Volume 47, Amer. Math. Soc. 1987

[19] S. DONKIN, On tilting modules for algebraic groups, *Math. Zeit.* **212**,(1993), 39–60

[20] S. DONKIN, Invariants of several matrices, *Invent. Math.* **110** (1992), 389–401

[21] S. DONKIN, On Schur algebras and related algebras IV : The blocks of the Schur algebras *J. Algebra* **168** (1994), 400–429

[22] S. DONKIN, On tilting modules and invariants for algebraic groups, *Representations of Algebras and Related Topics*,pp. 59–77, V. Dlab and L.L. Scott, (Ed.), Kluwer, Dordrecht/Boston/London 1994

[23] S. DONKIN, The *q*-Schur algebra, LMS Lecture Notes, CUP 1998

[24] S. DONKIN, Symmetric and exterior powers, linear source modules and representations of Schur superalgeras, *Proc. Lond. Math. Soc.* **83**, (2001), 647-680

[25] S. DONKIN, A note on the characters of the cohomology of the induced vector bundles on G/B in characteristic p, *J. Algebra* **258** (2002), 255-274

[26] S. DONKIN, Representations of Hecke algebras and Characters of Symmetric Groups, In: Studies in Memory of Issai Schur, Edited by Antony Joseph, Anna Melnikov and Rudolf Rentschler, *Progress in Mathematics* **210**, pp. 49-67, Birkhšer 2003

[27] S. DONKIN, On the cohomology of line bundles on the three dimensional flag variety, in characteristic p, in preparation

[28] S. DOTY, Polynomial representations, algebraic monoids, and Schur algebras of classical type, *Journal of Pure and Applied Algebra* **123** (1998), 165-199

[29] K. ERDMANN, Symmetric groups and quasi-hereditary algebras,*Finite Dimensional Algebras and Related Topics*,pp. 123–161, V. Dlab and L.L. Scott, (Ed.), Kluwer, Dordrecht/Boston/London 1994

[30] K. ERDMANN, Decomposition numbers for symmetric groups and composition factors of Weyl modules,*J. Algebra* **180** (1996), 316–320

256   S. Donkin

[31] MATTHEW FAYERS AND SINÉAD LYLE, Row and column theorems for homomorphisms between Specht modules, *Journal of Pure and Applied Algebra* **185** (2003), 147-164

[32] E. FRIEDLANDER AND B. PARSHALL, Support varieties for restricted Lie algebras, *Invent Math.* **86** (1986), 553-562

[33] J. A. GREEN, Locally finite representations, *J. Algebra* **41** (1976), 131-171

[34] J. A. GREEN, *Polynomial Representations of GL_n*, Lecture Notes in Mathematics **830**, Springer, Berlin/Heidelberg/New York 1980

[35] D. J. HEMMER AND D. K. NAKANO, Specht filtrations for Hecke algebras of type A, *J. Lond. Math. Soc.* **69** (2004), 623-638

[36] G. D. JAMES, The Representation Theory of the Symmetric Groups, *Lecture Notes in Mathematics* **682**, Springer 1978, Berlin/Heidelberg/New York

[37] G. D. JAMES, Trivial source modules for symmetric groups, *Arch. Math.* **41** (1983), 294-300

[38] J. C. JANTZEN, Über Darstellungen höherer Frobenius-Kerne halbeinfacher algebraischer Gruppen, *Math. Z.* **164** (1979), 271-292

[39] J. C. JANTZEN, *Representations of Algebraic Groups* (Second Edition), Mathematical Surveys and Monographs, **107**, American Mathematical Society, 2003

[40] J. C. JANZTEN, Support varieties of Weyl modules, *Bull. Lond. Math. Soc.* **19**, (1987),238-244

[41] W. VAN DER KALLEN, Longest weight vectors and excellent filtrations, *Math. Zeit* **201**, (1989), 19-31

[42] W. VAN DER KALLEN, Lectures on Frobenius splittings and B-modules, *Tata Institute of Fundamental Research*, Springer 1993

[43] A. S. KLESHSCHEV AND D. K. NAKANO, On comparing the cohomology of general linear groups and symmetric groups, *Pacific J. Math.* **201**, (2001),339-355

[44] I. G. MACDONALD, *Symmetric functions and Hall polynomials*, Second Edition, Oxford University Press, 1995

[45] O. MATHIEU, Tilting modules and their applications, In: *Analysis on Homogeneous Spaces and Representation Theory of Lie Groups*, pp.145-212, Advanced Studies in Pure Mathematics 26, 2000

[46] V. OSTRIK, Tensor ideals in the category of tilting modules, *Transform. Groups* **2**, (1997), 279-287

[47] C. PROCESI, The invariant theory of $n \times n$-matrices, *Advances in Mathematics* **19** (1976), 306-381

[48] I. Reiten, Tilting theory and homologically finite subcategories with applications to quasihereditary algebras, this volume.

[49] C. M. RINGEL, The category of modules with good filtrations over a quasi-hereditary algebra has almost split sequences, *Math. Zeitschrift* **208**(1991),209-225

Stephen Donkin
Department of Mathematics
University of York

Heslington, York YO10 5DD, England
sd510@york.ac.uk

10

Combinatorial aspects of the set of tilting modules

Luise Unger

1 Introduction

Let Λ be a basic artin algebra over a commutative artin ring R and let $\mathrm{mod}\,\Lambda$ be the category of finitely generated left Λ-modules. For a module $M \in \mathrm{mod}\,\Lambda$ we denote by $\mathrm{pd}\,M$ the projective dimension of M and by gl.dim Λ the global dimension of Λ.

A module $T \in \mathrm{mod}\Lambda$ is called a *tilting module* if

(i) the projective dimension of T is finite, and

(ii) $\mathrm{Ext}_\Lambda^i(\mathrm{T},\mathrm{T}) = 0$ for all $i > 0$, and

(iii) there is an exact sequence $0 \to {}_\Lambda\Lambda \to T^1 \to \cdots \to T^d \to 0$ with $T^i \in \mathrm{add}T$ for all $1 \leqslant i \leqslant d$.

Here addT denotes the category of direct sums of direct summands of T. We say that a tilting module is *basic* if in a direct sum decomposition of T the indecomposable direct summands of T occur with multiplicity one. Unless stated otherwise all tilting modules considered here will be assumed to be basic. We consider the set $\mathcal{T}_\Lambda$ of all tilting modules over Λ up to isomorphism.

It was observed by Ringel that $\mathcal{T}_\Lambda$ carries the structure of a simplicial complex whose geometric realization in some examples turned out to be an n-dimensional ball. Here $n + 1$ is the number of isomorphism classes of simple Λ-modules. During a ring theory conference in Antwerp 1987 he proposed to to study this complex. This was the starting point for the investigations of the combinatorial structure of $\mathcal{T}_\Lambda$.

The first article that dealt with the combinatorics of $\mathcal{T}_\Lambda$ was due to Riedtmann and Schofield [33], published in 1991. In this article Riedtmann and Schofield introduced three combinatorial setups: the partial order of tilting modules, the quiver of tilting modules and the simplicial complex of tilting modules. All these combinatorial structures were later studied and generalized by various authors. The aim of this article is to summarize the results in this context.

The notation and terminology introduced here will be fixed throughout the article. For unexplained representation-theoretical terminology we refer to [34] or [4].

2 The partial order of tilting modules

Following [3] we consider for a tilting module T the right perpendicular category

$$T^\perp = \{X \in \text{mod}\,\Lambda \mid \text{Ext}^i_\Lambda(T, X) = 0 \text{ for all } i > 0\}.$$

We define a partial order $\leqslant$ on $\mathcal{T}_\Lambda$ (compare [33]). For $T, T' \in \mathcal{T}_\Lambda$ we set $T \leqslant T'$ provided $T^\perp \subseteq T'^\perp$. Then $(\mathcal{T}_\Lambda, \leqslant)$ is a partially ordered set. The following proposition lists some elementary properties of $(\mathcal{T}_\Lambda, \leqslant)$.

Proposition 2.1.

(a) $T \leqslant T'$ if and only if $T \in T'^\perp$.

(b) $_\Lambda\Lambda$ is the unique maximal element.

(c) If Λ is Gorenstein, then the injective cogenerator $D\Lambda_\Lambda$ of modΛ is the unique minimal element.

(d) If $T \leqslant T'$ in $(\mathcal{T}_\Lambda, \leqslant)$ then pd$T \geqslant$ pdT'.

The partial order does not always admit a minimal element – an example where this situation occurs may be found in [27]. This raises the question under which circumstances minimal elements do exist and whether they are unique. This relates to a further concept in representation theory, namely to contravariantly finite subcategories of modΛ, introduced by Auslander and Smalø in [5]. Let $\mathcal{C}$ be a full subcategory of modΛ which is closed under direct sums, direct summands and isomorphisms. The subcategory $\mathcal{C}$ is called *contravariantly finite* in modΛ, if every $X \in$ modΛ has a right $\mathcal{C}$-approximation, i.e. there is a morphism $F_X \to X$ with $F_X \in \mathcal{C}$ such that the induced morphism $\mathrm{Hom}_\Lambda(C, F_X) \to \mathrm{Hom}_\Lambda(C, X)$ is surjective for all $C \in \mathcal{C}$. We denote by $\mathcal{P}^{<\infty}(\Lambda)$ the full subcategory of $\mathrm{mod}\Lambda$ of modules of finte projective dimensions. The answer to the question above was given in [23].

Theorem 2.2. *The partial order* $(\mathcal{T}_\Lambda, \leqslant)$ *contains a minimal element if and only if* $\mathcal{P}^{<\infty}(\Lambda)$ *is contravariantly finite in* $\mathrm{mod}\Lambda$. *Moreover, a minimal element is unique if it exists.*

3 The quiver of tilting modules

Riedtmann and Schofield related with $\mathcal{T}_\Lambda$ a quiver $\overrightarrow{\mathcal{K}_\Lambda}$ as follows. The vertices of $\overrightarrow{\mathcal{K}_\Lambda}$ are the elements of $\mathcal{T}_\Lambda$. For $T, T' \in \mathcal{T}_\Lambda$ we set $T' \to T$ in case $T' = M \oplus X$, $T = M \oplus Y$ with X, Y indecomposable and there is a non-split short exact sequence

$$0 \to X \to \widetilde{M} \to Y \to 0$$

with $\widetilde{M} \in \mathrm{addM}$. We call $\overrightarrow{\mathcal{K}_\Lambda}$ the *quiver of tilting modules* over Λ. This setup was used essentially in [39] and [42] and has been extended to infinite dimensional tilting modules in [8].

3.1 Examples

The quiver $\overrightarrow{\mathcal{K}_\Lambda}$ has a unique source, the projective tilting module $_\Lambda\Lambda$. Moreover, $\overrightarrow{\mathcal{K}_\Lambda}$ does not contain oriented cycles. We include some examples to stress further properties of $\overrightarrow{\mathcal{K}_\Lambda}$.

The first series of examples classifies the quivers of tilting modules over hereditary algebras with three isomorphism classes of sinple modules as it was given in [40].

If A is the path algebra of

$$\bullet \longrightarrow \bullet \longrightarrow \bullet \quad \text{or} \quad \bullet \longleftarrow \bullet \longrightarrow \bullet \quad \text{or} \quad \bullet \longrightarrow \bullet \longleftarrow \bullet$$

then $\overrightarrow{\mathcal{K}_A}$ equals

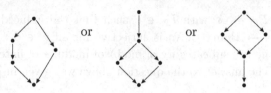

If A is a tame three-point quiver algebra, i.e. if A is the path algebra of the quiver ![triangle quiver], then the quiver of tilting modules decomposes into two connected components:

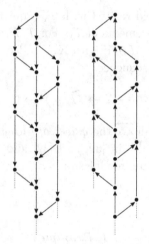

If A is a wild three-point quiver algebra, then $\overrightarrow{\mathcal{K}_A}$ decomposes into infinitely many connected components. All but finitely many of these components are binary trees without a source or sink, i.e. they are of the form

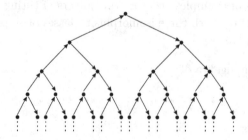

The following example from [33] shows that $\overrightarrow{\mathcal{K}_\Lambda}$ may be finite even for representation infinite algebras. Let Λ be given by the following quiver

with relations $\beta\alpha = \gamma\delta = 0$. Then gl.dim $\Lambda = 2$ and Λ is representation infinite. The quiver $\overrightarrow{\mathcal{K}_\Lambda}$ is

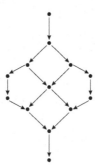

In general $\overrightarrow{\mathcal{K}(\Lambda)}$ may be rather complicated, even when it is finite. One measure for the complicatedness of a graph G is its genus $\gamma(G)$. This is the minimal genus of an orientable surface on which G can be embedded. It was proved in [43] that there are finite quivers of tilting modules of arbitrary genus.

Theorem 3.1. *For all integers $r \geqslant 0$ there is a representation finite, connected algebra Λ_r such that $\gamma(\overrightarrow{\mathcal{K}(\Lambda_r)}) = r$.*

The proof is constructive., namely Λ_1 is the path algebra of the quiver $\overrightarrow{\Delta}_1$

$$
\begin{array}{ccc}
& b & \\
\alpha \nearrow & \circ & \nwarrow \beta \\
a\, \circ & & \circ\, d \\
\gamma \searrow & \circ & \nearrow \delta \\
& c &
\end{array}
$$

bound by the relation $\alpha\beta = \gamma\delta$, and Λ_r for $r > 1$ is the path algebra of the quiver $\overrightarrow{\Delta}_r$

$$
\begin{array}{ccc}
& b & \quad 2 \leftarrow 3 \cdots \circ \leftarrow r \\
\alpha \nearrow & \circ & \nwarrow \beta \\
a\, \circ & & \circ\, d \\
\gamma \searrow & \circ & \nearrow \delta \quad \circ \\
& c & \quad 1
\end{array}
$$

bound by the relations $\alpha\beta = \gamma\delta$ and rad$^2 = 0$, i.e. the composition of two consecutive arrows in $\overrightarrow{\Delta}_r \backslash \{a\}$ is zero.

3.2 The Hasse diagram of the poset of tilting modules

Riedtmann and Schofield already observed in [33] a relation between the partial order $(\mathcal{T}_\Lambda, \leqslant)$ and the quiver $\overrightarrow{\mathcal{K}(\Lambda)}$. They raised the question whether the underlying graph $\mathcal{K}(\Lambda)$ of $\overrightarrow{\mathcal{K}(\Lambda)}$ is the Hasse diagram of $(\mathcal{T}_\Lambda, \leqslant)$. A positive answer to this question was given in [23].

Theorem 3.2. *Let $T, T' \in \mathcal{T}_\Lambda$ with $T \leqslant T'$ and $T \neq T'$. Then there exists $T'' \in \mathcal{T}_\Lambda$ with $T' \to T''$ in $\overrightarrow{\mathcal{K}(\Lambda)}$ such that $T \leqslant T''$. In particular, $\mathcal{K}(\Lambda)$ is the Hasse diagram of $(\mathcal{T}_\Lambda, \leqslant)$.*

A consequence of this theorem is that $\overrightarrow{\mathcal{K}(\Lambda)}$ has the Brauer-Thrall 1 property:

Corollary 3.3. *If $\overrightarrow{\mathcal{K}(\Lambda)}$ has a finite connected component $\mathcal{C}$ then $\overrightarrow{\mathcal{K}(\Lambda)} = \mathcal{C}$.*

A sink in $\overrightarrow{\mathcal{K}(\Lambda)}$ corresponds to a minimal element in $(\mathcal{T}_\Lambda, \leqslant)$. Hence combining this theorem with the theorem in section 2 we obtain the following consequences.

Corollary 3.4. *The quiver $\overrightarrow{\mathcal{K}(\Lambda)}$ contains at most one sink. It contains a sink if and only if $\mathcal{P}^{<\infty}(\Lambda)$ is contravariantly finite in modΛ.*

Corollary 3.5. *If $\overrightarrow{\mathcal{K}(\Lambda)}$ is finite, then $\mathcal{P}^{<\infty}(\Lambda)$ is contravariantly finite in modΛ.*

Corollary 3.6. *Let $T \in \mathcal{T}_\Lambda$ be a tilting module such that $End_\Lambda T$ is representation finite. Then $\mathcal{P}^{<\infty}(\Lambda)$ is contravariantly finite in modΛ.*

3.3 Connected components

We saw in section 3.2 that $\overrightarrow{\mathcal{K}(\Lambda)}$ is connected in case it is finite. The converse does not hold. In general very little is known about the connected components of $\overrightarrow{\mathcal{K}(\Lambda)}$ even in the case where Λ is the path algebra of a quiver $\overrightarrow{\Delta}$. We denote by Δ the underlying graph of $\overrightarrow{\Delta}$.

The following result taken from [39] deals with tame quiver algebras $k\overrightarrow{\Delta}$.

Theorem 3.7. *The following are equivalent for an affine diagram Δ.*

(a) $\overrightarrow{\mathcal{K}}_{mod\,k\overrightarrow{\Delta}}$ *is connected.*

(b) *There is a $k\overrightarrow{\Delta}$-tilting module T such that $End\,T$ is representation finite.*

(c) $\Delta \neq \tilde{\mathbb{A}}_{1,p}$.

Moreover, it is shown in [39] that $\overrightarrow{\mathcal{K}}_{mod\,k\overrightarrow{\Delta}}$ has precisely two connected components if Δ is of type $\tilde{\mathbb{A}}_{1,p}$.

It is conjectured (compare [39] and [24]) that $\overrightarrow{\mathcal{K}}_{mod\,k\overrightarrow{\Delta}}$ will have infinitely many connected components if Δ is a wild diagram and the number of vertices of Δ is greater than 2. This conjecture has been verified for three-point quiver algebras in [40].

3.4 The local structure

In this section we summarize the main results of [25] where the structure of $\overrightarrow{\mathcal{K}(\Lambda)}$ in the neighborhood of a vertex was studied. In this article the precise relationship between the number of neighbors of a vertex $T \in \overrightarrow{\mathcal{K}(\Lambda)}$ and the addT-coresolution of $_\Lambda\Lambda$ in the definition of a tilting module was established. Moreover, the number and length of paths leading to or starting in a given vertex was investigated.

Let us assume first that Λ is the path algebra of a quiver $\overrightarrow{\Delta}$. Let $s(T)$ respectively $e(T)$ be the number of arrows starting respectively ending at a vertex $T \in \overrightarrow{\mathcal{K}(\Lambda)}$. It is easy to see that $s(T) + e(T) \leqslant rkK_0(\Lambda)$, where $rkK_0(\Lambda)$ denotes the rank of the Grothendieck group of modΛ. We call T *saturated* if equality holds. The following result is contained in [41].

Theorem 3.8. *Let Λ be the path algebra of a quiver $\overrightarrow{\Delta}$. Then each connected component of $\overrightarrow{\mathcal{K}(\Lambda)}$ contains a non-saturated vertex.*

There is a purely combinatorial criterion to decide whether or not a vertex T ist saturated.

Proposition 3.9. *Let Λ be the path algebra of a quiver $\overrightarrow{\Delta}$. A vertex $T \in \overrightarrow{\mathcal{K}(\Lambda)}$ is saturated if and only if all entries in the dimension vector of T are at least 2.*

A different approach to compute the number of neighbors of a vertex T in $\overrightarrow{\mathcal{K}(\Lambda)}$ is as follows:

Let

$$0 \to {}_\Lambda\Lambda \to T^0 \to T^1 \to 0$$

and

$$0 \to T_1 \to T_0 \to D\Lambda_\Lambda \to 0$$

be minimal addT-(co)resolutions. For a Λ-module X we denote by $\delta(X)$ the number of pairwise non isomorphic indecomposable direct summands of X. The numbers $s(T)$ and $e(T)$ defined above may be interpreted as follows:

Proposition 3.10. *For a path algebra Λ with n isomophism classes of simple modules we have $s(T) = n - \delta(T_0) = \delta(T_1)$ and $e(T) = n - \delta(T^0) = \delta(T^1)$.*

Hence a module $T \in \overrightarrow{\mathcal{K}(\Lambda)}$ is saturated if and only if $n = \delta(T^0) + \delta(T_0)$. In particular, ${}_\Lambda\Lambda$ is never saturated.

Let us now assume that Λ is an arbitrary artin algebra. For $T \in \overrightarrow{\mathcal{K}(\Lambda)}$ we consider

$$0 \to {}_\Lambda\Lambda \to T^0 \to \cdots \to T^r \to 0$$

and

$$\cdots T_s \to \cdots \to T_0 \to D\Lambda_\Lambda \to 0$$

minimal addT-(co)resolutions.

For an indecomposable direct summand X of T choose $i(X)$ minimal such that X is a direct summand of $T^{i(X)}$. Note that each indecomposable direct summand X of T has to occur in the first exact sequence, hence $i(X)$ is well defined. If X occurs in the second exact sequence, we choose $j(X)$ minimal such that X is a direct summand of $T_{j(X)}$. Otherwise we set $j(X) = \infty$. Note that the validity of the generalized Nakayama conjecture will imply that $j(X) < \infty$. For more details about this connection we refer to the following section. With these notations the main result in [25] states:

Theorem 3.11. *For each indecomposable direct summand X of T there is a path $w(X)$ in $\overrightarrow{\mathcal{K}(\Lambda)}$ of length $i(X)$ ending in T and a path $u(X)$ of length $j(X)$ starting in T. These paths are pairwise disjoint.*

3.5 Complements to partial tilting modules

A direct summand M of a tilting module T is called a *partial tilting module*. We say that a basic module C is a complement to a partial tilting module T if $M \oplus C$ is a tilting module and addM $\cap$ addC $= 0$. In general, a partial tilting module admits many complements. In [22] very special complements were introduced which relate the structure of $\overrightarrow{\mathcal{K}(\Lambda)}$ with homological properties of modΛ.

A complement C to a partial tilting module is called a *source complement* if

$$(M \oplus C)^{\perp} = M^{\perp}.$$

It was shown in [21] that source complements to partial tilting modules do not always exist. Moreover, if M admits a source complement, then it is unique up to isomorphism.

Dually, we define for a partial tilting module the notion of a sink complement. A complement C to M is called a *sink complement* to M if

$$^{\perp}M \cap \mathcal{P}^{<\infty}(\Lambda) = {}^{\perp}(M \oplus C) \cap \mathcal{P}^{<\infty}(\Lambda).$$

The left perpendicular category $^{\perp}M$ is the full subcategory of modΛ with objects Y satisfying $\text{Ext}^{i}(Y, M) = 0$ for all $i > 0$. Again, sink complements need not exist, but they are unique in case they exist.

A partial tilting module M which admits an indecomposable complement is said to be an *almost complete partial tilting module*. Almost complete partial tilting modules always have a source complement. To be more precise, the following is basically contained in [10] or [17].

Theorem 3.12. *Let M be an almost complete partial tilting module. Then M admits a source complement X_0. Moreover, M has at most finitely many complements if and only if M admits a sink complement.*

If there are $r + 1$ non-isomorphic complements for some $r \geqslant 1$, then $r \leqslant fd(\Lambda)$ and there is a long exact sequence

$$0 \longrightarrow X_0 \longrightarrow M^1 \xrightarrow{f_1} M^2 \longrightarrow \cdots \longrightarrow M^{r-1} \xrightarrow{f_{r-1}} M^r \xrightarrow{f_r} X_r \longrightarrow 0$$

with $kerf_i = X_i$ for $1 \leqslant i \leqslant r$, $M^i \in addM$ for $1 \leqslant i \leqslant r$ and $X_0, \ldots, X_r$ complements to M. In particular, if $fd(\Lambda) < \infty$, then M has finitely many complements.

Here $fd(\Lambda) = \sup\{pd_\Lambda X \mid X \in \mathcal{P}^{<\infty}(\Lambda)\}$ is finitistic dimension of Λ. A classical conjecture (see for example [1], [2], [6], [30], [31] and [36]), the so called finitistic dimension conjecture states, that $fd(\Lambda)$ is always finite. This conjecture has been verified for some special classes of artin algebras, see for example [14], [15], [28], [29], [3], and also [37], [38].

Note that the complements $X_0, \ldots, X_r$ of the theorem above give rise to a path $M \oplus X_0 \to \cdots \to M \oplus X_r$ in $\overrightarrow{\mathcal{K}(\Lambda)}$.

It is well known that the finitistic dimension conjecture implies the generalized Nakayama conjecture which states that in a minimal projective resolution of $D\Lambda_\Lambda$ each indecomposable projective Λ-module occurs, or equivalently, for each simple Λ-modules S there is an integer t such that $\text{Ext}^t_\Lambda(D\Lambda_\Lambda, S) \neq 0$. We refer to [2] and [16] for related problems and the precise relationship between these homological conjectures.

There is a connection between the number of complements to almost complete partial tilting modules and the generalized Nakayama conjecture that was observed in [9] and [22]. To formulate this connection we need some further notation. We fix a complete set $\mathcal{S} = \{S_1, \ldots, S_n\}$ of representatives from the isomorphism classes of simple Λ-modules. If $S = S_i \in \mathcal{S}$ for some $1 \leqslant i \leqslant n$ we denote by $P(S)$ the projective cover of S and by $P_S = \bigoplus_{j \neq i} P(S_j)$. Note that P_S is an almost complete partial tilting module with source complement $P(S)$. The following result was proved in [22].

Theorem 3.13. *The following are equivalent:*

(1) *For each simple Λ-module S the category $^\perp P_S \cap \mathcal{P}^{<\infty}(\Lambda)$ is contravariantly finite.*

(2) *P_S has finitely many complements for all simple Λ-modules S.*

(3) *P_S has a sink complement for each simple Λ-module S.*

(4) *The generalized Nakayama conjecture holds for* Λ.

3.6 Generalizations to tilting objects

Let $\mathcal{H}$ be a connected hereditary k-category over an algebraically closed field k such that for all $X, Y \in \mathcal{H}$ we have that both $\mathrm{Hom}_{\mathcal{H}}(X, Y)$ and $\mathrm{Ext}^1_{\mathcal{H}}(X, Y)$ are finite dimensional k-vector spaces. Following [20] we call an object $T \in \mathcal{H}$ a *tilting object* provided $\mathrm{Ext}^1_{\mathcal{H}}(T, T) = 0$ and for $X \in \mathcal{H}$ with $\mathrm{Hom}_{\mathcal{H}}(T, X) = \mathrm{Ext}^1_{\mathcal{H}}(T, X) = 0$ we have that $X = 0$. For an object $X \in \mathcal{H}$ we denote by fac X the full subcategory of $\mathcal{H}$ whose objects are factor objects of objects in addX.

Let $\mathcal{H}$ be as above and assume that $\mathcal{H}$ contains a tilting object. It was proved by Happel in [19] that $\mathcal{H}$ is derived equivalent to the category of finite dimensional modules $\mathrm{mod}\, k\overrightarrow{\Delta}$ over the path algebra $k\overrightarrow{\Delta}$ of a finite quiver $\overrightarrow{\Delta}$ without oriented cycles or the category of coherent sheaves $\mathrm{coh}\mathbb{X}$ over a weighted projective line $\mathbb{X}$ (see for example [13]). In [18] a complete list of all categories derived equivalent to these two standard types is given.

In analogy to the algebra case one defines the set $\mathcal{T}_{\mathcal{H}}$ of all basic tilting objects over $\mathcal{H}$ up to isomorphism. The set $\mathcal{T}_{\mathcal{H}}$ is partially ordered in the following way. For $T, T' \in \mathcal{T}_{\mathcal{H}}$ we set $T \leqslant T'$ if fac $T \subseteq$ fac T'. The set $\mathcal{T}_{\mathcal{H}}$ together with this partial order is denoted by $(\mathcal{T}_{\mathcal{H}}, \leqslant)$. Moreover, one defines the quiver $\overrightarrow{\mathcal{K}_{\mathcal{H}}}$ of tilting objects as for the algebra case. The vertices of $\overrightarrow{\mathcal{K}_{\mathcal{H}}}$ are the elements of $\mathcal{T}_{\mathcal{H}}$ and for $T, T' \in \mathcal{T}_{\mathcal{H}}$ there is an arrow $T' \to T$ if $T' = M \oplus X$, $T = M \oplus Y$ with X, Y indecomposable and there is a short exact sequence

$$0 \to X \to \widetilde{M} \to Y \to 0$$

with $\widetilde{M} \in$ addM.

In [24] the partial order and the quiver of tilting object was investigated. It was shown that the underlying graph $\mathcal{K}_{\mathcal{H}}$ of $\overrightarrow{\mathcal{K}_{\mathcal{H}}}$ is the Hasse diagram of $(\mathcal{T}_{\mathcal{H}}, \leqslant)$. Moreover, it was proved that a vertex $T \in \overrightarrow{\mathcal{K}_{\mathcal{H}}}$ has precisely $n = \mathrm{rk}\, K_0(\mathcal{H})$ neighbors in $\overrightarrow{\mathcal{K}_{\mathcal{H}}}$ provided $\mathcal{H}$ does not contain nonzero projective objects. Here $K_0(\mathcal{H})$ denotes the Grothendieck group of $\mathcal{H}$.

In [24] also the question when $\overrightarrow{\mathcal{K}_{\mathcal{H}}}$ is connected is addressed. A complete answer is given in the cases where that $\mathcal{H}$ is derived equivalent to $\mathrm{mod}\, k\overrightarrow{\Delta}$ with $\overrightarrow{\Delta}$ not wild (i.e. $\overrightarrow{\Delta}$ is Dynkin or affine). If the underlying graph Δ of $\vec{\Delta}$ is Dynkin, then $\overrightarrow{\mathcal{K}_{\mathcal{H}}}$ is always connected. If it is affine, then $\overrightarrow{\mathcal{K}_{\mathcal{H}}}$ is connected unless $\mathcal{H} = \text{modk}\vec{\Delta}$ and $\Delta = \tilde{\mathbb{A}}_{1,n}$. This also covers the case that $\mathcal{H}$ is derived equivalent to $\text{coh}\,\mathbb{X}$ where $\mathbb{X}$ is a domestic weighted projective line. For the remaining cases we suggest the following conjecture as an answer.

Conjecture 3.14. *If $\mathcal{H}$ is derived equivalent to $\text{coh}\,\mathbb{X}$ and $\mathbb{X}$ is tubular or wild or to $\text{modk}\vec{\Delta}$ with $\vec{\Delta}$ wild, then $\overrightarrow{\mathcal{K}_{\mathcal{H}}}$ is connected if and only if $\mathcal{H}$ does not contain nonzero projective objects.*

Part of this conjecture was proved in [24]. It was shown that if $\mathcal{H}$ is derived equivalent to $\text{modk}\vec{\Delta}$ with $\vec{\Delta}$ wild, then $\overrightarrow{\mathcal{K}_{\mathcal{H}}}$ is connected if $\mathcal{H}$ does not contain nonzero projective objects. Investigations by Ringel [35] on tubular algebras in the non algebraically closed case show that a related general conjecture will fail.

4 The simplicial complex of tilting modules

Let $n + 1$ be the number of isomorphism classes of simple Λ modules. The set $\mathcal{T}_\Lambda$ forms a simplicial complex Σ_Λ as follows. The 0-simplices are the indecomposable direct summands of tilting modules $_\Lambda T \in \mathcal{T}_\Lambda$, and $\{T_0, \dots, T_r\}$ is an r-simplex if $\bigoplus_{i=0}^{n} T_i$ is a direct summand of a tilting modules $_\Lambda T \in \mathcal{T}_\Lambda$.

Note that we can recover the underlying graph $\mathcal{K}_\Lambda$ of $\overrightarrow{\mathcal{K}_\Lambda}$ from Σ_Λ and vice versa: The n-simplices correspond to the vertices, and there is an edge between two vertices if and only if the corresponding n-simplices contain a common $(n-1)$-simplex.

4.1 Examples

The set $\mathcal{T}_\Lambda$ is countable [17], hence Σ_Λ is a countable simplicial complex. Since a multiplicity free tilting module has $n + 1$ indecomposable direct summands, Σ_Λ is a simplicial complex of dimension n. Moreover, since all r-simplices are contained in an n-simples, Σ_Λ is a simplicial complex of pure dimension n.

We include some examples stressing further properties of Σ_Λ.

If Λ is the path algebra of the quiver $\bullet\!\!\longrightarrow\!\!\bullet\!\!\longrightarrow\!\!\bullet$ then Σ_Λ equals

If Λ is the path algebra of the quiver $\bullet\!\!\longrightarrow\!\!\bullet\!\!\longrightarrow\!\!\bullet$ modulo the ideal generated by the path of length 2, then Σ_Λ is of the form

In general Σ_Λ is not finite and not connected. As an example, if Λ is the path algebra of the quiver

then Σ_Λ decomposes into two $\mathbb{A}_\infty$-components.

The following example shows that Σ_Λ is not necessarily locally finite. Here Σ_Λ is the simplicial complex of tilting modules over the path algebra of :

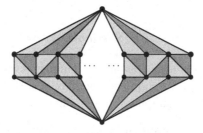

We will also be interested in the boundary of Σ_Λ. By definition the boundary $\delta\Sigma_\Lambda$ of a purely n-dimensional simplicial complex consists of the simplices contained in those $(n-1)$-simplices which lie in precisely one n-simplex. Clearly $\delta\Sigma_\Lambda$ is a simplicial complex of pure dimension $n-1$. The boundaries of the simplicial complexes considered above are

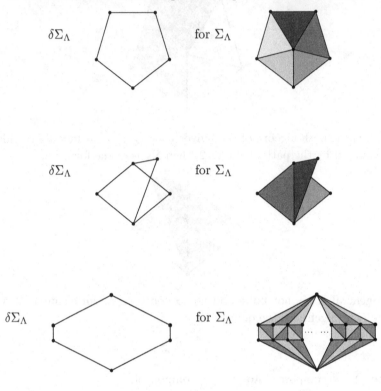

Note that $\delta\Sigma_\Lambda$ may be finite if Σ_Λ is infinite.

4.2 Shellability

For a countable simplicial complex of pure dimension n one has the notion of shellability: A simplicial complex Σ with these properties is called *shellable* if its n-simplices may be linearly ordered $\sigma_1, \sigma_2, \sigma_3, \dots$ such that for all $l > 1$ the simplicial subcomplex $\left(\bigcup_{i=1}^{l-1} \sigma_i \right) \cap \sigma_l$ is of pure dimension $n-1$. The linear order $\sigma_1, \sigma_2, \sigma_3, \dots$ is called a *shelling*.

The simplicial complex

is shellable – any linear order of its 2-simplices is a shelling.

We assume until the end of this section that Σ is a finite simplicial complex.

Shellability of Σ has surprising consequences in algebraic combinatorics. We briefly recall the facts. Let $\tau \in \Sigma$ be a simplex. The *link* of τ is the simplicial subcomplex $\mathrm{lk}(\tau) = \{\sigma \in \Sigma \mid \tau \cup \sigma \in \Sigma \text{ and } \tau \cap \sigma = \varnothing\}$. Let R be the ring of integers or a field. By $\tilde{H}_*(\Sigma, R)$ we denote the reduced simplicial homology of Σ with coefficients in R. The simplicial complex Σ is said to be *Cohen-Macaulay* over R if $\tilde{H}_i(\mathrm{lk}(\tau), R) = 0$ for all $\tau \in \Sigma$ and all $i < \dim \mathrm{lk}(\tau)$. The motivation for this terminology comes from another object studied in in algebraic combinatorics. Let $V = \{x_0, \ldots, x_r\}$ be the vertex set of Σ. Consider the polynomial ring $R[x_0, \ldots, x_r]$ and the ideal I_Σ generated by the square free monomials $x_{i_1} \cdots x_{i_s}$ such that $\{x_{i_1}, \ldots, x_{i_s}\} \notin \Sigma$. The quotient $R[\Sigma] = R[x_0, \ldots, x_r]/I_\Sigma$ is called the *face ring* or the *Stanley-Reisner ring* of Σ. A theorem of Reisner [32] states that Σ is Cohen-Macaulay over R if and only if $R[\Sigma]$ is a Cohen-Macauley ring. It was first observed by Hochster [26] that a shellable simplicial complex is Cohen-Macaulay over the integers. Examples for non-shellable simplicial Cohen-Macaulay complexes are given in [12].

Further consequences of shellability concern the geometric realization $\|\Sigma\|$ of Σ. For this connection one needs an invariant, the so called characteristic, associated with a shellable simplicial complex. To define this we introduce some further notation. For a given shelling of Σ and for all $l \geqslant 1$ we denote $\bigcup_{i=1}^{l} \sigma_i$ by Σ_l. For $l > 1$ one further defines the *restriction* $\mathcal{R}(\sigma_l)$ of an n-simplex σ_l as the set of vertices v of σ_l such that $\sigma_l \backslash \{v\}$ is contained in Σ_{l-1}. A shellable simplicial complex of pure dimension n is said to be *shellable of characteristic h* if h is the cardinality of the n-simplices σ satisfying $\mathcal{R}(\sigma) = \sigma$. The connection between this notion and the geometric realisation of a simplicial complex is given by a theorem of Björner [7]:

Theorem 4.1. *Let Σ be a shellable simplicial complex of pure dimension*

n and characteristic h. Then $\|\Sigma\|$ has the homotopy type of a wedge of h n-spheres. In particular, Σ is $(n-1)$-connected.

Recall that a purely n-dimensional simplicial complex is called $(n-1)$-connected if for all n-simplices σ and τ there is a sequence $\sigma = \sigma_1, \sigma_2, \ldots, \sigma_t = \tau$ of n-simplices such that for all $1 \leqslant i \leqslant t-1$ the simplices σ_i and σ_{i+1} have a common $(n-1)$-simplex.

Shellability of the simplicial complex Σ_Λ of tilting modules over Λ was studied in [39] and [42]. The main results state:

Theorem 4.2. *Let Σ_Λ be the simplicial complex of tilting modules over Λ.*

(a) *If Σ_Λ is finite, then it is shellable of characteristic 0. In particular, Σ_Λ is contractible.*

(b) *If Σ_Λ is finite, then $\delta\Sigma_\Lambda$ is shellable.*

Note that the characteristic of $\delta\Sigma_\Lambda$ may be arbitrarily large. If Λ_n is the path algebra of a linearly ordered $\mathbb{A}_{n+1}$ modulo the ideal generated by all paths of length two, then $\delta\Sigma_{\Lambda_n}$ is shellable of characteristic n.

4.3 Tilting modules of projective dimensions at most one

Much better results can be obtained provided Σ is a pseudomanifold, i.e. if Σ is $(n-1)$-connected and if every $(n-1)$-simplex is contained in at most two n-simplices. A result of Danaraj and Klee [11] states that for a shellable pseudomanifold Σ of dimension n its geometric realization is an n-ball or an n-sphere. The latter case occurs if and only if Σ has no boundary.

Let $\Sigma_\Lambda^{\leqslant 1}$ be the simplicial complex of tilting modules of projective dimensions at most one. This again is a countable simplicial complex of pure dimension n, where $n+1$ is the number of isomorphism classes of simple Λ-modules. In [33] Riedtmann and Schofield proved that $\Sigma_\Lambda^{\leqslant 1}$ is a shellable pseudomanifold provided it is finite. Moreover, $\Sigma_\Lambda^{\leqslant 1}$ always admits a boundary. Hence in combination with [11] the main result of [33] follows.

Theorem 4.3. *If $\Sigma_\Lambda^{\leqslant 1}$ is finite, then $\|\Sigma_\Lambda^{\leqslant 1}\|$ is an n-ball.*

It was proved in [42] that this result also extends to the boundaries of $\Sigma_\Lambda^{\leqslant 1}$.

Theorem 4.4. *If $\delta\Sigma_\Lambda^{\leqslant 1}$ is finite, then it is a pseudomanifold. In particular, $\|\delta\Sigma_\Lambda^{\leqslant 1}\|$ is an $(n-1)$-ball or an $(n-1)$-sphere.*

Both cases may occur. The example of Igusa, Smalø and Todorov [27] mentioned above yields an example where the boundary complex of $\Sigma_\Lambda^{\leqslant 1}$ is a ball.

4.4 Connected components

If the graph $\overrightarrow{\mathcal{K}(\Lambda)}$ of tilting modules is connected, then Σ_Λ is connected. In particular, Σ_Λ is connected in case it is finite. In general very little is known about the connected components of Σ_Λ. For quiver algebras Λ the situation is better understood. The following result was obtained in [41].

Theorem 4.5. *Let Λ be a path algebra of a quiver $\overrightarrow{\Delta}$ with n vertices. Then Σ_Λ has at most n connected components.*

To be more precise, it was shown in [41] that a connected component of Σ_Λ has to contain a simple Λ-module provided Λ is the path algebra of a quiver.

References

[1] M. Auslander, D. Buchsbaum: *Homological dimension in noetherian rings*, Proc. Nat. Acad. Sci. USA 42 (1956), 36-38.

[2] M. Auslander, I. Reiten: *On a generalized version of the Nakayama conjecture*, Proc. Amer. Math. Soc. 52, 1975, 69-74.

[3] M. Auslander, I. Reiten: *Applications of contravariantly finite subcategories*, Adv. Math. 86 (1991), 111-152.b

[4] M. Auslander, I. Reiten, S. Smalø: *Representation Theory of Artin Algebras,* Cambridge University Press (1995).

[5] A. Auslander, S. Smalø: *Preprojective modules over Artin algebras*, J. Algebra 66 (1980), 61-122.

[6] H. Bass: *Finitistic dimension and a homological generalization of semiprimary rings,* Trans. Amer. Math. Soc. 95 (1960) 466-488.

[7] A. Björner: *Some combinatorial and algebraic properties of Coxeter complexes and Tits buildings*, Adv. Math. (3) 52 (1984), 173-212.

[8] A. Buan, H. Krause, Ø. Solberg: *On the lattice of cotilting modules*, AMA Algebra Montp. Announc., 1 (2002).

[9] A. Buan, Ø. Solberg: *Relative cotilting and almost complete cotilting modules*, Proc. ICRA VIII, CMS Conference Proc. 24 (1998), 77-92.

[10] F. Coelho, D. Happel, L. Unger: *Complements to partial tilting modules*, J. Algebra 170 (1994), 184-205.

[11] G. Danaraj, V. Klee: *Shellings of spheres and polytopes*, Duke Math. J. 41 (1974), 443-451.

[12] G. Danaraj, V. Klee: *Which spheres are shellable?* Ann. Discrete Math.2 (1978), 33-52.

[13] W. Geigle, H. Lenzing: *A class of weighted projective curves arising in representation theory of finite dimensional algebras*, in: Singularities, representations of algebras and vector bundles, Springer Lecture Notes 1273 (1987) 265-297.

[14] E. Green, E. Kirkman, J. Kuzmanovich: *Finitistic dimension of finite dimensional monomial algebras*, J. Algebra 136 (1991), 37-51.

[15] E. Green, B. Zimmermann-Huisgen: *Finitistic dimension of artinian rings with vanishing radical cube*, Math. Zeitschrift 206 (1991), 505-526.

[16] D. Happel: *Reduction techniques for homological conjectures*, Tsukuba J. Math. 17 (1993), 115-130.

[17] D. Happel: *Selforthogonal modules*, Abelian groups and modules (Padova 1994), 257-276, Math. Appl., 343, Kluwer Acad. Publ., Dordrecht, 1995.

[18] D. Happel: *Quasitilted algebras*, Proc. ICRA VIII, CMS Conference Proc. 23 (1998), 55-83.

[19] D. Happel: *A characterization of hereditary categories with tilting object*, Invent. Math. 144 (2001) 381-398.

[20] D. Happel, I. Reiten, S. Smalø: *Tinting in abelian categories and quasitilted algebras*, Memoirs Amer. Math. Soc. 575 (1996).

[21] D. Happel, L. Unger: *Partial tilting modules and covariantly finite subcategories*, Comm. Algebra 22(5) (1994), 1723-1727.

[22] D. Happel, L. Unger: *Complements and the generalized Nakayama Conjecture*, Proc. ICRA VIII, CMS Conference Proc. 24 (1998), 293-310.

[23] D. Happel, L. Unger: *On a partial order of tilting modules*, (2001) to appear in Algebras and Representation Theory.

[24] D. Happel, L. Unger: *On the set of tilting objects in hereditary categories*, (2002) to appear in Proc. ICRA X.

[25] D. Happel, L. Unger: *On the quiver of tilting modules*, preprint (2003).

[26] M. Hochster: *Rings of invariants of tori, Cohen-Macaulay rings generated by monomials, and polytopes*, Ann. Math. (2) 96 (1972) 318-337.

[27] K. Igusa, S. Smalø, G. Todorov: *Finite projectivity and contravariantly finiteness*, Proc. Amer. Math. Soc. 109 (1990) 937-941.

[28] K. Igusa, G. Todorov: *On the finitistic global dimension conjecture for artin algebras*, preprint.

[29] K. Igusa, D. Zacharia: *Syzygy Pairs in a monomial algebra*, Proc. Amer. Math. Soc. 108 (1990), 601-604.

[30] J. P. Jans: *Some generalizations of finite projective dimension*, Ill. J. Math. 5 (1961), 334-344.

[31] R. J. Nunke: *Modules of extensions over Dedekind rings*, Ill. J. Math. 3 (1959), 22-242.

[32] G. Reisner: *Cohen-Macaulay quotients of polynomial rings*, Adv. Math. 21 (1976), 30-49.

[33] C. Riedtmann, A. Schofield: *On a simplicial complex associated with*

tilting modules, Comment. Math. Helv. 66 (1991), 70-78.

[34] C. M. Ringel: *Tame algebras and integral quadratic forms,* Springer Lecture Notes in Mathematics 1099, Heidelberg (1984).

[35] C. M. Ringel: *Tilting modules over tubular algebras,* in preparation.

[36] J. P. Serre: *Sur la dimension homologique des anneaux et des modules noethériens,* Proc. Intl. Symp. on Algebraic Number Theory, Tokyo 1955, 175-189.

[37] Ø. Solberg: *Infinite dimensional tilting modules over finite dimensional algebras,* this volume.

[38] J. Trlifaj: *Infinite dimensional tilting modules and cotorsion pairs,* this volume.

[39] L. Unger: *On the simplicial complex of exceptional modules,* Habilitationsschrift, Paderborn (1993).

[40] L. Unger: *The partial order of tilting modules over three-point-quiver-algebras,* Proceedings of ICRA VII, Canadian Mathematical Society Conference Proceedings **18** (1996) 671-679.

[41] L. Unger: *On the simplicial complex of tilting modules over quiver algebras,* Proc. London Math. Soc. (3) 73 (1996),27-46.

[42] L. Unger: *Shellability of simplicial complexes arising in representation theory,* Advances in Mathematics 144, (1999) 221-246.

[43] L. Unger, M. Ungruhe: *On the genus of the graph of tilting modules,* to appear in Beiträge zur Algebra und Geometrie.

Luise Unger
FernUniversität Hagen
Fachbereich Mathematik
D-58084 Hagen
Germany

11

Infinite dimensional tilting modules and cotorsion pairs

Jan Trlifaj[1]

Dedicated to Claus Michael Ringel on the occation of his 60th birthday.

Classical tilting theory generalizes Morita theory of equivalence of module categories. The basic property – existence of a maximal category equivalence represented by the tilting module – forces the module to be finitely generated, cf. [31, §2], [85].

However, some aspects of the classical theory can be extended to infinitely generated modules over arbitrary rings. In this paper, we investigate such an aspect in detail: the relation of tilting to approximations (preenvelopes and precovers) of modules. Then we show how infinitely generated tilting modules are employed for proving finitistic dimension conjectures in particular cases, and for characterizing Matlis localizations of commutative rings.

General existence theorems provide a big supply of approximations in the category Mod-R of all modules over an arbitrary ring R. However, the corresponding approximations may not be available in the subcategory of all finitely generated modules. So the usual sharp distinction between finitely and infinitely generated modules becomes unnatural, and even misleading.

Cotorsion pairs give a convenient tool for the study of module approximations. Tilting cotorsion pairs are defined as the cotorsion pairs induced by tilting modules. We present their characterization among all cotorsion pairs, and apply it to a classification of tilting classes in particular cases (e.g., over Prüfer domains). The point of the classification is that tilting classes coincide with classes of finite type. So we can replace

1 Supported by the research project MSM 0021620839.

J. Trlifaj

each infinitely generated tilting module by a set of finitely presented modules; in particular, the corresponding tilting class is axiomatizable in the language of the first order theory of modules.

Most of this paper is a survey of recent developments. We give complete definitions and statements of the results, but most proofs are omitted or replaced by outlines of the main ideas. For full details, we refer to the original papers listed in the references, or to the forthcoming monograph [55]. However, Theorems 3.5, 3.7, 4.14, and 4.15 are new, hence presented with full proofs.

In §1, we introduce cotorsion pairs and their relations to approximation theory of infinitely generated modules over arbitrary rings. In §2 and §3, we study infinitely generated tilting and cotilting modules, and characterize the induced tilting and cotilting cotorsion pairs. In §4, we deal with classes of finite and cofinite type, and with the classification of tilting and cotilting classes over particular rings. We also characterize Matlis localizations of commutative rings. Finally, §5 relates tilting approximations to the first and second finitistic dimension conjectures.

We start by fixing our notation. For an (associative, unital) ring R, Mod-R denotes the category of all (right R-) modules. mod-R denotes the subcategory of Mod-R formed by all modules possessing a projective resolution consisting of finitely generated modules. (If R is a right coherent ring then mod-R is just the category of all finitely presented modules).

Let $\mathcal{C}$ be a class of modules. For a cardinal κ, we denote by $\mathcal{C}^{<\kappa}$, and $\mathcal{C}^{\leq\kappa}$, the subclass of $\mathcal{C}$ consisting of the modules possessing a projective resolution containing only $< \kappa$-generated, and $\leq \kappa$-generated, modules, respectively. For example, mod-$R = (\text{Mod-}R)^{<\omega}$. Further, $\varinjlim \mathcal{C}$ denotes the class of all modules that are direct limits of modules from $\mathcal{C}$. (In general, $\varinjlim \mathcal{C}$ is not closed under direct limits, but it is in case $\mathcal{C} \subseteq$ mod-R. In that case, $\mathcal{C} = \varinjlim \mathcal{C} \cap$ mod-R provided $\mathcal{C}$ is closed under finite direct sums and direct summands.)

Let $n < \omega$. We denote by $\mathcal{P}_n$ ($\mathcal{I}_n$, $\mathcal{F}_n$) the class of all modules of projective (injective, flat) dimension $\leq n$. Further, $\mathcal{P}$ ($\mathcal{I}$, $\mathcal{F}$) denotes the class of all modules of finite projective (injective, flat) dimension. The injective hull of a module M is denoted by $E(M)$.

We denote by $\mathbb{Z}$ the ring of all integers, and by $\mathbb{Q}$ the field of all rational numbers. For a commutative domain R, Q denotes the quotient field of R.

For a left R-module N, we denote by $N^* = \mathrm{Hom}_{\mathbb{Z}}(N, \mathbb{Q}/\mathbb{Z})$ the *character module* of N. Note that N^* is a (right R-) module.

Let M be a module. Then M is a *dual module* provided that $M = N^*$ for a left R-module N. M is *pure-injective* provided that M is a direct summand in a dual module. M is (Enochs) *cotorsion* provided that $\mathrm{Ext}_R^1(F, M) = 0$ for each $F \in \mathcal{F}_0$. Notice that any dual module is pure-injective, and any pure-injective module is cotorsion (The converses do not hold in general; however, flat cotorsion modules over left coherent rings are pure-injective, [93]). The class of all pure-injective, and cotorsion, modules is denoted by $\mathcal{PI}$, and $\mathcal{EC}$, respectively.

A module M is *divisible* if $\mathrm{Ext}_R^1(R/rR, M) = 0$ for each $r \in R$, and *torsion-free* if $\mathrm{Tor}_1^R(M, R/Rr) = 0$ for each $r \in R$ (If R is a commutative domain, then these notions coincide with the classical ones). The class of all divisible and torsion-free modules is denoted by $\mathcal{DI}$ and $\mathcal{TF}$, respectively.

1 Cotorsion pairs and approximations of modules

Cotorsion pairs are analogs of (non-hereditary) torsion pairs, with Hom replaced by Ext. They were introduced by Salce (under the name "cotorsion theories") in [76]. The analogy with the well-known torsion pairs makes it possible to derive easily some basic notions and facts about cotorsion pairs. However, the main point concerning cotorsion pairs is their close relation to special approximations of modules: cotorsion pairs provide a homological tie between the dual notions of a special preenvelope and a special precover. This tie (discovered in [76], cf. 1.9.3) is a sort of remedy for the non-existence of a duality in Mod-R.

Before introducing cotorsion pairs, we define various Ext-orthogonal classes.

Let $\mathcal{C} \subseteq$ Mod-R. Define $\mathcal{C}^{\perp} = \bigcap_{1 \leqslant n < \omega} \mathcal{C}^{\perp_n}$ where $\mathcal{C}^{\perp_n} = \{M \in$ Mod-$R \mid \mathrm{Ext}_R^n(C, M) = 0$ for all $C \in \mathcal{C}\}$ for each $1 \leqslant n < \omega$. Dually, let $^{\perp}\mathcal{C} = \bigcap_{1 \leqslant n < \omega} {}^{\perp_n}\mathcal{C}$ where $^{\perp_n}\mathcal{C} = \{M \in$ Mod-$R \mid \mathrm{Ext}_R^n(M, C) = 0$ for all $C \in \mathcal{C}\}$ for each $1 \leqslant n < \omega$.

1.1. Cotorsion pairs. Let R be a ring. A *cotorsion pair* is a pair $\mathfrak{C} = (\mathcal{A}, \mathcal{B})$ of classes of modules such that $\mathcal{A} = {}^{\perp_1}\mathcal{B}$ and $\mathcal{B} = \mathcal{A}^{\perp_1}$. The class $\mathcal{A} \cap \mathcal{B}$ is called the *kernel* of $\mathfrak{C}$. The cotorsion pair $\mathfrak{C}$ is *hereditary* provided that $\mathrm{Ext}_R^i(A, B) = 0$ for all $A \in \mathcal{A}$, $B \in \mathcal{B}$ and $i \geqslant 2$.

Notice that $\mathfrak{C}$ is hereditary iff $\mathcal{A} = {}^{\perp}\mathcal{B}$ and $\mathcal{B} = \mathcal{A}^{\perp}$. The property of $\mathfrak{C}$ being hereditary can easily be expressed in terms of the properties of $\mathcal{A}$ and $\mathcal{B}$: $\mathfrak{C}$ is hereditary iff $\mathcal{A}$ is closed under kernels of epimorphisms iff $\mathcal{B}$ is closed under cokernels of monomorphisms.

Each module M in the kernel of a cotorsion pair $\mathfrak{C}$ is a *splitter*, that is, M satisfies $\mathrm{Ext}_R^1(M, M) = 0$. We will see that the kernel of $\mathfrak{C}$ in the tilting and cotilting cases plays an important role: it determines completely the classes $\mathcal{A}$ and $\mathcal{B}$. (This contrasts with what happens for torsion pairs: since $\mathrm{id}_M \in \mathrm{Hom}_R(M, M)$ for each module M, the "kernel" of any torsion pair is trivial.)

1.2. By changing the category, we could take a complementary point of view, working modulo the kernel rather than stressing its role. By a result of Beligiannis and Reiten [26], each complete hereditary cotorsion pair $\mathfrak{C} = (\mathcal{A}, \mathcal{B})$ in Mod-R determines a torsion pair, $(\underline{\mathcal{A}}, \underline{\mathcal{B}})$, in the stable module category $\underline{\mathrm{Mod}\text{-}R}$ (of Mod-R modulo the kernel of $\mathfrak{C}$), cf. 1.9.3. Consequently, special $\mathcal{A}$-precovers and special $\mathcal{B}$-preenvelopes are functorial modulo maps factoring through the kernel, cf. [66].

1.3. The class of all cotorsion pairs is partially ordered by inclusion in the first component: $(\mathcal{A}, \mathcal{B}) \leqslant (\mathcal{A}', \mathcal{B}')$ iff $\mathcal{A} \subseteq \mathcal{A}'$. The $\leqslant$-least cotorsion pair is $(\mathcal{P}_0, \mathrm{Mod}\text{-}R)$, the $\leqslant$-greatest $(\mathrm{Mod}\text{-}R, \mathcal{I}_0)$; these are the *trivial* cotorsion pairs.

The cotorsion pairs over a ring R form a complete lattice, $\mathfrak{L}_R$: given a sequence of cotorsion pairs $\mathcal{S} = ((\mathcal{A}_i, \mathcal{B}_i) \mid i \in I)$, the infimum of $\mathcal{S}$ in $\mathfrak{L}_R$ is $(\bigcap_{i \in I} \mathcal{A}_i, (\bigcap_{i \in I} \mathcal{A}_i)^{\perp_1})$, the supremum being $({}^{\perp_1}(\bigcap_{i \in I} \mathcal{B}_i), \bigcap_{i \in I} \mathcal{B}_i)$.

For any class of modules $\mathcal{C}$, there are two cotorsion pairs associated with $\mathcal{C}$: $({}^{\perp_1}\mathcal{C}, ({}^{\perp_1}\mathcal{C})^{\perp_1})$, called the cotorsion pair *generated* by $\mathcal{C}$, and $({}^{\perp_1}(\mathcal{C}^{\perp_1}), \mathcal{C}^{\perp_1})$, the cotorsion pair *cogenerated* by $\mathcal{C}$. If $\mathcal{C}$ has a representative set of elements $\mathcal{S}$, then the first cotorsion pair is generated by the single module $\prod_{S \in \mathcal{S}} S$, while the second is cogenerated by the single module $\bigoplus_{S \in \mathcal{S}} S$.

The existence of cotorsion pairs generated and cogenerated by any class of modules indicates that $\mathfrak{L}_R$ is a large class in general.

For example, the condition of all cotorsion pairs being trivial is extremely restrictive: by [86] and [43], for a right hereditary ring R, this condition holds iff $R = S$ or $R = T$ or R is the ring direct sum $S \boxplus T$, where S is semisimple artinian and T is Morita equivalent to a 2×2-matrix ring over a skew-field. As another example, consider the case of $R = \mathbb{Z}$: by [53], any partially ordered set embeds in $\mathfrak{L}_{\mathbb{Z}}$; in particular, $\mathfrak{L}_{\mathbb{Z}}$ is a proper class.

1.4. Replacing Ext by Tor in 1.1, we can define a *Tor-torsion pair* as the pair $(\mathcal{A}, \mathcal{B})$ where $\mathcal{A} = \{A \in \text{Mod-}R \mid \text{Tor}_1^R(A, B) = 0$ for all $B \in \mathcal{B}\}$ and $\mathcal{B} = \{B \in R\text{-Mod} \mid \text{Tor}_1^R(A, B) = 0$ for all $A \in \mathcal{A}\}$. Similarly to the case of cotorsion pairs, we can define Tor-torsion pairs generated (cogenerated) by a class of left (right) R-modules. Tor-torsion pairs over a ring R form a complete lattice; by 1.5.3 below, the cardinality of this lattice is $\leqslant 2^{2^{\kappa}}$ where $\kappa = \text{card}(R) + \aleph_0$.

The well-known Ext-Tor relations yield an embedding of the lattice of Tor-torsion pairs into $\mathfrak{L}_R$ as follows: a Tor-torsion pair $(\mathcal{A}, \mathcal{B})$ is mapped to the cotorsion pair $(\mathcal{A}, \mathcal{A}^{\perp_1})$. The latter cotorsion pair is easily seen to be generated by the class $\{B^* \mid B \in \mathcal{B}\}$. In this way, Tor-torsion pairs are identified with particular cotorsion pairs generated by classes of pure-injective modules.

The following lemma says that most of the classes of modules defined above occur as first or second components of cotorsion pairs cogenerated by sets:

Lemma 1.5. *Let R be a ring and $n < \omega$. Let $\kappa = \text{card}(R) + \aleph_0$.*

(1) *$\mathfrak{C} = (\mathcal{P}_n, (\mathcal{P}_n)^{\perp})$ is a hereditary cotorsion pair cogenerated by $\mathcal{P}_n^{\leqslant \kappa}$. If R is right noetherian then $\mathfrak{C}$ is cogenerated by $\mathcal{P}_n^{\leqslant \omega}$.*

(2) *Let $\mathfrak{C} = (\mathcal{A}, \mathcal{B})$ be a cotorsion pair generated by a class of pure-injective modules. Then $\mathfrak{C}$ is cogenerated by $\mathcal{A}^{\leqslant \kappa}$.*

(3) *Let $(\mathcal{A}, \mathcal{B})$ be a Tor-torsion pair. Then $(\mathcal{A}, \mathcal{A}^{\perp_1})$ is a cotorsion pair cogenerated by $\mathcal{A}^{\leqslant \kappa}$, and generated by $\{B^* \mid B \in \mathcal{B}\}$. In particular, $(\mathcal{F}_n, (\mathcal{F}_n)^{\perp})$ is a hereditary cotorsion pair cogenerated by $\mathcal{F}_n^{\leqslant \kappa}$.*

(4) *$(^{\perp}\mathcal{I}_n, \mathcal{I}_n)$ is a hereditary cotorsion pair cogenerated by $(^{\perp}\mathcal{I}_n)^{\leqslant \lambda}$ where λ is the least infinite cardinal such that each right ideal of R is λ-generated.*

(5) *Let R be a right noetherian ring. Then the cotorsion pair cogenerated by $\mathcal{I}_n$ is cogenerated by a set.*

(6) $(^{\perp_1}\mathcal{DI}, \mathcal{DI})$ *and* $(\mathcal{TF}, \mathcal{TF}^{\perp_1})$ *are cogenerated by sets of cardinality $\leqslant \kappa$.*

Proof. 1.[1] For $n = 0$, we apply the classical result of Kaplansky saying that each projective module is a direct sum of the countably generated ones. For $n \geqslant 1$, it suffices to prove that given a free resolution $\mathcal{R}$: $0 \to F_n \to \cdots \to F_0 \to M \to 0$ of $M \in \mathcal{P}_n$ with $F_i = R^{(A_i)}$ and $0 \neq x \in M$, there is a submodule $N \subseteq M$ and an exact subcomplex of $\mathcal{R}$, $0 \to G_n \to \cdots \to G_0 \to N \to 0$ where $G_i = R^{(B_i)}$, $B_i \subseteq A_i$, card$(B_i) \leqslant \kappa$ for all $i \leqslant n$, and $x \in N$. This is proved by a back and forth argument in $\mathcal{R}$, see [1]. The noetherian case is similar, cf. [80].

2. This is proved in [44].

3. The first statement follows by part 2. and by 1.4. The second is a particular case of the first one.

4. By Baer test lemma for injectivity, we have $M \in \mathcal{I}_n$ iff $\text{Ext}_R^1(N, M) = 0$ where N runs over a representative set of all n-th syzygies of cyclic modules.

5. Since R is right noetherian, there is a cardinal μ such that any injective module is a direct sum of $\leqslant \mu$-generated modules, and the proof proceeds in a dual way to 1., see [1].

6. The first cotorsion pair is cogenerated by the set $\{R/rR \mid r \in R\}$. The assertion concerning the second pair is a particular case of 3. $\quad\square$

The key property of cotorsion pairs is their relation to approximations of modules. The connection is via the notion of a special approximation, [93]:

1.6. Special approximations. Let R be a ring, M a module and $\mathcal{C}$ a class of modules. An R-homomorphism $f \colon M \to C$ is a *special $\mathcal{C}$-preenvelope* of M provided that f induces a short exact sequence $0 \to M \xrightarrow{f} C \to D \to 0$ with $C \in \mathcal{C}$ and $D \in {}^{\perp_1}\mathcal{C}$. $\mathcal{C}$ is a *special preenveloping class* if each module $M \in \text{Mod-}R$ has a special $\mathcal{C}$-preenvelope.

Dually, an R-homomorphism $g \colon C \to M$ is a *special $\mathcal{C}$-precover* of M provided that g induces a short exact sequence $0 \to B \to C \xrightarrow{g} M \to 0$

with $C \in \mathcal{C}$ and $B \in \mathcal{C}^{\perp_1}$. $\mathcal{C}$ is a *special precovering* class if each module $M \in$ Mod-R has a special $\mathcal{C}$-precover.

The terminology of 1.6 comes from the fact that special preenvelopes and precovers are special instances of the following more general notions, [46], [93]:

1.7. Let R be a ring, M a module, and $\mathcal{C}$ a class of modules. An R-homomorphism $f: M \to C$ with $C \in \mathcal{C}$ is a $\mathcal{C}$-*preenvelope* of M provided that for each $C' \in \mathcal{C}$ and each R-homomorphism $f': M \to C'$ there is an R-homomorphism $g: C \to C'$ such that $f' = gf$.

The $\mathcal{C}$-preenvelope f is a $\mathcal{C}$-*envelope* of M if f has the following minimality property: if g is an endomorphism of C such that $gf = f$ then g is an automorphism.

$\mathcal{C}$ is a *preenveloping (enveloping)* class provided that each module $M \in$ Mod-R has a $\mathcal{C}$-preenvelope (envelope).

The notions of a $\mathcal{C}$-*precover*, $\mathcal{C}$-*cover*, *precovering* class, and *covering* class are defined dually.

A preenvelope (precover) may be viewed as a kind of weak (co-) reflection [48]; however, we do not require the assignment $M \mapsto C$ ($C \mapsto M$) to be functorial or unique, cf. 1.2.

However, if a module M has a $\mathcal{C}$-envelope (cover) then the $\mathcal{C}$-envelope (cover) is easily seen to be uniquely determined up to isomorphism; morever the $\mathcal{C}$-envelope (cover) of M is isomorphic to a direct summand in any $\mathcal{C}$-preenvelope ($\mathcal{C}$-precover) of M, [93].

Classical examples of enveloping classes include $\mathcal{I}_0$ and $\mathcal{PI}$, see [38] and [92], and of covering classes, $\mathcal{P}_0$ in case R is a right perfect ring, and $\mathcal{TF}$ in case R is a domain, see [14] and [45]. We will have many more examples later in this section.

1.8. (i) The definitions above can be extended to the setting of an abitrary category $\mathcal{K}$ (in place of Mod-R) and its subcategory $\mathcal{C} \subseteq \mathcal{K}$. If $\mathcal{K} =$ mod-R, we say that $\mathcal{C}$ is *covariantly finite* (*contravariantly finite*) provided that $\mathcal{C}$ is preenveloping (precovering) in mod-R, cf. [13]. We refer to [72] for more on the role of covariantly and contravariantly finite subcategories of mod-R in tilting theory.

(ii) There is a related notion of a "Γ-separated cover" introduced by

Klingler and Levy in their classification of finitely generated modules over Dedekind-like rings, cf. [65]. For classical Dedekind-like rings, these "Γ-separated covers" turn out to be a particular sort of the covers defined in 1.7, though they do not fit any cotilting setting – see [68] for more details.

The following lemma connects cotorsion pairs to approximations of modules:

Lemma 1.9. *Let R be a ring, M a module, and $\mathfrak{C} = (\mathcal{A}, \mathcal{B})$ a cotorsion pair.*

   (1) [91]   *Assume M has a $\mathcal{B}$-envelope f. Then f is a special $\mathcal{B}$-preenvelope. So if $\mathcal{B}$ is enveloping then $\mathcal{B}$ is special preenveloping.*

   (2) [91]   *Assume M has a $\mathcal{A}$-cover f. Then f is a special $\mathcal{A}$-precover. So if $\mathcal{A}$ is covering then $\mathcal{A}$ is special precovering.*

   (3) [76]   *$\mathcal{A}$ is special precovering iff $\mathcal{B}$ is special preenveloping. In this case $\mathfrak{C}$ is called a* complete *cotorsion pair.*

Proof. 1. Since $\mathcal{I}_0 \subseteq \mathcal{B}$, f is monic, so there is a short exact sequence $0 \to M \xrightarrow{f} B \xrightarrow{g} C \to 0$. Take a short exact sequence $0 \to B' \to D \xrightarrow{h} C \to 0$ with $B' \in \mathcal{B}$. Considering the pull-back of g and h, and using the minimality of the map f, we obtain a splitting map for h, thus proving that $C \in \mathcal{A}$.

2. This is dual to 1.

3. Let M be a module. Consider a short exact sequence $0 \to M \to I \xrightarrow{f} J \to 0$ where $I \in \mathcal{I}_0$. Let $g \colon A \to J$ be a special $\mathcal{A}$-precover of J. Then the pull-back of g and f yields a special $\mathcal{B}$-preenvelope of M. The proof of the converse implication is dual. $\square$

The following example shows that in 1.9.3, we cannot claim that $\mathcal{A}$ is a covering class iff $\mathcal{B}$ is an enveloping one (however, by 1.11 below, the equivalence holds in case $\mathcal{A}$ is closed under direct limits):

Example 1.10. [22], [23], [24]   Let R be a commutative domain and $\mathfrak{C}$ be the cotorsion pair cogenerated by the quotient field Q. Matlis proved that $\mathfrak{C}$ is hereditary iff proj.dim$(Q) \leqslant 1$ (that is, R is a *Matlis domain*).

The class $\mathcal{B} = \{Q\}^{\perp_1}$ is the class of all *Matlis cotorsion* modules. Since

$\mathcal{B} = (\text{Mod-}Q)^{\perp_1}$, $\mathcal{B}$ is an enveloping class, [93]. For example, the $\mathcal{B}$-envelope of a torsion-free reduced module M coincides with the R-completion of M, cf. [50].

On the other hand, $\mathcal{A}$ (called the class of all *strongly flat* modules) is a covering class iff all proper factor-rings of R are perfect. For example, if R is a Prüfer domain then $\mathcal{A}$ is a covering class iff R is a Dedekind domain.

Cotorsion pairs $\mathfrak{C} = (\mathcal{A}, \mathcal{B})$ such that $\mathcal{A}$ is a covering class and $\mathcal{B}$ is an enveloping class are called *perfect*. By 1.9, each perfect cotorsion pair is complete. There is a sufficient condition for perfectness of complete cotorsion pairs due to Enochs. For a proof, we refer to [46] and [93]:

Theorem 1.11. *Let R be a ring, M a module, and $\mathfrak{C} = (\mathcal{A}, \mathcal{B})$ a cotorsion pair. Assume that $\mathcal{A}$ is closed under direct limits.*

(1) *If M has a $\mathcal{B}$-preenvelope then M has a $\mathcal{B}$-envelope.*

(2) *If M has an $\mathcal{A}$-precover then M has an $\mathcal{A}$-cover.*

In particular, $\mathfrak{C}$ is perfect iff $\mathfrak{C}$ is complete iff $\mathcal{A}$ is covering iff $\mathcal{B}$ is enveloping.

1.12. Let R be a ring, and $\mathcal{C}$ a subclass of mod-R closed under extensions and direct summands such that $R \in \mathcal{C}$. Let $\mathcal{D} = \varinjlim \mathcal{C}$. Then $\mathcal{C} = \mathcal{D} \cap \text{mod-}R$. Moreover, by [9], the Tor-torsion pair cogenerated by $\mathcal{C}$ has the form $(\mathcal{D}, \mathcal{E})$ for some $\mathcal{E} \subseteq R\text{-Mod}$.

By 1.5.3, there are two associated cotorsion pairs: $(\mathcal{A}, \mathcal{B})$ – the cotorsion pair cogenerated by $\mathcal{C}$, and $(\mathcal{D}, \mathcal{G})$ – the cotorsion pair generated by the class of all dual modules in $\mathcal{B}$. Clearly, $(\mathcal{A}, \mathcal{B}) \leqslant (\mathcal{D}, \mathcal{G})$.

Assume that R is an artin algebra. By [66], if $\mathcal{C}$ is resolving and contravariantly finite, then $(\mathcal{D}, \mathcal{G})$ is also generated by $\mathcal{H} = \mathcal{B} \cap \text{mod-}R$. So $(\mathcal{C}, \mathcal{H})$ is a complete hereditary cotorsion pair *in mod-R*. Moreover, $\mathcal{G} = \varinjlim \mathcal{H}$.

Let $\mathfrak{C} = (\mathcal{A}, \mathcal{B})$ be a complete cotorsion pair. It is an open problem whether $\mathcal{A}$ is a covering class iff $\mathcal{A}$ is closed under direct limits (The 'if' part is true by 1.11). 1.10 shows that $\mathcal{B}$ may be enveloping even if $\mathcal{A}$ is not closed under direct limits.

1.13. Invariants of modules.   Assume $\mathfrak{C} = (\mathcal{A}, \mathcal{B})$ is a perfect cotorsion pair. Then often the modules in the kernel, $\mathcal{K}$, of $\mathfrak{C}$ can be classified up to isomorphism by cardinal invariants. There are two ways to extend this classification:

a) Any module $A \in \mathcal{A}$ determines – by an iteration of $\mathcal{B}$-envelopes (of A, of the cokernel of the $\mathcal{B}$-envelope of A, etc.) – a long exact sequence all of whose members (except for A) belong to $\mathcal{K}$. This sequence is called the *minimal $\mathcal{K}$-coresolution* of A. The sequence of the cardinal invariants of the modules from $\mathcal{K}$ occuring in the coresolution provides for an invariant of A. In this way, the structure theory of the modules in $\mathcal{K}$ is extended to a structure theory of the modules in $\mathcal{A}$.

b) Dually, any module $B \in \mathcal{B}$ determines – by an iteration of $\mathcal{A}$-covers – a long exact sequence all of whose members (except for B) belong to $\mathcal{K}$, the *minimal $\mathcal{K}$-resolution* of B. This yields a sequence of cardinal invariants for any module $B \in \mathcal{B}$.

For specific examples to a) and b), we consider the case when R is a commutative noetherian ring:

If $\mathfrak{C} = (\text{Mod-}R, \mathcal{I}_0)$, then $\mathcal{K} = \mathcal{I}_0$, and by the classical theory of Matlis, each $M \in \mathcal{K}$ is determined up to isomorphism by the multiplicities of indecomposable injectives $E(R/p)$ (p a prime ideal of R) occuring in an indecomposable decomposition of M. The cardinal invariants of arbitary modules (in $\mathcal{A} = \text{Mod-}R$) constructed in a) are called the *Bass numbers*. A formula for their computation goes back to Bass: the multiplicity of $E(R/p)$ in the i-th term of the minimal injective coresolution of a module N is $\mu_i(p, N) = \dim_{k(p)} \text{Ext}^i_{R_p}(k(p), N_p)$ where $k(p) = R_p/\text{Rad}(R_p)$, and R_p and N_p is the localization of R and N at p, respectively, cf. [70].

If $\mathfrak{C} = (\mathcal{F}_0, \mathcal{EC})$, then $\mathcal{K}$ consists of the flat pure-injective modules: these are described by the ranks of the completions, T_p, of free modules over localizations R_p (p a prime ideal of R) occuring in their decomposition, [46]. The construction b) yields a sequence of invariants for any cotorsion module N. These invariants are called the *dual Bass numbers*. A formula for their computation is due to Xu [93]: the rank of T_p in the i-th term of the minimal flat resolution of N is $\pi_i(p, N) = \dim_{k(p)} \text{Tor}^{R_p}_i(k(p), \text{Hom}_R(R_p, N))$.

In view of 1.5, the following result says that most cotorsion pairs are complete, hence provide for approximations of modules.

For a module M and a class of modules $\mathcal{C}$, a *$\mathcal{C}$-filtration* of M is an increasing sequence of submodules of M, $(M_\alpha \mid \alpha < \sigma)$, such that $M_0 = 0$, $M = \bigcup_{\alpha<\sigma} M_\alpha$, $M_\alpha = \bigcup_{\beta<\alpha} M_\beta$ for all limit ordinals $\alpha < \sigma$, and $M_{\alpha+1}/M_\alpha$ is isomorphic to an element of $\mathcal{C}$ for each $\alpha < \sigma$. A module possessing a $\mathcal{C}$-filtration is called $\mathcal{C}$-*filtered*.

Theorem 1.14. *Let R be a ring and $\mathfrak{C} = (\mathcal{A}, \mathcal{B})$ a cotorsion pair cogenerated by a set of modules $\mathcal{S}$. Then $\mathfrak{C}$ is complete, and $\mathcal{A}$ is the class of all direct summands of all $\mathcal{S} \cup \{R\}$-filtered modules.*

Proof. [43] The core of the proof is a construction (by induction, using a push-out argument inspired by [52] in the non-limit steps), for each pair of modules, (M, N), of a short exact sequence $0 \to M \to B \to A \to 0$ such that A is $\{N\}$-filtered and $B \in \{N\}^{\perp_1}$. By assumption, $\mathfrak{C}$ is cogenerated by a single module, say N. For any module M, the short exact sequence above yields a special $\mathcal{B}$-preenvelope of M, proving that $\mathfrak{C}$ is complete.

For a module $X \in \mathcal{A}$, consider a short exact sequence $0 \to M \to F \to X \to 0$ with F free. Let $0 \to M \to B \to A \to 0$ be as above. The push-out of the maps $M \to F$ and $M \to B$ yields a split exact sequence $0 \to B \to G \to X \to 0$, and G is an extension of F by A, hence G is $\{N, R\}$-filtered. The converse is proved by induction on the length of the filtration.   □

1.14 was applied by Enochs to prove the flat cover conjecture: each module has a flat cover and a cotorsion envelope, [27]. This was generalized in [44] as follows:

Theorem 1.15. *Let R be a ring and $\mathfrak{C}$ be a cotorsion pair generated by a class of pure-injective modules. Then $\mathfrak{C}$ is perfect.*

Proof. By 1.5, $\mathfrak{C}$ is cogenerated by a set of modules. By 1.14, $\mathfrak{C}$ is a complete cotorsion pair. By a classical result of Auslander, the functor $\mathrm{Ext}^1_R(-, M)$ takes direct limits to the inverse ones for each pure-injective module M. In particular, $^{\perp_1}\{M\}$ is closed under direct limits. So 1.11 applies, proving that $\mathfrak{C}$ is perfect.   □

The flat cover conjecture is the particular case of 1.15 for $\mathfrak{C}$ generated by $\mathcal{PI}$. The importance of 1.15 for cotilting theory comes from the recent result of Šťovíček proving that any cotilting module is pure-injective (see Section 3).

For Dedekind domains, we can extend 1.15 further:

Theorem 1.16. *Let R be a Dedekind domain and $\mathfrak{C}$ be a cotorsion pair generated by a class of cotorsion modules. Then $\mathfrak{C}$ is perfect.*

Proof. Let C be a cotorsion module and $f\colon F \to C$ be its flat cover. Then F is flat and cotorsion, hence pure-injective. By [44], $^{\perp}C = {}^{\perp}F$. So $\mathfrak{C}$ is generated by a class of pure-injective modules, and 1.15 applies. $\square$

However, the possibility of extending 1.15 to larger classes of modules depends on the extension of set theory (ZFC) that we work in. Here, one uses the well-developed theory studying dependence of vanishing of Ext on set-theoretic assumptions. This theory originated in Shelah's solution of the Whitehead problem, but has many more applications [40].

In the positive direction, Gödel's axiom of constructibility (V = L) is useful, or rather its combinatorial consequence called Jensen's diamond principle $\Diamond$:

$\Diamond$ Let κ be a regular uncountable cardinal, E be a stationary subset in κ, and X be a set of cardinality κ such that $X = \bigcup_{\alpha<\kappa} X_\alpha$ where $(X_\alpha \mid \alpha < \kappa)$ is an increasing chain of subsets of X with $\mathrm{card}(X_\alpha) < \kappa$ for all $\alpha < \kappa$, and $X_\beta = \bigcup_{\gamma<\beta} X_\gamma$ for each limit ordinal $\beta < \kappa$.

Then there exist sets Y_α ($\alpha \in E$) such that $Y_\alpha \subseteq X_\alpha$ for all $\alpha \in E$, and moreover, for each $Z \subseteq X$, the set $\{\alpha \in E \mid Z \cap X_\alpha = Y_\alpha\}$ is stationary in κ.

(A subset $E \subseteq \kappa$ is *stationary* in κ if E has a non-empty intersection with each closed and unbounded subset of κ.)

The following recent result from [79] extends [44]. It is proved by induction, applying $\Diamond$ in regular cardinals, and Shelah's Singular Compactness Theorem in the singular ones:

Theorem 1.17. *Assume $\Diamond$. Let R be a ring and $\mathfrak{C} = (\mathcal{A}, \mathcal{B})$ a cotorsion pair generated by a set of modules. Assume that $\mathcal{A}$ is closed under pure submodules. Then $\mathfrak{C}$ is cogenerated by a set, and hence $\mathfrak{C}$ is complete.*

Theorem 1.17 is not provable in ZFC (+ GCH), see Theorems 1.19 and 1.20 below. However, it becomes a theorem in ZFC once we strengthen the assumptions on $\mathcal{A}$. This result, conjectured by Bazzoni in [17], has recently been proved in the following general form by Šťovíček [83] (the alternative (i)) and Šaroch [79] (the alternative (ii)):

Theorem 1.18. *Let R be a ring and $\mathfrak{C} = (\mathcal{A}, \mathcal{B})$ be a cotorsion pair such that $\mathcal{A}$ is closed under arbitrary direct products. Moreover, assume that either (i) $\mathcal{A}$ is closed under pure submodules, or (ii) $\mathfrak{C}$ is hereditary and $\mathcal{B} \subseteq \mathcal{I}_n$ for some $n < \omega$. Then $\mathfrak{C}$ is perfect.*

In the negative direction, Shelah's uniformization principle UP$^+$ is useful. Like Gödel's axiom of constructibility, UP$^+$ is relatively consistent with ZFC + GCH, but UP$^+$ and $\Diamond$ are mutually inconsistent.

UP^+ Let κ be a singular cardinal of cofinality ω. There is a stationary subset E in κ^+ consisting of ordinals of cofinality ω, and a ladder system $\mu = (\mu_\alpha \mid \alpha \in E)$ with the following uniformization property:

For each cardinal $\lambda < \kappa$ and each system of maps $h_\alpha \colon \mu_\alpha \to \lambda$ ($\alpha \in E$) there is a map $f \colon \kappa^+ \to \lambda$ such that for each $\alpha \in E$, f coincides with h_α in all but finitely many ordinals of the ladder μ_α.

(A *ladder system* $\mu = (\mu_\alpha \mid \alpha \in E)$ consists of ladders, the ladder μ_α being a strictly increasing countably infinite sequence of ordinals $< \alpha$ whose supremum is α.)

UP$^+$ can be used, for any non-right perfect ring R, to construct particular free modules $G \subseteq F$ such that $M = F/G$ is a non-projective module satisfying $\mathrm{Ext}^1_R(M, N) = 0$ for each module N with $\mathrm{card}(N) < \lambda$. The point is that in the particular setting, a homomorphism $\varphi \colon G \to N$ determines a system of maps h_α ($\alpha \in E$) as in the premise of UP$^+$. The uniformization map f can then be used to define a homomorphism $\psi \colon F \to N$ extending φ, thus giving $\mathrm{Ext}^1_R(M, N) = 0$.

The following is proved in [42] (cf. with 1.16):

Theorem 1.19. *Assume UP$^+$. Let R be a Dedekind domain with a countable spectrum, and $\mathfrak{C}$ a cotorsion pair generated by a set containing at least one non-cotorsion module. Then $\mathfrak{C}$ is not cogenerated by a set of modules.*

In the particular case of $R = \mathbb{Z}$, there is a stronger result by Eklof and Shelah [41], using a much stronger version of UP$^+$ which we do not state here, but just denote by SUP (SUP is also relatively consistent with ZFC + GCH, cf. [41]):

Theorem 1.20. *Assume SUP. Denote by* $\mathfrak{C} = (\mathcal{A}, \mathcal{B})$ *the cotorsion pair generated by* $\mathbb{Z}$. *Then* $\mathbb{Q}$ *does not have an* $\mathcal{A}$-*precover; in particular,* $\mathfrak{C}$ *is not complete.*

Notice that the class $\mathcal{A}$ in 1.20 is the well-known class of all *Whitehead groups*.

We finish this section by two open problems. *Let* R *be a ring and* $\mathfrak{C}$ *a cotorsion pair.*

1. *Is* $\mathfrak{C}$ *complete provided that* $\mathfrak{C}$ *is generated by a class of cotorsion modules?* This is true in the Dedekind domain case by 1.16. Notice that for right perfect rings, the question asks whether all cotorsion pairs are complete.

2. *Is the completeness of* $\mathfrak{C}$ *independent of ZFC in case* $\mathfrak{C}$ *is generated by a set containing at least one non-cotorsion module?* This is true when $R = \mathbb{Z}$ and $\mathfrak{C}$ is generated by $\mathbb{Z}$, cf. 1.17 and 1.20. The term "set" is important here, since by 1.5 and 1.14, in ZFC there are many complete cotorsion pairs $\mathfrak{C} = (\mathcal{A}, \mathcal{B})$ with $\mathcal{B}$ containing non-cotorsion modules.

2 Tilting cotorsion pairs

In this section, we investigate relations between tilting and approximation theory of modules. For this purpose, it is convenient to work with a rather general definition of a tilting module. Our definition allows for infinitely generated modules, and also modules of finite projective dimension > 1.

2.1. Tilting modules.   Let R be a ring. A module T is *tilting* provided that

(1) proj.dim$(T) < \infty$;

(2) $\operatorname{Ext}_R^i(T, T^{(\kappa)}) = 0$ for any cardinal κ and any $i \geqslant 1$;

(3) There are $k < \omega$, $T_i \in \mathrm{Add}(T)$ $(i \leqslant k)$, and an exact sequence

$$0 \to R \to T_0 \to \cdots \to T_k \to 0.$$

Here, $\mathrm{Add}(T)$ denotes the class of all direct summands of arbitrary direct sums of copies of the module T.

Let $n < \omega$. Tilting modules of projective dimension $\leqslant n$ are called *n-tilting*. A class of modules $\mathcal{C}$ is *n-tilting* if there is an n-tilting module T such that $\mathcal{C} = \{T\}^{\perp}$.

A cotorsion pair $\mathfrak{C} = (\mathcal{A}, \mathcal{B})$ is *n-tilting* provided that $\mathcal{B}$ is an n-tilting class.

Notice that the notions above do not change when replacing the tilting module T by the tilting module $T^{(\kappa)}$ $(\kappa > 1)$. It is convenient to define an equivalence of tilting modules as follows: T is *equivalent* to T' provided that the induced tilting classes coincide: $\{T\}^{\perp} = \{T'\}^{\perp}$ (This is also equivalent to $\mathrm{Add}(T) = \mathrm{Add}(T')$.)

Clearly, 0-tilting modules coincide with the projective generators. Finite dimensional tilting modules over artin algebras have been studied in great detail - we refer to [4], [72] and [90] in this volume for much more on this classical case. Now, we give several examples of infinitely generated 1-tilting modules:

2.2. Fuchs tilting modules.   [50], [51] Let R be a commutative domain, and S a multiplicative subset of R. Let $\delta_S = F/G$ where F is the free module with the basis given by all sequences $(s_0, \ldots, s_n)$ where $n \geqslant 0$, and $s_i \in S$ for all $i \leqslant n$, and the empty sequence $w = ()$. The submodule G is generated by the elements of the form $(s_0, \ldots, s_n)s_n - (s_0, \ldots, s_{n-1})$ where $0 < n$ and $s_i \in S$ for all $1 \leqslant i \leqslant n$, and of the form $(s)s - w$ where $s \in S$.

The module $\delta = \delta_{R \setminus \{0\}}$ was introduced by Fuchs. Facchini [47] proved that δ is a 1-tilting module. The general case of δ_S comes from [51]: the module δ_S is a 1-tilting module, called the *Fuchs tilting module*. The corresponding 1-tilting class is $\{\delta_S\}^{\perp} = \{M \in \mathrm{Mod}\text{-}R \mid Ms = M \text{ for all } s \in S\}$, the class of all *S-divisible modules*. If R is a Prüfer domain or a Matlis domain, then the 1-tilting cotorsion pair cogenerated by δ is $(\mathcal{P}_1, \mathcal{DI})$.

J. Trlifaj

Example 2.3. [5] Let R be a commutative 1-Gorenstein ring. Let P_0 and P_1 denote the set of all prime ideals of height 0 and 1, respectively. By a classical result of Bass, the minimal injective coresolution of R has the form

$$0 \to R \to \bigoplus_{q \in P_0} E(R/q) \xrightarrow{\pi} \bigoplus_{p \in P_1} E(R/p) \to 0.$$

Consider a subset $P \subseteq P_1$. Put $R_P = \pi^{-1}(\bigoplus_{p \in P} E(R/p))$. Then $T_P = R_P \oplus \bigoplus_{p \in P} E(R/p)$ is a 1-tilting module, the corresponding 1-tilting class being $\{T_P\}^\perp = \{M \mid \operatorname{Ext}^1_R(E(R/p), M) = 0$ for all $p \in P\}$. In particular, if R is a Dedekind domain then $\{T_P\}^\perp = \{M \mid \operatorname{Ext}^1_R(R/p, M) = 0$ for all $p \in P\} = \{M \mid pM = M$ for all $p \in P\}$.

In his classical work [74], Ringel discovered analogies between modules over Dedekind domains and tame hereditary algebras. The analogies extend to the setting of infinite dimensional tilting modules:

2.4. Ringel tilting modules. [74], [75] Let R be a connected tame hereditary algebra over a field k. Let G be the generic module. Then $S = \operatorname{End}(G)$ is a skew-field and dim $_S G = n < \omega$. Denote by $\mathcal{T}$ the set of all tubes. If $\alpha \in \mathcal{T}$ is a homogenous tube, we denote by R_α the corresponding Prüfer module. If $\alpha \in \mathcal{T}$ is not homogenous, denote by R_α the direct sum of all Prüfer modules corresponding to the rays in α. Then there is an exact sequence

$$0 \to R \to G^{(n)} \xrightarrow{\pi} \bigoplus_{\alpha \in \mathcal{T}} R_\alpha^{(\lambda_\alpha)} \to 0$$

where $\lambda_\alpha > 0$ for all $\alpha \in \mathcal{T}$.

Let $P \subseteq \mathcal{T}$. Put $R_P = \pi^{-1}(\bigoplus_{\alpha \in P} R_\alpha^{(\lambda_\alpha)})$. Then $T_P = R_P \oplus \bigoplus_{\alpha \in P} R_\alpha$ is a 1-tilting module, called the *Ringel tilting module*. The corresponding 1-tilting class is the class of all modules M such that $\operatorname{Ext}^1_R(N, M) = 0$ for all (simple) regular modules $N \in P$. In particular, if $P \neq P' \subseteq \mathcal{T}$, then the 1-tilting modules T_P and $T_{P'}$ are not equivalent.

2.5. Lukas tilting modules. [64], [69] Let R be a connected wild hereditary algebra over a field k. Denote by τ the Auslander-Reiten translation, and by $\mathcal{R}$ the class of all *Ringel divisible modules*, that is, of all modules D such that $\operatorname{Ext}^1_R(M, D) = 0$ for each regular module M.

Let M be any regular module. Then for each finite dimensional module N, Lukas constructed an exact sequence $0 \to N \to A_M \to B_M \to 0$ where $A_M \in M^{\perp}$ and B_M is a finite direct sum of copies of $\tau^n M$ for some $n < \omega$. Letting $C_M = \{\tau^m M \mid m < \omega\}$, we can iterate this construction (for $N = R$, $N = A_M$, etc.) and get an exact sequence $0 \to R \to C_M \to D_M \to 0$ where D_M has a countable C_M-filtration. Then $T_M = C_M \oplus D_M$ is a 1-tilting module, called the *Lukas tilting module*. The corresponding 1-tilting class is $\mathcal{R}$ (so in contrast to 2.4, T_M and $T_{M'}$ are equivalent for all regular modules M and M').

Next, we consider a simple example of an infinitely generated n-tilting module. In §5, we will see that this example is related to the validity of the first finitistic dimension conjecture for Iwanaga-Gorenstein rings.

A ring R is called *Iwanaga-Gorenstein* provided that R is left and right noetherian and the left and right injective dimensions of the regular module are finite, [46]. In this case, inj.dim$(R_R) =$ inj.dim$(_R R) = n$ for some $n < \omega$, and R is called *n-Gorenstein*. Notice that 0-Gorenstein rings coincide with the quasi-Frobenius rings.

Example 2.6. Let R be an n-Gorenstein ring. Let

$$0 \to R \to E_0 \to \cdots \to E_n \to 0$$

be the minimal injective coresolution of R. Then $T = \bigoplus_{i \leqslant n} E_i$ is an n-tilting module. The only non-trivial fact needed for this is that $\mathcal{P} = \mathcal{P}_n = \mathcal{I}_n = \mathcal{I} \ (= \mathcal{F}_n = \mathcal{F})$ for any n-Gorenstein ring, cf. [46, §9].

For any tilting cotorsion pair $\mathfrak{C} = (\mathcal{A}, \mathcal{B})$, there is a close relation among the classes $\mathcal{A}$, $\mathcal{B}$, and the kernel of $\mathfrak{C}$:

Lemma 2.7. *Let R be a ring and $\mathfrak{C} = (\mathcal{A}, \mathcal{B})$ a tilting cotorsion pair. Let T be an n-tilting module with $\{T\}^{\perp} = \mathcal{B}$. Then*

  (1) *$\mathfrak{C}$ is hereditary and complete. Moreover, $\mathfrak{C} \leqslant (\mathcal{P}_n, \mathcal{P}_n^{\perp})$, and the kernel of $\mathfrak{C}$ equals $Add(T)$.*

  (2) *$\mathcal{A}$ coincides with the class of all modules M such that there is an exact sequence*

$$0 \to M \to T_0 \to \cdots \to T_n \to 0$$

where $T_i \in Add(T)$ for all $i \leqslant n$.

(3) Let $0 \to F_n \to \cdots \to F_0 \to T \to 0$ be a free resolution of T and let $\mathcal{S} = \{S_i \mid i \leqslant n\}$ be the corresponding set of syzygies of T. Then $\mathcal{A}$ coincides with the class of all direct summands of all $\mathcal{S}$-filtered modules.

(4) $\mathcal{B}$ coincides with the class of all modules N such that there is a long exact sequence

$$\cdots \to T_{i+1} \to T_i \to \cdots \to T_0 \to N \to 0$$

where $T_i \in Add(T)$ for all $i < \omega$. In particular, $\mathcal{B}$ is closed under arbitrary direct sums.

Proof. 1. The first claim follows from $\mathcal{B} = \{T\}^{\perp}$ by 1.14, the second is clear from $T \in \mathcal{P}_n$. The last claim is proved in [2].

2. Since $\mathcal{A}$ is closed under kernels of monomorphisms, any M possessing such exact sequence is in $\mathcal{A}$. Conversely, we obtain the desired sequence by an iteration of special $\mathcal{B}$-preenvelopes (of M etc.). The fact that we can stop at n follows from proj.dim$(T) \leqslant n$.

3. This follows by the characterization of $\mathcal{A}$ given in 1.14, since $\mathfrak{C}$ is cogenerated by $\bigoplus_{i \leqslant n} S_i$.

4. If $N \in \mathcal{B}$ then the long exact sequence can be obtained by an iteration of special $\mathcal{A}$-precovers (of N etc.). The converse uses proj.dim$(T) \leqslant n$ once again. $\qquad\qquad\qquad\qquad\qquad\qquad\qquad\qquad\qquad\qquad\qquad\square$

We arrive at the characterization of tilting cotorsion pairs in terms of approximations. We start with the case of $n = 1$ treated in [7] and [84]:

Theorem 2.8. *Let R be a ring.*

(1) *A class of modules C is 1-tilting iff C is a special preenveloping torsion class.*

(2) *Let $\mathfrak{C} = (\mathcal{A}, \mathcal{B})$ be a cotorsion pair. Then $\mathfrak{C}$ is 1-tilting iff $\mathcal{A} \subseteq \mathcal{P}_1$ and $\mathcal{B}$ is closed under arbitrary direct sums.*

Proof. 1. Since $\{T\}^{\perp}$ is closed under homomorphic images and extensions for any module T with proj.dim$(T) \leqslant 1$, the only-if part is a consequence of parts 1. and 4. of 2.7 (for $n = 1$). For the if-part, we consider a special $\mathcal{B}$-preenvelope of R; this yields an exact sequence $0 \to R \to B \to A \to 0$ with $B \in \mathcal{B}$ and $A \in \mathcal{A}$. Then $T = A \oplus B$ is a 1-tilting module with $\{T\}^{\perp} = \mathcal{C}$, cf. [7].

2. The only-if part follows directly from parts 1. and 4. of 2.7. For the if-part, we first note that $\mathfrak{C}$ is complete by [19]. Further, $\mathcal{B}$ is closed under homomorphic images and extensions since $\mathcal{B} = \mathcal{A}^{\perp_1}$ and $\mathcal{A} \subseteq \mathcal{P}_1$. So $\mathcal{B}$ is a torsion class, and part 1. applies. $\qquad \square$

We stress that the special approximations induced by 1-tilting modules may not have minimal versions in general (compare this with 3.10.1 below). For example, if R is a Prüfer domain and δ is the Fuchs tilting module from 2.2 then special $\{\delta\}^{\perp}$-preenvelopes coincide with special divisible preenvelopes (and also with special FP-injective preenvelopes). However, if proj.dim$(Q) \geqslant 2$ then the regular module R does not have a divisible envelope (and so it does not have an FP-injective envelope), see [87].

The characterization in the general case is due to Angeleri Hügel and Coelho [2], with a recent improvement from [84]:

Theorem 2.9. *Let R be a ring and $\mathfrak{C} = (\mathcal{A}, \mathcal{B})$ be a cotorsion pair. Let $n < \omega$. Then $\mathfrak{C}$ is n-tilting iff $\mathfrak{C}$ is hereditary (and complete), $\mathcal{A} \subseteq \mathcal{P}_n$, and $\mathcal{B}$ is closed under arbitrary direct sums.*

Proof. The only-if part follows by 2.7. For the if-part, we first note that $\mathfrak{C}$ is complete by [84]. Consider the iteration of special $\mathcal{B}$-preenvelopes of R, of Coker(f) (where f is a special $\mathcal{B}$-preenvelope of R), etc. By assumption, this yields a finite $(\mathcal{A} \cap \mathcal{B})$-coresolution of R, $0 \to R \to T_0 \to \cdots \to T_n \to 0$. Then $T = \bigoplus_{i \leqslant n} T_i$ is n-tilting with $\{T\}^{\perp} = \mathcal{B}$, cf. [2]. $\qquad \square$

2.10. (cf. [81]) Many authors define a partial tilting module P as the module satisfying the first two conditions of 2.1 (for P). However, in general, these two conditions are not sufficient for existence of a *complement* of P (= a module P' such that $T = P \oplus P'$ is tilting and $\{P\}^\perp = \{T\}^\perp$). For a counter-example, consider $R = \mathbb{Z}$ and $P = \mathbb{Q}$; then $\{P\}^\perp$ is the class of all cotorsion groups which is not closed under arbitrary direct sums.

The extra condition (E): "$\{P\}^\perp$ is closed under arbitrary direct sums" is clearly necessary for the existence of a complement of P. We define a *partial n-tilting* module P as a module of projective dimension $\leqslant n$ satisfying (E), and $\operatorname{Ext}^i_R(P, P) = 0$ for all $0 < i < \omega$. Then a complement of P always exists in Mod-R by 2.9: $\{P\}^\perp$ is an n-tilting class with an n-tilting module T such that $\{T\}^\perp = \{P\}^\perp$, so T is a complement of P, cf. [3]. Condition (E) is of course redundant in case $P \in$ mod-R.

Let P be a finitely presented partial 1-tilting module. If R is an artin algebra then P has a finitely presented complement by a classical result of Bongartz, cf. [90]. However, a finitely presented complement of P may not exists even if R is a hereditary noetherian domain, cf. [35].

Rickard and Schofield constructed artin algebras and finitely presented partial 2-tilting modules with no finitely presented complements, cf. [90].

3 Cotilting cotorsion pairs

In this section, we consider the dual case of cotilting modules and cotilting cotorsion pairs.

Similarly as tilting modules, the cotilting ones have first appeared in the representation theory of finite dimensional k-algebras. There, the finite dimensional cotilting modules coincide with the k-duals of the finite dimensional tilting modules, and the theory is obtained by applying the k-duality.

1-cotilting modules over general rings are closely related to dualities (see [33] in this volume for more details). In §4, we will show that there is an explicit homological duality between arbitrary tilting modules and classes on one hand, and cotilting modules and classes of cofinite type on the other hand. The adjective "of cofinite type" is essential here: Bazzoni [18] proved that there exist 1-cotilting modules not equivalent to duals of any 1-tilting modules, cf. 4.16.

The problem of the cotilting setting is that the dual of the key approximation construction 1.14 is not available in ZFC: by 1.20, there is an extension of ZGC + GCH with a cotorsion pair $\mathfrak{C}$ generated by a set such that $\mathfrak{C}$ is not complete. This has recently been overcome by Šťovíček who proved (in ZFC) that all cotilting modules are pure-injective (that is, they are direct summands in dual modules, see 3.3 below), so 1.15 applies and gives the desired covers and envelopes.

3.1. Cotilting modules. Let R be a ring. A module C is *cotilting* provided that

(1) inj.dim$(C) < \infty$;

(2) $\text{Ext}_R^i(C^\kappa, C) = 0$ for any cardinal κ and any $i \geq 1$;

(3) There are $k < \omega$, $C_i \in \text{Prod}(C)$ $(i \leq k)$, and an exact sequence

$$0 \to C_k \to \cdots \to C_0 \to W \to 0,$$

where W is an injective cogenerator for Mod-R, and $\text{Prod}(C)$ denotes the class of all direct summands of arbitrary direct products of copies of the module C.

Let $n < \omega$. Cotilting modules of injective dimension $\leq n$ are called *n-cotilting*. A class of modules $\mathcal{C}$ is *n-cotilting* if there is an n-cotilting module C such that $\mathcal{C} = {}^\perp\{C\}$. A cotorsion pair $\mathfrak{C} = (\mathcal{A}, \mathcal{B})$ is *n-cotilting* provided that $\mathcal{A}$ is an n-cotilting class.

The equivalence of cotilting modules is defined as follows: C is *equivalent* to C' provided that the induced cotilting classes coincide: ${}^\perp\{C\} = {}^\perp\{C'\}$ (that is, $\text{Prod}(C) = \text{Prod}(C')$.)

0-cotilting modules coincide with the injective cogenerators. In 4.12 below, we will see that any resolving subclass of $\mathcal{P}_n^{<\omega}$ yields an n-cotilting class (of left R-modules), so there is a big supply of n-cotilting modules for $n \geq 1$ in general.

We will need the following version of a characterization of cotilting modules by Bazzoni [16]. It generalizes the case of $n = 1$ from [32].

Lemma 3.2. *Let R be a ring, C a module, and $0 < n < \omega$. Then C is n-cotilting iff ${}^\perp\{C\}$ coincides with the class, $\text{Cog}_n(C)$, of all modules*

M possesing an exact sequence $0 \to M \to C_0 \to \cdots \to C_n$ where κ is a cardinal and $C_i = C^\kappa$ for all $i \leqslant n$.

A class $\mathcal{C}$ of modules is *definable* provided that $\mathcal{C}$ is closed under arbitrary direct products, direct limits, and pure submodules, [37]. (Definability implies axiomatizability: definable classes are axiomatized by equality to 1 of certain of the Baur-Garavaglia-Monk invariants. Definable classes of modules correspond bijectively to closed sets of indecomposable pure-injective modules, cf. [37] and [71].)

The following result was first proved in the particular case of $R = \mathbb{Z}$ in [54], and for R a Dedekind domain in [44], as a consequence of the classification of all cotilting modules in these cases. Bazzoni made a crucial step towards a general proof by proving the result for any ring R and $n = 1$ in [15]. An extension to arbitrary n, but for R countable, appeared in [20]. The general case was finally proved by Šťovíček [83] (by induction on n, applying 1.18(i) and [17] in the inductive step):

Theorem 3.3. [83], [17] *Let R be a ring, $n < \omega$, and C be an n-cotilting module. Then C is pure-injective, and $^\perp\{C\}$ is a definable class.*

Being definable, n-cotilting classes are completely characterized by the indecomposable pure-injective modules they contain, cf. [71].

3.4. We now introduce (almost) rigid systems in order to characterize cotilting modules and the corresponding cotilting classes:

Let $n < \omega$. Consider a set $\mathcal{S} = \{M_\alpha \mid \alpha < \kappa\}$ of modules such that each M_α $(\alpha < \kappa)$ is pure-injective with inj.dim$(M_\alpha) \leqslant n$, and Ext$_R^i(M_\alpha, M_\beta) = 0$ for all $\alpha, \beta < \kappa$ and $1 \leqslant i \leqslant n$ (So in particular, each M_α is a splitter.) Then $\mathcal{S}$ is an *n-rigid system* if all the elements of $\mathcal{S}$ are indecomposable. $\mathcal{S}$ is *almost n-rigid* if M_0 is superdecomposable, and all M_α $(0 < \alpha < \kappa)$ are indecomposable.

Theorem 3.5. *Let R be a ring, $n < \omega$, and C an n-cotilting module. Then there is an almost n-rigid system $\mathcal{S}$ such that $C' = \prod_{M \in \mathcal{S}} M$ is an n-cotilting module equivalent to C.*

Proof. By a result of Fisher [63, 8.28], the pure-injective module C is of the form $C = M_0 \oplus E$ where M_0 is zero or superdecomposable, and E is zero or a pure-injective hull of a direct sum of indecomposable pure-injective modules, $E = PE(\bigoplus_{0 < \alpha < \kappa} M_\alpha)$. Then E is a direct summand in $P = \prod_{0 < \alpha < \kappa} M_\alpha$, and P is a pure submodule, hence a direct summand, in E^κ. Put $C' = M_0 \oplus P$. Then $^\perp\{C\} = {}^\perp\{C'\}$, and also $\mathrm{Cog}_n(C) = \mathrm{Cog}_n(C')$, so 3.2 gives that C' is an n-cotilting module equivalent to C. Then $\mathcal{S} = \{M_\alpha \mid \alpha < \kappa\}$ is an almost n-rigid system. $\qquad\square$

3.6. Assume there are no superdecomposable pure-injective modules (i.e., in the terminology of Jensen and Lenzing [63, §8], R has sufficiently many algebraically compact indecomposable modules). Then the system $\mathcal{S}$ in 3.5 is n-rigid. So it only remains to determine which of the n-rigid systems indeed yield n-cotilting modules.

This occurs when R is a Dedekind domain, or a tame hereditary algebra, for example; in fact, in these cases the structure of indecomposable pure-injective modules is well-known, see [37] and [63].

In the Dedekind domain case, 1-rigid systems contain no finitely generated modules. It follows from 3.5 that up to equivalence, cotilting modules are of the form $C_P = \prod_{p \in P} J_p \oplus \bigoplus_{q \in \mathrm{Spec}(R) \backslash P} E(R/q)$ where $0 \notin P \subseteq \mathrm{Spec}(R)$, and J_p denotes the completion of the localization of R at p, cf. 4.14 and 4.17 below.

For the case of tame hereditary algebras, we refer to [28], [29], and [81].

Notice that by 3.7 below, in the right artinian case, each 1-rigid system yields a partial 1-cotilting module in the sense of 3.12.

In the noetherian case, there is more to say for $n = 1$. We can characterize 1-cotilting classes in terms of 1-rigid systems (for a different description, in terms of torsion-free classes in mod-R, see 3.11 below):

Theorem 3.7. *Let R be a right noetherian ring.*

If $\mathcal{C}$ is a 1-cotilting class then there is a 1-rigid system $\mathcal{S}$ such that $\mathcal{C} = \bigcap_{M \in \mathcal{S}} {}^\perp\{M\}$.

Conversely, if R is right artinian and $\mathcal{S}$ a 1-rigid system then $\bigcap_{M \in \mathcal{S}} {}^\perp\{M\}$ is a 1-cotilting class.

Proof. Let C be a 1-cotilting module such that $\mathcal{C} = {}^\perp\{C\}$. By a result of Ziegler, C is elementarily equivalent to a pure-injective envelope of a direct sum of indecomposable pure-injective modules, hence to a direct product of indecomposable pure-injective modules, $E = \prod_{\alpha < \kappa} M_\alpha$, cf. [71]. In particular, E is a direct summand in an ultrapower of C. Since any ultrapower of C is isomorphic to a direct limit of products of copies of C, 3.3 yields $E \in \mathcal{C}$. For right noetherian rings, Baer test lemma shows that $\mathcal{I}_1$ is definable, so $E \in \mathcal{I}_1$ because E is elementarily equivalent to $C \in \mathcal{I}_1$.

Since $\{A\}^{\perp_1}$ is definable for each finitely presented module A, we have $C \in \{A\}^{\perp_1}$ iff $E \in \{A\}^{\perp_1}$. By a classical result of Auslander, $\mathrm{Ext}^1_R(-, I)$ takes direct limits into inverse ones for any pure-injective module I. Since R is right noetherian, it follows that $\mathcal{C} = {}^{\perp_1}\{C\} = {}^{\perp_1}\{E\}$. In particular, E is a pure-injective splitter of injective dimension ≤ 1, so the modules M_α form a 1-rigid system.

Conversely, by [28], ${}^{\perp}\{M\}$ is closed under arbitrary direct products for any $M \in \mathcal{S}$. Let $P = \prod_{M \in \mathcal{S}} M$. By 1.15 and 3.10, ${}^{\perp}\{P\}$ is a 1-cotilting class. □

There is a similar result for Prüfer domains

Theorem 3.8. [18] *Let R be a Prüfer domain. Then each cotilting module C has injective dimension ≤ 1, and C is equivalent to a cotilting module which is a direct product of indecomposable pure-injective modules. In particular, there is a 1-rigid system $\mathcal{S}$ such that $\mathcal{C} = \bigcap_{M \in \mathcal{S}} {}^{\perp}\{M\}$.*

Now, we turn to relations between cotilting modules and approximations. Except for part 3., the dual of 2.7 holds true – a proof making use of [11] appears in [2]. In view of 3.3, one can also proceed directly, by dualizing the proof of 2.7 with help of 1.15:

Lemma 3.9. *Let R be a ring and $\mathfrak{C} = (\mathcal{A}, \mathcal{B})$ be a cotilting cotorsion pair. Let C be an n-cotilting module with ${}^{\perp}\{C\} = \mathcal{A}$. Then*

(1) *$\mathfrak{C}$ is hereditary and complete. Moreover, $({}^{\perp}\mathcal{I}_n, \mathcal{I}_n) \leq \mathfrak{C}$, and the kernel of $\mathfrak{C}$ equals $\mathrm{Prod}(C)$.*

(2) *$\mathcal{A}$ coincides with the class of all modules M such that there is a long exact sequence*

$$0 \to M \to C_0 \to \cdots \to C_i \to C_{i+1} \to \cdots$$

where $C_i \in \mathrm{Prod}(C)$ for all $i < \omega$. In particular, $\mathcal{A}$ is closed under arbitrary direct products.

(3) $\mathcal{B}$ coincides with the class of all modules N such that there is an exact sequence

$$0 \to C_n \to \cdots \to C_0 \to N \to 0$$

where $C_i \in Prod(C)$ for all $i \leqslant n$.

Theorem 3.10. *Let R be a ring.*

(1) *A class of modules $\mathcal{C}$ is 1-cotilting iff $\mathcal{C}$ is a covering torsion-free class.*

(2) *Let $\mathfrak{C} = (\mathcal{A}, \mathcal{B})$ be a cotorsion pair, and $n < \omega$. Then $\mathfrak{C}$ is n-cotilting iff $\mathfrak{C}$ is hereditary (and perfect), $\mathcal{B} \subseteq \mathcal{I}_n$, and $\mathcal{A}$ is closed under arbitrary direct products.*

Proof. The proof is dual to the proofs of 2.8.1 and 2.9, using 1.15, 1.18(ii), and 3.3. $\qquad\qquad\square$

So 1-cotilting classes coincide with those torsion-free classes $\mathcal{C}$ that are covering. If R is right noetherian then $\mathcal{C}$ is completely determined by its subclass $\mathcal{C} \cap$ mod-R, and the latter is characterized as a torsion-free class in mod-R containing R. More precisely, we have

Theorem 3.11. [28] *Let R be a right noetherian ring. There is a bijective correspondence between 1-cotilting classes of modules, $\mathcal{C}$, and torsion-free classes, $\mathcal{E}$, in mod-R containing R. The correspondence is given by the mutually inverse assignments $\mathcal{C} \mapsto \mathcal{C} \cap$ mod-R and $\mathcal{E} \mapsto \varinjlim \mathcal{E}$.*

Proof. If $\mathcal{C}$ is a 1-cotilting class, then clearly $\mathcal{C} \cap$ mod-R is a torsion-free class in mod-R containing R.

Conversely, given $\mathcal{E}$ as in the claim, let $\mathcal{C} = \varinjlim \mathcal{E}$. By [36], $\mathcal{C}$ is a torsion-free class in Mod-R. Since $R \in \mathcal{E}$, by 1.12, there is a Tor-torsion pair of the form $(\mathcal{C}, \mathcal{D})$. By 1.5.3 and 1.15, $\mathcal{C}$ is a covering class. By 3.10.1, $\mathcal{C}$ is 1-cotilting.

Now, $\mathcal{E} = \varinjlim \mathcal{E} \cap$ mod-R. Conversely, given a 1-cotilting class $\mathcal{C}$, each $M \in \mathcal{C}$ is a directed union of the system of its finitely presented submodules, $\{M_i \mid i \in I\}$ (because R is right noetherian). Since $\mathcal{C}$ is 1-cotilting,

$M_i \in \mathcal{C}$ for each $i \in I$. So $\mathcal{C} = \varinjlim(\mathcal{C} \cap \text{mod-}R)$, and the assignments are mutually inverse. $\square$

We note that the corresponding result to 3.11 does not hold for 1-tilting classes. Namely, given a right noetherian ring R and a 1-tilting (torsion) class $\mathcal{T}$ in Mod-R, the class $\mathcal{T} \cap \text{mod-}R$ is certainly a torsion class in mod-R. Let $\mathcal{C} = \varinjlim(\mathcal{T} \cap \text{mod-}R)$. By [36], $\mathcal{C}$ is a torsion class in Mod-R contained in $\mathcal{T}$. However, $\mathcal{C}$ is not 1-tilting in general: if R is an artin algebra and $\mathcal{T} = (\mathcal{P}_1^{<\omega})^{\perp}$, then $\mathcal{C}$ is closed under arbitrary direct products iff $\mathcal{P}_1^{<\omega}$ is contravariantly finite. The latter fails for the IST-algebra [61], for example.

(However, if R is an artin algebra and $\mathcal{C}$ a 1-tilting class, there is a way of reconstructing $\mathcal{C}$ from $\mathcal{C} \cap \text{mod-}R$, see 4.3 below.)

3.12. (cf. [81])   Define a *partial 1-cotilting* module P as a splitter of injective dimension $\leqslant 1$ satisfying the extra condition of $^{\perp}\{P\}$ being closed under arbitrary direct products. Then P has a complement in the sense that there is a module P' such that $C = P \oplus P'$ is 1-cotilting and $^{\perp}\{P\} = {}^{\perp}\{C\}$. This follows from 3.10 and [86, §6]. (Note that P is pure-injective by 3.3.) By [28], the extra condition is redundant when P is pure-injective and R is right artinian.

We finish this section by an open problem:

Let R be a ring. Does 3.5 hold in the stronger form, with n-rigid systems replacing the almost n-rigid ones?

The answer is affirmative for R von Neumann regular by 3.3 (since then pure-injective modules coincide with the injective ones), for R right noetherian and right hereditary by 3.7, and for R a Prüfer domain by 3.8.

4 Finite type, duality, and some examples

The notions of a tilting and cotilting module are formally dual. The duality can be made explicit using the notions of a module of finite and cofinite type.

4.1. Modules of finite type. Let R be a ring.

(1) Let $\mathcal{C}$ be a class of modules. Then $\mathcal{C}$ is of *finite type* (*countable type*) provided there exist $n < \omega$ and a subset $\mathcal{S} \subseteq \mathcal{P}_n^{<\omega}$ ($\mathcal{S} \subseteq \mathcal{P}_n^{\leqslant\omega}$) such that $\mathcal{C} = \mathcal{S}^\perp$.

(2) Let M be a module. Then M is of *finite type* (*countable type*, *definable*) provided the class $\{M\}^\perp$ is of finite type (countable type, definable).

The key fact is that tilting modules (classes) are exactly the modules (classes) of finite type. One implication is easy to prove:

Lemma 4.2. [5] *Let R be a ring and $\mathcal{C}$ be a class of modules of finite type. Then $\mathcal{C}$ is tilting (and definable).*

Proof. By assumption, there are $n < \omega$ and a set $\mathcal{S} \subseteq \mathcal{P}_n^{<\omega}$ such that $\mathcal{C} = \mathcal{S}^\perp$.

By a classical result of Brown, the covariant functor $\mathrm{Ext}_R^n(M, -)$ commutes with direct limits for each $n \geqslant 0$ and each $M \in \mathrm{mod}\text{-}R$. Also, it is easy to see that $\{N\}^{\perp_1}$ is closed under pure submodules for any finitely presented module N. It follows that $\mathcal{C}$ is definable.

Let $\mathfrak{C} = (\mathcal{A}, \mathcal{C})$ be the cotorsion pair generated by $\mathcal{C}$. By 1.14, $\mathcal{A} \subseteq \mathcal{P}_n$, so 2.9 gives that $\mathfrak{C}$ is a tilting cotorsion pair. That is, $\mathcal{C}$ is a tilting class. $\qquad\square$

4.2 says that there is a rich supply of tilting classes in general: any subset $\mathcal{S} \subseteq \mathcal{P}_n^{<\omega}$ (for some $n < \omega$) determines one. A more precise general description appears in 4.12 below; for artin algebras, there is also the following analog of 3.11:

Theorem 4.3. [64], [21] *Let R be an artin algebra. There is a bijective correspondence between 1-tilting classes, $\mathcal{C}$, and torsion classes, $\mathcal{T}$, in $\mathrm{mod}\text{-}R$ containing all finitely generated injective modules. The correspondence is given by the mutually inverse assignments $\mathcal{C} \mapsto \mathcal{C} \cap \mathrm{mod}\text{-}R$, and $\mathcal{T} \mapsto \mathrm{Ker\,Hom}_R(-, \mathcal{F})$ where $(\mathcal{T}, \mathcal{F})$ is a torsion pair in $\mathrm{mod}\text{-}R$.*

The proof of the converse of 4.2 is much more involved. First, we note the following characterization of finite type:

Lemma 4.4. [5] *Let R be a ring and T be a tilting module. Let $\mathcal{B} = \{T\}^{\perp}$, and $(\mathcal{A}, \mathcal{B})$ be the corresponding tilting cotorsion pair. Then T is of finite type iff T is definable and $T \in \varinjlim \mathcal{A}^{<\omega}$.*

For $n = 1$, the last condition of 4.4 holds by the following lemma:

Lemma 4.5. *Let R be a ring and M be a module of projective dimension $\leqslant 1$. Let $(\mathcal{A}, \mathcal{B})$ be the cotorsion pair cogenerated by M. Then $M \in \varinjlim \mathcal{A}^{<\omega}$.*

Proof. Since $M \in \mathcal{P}_1$, there is an exact sequence $0 \to F \subseteq G \to M \to 0$ where F and G are free modules. Let $\{x_\alpha \mid \alpha < \kappa\}$ and $\{y_\beta \mid \beta < \lambda\}$ be a free basis of F and G, respectively. W.l.o.g., κ is infinite. For each finite subset $S \subseteq \kappa$ let S' be the least (finite) subset of λ such that $F_S = \bigoplus_{\alpha \in S} x_\alpha R \subseteq G_S = \bigoplus_{\beta \in S'} y_\beta R$. Then F is a directed union of its summands of the form F_S where S runs over all finite subsets of κ. Let $M_S = G_S/F_S$. Then $M_S \in \mathcal{P}_1^{<\omega}$, and $M = P \oplus H$ where P is free and $H = \varinjlim_S M_S$.

It suffices to prove that $M_S \in \mathcal{A}^{<\omega}$ for each finite subset $S \subseteq \kappa$: Take an arbitrary $N \in \mathcal{B} = \{M\}^{\perp}$. Then any homomorphism from F to N extends to G. Let φ be a homomorphism from F_S to N. Since F_S is a direct summand in F, φ extends to F, hence to G, and G_S. It follows that $N \in \{M_S\}^{\perp}$, so $M_S \in \mathcal{A}^{<\omega}$, and $H \in \varinjlim \mathcal{A}^{<\omega}$. $\square$

In general, 4.5 fails for modules of projective dimension $n > 1$. By [80], for each $n > 1$, there is an artin algebra R such that $\mathcal{P} = \mathcal{P}_n$, but $\mathcal{P}^{<\omega} = \mathcal{P}_1^{<\omega}$. So $\varinjlim \mathcal{P}_1^{<\omega} = \mathcal{P}_1$, and 4.5 fails for the cotorsion pair $(\mathcal{P}_n, (\mathcal{P}_n)^{\perp})$, cf. 1.5.1.

Next step for the converse of 4.2 was done in [19], where it was proved that all 1-tilting modules are of countable type. Using this, Bazzoni and Herbera proved in [21] that all 1-tilting modules are definable, so the converse of 4.2 holds for $n = 1$ by 4.4 and 4.5.

For $n \geqslant 1$, Šťovíček and Trlifaj proved in [84] that all n-tilting modules are of countable type. Their proof – generalizing the case of $n = 1$ from [19] – makes essential use of the set-theoretic methods developed by Eklof, Fuchs and Shelah for the structure theory of so called Baer modules, cf. [39], [40].

Building on [21] and [84], Bazzoni and Šťovíček finally proved the converse of 4.2 in full generality:

Theorem 4.6. [25] *Let R be a ring, $n < \omega$, and T be an n-tilting module. Then T is of finite type.*

4.7. Let R be a ring, $n < \omega$, T be an n-tilting module, and $(\mathcal{A}, \mathcal{B})$ be the corresponding n-tilting cotorsion pair. Then T is $\mathcal{A}^{\leq \omega}$-filtered by [84].

Though $T \in \varinjlim \mathcal{A}^{<\omega}$ by 4.4, T need not be $\mathcal{A}^{<\omega}$-filtered (for example, if R is a ring possesing a countably generated projective module P which is not a direct sum of finitely generated modules, then $T = P \oplus R$ is 0-tilting, but T is not a direct sum of modules in $\mathcal{P}_0^{<\omega}$). However, T is always equivalent to a $\mathcal{A}^{<\omega}$-filtered n-tilting module T' (where $T' = \bigoplus_{i \leq n} T_i$ can be obtained by the following iteration of special $(\mathcal{A}^{<\omega})^\perp$-preenvelopes with $\mathcal{A}^{<\omega}$-filtered cokernels: $\mu_0 \colon R \to T_0$, $\mu_1 \colon R/T_0 \to T_1$ etc.).

The counterpart of a tilting (right R-) module (of finite type) is a cotilting left R-module of cofinite type:

4.8. Modules of cofinite type. Let R be a ring. Let $\mathcal{C} \subseteq R$-Mod. Then $\mathcal{C}$ is of *cofinite type* provided that there exist $n < \omega$ and a subset $\mathcal{S} \subseteq \mathcal{P}_n^{<\omega}$ such that $\mathcal{C} = \mathcal{S}^\intercal$, where $\mathcal{S}^\intercal = \{M \in R\text{-Mod} \mid \operatorname{Tor}_i^R(S, M) = 0 \text{ for all } S \in \mathcal{S} \text{ and all } 0 < i \leq n\}$.

Let C be a left R-module. Then C is of *cofinite type* provided that the class $^\perp\{C\}$ is of cofinite type.

Applying 3.10, we can dualize 4.2:

Lemma 4.9. *Let R be a ring and $\mathcal{C}$ be a class of left R-modules of cofinite type. Then $\mathcal{C}$ is cotilting (and definable).*

4.5 yields a characterization of 1-cotilting classes of cofinite type:

Lemma 4.10. *Let R be a ring and $\mathcal{C}$ be a class of left R-modules. Then $\mathcal{C}$ is 1-cotilting of cofinite type iff there is a module $M \in \mathcal{P}_1$ such that $\mathcal{C} = \{M\}^\intercal$.*

Proof. For the only-if part, consider $\mathcal{S} \subseteq \mathcal{P}_n^{<\omega}$ such that $\mathcal{S}^\intercal = \mathcal{C}$. Put $M = \bigoplus_{S \in A} S$ where A is a representative set of elements of $\mathcal{S}$. Then

$\mathcal{C} = \{M\}^{\mathsf{T}} = {}^{\perp}\{M^*\}$, so $M^* \in \mathcal{I}_1$ (because $\mathcal{C}$ is 1-cotilting). It follows that $M \in \mathcal{F}_1$. So $S \in \mathcal{F}_1 \cap \text{mod-}R = \mathcal{P}_1^{<\omega}$ for each $S \in A$, and $M \in \mathcal{P}_1$.

For the if part, let $(\mathcal{A}, \mathcal{B})$ be the cotorsion pair cogenerated by M. Since $\mathcal{A} \subseteq \mathcal{P}_1$, it suffices to show that $\{M\}^{\mathsf{T}} = (\mathcal{A}^{<\omega})^{\mathsf{T}}$. By 4.5, $(\mathcal{A}^{<\omega})^{\mathsf{T}} \subseteq \{M\}^{\mathsf{T}}$. Conversely, let $N \in \{M\}^{\mathsf{T}}$. Then $N^* \in \mathcal{B}$, so $N \in \mathcal{A}^{\mathsf{T}} \subseteq (\mathcal{A}^{<\omega})^{\mathsf{T}}$. □

The bijective correspondence between tilting classes, and cotilting classes of cofinite type, is mediated by resolving subclasses of mod-R. It is analogous to the classical characterization of cotilting classes in mod-R over artin algebras due to Auslander and Reiten [12].

Definition 4.11. Let R be a ring and $\mathcal{S} \subseteq \text{mod-}R$. Then $\mathcal{S}$ is *resolving* provided that $\mathcal{P}_0^{<\omega} \subseteq \mathcal{S}$, $\mathcal{S}$ is closed under direct summands and extensions, and $\mathcal{S}$ is closed under kernels of epimorphisms.

Notice that a subclass $\mathcal{S} \subseteq \mathcal{P}_1^{<\omega}$ is resolving iff $\mathcal{S}$ is closed under extensions and direct summands, and $R \in \mathcal{S}$.

Theorem 4.12. [5], [25] *Let R be a ring and $n < \omega$. There is a bijective correspondence among*

- *n-tilting classes in Mod-R,*
- *resolving subclasses of $\mathcal{P}_n^{<\omega}$,*
- *n-cotilting classes of cofinite type in R-Mod.*

Proof. Given an n-tilting class $\mathcal{T} \subseteq \text{Mod-}R$, we put $\mathcal{S} = {}^{\perp}\mathcal{T} \cap \text{mod-}R$; conversely, given a resolving subclass $\mathcal{S} \subseteq \mathcal{P}_n^{<\omega}$, we let $\mathcal{T} = \mathcal{S}^{\perp}$. These assignments are mutually inverse. Similarly, given an n-cotilting class of cofinite type $\mathcal{C} \subseteq R$-Mod, we let $\mathcal{S} = {}^{\mathsf{T}}\mathcal{C} \cap \text{mod-}R$; conversely, $\mathcal{C} = \mathcal{S}^{\mathsf{T}}$. For more details, we refer to [5]. □

Moreover, if T is an n-tilting module then T^* is an n-cotilting left R-module of cofinite type; in the correspondence of 4.12, the n-tilting class $\{T\}^{\perp}$ corresponds to the n-cotilting class of cofinite type ${}^{\perp}\{T^*\} = \{T\}^{\mathsf{T}}$, cf. [5].

Lemma 4.13. [5] *Let R be a left noetherian ring and C be a 1-cotilting left R-module. Then ${}^{\perp}\{C\} = \{C^*\}^{\mathsf{T}}$.*

Proof. By 3.3, C is pure-injective, so C is a direct summand in C^{**}. In particular, $\{C^*\}^{\intercal} = {}^{\perp}\{C^{**}\} \subseteq {}^{\perp}\{C\}$. Conversely, take $M \in R\text{-mod}$. If $\operatorname{Ext}_R^1(M, C) = 0$, then the Ext-Tor relations yield $\operatorname{Tor}_1^R(C^*, M) = 0$. Since R is left noetherian, if $N \in {}^{\perp}\{C\}$ then N is a directed union, $N = \bigcup_{i \in I} N_i$, of submodules of N such that $N_i \in {}^{\perp}\{C\} \cap R\text{-mod}$ for all $i \in I$. So $N_i \in \{C^*\}^{\intercal}$. Since Tor commutes with direct limits, we have $N \in \{C^*\}^{\intercal}$. This proves that ${}^{\perp}\{C\} = \{C^*\}^{\intercal}$. $\qquad \square$

4.10 and 4.13 yield a partial converse of 4.9:

Theorem 4.14. *Let R be a left noetherian ring. Assume that $\mathcal{F}_1 = \mathcal{P}_1$ (this holds when R is (i) right perfect or (ii) right hereditary or (iii) 1-Gorenstein, for example). Then every 1-cotilting left R-module is of cofinite type.*

Proof. Let C be a 1-cotilting left R-module. Then $C^* \in \mathcal{F}_1 = \mathcal{P}_1$. By 4.13, ${}^{\perp}\{C\} = \{C^*\}^{\intercal}$. The latter class is of cofinite type by 4.10. $\qquad \square$

4.14 applies to the left artinian case:

4.15. 1-tilting and 1-cotilting classes over artinian rings. Let R be a left artinian ring. Then 1-cotilting classes of left R-modules are of cofinite type, hence coincide with the classes of the form $\{M \in R\text{-Mod} \mid \operatorname{Tor}_1^R(S, M) = 0 \text{ for all } S \in \mathcal{S}\}$ for some $\mathcal{S} \subseteq \mathcal{P}_1^{<\omega}$. Moreover, by 4.12, these classes correspond bijectively to the classes $\mathcal{S}'$ closed under extensions and direct summands, and satisfying $\mathcal{P}_0^{<\omega} \subseteq \mathcal{S}' \subseteq \mathcal{P}_1^{<\omega}$. By 3.11, they also correspond to torsion-free classes in $R\text{-mod}$ containing R.

If R is an artin algebra, then 1-tilting classes correspond bijectively to torsion classes in mod-R containing all finitely generated injective modules by 4.3. Moreover, 1-cotilting left R-modules coincide (up to equivalence) with duals of 1-tilting modules.

In contrast with 4.6, the converse of 4.9 does not hold in general: there exist Prüfer domains with 1-cotilting modules that are not of cofinite type. We are going to discuss the Prüfer and Dedekind domain cases in more detail:

4.16. Tilting and cotilting classes over Prüfer domains. [18], [19], [77], [78] Let R be a Prüfer domain. Then all tilting modules have projective dimension $\leqslant 1$. Moreover, for each 1-tilting class, $\mathcal{T}$, there is a set, $\mathcal{E}$, of non-zero finitely generated (projective) ideals of R such that $\mathcal{T}$ consists of all modules M satisfying $IM = M$ for all $I \in \mathcal{E}$ (or, equivalently, $\text{Ext}^1_R(R/I, M) = 0$ for all $I \in \mathcal{E}$).

Moreover, tilting classes correspond bijectively to finitely generated localizing systems, $\mathcal{I}$, of R in the sense of [49, §5.1]. (A multiplicatively closed filter $\mathcal{I}$ of non-zero ideals of R is a *finitely generated localizing system* provided that $\mathcal{I}$ contains a basis consisting of finitely generated ideals; by [49, 5.1], finitely generated localizing systems correspond bijectively to overrings of R.) Given such system $\mathcal{I}$, the corresponding 1-tilting class $\mathcal{T}_\mathcal{I}$ consists of all modules M satisfying $IM = M$ for all $I \in \mathcal{I}$.

The appropriate generalization of the Fuchs tilting module δ_S from 2.2 to the present setting is obtained by replacing finite sequences of elements of S by finite sequences of finitely generated (invertible) ideals in $\mathcal{I}$. The resulting 1-tilting module, $\delta_\mathcal{I}$, generates $\mathcal{T}_\mathcal{I}$. This yields a classification of tilting modules over Prüfer domains up to equivalence – for more details, we refer to [78].

Since the weak global dimension of a Prüfer domain is $\leqslant 1$, all pure-injective modules have injective dimension $\leqslant 1$, in particular, all cotilting modules have injective dimension $\leqslant 1$. By (the proof of) 4.12, the cotilting classes of cofinite type coincide with the classes of the form $\{M \mid \text{Tor}^R_1(M, R/I) = 0$ for all $I \in \mathcal{I}\}$ where $\mathcal{I}$ is a finitely generated localizing system.

By [18], the structure of cotilting modules can be reduced to the valuation domain case: if C is a cotilting module then C is equivalent to the direct product $\prod_m C_m$ where m runs over all maximal ideals of R, and the localization of C at m, C_m, is a cotilting R_m-module. By Theorem 3.8, we can further reduce to products of indecomposable pure-injective R_m-modules (the latter are known to be isomorphic to injective envelopes of the uniserial R_m-modules R_m/I where $I \neq R_m$ is an ideal of R_m, or to pure-injective envelopes of the uniserial R_m-modules of the form J/I where $0 \subseteq I \subsetneq J \subseteq Q_m$, [50, XIII.5]).

If R is a valuation domain then all cotilting modules are of cofinite type iff R is *strongly discrete* (that is, R contains no non-zero idempotent prime ideals). Indeed, if L is a non-zero idempotent prime ideal in R, denote by $\mathcal{C}_L$ the class of all modules M such that the annihilator of x

is either 0 or L, for each $0 \neq x \in M$. Then $\mathcal{C}_L$ is 1-cotilting, but not of cofinite type. For more details, we refer to [18].

A complete description is available for Dedekind domains:

4.17. Tilting and cotilting modules over Dedekind domains.
[19] Let R be a Dedekind domain. By 2.3, for each set of maximal ideals, P, there is a tilting module $T_P = R_P \oplus \bigoplus_{p \in P} E(R/p)$ with the corresponding tilting class $\{T_P\}^\perp = \{M \mid pM = M \text{ for all } p \in P\}$. Since localizing systems of ideals of R are determined by their prime ideals, by 4.16, any tilting module T is equivalent to T_P for a set of maximal ideals P, cf. [19]. (In the particular case when $R = \mathbb{Z}$, and R is a small Dedekind domain, this result was proved assuming V = L in [54] and [89], respectively).

By 4.12, cotilting classes of cofinite type are exactly the classes of the form $\mathcal{C}_P = \{M \mid \operatorname{Tor}_1^R(M, R/p) = 0 \text{ for all } p \in P\}$ for a set, P, of maximal ideals of R. Moreover, $\mathcal{C}_P = {}^\perp\{C_P\}$ where $C_P = \prod_{p \in P} J_p \oplus \bigoplus_{q \in \operatorname{Spec}(R) \setminus P} E(R/q)$ is a cotilting module. (Here, J_p denotes the completion of the localization of R at p).

By 4.14 (or 3.6), all cotilting classes are of the form $\mathcal{C}_P$, and all cotilting modules are equivalent to the modules of the form C_P, for a set, P, of maximal ideals of R, cf. [44].

The analogies between modules over Dedekind domains and tame hereditary algebras (cf. 2.3 and 2.4) extend also to the tilting and cotilting setting, see [28], [29], and [81], for more details.

Finally, we present a recent application of 1-tilting modules to decomposition theory over commutative rings:

4.18. Matlis localizations. [6] Let R be a commutative ring, Σ the monoid of all regular elements (= non-zero-divisors) in R, and S a submonoid of Σ. Consider the localization RS^{-1} of R at S. The ring RS^{-1} is a *Matlis localization* of R provided that RS^{-1} has projective dimension $\leqslant 1$ as an R-module.

Assume R is a commutative domain. Then $Q = R\Sigma^{-1}$, so Q is a Matlis localization iff R is a Matlis domain in the sense of 1.10. Lee, extending earlier work of Kaplansky and Hamsher, characterized Matlis domains by the property that Q/R decomposes into a direct sum of countably generated R-modules, [67]. Lee's result was extended by Fuchs and Salce to the case of an arbitrary submonoid S of Σ, [51].

Infinite dimensional tilting theory makes it possible to extend this characterization further, to the setting of arbitrary commutative rings:

Theorem 4.19. [6] *Let R be a commutative ring and S be a submonoid of Σ. The following conditions are equivalent:*

(1) RS^{-1} *is a Matlis localization.*

(2) $T_S = RS^{-1} \oplus RS^{-1}/R$ *is a 1-tilting module.*

(3) RS^{-1}/R *decomposes into a direct sum of countably presented R-modules.*

Under these conditions, the 1-tilting class generated by T_S equals $\{M \in \text{Mod-}R \mid Ms = M$ for all $s \in S\}$, the class of all S-divisible modules (cf. 2.2). Condition (3) can be viewed as a stronger form of countable type going in a different direction than finite type: when computing the 1-tilting class $T_S^\perp = (RS^{-1}/R)^\perp$, we can replace the single (infinitely generated) module RS^{-1}/R by a set of countably presented modules which moreover form a direct sum decomposition of RS^{-1}/R.

We finish this section by an open problem:

4.20. *Characterize the rings R such that all cotilting left R-modules are equivalent to duals of tilting (right R-) modules.*

This property holds for all left noetherian right hereditary rings, and all 1-Gorenstein rings, by 4.14, and (trivially) for all von Neumann regular rings, but fails for any non-strongly discrete valuation domain by 4.16.

5 Tilting modules and the finitistic dimension conjectures

Let R be a ring and $\mathcal{C}$ be a class of modules. The $\mathcal{C}$-*dimension* of R is defined as the supremum of projective dimensions of all modules in $\mathcal{C}$.

If $\mathcal{C} = \text{Mod-}R$ then the $\mathcal{C}$-dimension is called the (right) *global dimension* of R; if $\mathcal{C} = \mathcal{P}$, it is called the *big finitistic dimension* of R. If $\mathcal{C}$ is the class of all finitely generated modules in $\mathcal{P}$ then the $\mathcal{C}$-dimension is called the *little finitistic dimension* of R. These dimensions are denoted by gl.dim(R), Fin.dim(R), and fin.dim(R), respectively.

Clearly, fin.dim(R) $\leqslant$ Fin.dim(R) $\leqslant$ gl.dim(R) for any ring R. Moreover, if R has finite global dimension, then gl.dim(R) is attained on cyclic modules, so all the three dimensions coincide.

If R has infinite global dimension, then the finitistic dimensions take the role of the global dimension to provide a fine measure of complexity of the module category. For example, if $R = \mathbb{Z}_{p^n}$ for a prime integer p and $n > 1$, then R has infinite global dimension, but both finitistic dimensions are 0; they certainly reflect better the fact that R is of finite representation type.

In [14], Bass considered the following assertions

(I) fin.dim(R) $=$ Fin.dim(R)

(II) fin.dim(R) is finite

and proposed to investigate the validity of these assertions in dependence on the structure of the ring R. Later, (I) and (II) became known as the *first*, and the *second, finitistic dimension conjecture*, respectively.

For R commutative and noetherian, Bass, Raynaud and Gruson proved that Fin.dim(R) coincides with the Krull dimension of R, so classical examples of Nagata can be used to provide counter-examples to the assertion (II). In case R is commutative local noetherian, Auslander and Buchweitz proved that fin.dim(R) coincides with the depth of R, so (I) holds iff R is a Cohen-Macaulay ring.

Assume that R is right artinian. Then the validity of (II) is still an open problem. However, Huisgen-Zimmermann proved that (I) need not hold even for monomial finite dimensional algebras, [57]. Smalø constructed, for any $1 < n < \omega$, examples of finite dimensional algebras with fin.dim(R) $= 1$ and Fin.dim(R) $= n$, [80].

However, there are many positive results available: (II) was proved for all monomial algebras in [56], for algebras of representation dimension $\leqslant 3$ in [62], etc.

(I) and (II) were proved for all algebras such that $\mathcal{P}^{<\omega}$ is contravariantly finite in [12] and [60]. In this section, we use tilting approximations to give a simple proof of the latter result. Then we prove (I) for all Iwanaga-Gorenstein rings.

In the rest of this section, R denotes a right noetherian ring, and $\mathfrak{C} = (\mathcal{A}, \mathcal{B})$ the cotorsion pair cogenerated by $\mathcal{P}^{<\omega}$. By 1.14, $\mathfrak{C}$ is complete and hereditary; moreover, $\mathcal{P}^{<\omega} = \mathcal{A} \cap \mathrm{mod}\text{-}R$.

The basic relation between tilting approximations and the finitistic dimension conjectures comes from [8]:

Theorem 5.1. *Let R be a right noetherian ring. Then (II) holds iff $\mathfrak{C}$ is a tilting cotorsion pair. Moreover, if T is a tilting module such that $\{T\}^{\perp} = \mathcal{B}$, then $\mathrm{fin.dim}(R) = \mathrm{proj.dim}(T)$.*

Proof. Assume $\mathrm{fin.dim}(R) = n < \omega$. Then $\mathcal{P}^{<\omega} \subseteq \mathcal{P}_n$, so $\mathcal{B}$ is of finite type, and $\mathfrak{C}$ is a tilting cotorsion pair by 4.2. Conversely, if $\mathfrak{C}$ is n-tilting then $\mathcal{P}^{<\omega} \subseteq \mathcal{P}_n$, so (II) holds. Since $\mathrm{fin.dim}(R)$ is the least m such that $\mathcal{A} \subseteq \mathcal{P}_m$, we infer that $\mathrm{fin.dim}(R) = \mathrm{proj.dim}(T)$. $\square$

A dual version of 5.1 for artin algebras appears in [30].

5.2. The tilting module T in 5.1 is unique up to equivalence, and it is clearly of finite type. In principle, T can be constructed as in the proof of 2.9: that is, by an iteration of special $\mathcal{B}$-preenvelopes of R etc. yielding an $\mathrm{Add}(T)$-coresolution of R, $0 \to R \to T_0 \to \cdots \to T_n \to 0$, and giving $T = \bigoplus_{i \leqslant n} T_i$. However, little is known of the (definable) class $\mathcal{B}$ in general, so this construction is of limited use. (The construction works fine for $\mathrm{gl.dim}(R) < \infty$. Then $\mathcal{B} = \mathcal{I}_0$, so the $\mathrm{Add}(T)$-coresolution above can be taken as the minimal injective coresolution of R.)

In the artinian case, we can compute $\mathrm{fin.dim}(R)$ using $\mathcal{A}$-approximations of all the (finitely many) simple modules. This is proved in [88], generalizing [12]:

Theorem 5.3. *Let R be a right artinian ring and $\{S_0, \ldots S_m\}$ be a representative set of all simple modules. For each $i \leqslant m$, take a special $\mathcal{A}$-preenvelope of S_i, $f_i \colon A_i \to S_i$. Then $\mathrm{fin.dim}(R) = \max_{i \leqslant m} \mathrm{proj.dim}(A_i)$.*

Moreover, all the modules A_i ($i \leqslant m$) can be taken finitely generated iff

$\mathcal{P}^{<\omega}$ *is contravariantly finite. In this case (II) holds true, since* $\mathcal{P}^{<\omega} = \mathcal{A} \cap \text{mod-}R$.

Now, we relate pure-injectivity properties of the tilting module T from 5.1 to closure properties of the class $\mathcal{A}$.

A module M is *pure-split* if all pure submodules of M are direct summands; M is $\sum$-*pure-split* iff all modules in Add(M) are pure-split. For example, any $\sum$-pure-injective module is $\sum$-pure-split, [59].

A module M is *product complete* if $\text{Prod}(M) \subseteq \text{Add}(M)$. Any product complete module is $\sum$-pure-injective, [63].

The following is proved in [8] and [9]:

Lemma 5.4. *Let R be a right noetherian ring satisfying (II). Let T be the tilting module from 5.1. Then*

   (1) *T is $\sum$-pure-split iff $\mathcal{A}$ is closed under direct limits.*

   (2) *T is product complete iff $\mathcal{A}$ is closed under products iff $\mathcal{A}$ is definable.*

   (3) *$\mathcal{A} = \mathcal{P}$ iff Add(T) is closed under cokernels of monomorphisms.*

5.5. The condition $\mathcal{A} = \mathcal{P}$ implies (I), since any module of finite projective dimension is then a direct summand in a $\mathcal{P}^{<\omega}$-filtered module, by 1.14. In fact, when proving the first finitistic dimension conjectures in 5.6 and 5.8 below, we always prove that $\mathcal{A} = \mathcal{P}$. However, (I) may hold even if $\mathcal{A} \subsetneqq \mathcal{P}$, see [9].

Theorem 5.6. [8] *Let R be an artin algebra such that (II) holds. Let T be the tilting module from 5.1. Then T can be taken finitely generated iff $\mathcal{P}^{<\omega}$ is contravariantly finite. In this case, (I) holds.*

Proof. If $\mathcal{P}^{<\omega}$ is contravariantly finite, then $\mathcal{B}^{<\omega}$ is covariantly finite (by a version of 1.9.3 in mod-R). As in the proof of 2.9, an iteration of the $\mathcal{B}^{<\omega}$-envelopes of R etc. yields an Add(T)-coresolution of R, $0 \to R \to T_0 \to \cdots \to T_n \to 0$. Then $T' = \bigoplus_{i \leqslant n} T_i$ is a finitely generated tilting module equivalent to T. The converse implication follows from [12].

If T is finitely generated then T is $\sum$-pure injective, and [8] gives that Add(T) is closed under cokernels of monomorphisms. By 5.4.3, $\mathcal{A} = \mathcal{P}$, so (I) holds true. $\square$

5.3 and 5.6 now give

Corollary 5.7. [12], [60] *Let R be an artin algebra such that $\mathcal{P}^{<\omega}$ is contravariantly finite. Then (I) and (II) hold for R.*

Note that all right serial artin algebras satisfy the assumption of 5.7, see [58]. However, there are finite dimensional algebras R with fin.dim$(R) =$ Fin.dim$(R) = 1$ such that $\mathcal{P}^{<\omega}$ is not contravariantly finite, for example the IST-algebra [61]; for those algebras, T is an infinitely generated 1-tilting module. For an explicit computation of T for the IST-algebra, we refer to [82].

Finally, we turn to Iwanaga-Gorenstein rings (see 2.6). Let $n < \omega$ and R be n-Gorenstein. Then $\mathcal{P} = \mathcal{I} = \mathcal{P}_n = \mathcal{I}_n$. In particular, there exist cotorsion pairs $\mathfrak{D} = (\mathcal{P}, \mathcal{GI})$ and $\mathfrak{E} = (\mathcal{GP}, \mathcal{I})$. The modules in $\mathcal{GI}$ are called *Gorenstein injective*, the ones in $\mathcal{GP}$ *Gorenstein projective*. The kernel of $\mathfrak{D}$ equals $\mathcal{I}_0$, the kernel of $\mathfrak{E}$ is $\mathcal{P}_0$, cf. [46]. Clearly, Fin.dim$(R) = n$, so (II) holds.

By [5], also (I) holds:

Theorem 5.8. *Let R be an Iwanaga-Gorenstein ring. Then (I) holds true. Moreover, the tilting module T from 5.1 can be taken of the form $T = \bigoplus_{i \leqslant n} I_i$ where $0 \to R \to I_0 \to \cdots \to I_n \to 0$ is the minimal injective coresolution of R.*

Proof. By 1.5.1, the cotorsion pair $\mathfrak{D} = (\mathcal{P}, \mathcal{GI})$ is of countable type. By [19], for each $C \in \mathcal{P}^{<\omega}$ there is a $\mathcal{P}^{<\omega}$-filtered module D such that $D = C \oplus P$ where $P \in \mathcal{P}_0$. So $C \in \mathcal{A}$, that is, $\mathcal{A} = \mathcal{P}$, and (I) holds. Since the minimal $\mathcal{GI}$-coresolution of R is actually its minimal injective coresolution, and $\mathfrak{E} = \mathfrak{D}$, T can be taken as claimed by 5.2. □

If R in 5.8 is an artin algebra, then T is finitely generated. So by 5.6, Iwanaga-Gorenstein artin algebras give yet another example of algebras with $\mathcal{P}^{<\omega}$ contravariantly finite, [12].

References

[1] S.T. ALDRICH, E. ENOCHS, O. JENDA, AND L. OYONARTE, *Envelopes and covers by modules of finite injective and projective dimensions*, J. Algebra **242** (2001), 447-459.

[2] L. ANGELERI HÜGEL AND F. COELHO, *Infinitely generated tilting modules of finite projective dimension*, Forum Math. **13** (2001), 239-250.

[3] L. ANGELERI HÜGEL AND F. COELHO, *Infinitely generated complements to partial tilting modules*, Math. Proc. Camb. Phil. Soc. **132** (2002), 89-96.

[4] L. ANGELERI HÜGEL, D. HAPPEL AND H. KRAUSE, *Basic results of classical tilting theory*, this volume.

[5] L. ANGELERI HÜGEL, D. HERBERA, AND J. TRLIFAJ, *Tilting modules and Gorenstein rings*, to appear in Forum Math. **18** (2006), pp. 217-235.

[6] L. ANGELERI HÜGEL, D. HERBERA, AND J. TRLIFAJ, *Divisible modules and localization*, J. Algebra **294** (2005), 519-551.

[7] L. ANGELERI HÜGEL, A. TONOLO, AND J. TRLIFAJ, *Tilting preenvelopes and cotilting precovers*, Algebras and Repres. Theory **4** (2001), 155-170.

[8] L. ANGELERI HÜGEL AND J. TRLIFAJ, *Tilting theory and the finitistic dimension conjectures*, Trans. Amer. Math. Soc. **354** (2002), 4345-4358.

[9] L. ANGELERI HÜGEL AND J. TRLIFAJ, *Direct limits of modules of finite projective dimension*, in Rings, Modules, Algebras, and Abelian Groups, LNPAM **236**, M. Dekker (2004), 27-44.

[10] M. AUSLANDER, *Functors and morphisms determined by objects*, in Representation Theory of Algebras, LNPAM **37**, M. Dekker (1978), 1-244.

[11] M. AUSLANDER AND R. BUCHWEITZ, *Homological Theory of Cohen-Macaulay Approximations*, Mem.Soc.Math. de France **38** (1989), 5–37.

[12] M. AUSLANDER AND I. REITEN, *Applications of contravariantly finite subcategories*, Adv.Math. **86** (1991), 111–152.

[13] M. AUSLANDER AND S. O. SMALØ, *Preprojective modules over artin algebras*, J. Algebra **66** (1980), 61-122.

[14] H. BASS, *Finitistic dimension and a homological generalization of semiprimary rings*, Trans. Amer. Math. Soc. **95** (1960), 466-488.

[15] S. BAZZONI, *Cotilting modules are pure-injective*, Proc. Amer. Math. Soc. **131** (2003), 3665-3672.

[16] S. BAZZONI, *A characterization of n-cotilting and n-tilting modules*, J. Algebra **273** (2004), 359-372.

[17] S. BAZZONI, *n-cotilting modules and pure-injectivity*, Bull. London Math. Soc. **36** (2004), 599-612.

[18] S. BAZZONI, *Cotilting and tilting modules over Prüfer domains*, to appear in Forum Math.

[19] S. BAZZONI, P. C. EKLOF, AND J. TRLIFAJ, *Tilting cotorsion pairs*, to appear in Bull. London Math. Soc. **37** (2005), 683-696.

[20] S. BAZZONI, R. GÖBEL, AND L. STRÜNGMANN, *Pure injectivity of n-cotilting modules: the Prüfer and the countable case*, Archiv d. Math. **84** (2005), 216-224.

[21] S. BAZZONI AND D. HERBERA, *One dimensional tilting modules are of finite type*, preprint.

[22] S. BAZZONI AND L. SALCE, *Strongly flat covers*, J. London Math. Soc. **66** (2002), 276-294.

[23] S. BAZZONI AND L. SALCE, *On strongly flat modules over integral domains*, Rocky Mountain J. Math. **34** (2004), 417-439.

[24] S. BAZZONI AND L. SALCE, *Almost perfect domains*, Colloq. Math. **95** (2003), 285-301.

[25] S.BAZZONI AND J. ŠŤOVÍČEK, *All tilting modules are of finite type*, preprint.

[26] A. BELIGIANNIS AND I. REITEN, **Homological and homotopical aspects of torsion theories**, to appear in Mem. Amer. Math. Soc.

[27] L. BICAN, R. EL BASHIR, AND E. ENOCHS, *All modules have flat covers*, Bull. London Math. Soc. **33** (2001), 385–390.

[28] A.B. BUAN AND H. KRAUSE, *Cotilting modules over tame hereditary algebras*, Pacific J. Math. **211** (2003), 41-60.

[29] A.B. BUAN AND H. KRAUSE, *Tilting and cotilting for quivers of type $\tilde{A}_n$*, J. Pure Appl. Algebra **190** (2004), 1-21.

[30] A.B. BUAN, H. KRAUSE, AND O. SOLBERG, *On the lattice of cotilting modules*, Algebra Montpellier Announcements.

[31] R.R. COLBY AND K.R. FULLER, **Equivalence and Duality for Module Categories**, Cambridge Tracts in Math. 161, Cambridge Univ. Press, Cambridge 2004.

[32] R.COLPI, G.D'ESTE, AND A.TONOLO, *Quasi-tilting modules and counter equivalences*, J. Algebra **191** (1997), 461-494.

[33] R. COLPI AND K.R. FULLER, *Cotilting dualities*, this volume.

[34] R.COLPI, A.TONOLO, AND J.TRLIFAJ, *Partial cotilting modules and the lattices induced by them*, Comm. Algebra **25** (1997), 3225–3237.

[35] R.COLPI AND J.TRLIFAJ, *Tilting modules and tilting torsion theories*, J. Algebra **178** (1995), 614–634.

[36] W. CRAWLEY-BOEVEY, *Locally finitely presented additive categories*, Comm. Algebra **22** (1994), 1644–1674.

[37] W. CRAWLEY-BOEVEY, *Infinite dimensional modules in the representation theory of finite dimensional algebras*, CMS Conf. Proc. **23** (1998), 29–54.

[38] B.ECKMANN AND A.SCHOPF, *Über injective Moduln*, Arch. Math. (Basel) **4** (1953), 75–78.

[39] P. C. EKLOF, L. FUCHS AND S. SHELAH, *Baer modules over domains*, Trans. Amer. Math. Soc. **322** (1990), 547–560.

[40] P. C. EKLOF AND A. H. MEKLER, **Almost Free Modules**, Revised Ed., North-Holland Math. Library, Elsevier, Amsterdam 2002.

[41] P. C. EKLOF AND S. SHELAH, *On the existence of precovers*, Illinois J. Math. **47** (2003), 173-188.

[42] P. C. EKLOF, S. SHELAH, AND J.TRLIFAJ, *On the cogeneration of cotorsion pairs*, J. Algebra **277** (2004), 572–578.

[43] P. C. EKLOF AND J. TRLIFAJ, *How to make Ext vanish*, Bull. London Math. Soc. **33** (2001), 31–41.

[44] P. C. EKLOF AND J. TRLIFAJ, *Covers induced by Ext*, J. Algebra **231** (2000), 640–651.

[45] E. ENOCHS, *Torsion free covering modules*, Proc. Amer. Math. Soc. **14** (1963), 884-889.

[46] E. ENOCHS AND O. JENDA, **Relative Homological Algebra**, de Gruyter Expos. in Math. **30**, de Gruyter, Berlin 2000.

[47] A. FACCHINI, *A tilting module over commutative integral domains*, Comm. Algebra **15** (1987), 2235-2250.

[48] C. FAITH, **Algebra: Rings, Modules and Categories I**, Springer, New York 1973.

[49] M. FONTANA, J.A. HUCKABA, AND I.J. PAPICK, **Prüfer domains**, M.Dekker, New York 1997.

[50] L. FUCHS AND L.SALCE, **Modules over Non-Noetherian Domains**, AMS, Providence 2001.

[51] L. FUCHS AND L.SALCE, *S-divisible modules over domains*, Forum Math. **4**(1992), 383-394.

[52] R. GÖBEL AND S. SHELAH, *Cotorsion theories and splitters*, Trans. Amer. Math. Soc. **352** (2000), 5357-5379.

[53] R. GÖBEL, S. SHELAH, AND S.L. WALLUTIS, *On the lattice of cotorsion theories*, J. Algebra **238** (2001), 292-313.

[54] R. GÖBEL AND J.TRLIFAJ, *Cotilting and a hierarchy of almost cotorsion groups*, J. Algebra **224** (2000), 110-122.

[55] R. GÖBEL AND J. TRLIFAJ, **Approximations and Endomorphism Algebras of Modules**, preprint of a monograph.

[56] L. L. GREEN, E. E. KIRKMAN, AND J. J. KUZMANOVICH, *Finitistic dimension of finite dimensional monomial algebras*, J. Algebra **136** (1991), 37-51.

[57] B. HUISGEN ZIMMERMANN, *Homological domino effects and the first finitistic dimension conjecture*, Invent. Math., **108** (1992), 369-383.

[58] B. HUISGEN ZIMMERMANN, *Syzigies and homological dimensions of left serial rings*, Methods in Module Theory, LNPAM **140**, M.Dekker, New York 1993, 161-174.

[59] B. HUISGEN ZIMMERMANN, *Purity, algebraic compactness, direct sum decompositions, and representation type*, Trends in Mathematics, Birkhäuser, Basel 2000, 331-368.

[60] B. HUISGEN ZIMMERMANN AND S. SMALØ, *A homological bridge between finite and infinite dimensional representations*, Algebras and Repres. Theory **4** (2001), 155-170.

[61] K. IGUSA, S. O. SMALØ AND G. TODOROV, *Finite projectivity and contravariant finiteness*, Proc. Amer. Math. Soc. **109** (1990), 937-941.

[62] K. IGUSA AND G. TODOROV, *On the finitistic global dimension conjecture for artin algebras*, in Representations of algebras and related topics, Fields Inst. Comm. **45** (2005).

[63] C. JENSEN AND H. LENZING, **Model Theoretic Algebra**, ALA 2, Gordon & Breach, Amsterdam 1989.

[64] O. KERNER AND J. TRLIFAJ, *Tilting classes over wild hereditary algebras*, J. Algebra **290** (2005), 538-556.

[65] L. KLINGLER AND L. LEVY, **Representation Type of Commutative Noetherian Rings III: Global Wildness and Tameness**, Memoir Amer. Math. Soc., Vol. **832** (2005).

[66] H. KRAUSE AND O. SOLBERG, *Applications of cotorsion pairs*, J. London Math. Soc. **68** (2003), 631-650

[67] S.B. LEE, *On divisible modules over domains*, Arch. Math. **53** (1989), 259-262.

[68] L. LEVY AND J. TRLIFAJ, *Γ-separated covers*, to appear as chapter 1 in *Abelian Groups, Rings, and Modules*, CRC Press, New York 2005.

[69] F. LUKAS, *Infinite-dimensional modules over wild hereditary algebras*, J. London Math. Soc. **44** (1991), 401-419.

[70] M. MATSUMURA, **Commutative Ring Theory**, 5th ed., CSAM **8**, Cambridge Univ. Press, Cambridge 1994.

[71] M. PREST, **Model Theory and Modules**, LMSLN **130**, Cambridge Univ. Press, Cambridge 1988.

[72] I. REITEN, *Tilting theory and homologically finite subcategories with*

applications to quasi-hereditary algebras, in this volume.

[73] I. REITEN AND C.M. RINGEL, *Infinite dimensional representations of canonical algebras*, to appear in Canad. J. Math.

[74] C.M. RINGEL, *Infinite dimensional representations of finite dimensional hereditary algebras*, Symposia Math. **23** (1979), Academic Press, 321–412.

[75] C.M. RINGEL, *Infinite length modules. Some examples as introduction*, Trends in Mathematics, Birkhäuser, Basel 2000, 1-74

[76] L. SALCE, *Cotorsion theories for abelian groups*, Symposia Math. **23** (1979), Academic Press, 11–32.

[77] L. SALCE, *Tilting modules over valuation domains*, Forum Math. **16** (2004), 539–552.

[78] L. SALCE, *F-divisible modules and 1-tilting modules over Prüfer domains*, J. Pure Appl. Algebra **199** (2005), 245-259.

[79] J. ŠAROCH AND J. TRLIFAJ, *Completeness of cotorsion pairs*, to appear in Forum Math.

[80] S. SMALØ, *Homological differences between finite and infinite dimensional representations of algebras*, Trends in Mathematics, Birkhäuser, Basel 2000, 425-439.

[81] O. SOLBERG, *Infinite dimensional tilting modules over finite dimensional algebras*, this volume.

[82] J. ŠŤOVÍČEK, *Tilting modules over an algebra by Igusa, Smalø and Todorov*, preprint.

[83] J. ŠŤOVÍČEK, *All n-cotilting modules are pure-injective*, to appear in Proc. Amer. Math. Soc.

[84] J. ŠŤOVÍČEK AND J. TRLIFAJ, *All tilting modules are of countable type*, preprint.

[85] J. TRLIFAJ, *Every *-module is finitely generated*, J. Algebra **169** (1994), 392-398.

[86] J. TRLIFAJ, *Whitehead test modules*, Trans. Amer. Math. Soc. **348** (1996), 1521-1554.

[87] J. TRLIFAJ, *Cotorsion theories induced by tilting and cotilting modules*, Abelian Groups, Rings and Modules, Contemporary Math. **273** (2001), 285-300.

[88] J. TRLIFAJ, *Approximations and the little finitistic dimension of artinian rings*, J. Algebra **246** (2001), 343-355.

[89] J. TRLIFAJ AND S.L. WALLUTIS, *Tilting modules over small Dedekind domains*, J. Pure Appl. Algebra **172** (2002), 109-117. Corrigendum: **183** (2003), 329-331.

[90] L. UNGER, *Combinatorial aspects of the set of tilting modules*, this volume.

[91] T. WAKAMATSU, *Stable equivalence of self-injective algebras and a generalization of tilting modules*, J. Algebra **134** (1990), 298-325.

[92] R. WARFIELD, *Purity and algebraic compactness for modules*, Pacific J. Math. **28** (1969), 699–719.

[93] J. XU, **Flat Covers of Modules**, Lecture Notes in Math. 1634, Springer-Verlag, New York 1996.

Jan Trlifaj

Charles University, Faculty of Mathematics and Physics,
Department of Algebra, Sokolovská 83, 186 75 Prague 8
Czech Republic

12

Infinite dimensional tilting modules over finite dimensional algebras

Øyvind Solberg

Introduction

The theory for tilting and cotilting modules has its roots in the representation theory of finite dimensional algebras (artin algebras) generalizing Morita equivalence and duality. First through reflection functors studied in [16] and a module theoretic interpretation of these in [8], tilting modules of projective dimension at most one got an axiomatic description in [18, 41]. Among others, [6, 17, 59] developed this theory further. These concepts were generalized in [40, 52] to tilting modules of arbitrary finite projective dimension. In the seminal paper [9] tilting and cotilting modules were characterized by special subcategories of the category of finitely presented modules. This paper started among other things the close connections between tilting and cotilting theory and homological conjectures studied further in [23, 42, 44].

Generalizations of tilting modules of projective dimension at most one to arbitrary associative rings have been considered in [4, 27, 32, 51]. As tilting and cotilting modules in this general setting is not necessarily linked by applying a duality, a parallel development of cotilting modules were pursued among others in [25, 24, 26, 28, 29, 30, 31]. Definitions of tilting and cotilting modules of arbitrary projective and injective dimension were introduced in [3] and [61], where the definition introduced in [3] being the most widely used now.

As the theory of tilting and cotilting modules has its origin in the representation theory of finite dimensional algebras, the current research on tilting and cotilting modules over more general rings seems also to return to finiteness conditions by considering modules of finite, cofinite and countable type. In this paper we consider finite dimensional algebras or artin algebras Λ, and review some of the properties and the structure of all tilting and cotilting modules. Of particular interest is the information these can give on the representation theory of finitely presented modules over Λ.

Let Λ be an artin algebra, and let $\operatorname{Mod}\Lambda$ denote the category of all left Λ-modules. Our aim is to show that generalizing the characterization of finitely presented tilting and cotilting modules over Λ to arbitrary tilting and cotilting modules in $\operatorname{Mod}\Lambda$ gives many of the results obtained for these modules. After recalling definitions and some basic results in section 1, the next section gives analogues in $\operatorname{Mod}\Lambda$ of all the known characterizations of finitely presented tilting and cotilting modules given in [9]. Section 3 is devoted to discussing the finitistic dimension conjectures. Here the little finitistic dimension of Λ, when finite, is shown to be obtained as the projective dimension of a tilting module in $\operatorname{Mod}\Lambda$, and we investigate when the big finitistic dimension is given by the projective dimension (the injective dimension) of a tilting (cotilting) module. We proceed in the next section to show that all finitely presented partial tilting and cotilting modules, when viewed as partial tilting and cotilting modules in $\operatorname{Mod}\Lambda$ always have a complement, contrary to within finitely presented modules. Moreover, the classical completion result for all finitely presented partial tilting and cotilting modules of projective and injective dimension at most one, respectively, is generalized. We end by giving the classification of all cotilting modules over a tame hereditary algebra.

1 Definitions and preliminaries

In this section we recall the definitions and the preliminary results that we shall use throughout the paper.

Let Λ be a ring. Denote by $\operatorname{mod}\Lambda$ the full subcategory of $\operatorname{Mod}\Lambda$ consisting of all finitely presented Λ-modules. For a module M in $\operatorname{Mod}\Lambda$ we define the full subcategory (i) $\operatorname{add}M$ to be the direct summands of all finite coproducts of copies of M, (ii) $\operatorname{Add}M$ to be the direct summands of arbitrary coproducts of copies of M, and (iii) $\operatorname{Prod}M$ to be the direct summands of arbitrary products of copies of M. For M in $\operatorname{mod}\Lambda$ let $\operatorname{Sub}(M)$ be the full subcategory of $\operatorname{mod}\Lambda$ consisting of all submodules of a finite coproducts of copies of M. For a subcategory $\mathcal{C}$ of $\operatorname{Mod}\Lambda$

the full subcategory $\{Y \in \operatorname{Mod}\Lambda \mid \operatorname{Ext}_\Lambda^i(\mathcal{C}, Y) = (0) \text{ for all } i > 0\}$ is denoted by $\mathcal{C}^\perp$. Dually $^\perp\mathcal{C}$ denotes the full subcategory of $\operatorname{Mod}\Lambda$ given by $\{X \in \operatorname{Mod}\Lambda \mid \operatorname{Ext}_\Lambda^i(X, \mathcal{C}) = (0) \text{ for all } i > 0\}$. The full subcategory of $\operatorname{Mod}\Lambda$ consisting of all modules X with a finite resolution

$$0 \to C_n \to C_{n-1} \to \cdots \to C_1 \to C_0 \to X \to 0$$

for some $n \geq 0$ and with C_i in $\mathcal{C}$ is denoted by $\hat{\mathcal{C}}$. If all modules in $\operatorname{Mod}\Lambda$ is in $\hat{\mathcal{C}}$ with the length n bounded by some number $N < \infty$, we say that the resolution dimension $\operatorname{resdim}_\mathcal{C}(\operatorname{Mod}\Lambda)$ of $\operatorname{Mod}\Lambda$ is finite, and otherwise it is infinite. Dually the full subcategory of $\operatorname{Mod}\Lambda$ consisting of all modules Y with a finite coresolution

$$0 \to Y \to C^0 \to C^1 \to \cdots \to C^{n-1} \to C^n \to 0$$

for some $n \geq 0$ and with C^i in $\mathcal{C}$ for all i is denoted by $\check{\mathcal{C}}$. The injective dimension and the projective dimension of a module X are denoted by $\operatorname{id}_\Lambda X$ and $\operatorname{pd}_\Lambda X$ respectively.

Now we give the definitions of a tilting and a cotilting module over an arbitrary ring Λ from [3]. Recall that a Λ-module T is a *tilting module* if

(T1) $\operatorname{pd}_\Lambda T < \infty$;

(T2) $\operatorname{Ext}_\Lambda^i(T, \amalg T) = (0)$ for all $i > 0$ and all coproducts $\amalg T$ of copies of T;

(T3) there exists a long exact sequence $0 \to \Lambda \to T^0 \to T^1 \to \cdots \to T^{n-1} \to T^n \to 0$ with T^i in $\operatorname{Add} T$ for all $i = 0, 1, \ldots, n$.

A module T is called a *partial tilting module* if T satisfies the conditions (T1) and (T2). If T is a partial tilting module, then a module T' such that $T \amalg T'$ is a tilting module is called a *complement* of T. A Λ-module T is a *cotilting module* if

(C1) $\operatorname{id}_\Lambda T < \infty$;

(C2) $\operatorname{Ext}_\Lambda^i(\prod T, T) = (0)$ for all $i > 0$ and all products $\prod T$ of copies of T;

(C3) there exists an injective generator I and a long exact sequence $0 \to T_n \to \cdots \to T_1 \to T_0 \to I \to 0$ with T_i in $\operatorname{Prod} T$ for all $i = 0, 1, \ldots, n$.

A module T is called a *partial cotilting module* if T satisfies the conditions (C1) and (C2). Dually we also define complements of partial cotilting modules. Also recall that a module T in $\operatorname{mod} \Lambda$ is a tilting or a cotilting module if $\operatorname{Add} T$ and $\operatorname{Prod} T$ are replaced by $\operatorname{add} T$ and arbitrary coproducts and products are replaced by finite coproducts in the above definitions. In particular, a tilting or cotilting module in $\operatorname{mod} \Lambda$ is still a tilting or cotilting module in $\operatorname{Mod} \Lambda$, respectively.

We denote by $\mathcal{I}^\alpha(\operatorname{Mod} \Lambda)$ and $\mathcal{P}^\alpha(\operatorname{Mod} \Lambda)$ for α in $\mathbb{N} \cup \{\infty\}$ the full subcategories of $\operatorname{Mod} \Lambda$ consisting of all modules X with $\operatorname{id}_\Lambda X < \alpha$ and all modules Y with $\operatorname{pd}_\Lambda Y < \alpha$, respectively. The categories $\mathcal{I}^\alpha(\operatorname{Mod} \Lambda)$ and $\mathcal{P}^\alpha(\operatorname{Mod} \Lambda)$ are examples of coresolving and resolving categories, that is, a subcategory $\mathcal{X}$ of $\operatorname{Mod} \Lambda$ is called *resolving* if $\mathcal{X}$ contains all projective modules and is closed under kernels of epimorphisms, direct summands and extensions. Coresolving is defined dually.

The *big finitistic dimension* $\operatorname{Findim} \Lambda$ of Λ is given by

$$\sup\{\operatorname{pd}_\Lambda X \mid X \in \operatorname{Mod} \Lambda \text{ with } \operatorname{pd}_\Lambda X < \infty\}.$$

The *little finitistic dimension* $\operatorname{findim} \Lambda$ of Λ is defined as

$$\sup\{\operatorname{pd}_\Lambda X \mid X \in \operatorname{mod} \Lambda \text{ with } \operatorname{pd}_\Lambda X < \infty\}.$$

Let $\mathcal{X}$ be a class of Λ-modules. For a given Λ-module C, a map $\varphi \colon X \to C$ is a *right $\mathcal{X}$-approximation* of C if X is in $\mathcal{X}$ and $\operatorname{Hom}_\Lambda(\mathcal{X}, X) \xrightarrow{\operatorname{Hom}_\Lambda(\mathcal{X}, \varphi)} \operatorname{Hom}_\Lambda(\mathcal{X}, C)$ is surjective. The approximation φ is *minimal* if every endomorphism $f \colon X \to X$ with $\varphi f = \varphi$ is an isomorphism. If every Λ-module has a right $\mathcal{X}$-approximation, then $\mathcal{X}$ is called *contravariantly finite* in $\operatorname{Mod} \Lambda$.

One crucial result with respect to approximations is the Wakamatsu's Lemma [65]. It says the following.

Lemma 1.1. [9, Lemma 1.3] *Let $\mathcal{X}$ be a class of Λ-modules closed under extensions, and let $0 \to Y \to X \xrightarrow{\varphi} C \to 0$ be an exact sequence of Λ-modules.*

(a) *If φ is a minimal right $\mathcal{X}$-approximation, then $\operatorname{Ext}^1_\Lambda(\mathcal{X}, Y) = (0)$.*

(b) *If $\operatorname{Ext}^1_\Lambda(\mathcal{X}, Y) = (0)$ and X is in $\mathcal{X}$, then φ is a right $\mathcal{X}$-approximation.*

If an right $\mathcal{X}$-approximation $\varphi\colon X \rightarrow C$ is surjective and $\text{Ext}^1_\Lambda(\mathcal{X}, \text{Ker}\,\varphi) = (0)$, the approximation is called *special*.

We leave it to the reader to define the dual concepts of all the above notions for right approximations in ModΛ and the corresponding notions in mod Λ.

2 The subcategory correspondence

Let Λ be an artin algebra. In Theorem 5.5 of the paper [9] tilting and cotilting modules in mod Λ were characterized by certain subcategories of mod Λ. This characterization has lead to many fruitful applications as reviewed in [53]. So it is natural to ask if there is a similar correspondence for arbitrary tilting and cotilting modules over Λ. This section is devoted to discussing such correspondences. For a closely related investigation of the associated cotorsion pairs see [62].

First we recall the characterization of equivalence classes of tilting and cotilting modules in mod Λ given in [9]. Here we say that two tilting or cotilting modules T and T' in mod Λ are *equivalent* if add T = add T'. Note that in the following result all modules and subcategories are in mod Λ.

Theorem 2.1. [9, Theorem 5.5] *Let T be a module.*

(a) $T \mapsto {}^\perp T$ *gives a one-to-one correspondence between equivalence classes of cotilting modules and contravariantly finite resolving subcategories $\mathcal{X}$ of* mod Λ *with $\widehat{\mathcal{X}} =$ mod Λ. The inverse is given by $\mathcal{X} \mapsto \mathcal{X} \cap \mathcal{X}^\perp$.*

(b) $T \mapsto \widetilde{\text{add}\,T}$ *gives a one-to-one correspondence between equivalence classes of cotilting modules and covariantly finite coresolving subcategories $\mathcal{Y}$ of $\mathcal{I}^\infty(\text{mod}\,\Lambda)$. The inverse is given by $\mathcal{Y} \mapsto {}^\perp\mathcal{Y} \cap \mathcal{Y}$.*

(c) $T \mapsto T^\perp$ *gives a one-to-one correspondence between equivalence classes of tilting modules and covariantly finite coresolving subcategories $\mathcal{Y}$ of with $\check{\mathcal{Y}} =$ mod Λ. The inverse is given by $\mathcal{Y} \mapsto {}^\perp\mathcal{Y} \cap \mathcal{Y}$.*

(d) $T \mapsto \widetilde{\text{add}\,T}$ *gives a one-to-one correspondence between equivalence classes of tilting modules and contravariantly finite re-*

solving subcategories $\mathcal{X}$ of $\mathcal{P}^\infty(\mathrm{mod}\,\Lambda)$. The inverse is given by $\mathcal{X} \mapsto \mathcal{X} \cap \mathcal{X}^\perp$.

The usual duality $D\colon \mathrm{mod}\,\Lambda \to \mathrm{mod}\,\Lambda^{\mathrm{op}}$ maps tilting modules to cotilting modules and vice versa. So a result for one case automatically yields a result for the other. Given this remark it is superfluous to give the correspondences both for tilting and cotilting modules. However since a result for tilting modules in $\mathrm{Mod}\,\Lambda$ can not be translated to a result about cotilting modules by applying the duality D, we have chosen to give all correspondences in order to clearly see the relationship with the analogous correspondences in $\mathrm{Mod}\,\Lambda$.

We want to present analogues of the above characterizations for tilting and cotilting modules in $\mathrm{Mod}\,\Lambda$. As infinitely generated tilting and cotilting modules T are defined in terms $\mathrm{Add}\,T$ and $\mathrm{Prod}\,T$ the following are natural definitions. Two tilting (or cotilting) modules T and T' in $\mathrm{Mod}\,\Lambda$ are *equivalent* if $\mathrm{Add}\,T = \mathrm{Add}\,T'$ (or $\mathrm{Prod}\,T = \mathrm{Prod}\,T'$).

We first discuss the situation for cotilting modules over Λ. A first approximation to such a characterization was found in [3], and it reads as follows (this result is true for any ring).

Theorem 2.2. [3, Theorem 4.2] *Let $\mathcal{X}$ be class of modules in $\mathrm{Mod}\,\Lambda$ closed under kernels of epimorphisms and such that $\mathcal{X} \cap \mathcal{X}^\perp$ is closed under products. The following are equivalent.*

  (a) *There exists a cotilting module T with $\mathrm{id}_\Lambda T \leqslant n$ such that $\mathcal{X} = {}^\perp T$;*

  (b) *Every left Λ-module has a special $\mathcal{X}$-approximation and all modules Y in $\mathcal{X}^\perp$ have $\mathrm{id}_\Lambda Y \leqslant n$.*

From our view point and for our purposes this is not a true generalization of the characterization given of tilting and cotilting modules in [9], since it involve both a category and its Ext-orthogonal category. A characterization of the equivalence classes of cotilting modules in terms of subcategories of $\mathrm{Mod}\,\Lambda$ is given as follows.

Theorem 2.3. [49, Theorem 5.6] *Let T be a module in $\mathrm{Mod}\,\Lambda$. The map $T \mapsto {}^\perp T$ gives a one-to-one correspondence between equivalence classes of cotilting modules over Λ and resolving subcategories $\mathcal{X}$ of $\mathrm{Mod}\,\Lambda$ closed under products with $\widehat{\mathcal{X}} = \mathrm{Mod}\,\Lambda$, such that every Λ-module has a special right $\mathcal{X}$-approximation. The inverse is given by $\mathcal{X} \mapsto \mathcal{X} \cap \mathcal{X}^\perp$.*

Remark 2.4. This result is in fact true for any ring R by replacing $\widehat{\mathcal{X}} =$ Mod R by resdim$_{\mathcal{X}}$(Mod R) $< \infty$ as formulated in [49]. So we see that the only real difference with Theorem 2.1 (a) is that the corresponding category should be closed under products, since in mod Λ any module having a right $\mathcal{X}$-approximation with $\mathcal{X}$ in mod Λ has a minimal right approximation. In addition, when the category $\mathcal{X}$ is extension closed, then by Wakamatsu's Lemma a minimal approximation is special.

It is natural to ask if all the known characterizations of tilting and cotilting modules in mod Λ have counterparts in Mod Λ. We first consider an analogue of Theorem 2.1 (b).

Proposition 2.5. *Let T be a module in* Mod Λ. *The map $T \mapsto \widehat{\mathrm{Prod}\,T}$ gives a one-to-one correspondence between equivalence classes of cotilting modules over Λ and coresolving subcategories $\mathcal{Y}$ of* Mod Λ

(i) *closed under products,*

(ii) *contained in $\mathcal{I}^{\infty}$(Mod Λ),*

(iii) *every Λ-module has a special left $\mathcal{Y}$-approximation,*

(iv) *if $0 \to \Lambda/\mathfrak{r} \to Y^{\Lambda/\mathfrak{r}} \to X^{\Lambda/\mathfrak{r}} \to 0$ is a special left $\mathcal{Y}$-approximation, then the category $\{X \in \mathrm{Mod}\,\Lambda \mid \mathrm{Ext}^1_{\Lambda}(X, Y^{\Lambda/\mathfrak{r}}) = (0)\}$ is closed under products.*

Proof. Assume first that T is a cotilting module in Mod Λ. We claim that $\widehat{\mathrm{Prod}\,T} = ({}^{\perp}T)^{\perp}$. Using that Prod T is contained in $({}^{\perp}T)^{\perp}$, we conclude that $({}^{\perp}T)^{\perp}$ contains $\widehat{\mathrm{Prod}\,T}$. Taking a sequence $\cdots \to X_2 \xrightarrow{f_2} X_1 \xrightarrow{f_1} X_0 \xrightarrow{f_0} Z \to 0$ of special right ${}^{\perp}T$-approximations of a module Z in $({}^{\perp}T)^{\perp}$, the kernels Ker f_i are in $({}^{\perp}T)^{\perp}$ and X_i are in Prod T. Since $({}^{\perp}T)^{\perp}$ is contained in $\mathcal{I}^{n+1}$(Mod Λ) for some $n < \infty$, the module Ker f_{n-1} is in Prod T. Hence Z is in $\widehat{\mathrm{Prod}\,T}$ and $({}^{\perp}T)^{\perp} = \widehat{\mathrm{Prod}\,T}$, which clearly is closed under products. Then by [9, Proposition 1.8] the subcategory $\widehat{\mathrm{Prod}\,T}$ is coresolving where all modules have a special left $\widehat{\mathrm{Prod}\,T}$-approximation.

Dual to Theorem 3.1 in [49] any module in $\widehat{\mathrm{Prod}\,T}$ is a direct factor of a module having a finite filtration in products of copies of $Y^{\Lambda/\mathfrak{r}}$. Then it is easy to see that $\{X \in \mathrm{Mod}\,\Lambda \mid \mathrm{Ext}^1_{\Lambda}(X, Y^{\Lambda/\mathfrak{r}}) = (0)\} = {}^{\perp}T$ and therefore closed under products.

Conversely, let $\mathcal{Y}$ be a subcategory of Mod Λ satisfying the properties (i)–(iv). Then $^\perp \mathcal{Y}$ is resolving and all modules have a special right $^\perp \mathcal{Y}$-approximation. Dual to as in the proof of Theorem 3.1 in [49] a module is in $\mathcal{Y}$ if and only if it is a direct factor of a module having a finite filtration in products of copies of $Y^{\Lambda/\mathfrak{r}}$. Hence $\mathcal{Y}$ is contained in $\mathcal{I}^{n+1}(\mathrm{Mod}\,\Lambda)$ for some $n < \infty$. Then we infer that $\{X \in \mathrm{Mod}\,\Lambda \mid \mathrm{Ext}^1_\Lambda(X, Y^{\Lambda/\mathfrak{r}}) = (0)\} = {}^\perp \mathcal{Y}$ and that $\widehat{{}^\perp \mathcal{Y}} = \mathrm{Mod}\,\Lambda$. Then $^\perp \mathcal{Y} = {}^\perp T$ for some cotilting module T with $\mathrm{Prod}\,T = {}^\perp \mathcal{Y} \cap \mathcal{Y}$. Furthermore by the above we have that $\mathcal{Y} = \widehat{\mathrm{Prod}\,T}$. $\square$

Now we give the analogues of (c) and (d) in Theorem 2.1. The proofs are dual to those of Theorem 2.3 and Proposition 2.5.

Proposition 2.6. *Let T be a module in $\mathrm{Mod}\,\Lambda$.*

(a) *The map $T \mapsto T^\perp$ gives a one-to-one correspondence between equivalence classes of tilting modules over Λ and coresolving subcategories $\mathcal{Y}$ of $\mathrm{Mod}\,\Lambda$ closed under coproducts with $\breve{\mathcal{Y}} = \mathrm{Mod}\,\Lambda$, such that every Λ-module has a special left $\mathcal{Y}$-approximation.*

(b) *The map $T \mapsto \widetilde{\mathrm{Add}\,T}$ gives a one-to-one correspondence between equivalence classes of tilting modules over Λ and resolving subcategories $\mathcal{X}$ of $\mathrm{Mod}\,\Lambda$*

    (i) *closed under coproducts,*

    (ii) *contained in $\mathcal{P}^\infty(\mathrm{Mod}\,\Lambda)$,*

    (iii) *every Λ-module has a special right $\mathcal{X}$-approximation,*

    (iv) *if $0 \to Y_{\Lambda/\mathfrak{r}} \to X_{\Lambda/\mathfrak{r}} \to \Lambda/\mathfrak{r} \to 0$ is a special right $\mathcal{X}$-approximation, then the category $\{Y \in \mathrm{Mod}\,\Lambda \mid \mathrm{Ext}^1_\Lambda(X_{\Lambda/\mathfrak{r}}, Y) = (0)\}$ is closed under coproducts.*

In [9] the subcategory $^\perp T$ of $\mathrm{mod}\,\Lambda$ is shown not only to be contravariantly finite for a cotilting module T in $\mathrm{mod}\,\Lambda$, but also to be covariantly finite (see [9, Corollary 5.10]). Using subcategories of $\mathrm{Mod}\,\Lambda$ a resolving and contravariantly finite subcategory of $\mathrm{mod}\,\Lambda$ is shown also to be covariantly finite in [48, Corollary 2.6]. Is something similar true for subcategories of $\mathrm{Mod}\,\Lambda$?

We end the discussion on arbitrary tilting and cotilting modules by considering the pure-injectivity of cotilting modules. Recall that a module M is *pure-injective* if the natural map $M \to D^2(M)$ splits [47].

It was conjectured that all cotilting modules are pure-injective. This was first proven for cotilting modules of injective dimension at most one (see [11]). Recently, all cotilting modules are shown to be pure-injective (see [60]). Before this result was known, pure-injectivity of a cotilting module was characterized (see [12] and [49, Theorem 5.7]). Using [60] they now become additional properties of a cotilting module (true for any ring).

Theorem 2.7. *Let T be a cotilting module. Then the following assertions hold.*

(a) $^{\perp}T$ *is closed under pure factor modules,*

(b) $^{\perp}T$ *is closed under pure submodules,*

(c) $^{\perp}T$ *is closed under filtered colimits,*

(d) *every Λ-module has a minimal right $^{\perp}T$-approximation and* $D^2(^{\perp}T) \subseteq {}^{\perp}T$,

(e) T *is pure-injective.*

As already indicated by the above the theory for tilting and cotilting modules is in particular rich both for modules in $\mathrm{mod}\,\Lambda$ and $\mathrm{Mod}\,\Lambda$ when restricted to projective and injective dimension at most one, respectively. In particular for $\mathrm{mod}\,\Lambda$ every partial tilting module of projective dimension at most one can be completed to a tilting module (see [17]). This question we address in section 4 for tilting and cotilting modules in $\mathrm{Mod}\,\Lambda$. Furthermore, all torsion pairs $(\mathcal{T},\mathcal{F})$ in $\mathrm{mod}\,\Lambda$ with $\mathcal{F} = \mathrm{Sub}(M)$ for some M in $\mathrm{mod}\,\Lambda$ and with Λ in $\mathcal{F}$ are in one-to-one correspondence with cotilting modules T of injective dimension at most one [10] ($\mathcal{F} = \mathrm{Sub}(T)$). However there are torsion pairs in $\mathrm{mod}\,\Lambda$ not induced by a cotilting (or tilting) module in $\mathrm{mod}\,\Lambda$. For example, for the Kronecker algebra let $\mathcal{F}$ be the additive closure of the preprojective modules and any proper non-empty set of tubes in the Auslander-Reiten quiver. But in [19] it is shown that all torsion pairs $(\mathcal{T},\mathcal{F})$ with Λ in $\mathcal{F}$ are induced by cotilting modules of injective dimension at most one in $\mathrm{Mod}\,\Lambda$. We end this section by explaining this correspondence (true for left noetherian rings). For a subcategory $\mathcal{C}$ in $\mathrm{mod}\,\Lambda$ denote by $\varinjlim \mathcal{C}$ the full additive subcategory of all direct summands of the filtered colimits of modules in $\mathcal{C}$.

Theorem 2.8. *There is a bijection between torsion pairs $(\mathcal{T},\mathcal{F})$ in* mod Λ *with Λ in $\mathcal{F}$ and equivalence classes of cotilting modules of injective dimension at most one in* Mod Λ.

A cotilting module T of injective dimension at most one corresponding to a torsion pair $(\mathcal{T}, \mathcal{F})$ satisfies Prod $T = \varinjlim \mathcal{F} \cap (\varinjlim \mathcal{F})^{\perp}$ *and* $^{\perp}T \cap$ mod $\Lambda = \mathcal{F}$.

Proof. We sketch the proof of one direction to illustrate the use of the results in this section. Let $(\mathcal{T}, \mathcal{F})$ be a torsion pair in mod Λ with Λ in $\mathcal{F}$. By [33] $(\varinjlim \mathcal{T}, \varinjlim \mathcal{F})$ is a torsion pair in Mod Λ. Note that any torsion free class of a torsion pair in Mod Λ containing Λ is a resolving subcategory. Hence $\varinjlim \mathcal{F}$ is resolving and closed under products. Since a first syzygy of any module in Mod Λ is in $\varinjlim \mathcal{F}$, we have that $\widehat{\varinjlim \mathcal{F}} =$ Mod Λ and $(\varinjlim \mathcal{F})^{\perp}$ is contained in $\mathcal{I}^2(\text{Mod } \Lambda)$. Every Λ-module has a minimal (special) right $\varinjlim \mathcal{F}$-approximation by [19, Lemma 1.3] or [49, Theorem 2.6], since $\varinjlim \mathcal{F}$ is resolving and closed under products and filtered colimits. Hence by Theorem 2.3 and 2.7 we have $\varinjlim \mathcal{F} \cap (\varinjlim \mathcal{F})^{\perp} = \text{Prod } T$ for some pure-injective cotilting module of injective dimension at most one in Mod Λ. $\square$

A different characterization of these torsion pairs when the ring is left or right artinian can be found in [62, Theorem 3.7]. For related information and examples over concealed canonical algebras see [54].

3 The finitistic dimension conjectures

The main interest in this section is the finitistic conjectures for an artin algebra, that is, when is Findim Λ = findim Λ and when is findim Λ finite. The first of these conjectures was disproved by in [43], and the second one has been proven for

   (i) monomial algebras in [39] (again in [46]),

   (ii) radical cube zero (and even more generally for algebras with Loewy length $2n + 1$ and $\Lambda/\mathfrak{r}^n$ of finite representation type, see [34]) in [38] ([45]),

   (iii) when $\mathcal{P}^{\infty}(\text{mod } \Lambda)$ is contravariantly finite in [9],

   (iv) when the representation dimension of Λ is at most 3 in [45] (all special biserial algebras has representation dimension at most 3, see [35]).

None of the above proofs of the finiteness of findim Λ directly involve a tilting or a cotilting module. The results on relations between these conjectures and tilting and cotilting modules are more in the direction of finding test classes of modules for these conjectures, and if the dimensions are finite to show that they are obtained as the projective or the injective dimension of a tilting or cotilting module, respectively. For related results see [62].

First we discuss how findim Λ is obtained as the projetive dimension of a tilting module in Mod Λ when it is finite.

Proposition 3.1. [5, Proposition 2.3] *For any positive integer n there is a tilting module T_n with projective dimension at most n and $T^{\perp} = \mathcal{P}^{n+1}(\operatorname{mod} \Lambda)^{\perp}$. If* findim $\Lambda \geqslant n$, *then T_n has projective dimension n.*

Proof. The category $\mathcal{Y} = \mathcal{P}^{n+1}(\operatorname{mod} \Lambda)^{\perp}$ is clearly coresolving. Since $\mathcal{P}^{n+1}(\operatorname{mod} \Lambda)$ consists only of finitely presented modules, $\mathcal{Y}$ is closed under coproducts. Every Λ-module has a special left $\mathcal{Y}^{\perp}$-approximation by [37, Theorem 10]. It is also clear that $\breve{\mathcal{Y}} = \operatorname{Mod} \Lambda$. By Proposition 2.6 there exists a tilting module T_n such that $T_n^{\perp} = \mathcal{Y}$. The last claim is left to the reader. $\qquad \square$

This immediately gives the following relationship between the finitistic dimensions Findim Λ and findim Λ (for the second statement see [21, Corollary 2.2]).

Corollary 3.2.

(a) Findim $\Lambda \geqslant \sup\{\operatorname{pd}_{\Lambda} T \mid T \text{ tilting module}\} \geqslant$ findim Λ.

(b) Findim $\Lambda \geqslant \sup\{\operatorname{id}_{\Lambda^{\mathrm{op}}} T \mid T \text{ cotilting } \Lambda^{\mathrm{op}}\text{-module}\} \geqslant$ findim Λ.

In view of this result an obvious question is: When Findim Λ is finite or findim Λ is finite, does there exist tilting or cotilting modules with projective or injective dimensions equal to one of the finitistic dimensions? It seems to be unknown in general if Findim Λ always can be obtained in this way when it is finite. But we give some sufficient conditions later.

It was first proved in [9] that if $\mathcal{P}^{\infty}(\operatorname{mod} \Lambda)$ is contravariantly finite, then findim Λ is finite. It is easy to show using Theorem 2.1 (d) that then findim Λ is obtained as the projective dimension of a finitely presented tilting module.

We show next that the finitistic dimension findim Λ always is obtained by the projective dimension of a tilting module when it is finite.

Theorem 3.3. *The following are equivalent.*

(a) findim $\Lambda < \infty$.

(b) *there exists a tilting module T such that $T^\perp = \mathcal{P}^\infty(\mathrm{mod}\,\Lambda)^\perp$.*

(c) $\widetilde{\mathcal{P}^\infty(\mathrm{mod}\,\Lambda)}^\perp = \mathrm{Mod}\,\Lambda$.

If any of these are true, then findim $\Lambda = \mathrm{pd}_\Lambda T$.

The equivalence of (a) and (b) is given for left noetherian rings in [5, Theorem 2.6]. The equivalence of (b) and (c) is an immediate consequence of Proposition 2.6 noting that $\mathcal{P}^\infty(\mathrm{mod}\,\Lambda)^\perp$ always is coresolving and closed under coproducts, and every Λ-module has a special left $\mathcal{P}^\infty(\mathrm{mod}\,\Lambda)^\perp$-approximation.

As a consequence of the above result Findim Λ is obtained as the projective dimension of a tilting module T whenever Findim Λ and findim Λ are equal. When $\mathcal{P}^\infty(\mathrm{mod}\,\Lambda)$ is contravariantly finite in mod Λ then Findim Λ and findim Λ are shown to be equal in [44] (for an alternative proof see [5, Corollary 4.3]). Hence this gives a situation where also Findim Λ is realized as the projective dimension of a tilting module. Next we give a new criterion for when Findim Λ is obtained as the projective dimension of a tilting module.

Proposition 3.4. (a) *Assume that* Findim $\Lambda < \infty$. *The category $\mathcal{P}^\infty(\mathrm{Mod}\,\Lambda)^\perp$ is closed under coproducts if and only if there exists a tilting module T such that $\widehat{\mathrm{Add}\,T} = \mathcal{P}^\infty(\mathrm{Mod}\,\Lambda)$. In this case* Findim $\Lambda = \mathrm{pd}_\Lambda T$.

(b) *Assume that* Findim $\Lambda^{\mathrm{op}} < \infty$. *The category $^\perp\mathcal{I}^\infty(\mathrm{Mod}\,\Lambda)$ is closed under products if and only if there exists a cotilting module T such that $\widehat{\mathrm{Prod}\,T} = \mathcal{I}^\infty(\mathrm{Mod}\,\Lambda)$. In this case* Findim $\Lambda^{\mathrm{op}} = \mathrm{id}_\Lambda T$.

When Findim $\Lambda = n$ is finite, then the subcategory $\mathcal{P}^\infty(\mathrm{Mod}\,\Lambda)$ is equal to $\{X \in \mathrm{Mod}\,\Lambda \mid \mathrm{Ext}_\Lambda^1(X, \Omega_\Lambda^{-n}(\Lambda/\mathfrak{r})) = (0)\}$. This subcategory is resolving and every Λ-module has a minimal (special) right $\mathcal{P}^\infty(\mathrm{Mod}\,\Lambda)$-

approximation by [36, Corollary 10]. The condition (iv) in Proposition 2.6 amounts to saying that $\mathcal{P}^{\infty}(\operatorname{Mod}\Lambda)^{\perp}$ is closed under coproducts. Similar arguments are used in (b), hence the above result is then a direct consequence of Proposition 2.5 and Proposition 2.6. For related results see also [5, Section 3].

The categories $\mathcal{P}^n(\operatorname{Mod}\Lambda)$ are proven to be contravariantly finite for any $n \geqslant 1$ in [1]. Then using similar arguments as in [9, Corollary 3.10] one can show that the following are equivalent (i) $\operatorname{Findim}\Lambda$ is finite, (ii) $\mathcal{P}^{\infty}(\operatorname{Mod}\Lambda)$ is contravariantly finite and (iii) $\mathcal{P}^{\infty}(\operatorname{Mod}\Lambda)$ is closed under coproducts (see [49, Corollary 2.7]).

We end this section by characterizing the cotilting module in Proposition 3.4 (b). To this end we use the characterization of cotilting modules to define further structures on the class of cotilting modules.

Let $\operatorname{Cotilt}\Lambda$ denote the set of all equivalence classes of cotilting modules in $\operatorname{Mod}\Lambda$ (shown to be a set for any ring in [21]). We can partially order the cotilting modules according to the Hasse diagram of the subcategories $^{\perp}T$ for T in $\operatorname{Cotilt}\Lambda$. Such a Hasse diagram of the subcategories $^{\perp}T$ in $\operatorname{mod}\Lambda$ for the set of all (co)tilting modules T in $\operatorname{mod}\Lambda$ has been considered in various contexts in [56, 64] (also see [63]). The Hasse diagram of $\operatorname{Cotilt}\Lambda$ was considered in [21, 22], and it was shown to be a lattice in [21].

The characterization of cotilting modules in Theorem 2.3 imply that for any subset $\{T_i\}_{i\in I}$ of $\operatorname{Cotilt}\Lambda$ with $\sup\{\operatorname{id}_{\Lambda}T_i\}_{i\in I} < \infty$, there exists a cotilting module T in $\operatorname{Cotilt}\Lambda$ such that $^{\perp}T = \bigcap_{i\in I} {}^{\perp}T_i$. Moreover, $\operatorname{id}_{\Lambda}T = \sup\{\operatorname{id}_{\Lambda}T_i\}_{i\in I}$. In particular, when $\operatorname{Findim}\Lambda^{\mathrm{op}}$ is finite there is a unique minimal cotilting module $T_{\min}$ in $\operatorname{Cotilt}\Lambda$ with $\operatorname{id}_{\Lambda}T = \sup\{\operatorname{id}_{\Lambda}T\}_{T\in\operatorname{Cotilt}\Lambda}$. It is unknown whether or not $\operatorname{Findim}\Lambda^{\mathrm{op}} = \operatorname{id}_{\Lambda}T_{\min}$. We want to compare $T_{\min}$ to the module defined in the following lemma (see [21, Lemma 3.1 and 3.2]).

Lemma 3.5. *Suppose that* $\operatorname{Findim}\Lambda^{\mathrm{op}}$ *is finite.*

(a) *There exists a Λ-module T such that*
$$^{\perp}\mathcal{I}^{\infty}(\operatorname{Mod}\Lambda) \cap \mathcal{I}^{\infty}(\operatorname{Mod}\Lambda) = \operatorname{Add}T.$$

(b) $\mathcal{I}^{\infty}(\operatorname{Mod}\Lambda) = \widehat{\operatorname{Add}T}$, *where all the resolutions in* $\operatorname{Add}T$ *can be chosen to have length at most* $\operatorname{Findim}\Lambda^{\mathrm{op}}$.

(c) $\operatorname{id}_\Lambda T = \operatorname{Findim}\Lambda^{\mathrm{op}}$.

Denote the module T in the above lemma by T_{inj}. When $\operatorname{Findim}\Lambda^{\mathrm{op}}$ is finite it is immediate that (i) $\operatorname{id}_\Lambda T_{\mathrm{inj}} < \infty$, (ii) $\operatorname{Ext}^i_\Lambda(T_{\mathrm{inj}}, T_{\mathrm{inj}}) = (0)$ for all $i > 0$ and (iii) there exists a long exact sequence

$$0 \to T_n \to T_{n-1} \to \cdots \to T_1 \to T_0 \to D(\Lambda) \to 0$$

with T_i in $\operatorname{Add} T_{\mathrm{inj}}$ for all i. So T_{inj} is very close to being a cotilting module, but nevertheless if $\operatorname{Findim}\Lambda^{\mathrm{op}}$ is finite, then it is obtained on a selforthogonal module. Moreover, we have $^\perp\operatorname{Add} T_{\mathrm{inj}} = {}^\perp\mathcal{I}^\infty(\operatorname{Mod}\Lambda)$.

The following result points out a relationship between the modules T_{inj} and T_{min} and the dimensions $\operatorname{id}_\Lambda T_{\mathrm{inj}}$ and $\operatorname{id}_\Lambda T_{\mathrm{min}}$ when $\operatorname{Findim}\Lambda^{\mathrm{op}}$ is finite (see [21, Theorem 3.3]). Recall that a modules M in $\operatorname{Mod}\Lambda$ is called *product complete* if $\operatorname{Add} M = \operatorname{Prod} M$, and it is called Σ-*pure-injective* if every coproduct $M^{(\alpha)}$ is pure-injective.

Theorem 3.6. *Suppose that $\operatorname{Findim}\Lambda^{\mathrm{op}}$ is finite. Then the following are equivalent.*

(a) $^\perp\mathcal{I}^\infty(\operatorname{Mod}\Lambda) = {}^\perp T_{\mathrm{min}}$,

(b) $^\perp\mathcal{I}^\infty(\operatorname{Mod}\Lambda)$ *is closed under products,*

(c) T_{inj} *is product complete,*

(d) T_{inj} *is a Σ-pure-injective cotilting module.*

Moreover, when one of these conditions holds T_{min} and T_{inj} are equivalent cotilting modules and $\operatorname{Findim}\Lambda^{\mathrm{op}} = \operatorname{id}_\Lambda T_{\mathrm{min}}$.

4 Complements of tilting and cotilting modules

As already mentioned, a classical result in the theory of tilting modules in $\operatorname{mod}\Lambda$ says that for any partial tilting module T of projective dimension at most one, there exists a module T' in $\operatorname{mod}\Lambda$ such that $T \amalg T'$ is a tilting module with $T^\perp = (T \amalg T')^\perp$ (the Bongartz construction or complement). For a partial tilting module T of projective dimension at least two in $\operatorname{mod}\Lambda$, such a complement T' in $\operatorname{mod}\Lambda$ does not always exist as shown in [55]. This naturally gives rise to at least two questions: (1) Does the Bongartz construction of a complement generalize to partial tilting (or cotilting) modules of projective (injective) dimension at most one in Mod Λ? (2) Does any partial tilting or cotilting module T in mod Λ have a complement when considered as a partial tilting or cotilting module in Mod Λ? This section is devoted to discussing these questions. The answer to the second question is always yes, and the first question is answered in full for tilting and cotilting modules. For related results see [62].

First we discuss a criterion for existence of complements of partial tilting and cotilting modules. The following is an easy consequence of our characterizations of tilting and cotilting modules in terms of subcategories of Mod Λ.

Proposition 4.1. [49, Proposition 6.1]

(a) *Let T be a partial tilting module. Then T has a complement if and only if $T^{\perp}$ contains a coresolving subcategory $\mathcal{Y}$ containing $\mathrm{Add}\,T$ and closed under coproducts with $\breve{\mathcal{Y}} = \mathrm{Mod}\,\Lambda$, such that every Λ-module has a special left $\mathcal{Y}$-approximation.*

(b) *Let T be a partial cotilting module. Then T has a complement if and only if $^{\perp}T$ contains a resolving subcategory $\mathcal{X}$ containing $\mathrm{Prod}\,T$ and closed under products with $\hat{\mathcal{X}} = \mathrm{Mod}\,\Lambda$, such that every Λ-module has a special right $\mathcal{X}$-approximation.*

Recall that for an extension closed subcategory $\mathcal{X}$ of Mod Λ a module M in $\mathcal{X}$ is called Ext-*injective* in $\mathcal{X}$ if $\mathrm{Ext}^1_{\Lambda}(\mathcal{X}, M) = (0)$ and Ext-*projective* in $\mathcal{X}$ if $\mathrm{Ext}^1_{\Lambda}(M, \mathcal{X}) = (0)$. Note that if T is a partial tilting module, then an Ext-projective complement T' in $T^{\perp}$ correspond to the Bongartz complement as $T^{\perp} = (T \amalg T')^{\perp}$. As a further application of our characterizations of tilting and cotilting modules we obtained the following. The different proofs are given in [2, Theorem 2.1] and [49, Theorem 6.2].

Theorem 4.2.   (a) *Let T be a partial tilting module. Then T has a complement which is Ext-projective in $T^{\perp}$ if and only if $T^{\perp}$ is closed under coproducts.*

(b) *Let T be a partial cotilting module. Then T has a complement which is Ext-injective in $^{\perp}T$ if and only if $^{\perp}T$ is closed under products and each Λ-module has a special right $^{\perp}T$-approximation.*

Here we see that the statements of the results for tilting and cotilting are not dual. The explanation for this is the following. By [37,

Theorem 10] every Λ-module has a special left $M^{\perp}$-approximation for any Λ-module M, while it is unknown if every Λ-module has a special right $^{\perp}M$-approximation for any Λ-module M. However, "duality" is restored if we assume that T is pure-injective, since by [36, Corollary 10] every Λ-module has a minimal right $^{\perp}M$-approximation whenever M is pure-injective. In view of the fact that all cotilting modules are pure-injective (see [60]), for a partial cotilting module to have a complement it must be pure-injective. In fact, we have the following, where the equivalence of (a) and (c) can be found in [2, p. 93].

Proposition 4.3. [49, Corollary 6.3] *Let T be a pure-injective partial cotilting module. The following are equivalent.*

(a) *T admits a complement which is Ext-injective in $^{\perp}T$,*

(b) *T admits a pure-injective complement which is Ext-injective in $^{\perp}T$,*

(c) *$^{\perp}T$ is closed under products.*

Let us proceed by addressing the first question for cotilting modules: (1) Does the Bongartz construction of a complement generalize to cotilting modules of injective dimension at most one in Mod Λ? In [11] any cotilting module of injective dimension at most one is shown to be pure-injective. Hence a necessary assumption for being able to answer question (1) affirmatively is to start with a pure-injective partial cotilting module of injective dimension at most one. This is also shown to be sufficient in [19, Corollary 1.12].

Proposition 4.4. *Suppose that T is a pure-injective partial cotilting module of injective dimension at most one. Then there exists a pure-injective module T' such that $T \amalg T'$ is a pure-injective cotilting module.*

Proof. Let M be a pure-injective module of injective dimension at most one. It is shown in [19, Corollary 1.10] that $^{\perp}M$ is closed under products. The claim then follows immediately from the above result. $\square$

It follows from [19, Lemma 1.11] that the module T' found in the above proposition corresponds to the Bongartz construction.

Now we address the question (1) for partial tilting modules. This involves the following notions. Recall that a class of modules $\mathcal{X}$ is said to be of *finite (countable) type* if $\mathcal{X} = \mathcal{S}^{\perp}$ for some subset $\mathcal{S}$ of finitely presented (countably generated) modules in $\mathcal{P}^{n+1}(\mathrm{Mod}\,\Lambda)$ for some $n < \infty$. A module M is of *finite* or *countable* type if the subcategory $M^{\perp}$ is of finite or countable type, respectively. For further details and results about these notions for tilting modules see [62].

In [13] all tilting modules of projective dimension at most one are shown to be of countable type. But even more is true, recently these tilting modules were shown to be of finite type (see [14]), and all tilting modules were shown to have the same property in [15]. So to answer question (1) for partial tilting modules T of projective dimension at most one, the assumption of T being of finite type is necessary. But this is also sufficient due to the following fact. If M is a module of finite type, then $M^{\perp}$ is closed under products, filtered colimits and pure submodules, in particular closed under coproducts. As a consequence of this and Theorem 4.2 (a) we obtain the following.

Proposition 4.5. *Suppose that T is a partial tilting module of finite type (and of projective dimension at most one). Then there exists a module T' such that $T \amalg T'$ is a tilting module of finite type (and of projective dimension at most one).*

We end this section by answering the second question: Do any partial tilting or cotilting module T in $\mathrm{mod}\,\Lambda$ have a complement when considered as a partial tilting or cotilting module in $\mathrm{Mod}\,\Lambda$? Since $T^{\perp}$ is closed under coproducts and $^{\perp}T$ is closed under products (actually definable) when T is finitely presented, the following result is an immediate consequence of Theorem 4.2 (see [2, Corollary 2.2] and [49, Corollary 6.4]).

Proposition 4.6. *Let T be a finitely presented Λ-module.*

(a) *If T is a partial tilting module, then T has a complement which is Ext-projective in $T^{\perp}$.*

(b) *If T is a partial cotilting module, then T has a complement which is pure-injective and Ext-injective in $^{\perp}T$.*

5 Classification of all cotilting modules

For an artin algebra Λ of finite representation type all modules in $\text{Mod}\,\Lambda$ is isomorphic to a direct sum of indecomposable finitely presented modules (see [7, 58]). Therefore every tilting or cotilting module is equivalent to a finitely presented tilting or cotilting module, hence there is only a finite number of them.

For algebras of infinite representation type a complete classification of all cotilting modules seems only to be known for tame hereditary algebras ([19, 20]). The aim of this section is to give this classification. For information on cotilting modules over concealed canonical algebras see [54, Section 10].

Assume throughout that Λ is a finite dimensional tame hereditary algebra. We start by reviewing some facts about these algebras (see [57]). The category of finitely presented regular modules is denoted by $\mathcal{R}$, which are all the modules occurring in the tubes of the Auslander-Reiten quiver of Λ. This category is an abelian category, and $\mathbb{P}$ denotes the set of isomorphism classes of all simple objects in $\mathcal{R}$. Two elements S and S' in $\mathbb{P}$ are said to be equivalent if $\text{Ext}^1_\Lambda(S, S') \neq (0)$ or $\text{Ext}^1_\Lambda(S', S) \neq (0)$. Take the transitive closure of this relation in $\mathbb{P}$, and let $[S]$ denote the equivalence class of S. For each S in $\mathbb{P}$ there are unique indecomposable objects S_n and S_{-n} of length n in $\mathcal{R}$ such that $\text{Hom}_\Lambda(S, S_n) \neq (0)$ and $\text{Hom}_\Lambda(S_{-n}, S) \neq (0)$. Moreover there are chains of monomorphisms $S = S_1 \to S_2 \to \cdots$ and chains of epimorphisms $\cdots \to S_{-2} \to S_{-1} = S$ for each S in $\mathbb{P}$. The colimit $\varinjlim S_n = S_\infty$ is the corresponding *Prüfer module*, and the inverse limit $\varprojlim S_{-n} = S_{-\infty}$ is the *adic* module. Finally there is a unique generic module G, that is, G is indecomposable of infinite length and has finite length over $\text{End}_\Lambda(G)$.

Given a module M in $\text{Mod}\,\Lambda$, denote by $\text{indec}\,M$ the set of all the isoclasses of indecomposable direct summands of M. If M is pure-injective, then there is a unique family $\{M_i\}_{i \in I}$ of modules in $\text{indec}\,M$ such that M is the pure-injective envelope of $\amalg_{i \in I} M_i$. With these preliminaries we can give the classification of all cotilting modules over Λ. Recall that by [11] all cotilting modules of injective dimension at most one are pure-injective, so that all cotilting modules over Λ are pure-injective.

Theorem 5.1. [19, Theorem 3.9] *Let Λ be a tame hereditary algebra, and let T be a pure-injective Λ-module.*

(a) *Suppose that all indecomposable direct summands of T are finitely presented. Then T is a cotilting module if and only if* card(indec T) *equals the number of non-isomorphic simple Λ-modules and* $\operatorname{Ext}_{\Lambda}^{1}(T',T'') = (0)$ *for all T' and T'' in* indec T.

(b) *Suppose there is an indecomposable direct summand of T which is not finitely presented. Then T is a cotilting module if and only if the following holds:*

    (i) *Each M in* indec T *is either generic or of the form S_n for some S in $\mathbb{P}$ and some n in $\mathbb{N} \cup \{-\infty, \infty\}$.*

    (ii) *For each S in $\mathbb{P}$, the set I_S consisting of all the non-isomorphic modules M in* indec T *with $M \simeq S'_n$ for some n in $\mathbb{N} \cup \{-\infty, \infty\}$ and some S' in $[S]$ satisfies* card I_S = card$[S]$ *and* $\operatorname{Ext}_{\Lambda}^{1}(T', T'') = (0)$ *for all T' and T'' in I_S.*

(c) *Two cotilting modules T_1 and T_2 are equivalent if and only if* indec$(T_1 \amalg G)$ = indec$(T_2 \amalg G)$.

The cotilting modules occurring in part (b) are given by for every tube choosing the rank of the tube indecomposable non-isomorphic modules totally from the tube, its Prüfer module or its adic module. For a tube of rank one there are only two choices, the Prüfer or the adic module. A complete characterization of the different possible choices in a tube is given in [20].

References

[1] Aldrich, S. T., Enochs, E. E., Jenda, O. M. G., Oyonarte, L., *Envelopes and covers by modules of finite injective and projective dimension*, J. Algebra 242 (2001), 447–459.

[2] Angeleri-Hügel, L., Coelho, F. U., *Infinitely generated complements to partial tilting modules*, Math. Proc. Cambridge Philos. Soc. 132 (2002), no. 1, 89–96.

[3] Angeleri-Hügel, L., Coelho, F. U., *Infinitely generated tilting modules of finite projective dimension*, Forum Math. 13 (2001), no. 2, 239–250.

[4] Angeleri-Hügel, L., Tonolo, A., Trlifaj, J., *Tilting preenvelopes and cotilting precovers*, Algebras and Representation Theory, 4 (2001), 155–170.

[5] Angeleri-Hügel, L., Trlifaj, J., *Tilting theory and the finitistic dimension conjectures*, Trans. Amer. Math. Soc. 354 (2002), no. 11, 4345–4358.

[6] Assem, I., *Tilting theory - an introduction*, Topics in Algebra, Banach Center Publications 26 PWN, Warsaw (1990), 127–180.

[7] Auslander, M., *Representation theory of artin algebras* II, Comm. Algebra 1 (1974), 269–310. 1976.

[8] Auslander, M., Platzeck, M. I., Reiten, I., *Coxeter functors without diagrams*, Trans. Amer. Math. Soc. 250 (1979), 1–12.

[9] Auslander, M., Reiten, I., *Applications of contravariantly finite subcategories*, Adv. Math. 86 (1991) no. 1, 111–152.

[10] Auslander, M., Smalø, S. O., *Almost split sequences in subcategories*, J. Algebra 69 (1981), no. 2, 426–454. Addendum: J. Algebra 71 (1981), no. 2, 592–594.

[11] Bazzoni, S., *Cotilting modules are pure-injective*, Proc. Amer. Math. Soc. 131 (2003), no. 12, 3665–3672.

[12] Bazzoni, S., *n-cotilting modules and pure-injectivity*, Bull. London Math. Soc., 36 (2004), no. 5, 599–612.

[13] Bazzoni, S., Eklof, P., Trlifaj, J., *Tilting cotorsion pairs*, Bull. London Math. Soc., 37 (2005), 683–696.

[14] Bazzoni, S., Herbera, D., *One dimensional tilting modules are of finite types*, preprint.

[15] Bazzoni, S., Stovicek, J., *All tilting modules are of finite type*, preprint

[16] Bernstein, I., Gelfand, I. M., Ponomarev, V. A., *Coxeter functors and Gabriel's theorem*, Usp. Mat. Nauk 28 (1973), 19–23.

[17] Bongartz, K., *Tilted algebras*, Representations of algebras (Puebla, 1980), Lecture Notes in Math. 904, Springer, Berlin-New York (1981), 26–38.

[18] Brenner, S., Butler, M. C. R., *Generalizations of the Bernstein-Gelfand-Ponomarev reflection functors*, Lecture Notes in Math. 832, Springer-Verlag, Berlin, 1980, 103–169.

[19] Buan, A. B., Krause, H., *Cotilting modules over tame hereditary algebras*, Pacific J. Math. 211 (2003), no. 1, 41–59.

[20] Buan, A. B., Krause, H., *Tilting and cotilting for quivers of type $\tilde{A}_n$*, J. Pure Applied algebra, 190 (2004), no. 1–3, 1–21.

[21] Buan, A. B., Krause, H., Solberg, Ø., *On the lattice of cotilting modules*, AMA Algebra Montp. Announc. 2002, Paper 2, 6 pp.

[22] Buan, A. B., Solberg, Ø., *Limits of pure-injective cotilting modules*, J. Algebra and Representation Theory, to appear.

[23] Buan, A. B., Solberg, Ø., *Relative cotilting theory and almost complete cotilting modules*, Algebras and modules II, CMS Conf. Proc. 24 (1998), 77–92.

[24] Colby, R. R., *A cotilting theorem for rings*, Methods in Module Theory, M. Dekker, New York, 1993, pp. 33–37.

[25] Colby, R. R., *A generalization of Morita duality and the tilting theorem*, Comm. Algebra 17 (1989), 1709–1722.

[26] Colby, R. R., Fuller, K. R., *Tilting and serially tilted algebras*, Comm. Algebra 23 (1995), 1585–1616.

[27] Colby, R. R., Fuller, K. R., *Tilting and torsion theory counter equivalences*, Comm. Algebra 23 (1995), 4833-4849.

[28] Colpi, R., *Cotilting bimodules and their dualities*, Proc. Euroconf. Murcia '98, LNPAM 210, M. Dekker, New York 2000, 81–93.

[29] Colpi, R., D'Este, G., Tonolo, A., *Quasi-tilting modules and counter equivalences*, J. Algebra 191 (1997), 461–494.

[30] Colpi, R., Fuller, K. R., *Cotilting modules and bimodules*, Pacific J. Math. 192 (2) (2000), 275–291.

[31] Colpi, R. R., Tonolo, A., Trlifaj, J., *Partial cotilting modules and the lattice induced by them*, Comm. Algebra 25 (1997), 3225–3237.

[32] Colpi, R. R., Trlifaj, J., *Tilting modules and tilting torsion theories*, J. Algebra 178 (1995), 614–634.

[33] Crawley-Boevey, W., *Locally finitely presented additive categories*, Comm. Algebra 22 (1994), 1644–1674.

[34] Dräxler, P., Happel, D., *A proof of the generalized Nakayama conjecture for algebras with $J^{2l+l} = 0$ and A/J^l representation finite*, J. Pure Appl. Algebra 78 (1992), no. 2, 161–164.

[35] Erdmann, K., Holm, T., Iyama, O., Schröer, J., *Radical embeddings and representation dimension*, Adv. Math., 185 (2005), no. 1, 159–177.

[36] Eklof, P., Trlifaj, J., *Covers induced by* Ext, J. Algebra 231 (2000), 640–651.

[37] Eklof, P., Trlifaj, J., *How to make* Ext *vanish*, Bull. London Math. Soc., 33 (2001), 41–51.

[38] Green, E. L., Huisgen-Zimmermann, B., *Finitistic dimension of artinian rings with vanishing radical cube*, Math. Z. 206 (1991), no. 4, 505–526.

[39] Green, E. L., Kirkman, E.., Kuzmanovich, J., *Finitistic dimensions of finite dimensional monomial algebras*, J. Algebra 136 (1991), no. 1, 37–50.

[40] Happel, D., *Triangulated categories in the representation theory of finite-dimensional algebras*, London Math. Soc. Lecture Notes Series, 119. Cambridge Univ. Press (1988).

[41] Happel, D., Ringel, C. M., *Tilted algebras*, Trans. Amer. Math. Soc. 274 (1982), 399–443.

[42] Happel, D., Unger, L., *Complements and the Generalized Nakayama conjecture*, Algebras and modules II, CMS Conf. Proc. 24 (1998), 293–310.

[43] Huisgen-Zimmermann, B., *Homological domino effects and the first finitistic dimension conjecture*, Invent. Math. 108 (1992) 369–383.

[44] Huisgen-Zimmermann, B., Smalø, S. O., *A homological bridge between finite and infinite dimensional representations*, Algebras and Representation Theory 1 (1998), 169–188.

[45] Igusa, K., Todorov, G., *On the finitistic global dimension conjecture for artin algebras*, preprint.

[46] Igusa, K., Zacharia, D., *Syzygy pairs in a monomial algebra*, Proc. Amer. Math. Soc. 108 (1990), no. 3, 601–604.

[47] Jensen, C. U., Lenzing, H., *Model-theoretic algebra with particular emphasis on fields, rings, modules*, Algebra, Logic and Applications, 2. Gordon and Breach Science Publishers, New York, 1989, xiv+443 pp.

[48] Krause, H., Solberg, Ø., *Application of cotorsion pairs*, J. London Math. Soc. (2) 68 (2003), no. 3, 631–650.

[49] Krause, H., Solberg, Ø., *Filtering modules of finite projective dimension*, Forum Math. 15 (2003), no. 3, 377-393.

[50] Mantese, F., PhD-thesis, University of Padova, 2003.

[51] Menini, C., Orsatti, A., *Representable equivalences between categories of modules and applications*, Rend. Sem. Math. Univ. Padova 82 (1989), 203–231.

[52] Miyashita, Y., *Tilting modules of finite projective dimension*, Math. Z. 193 (1986), 113–146.

[53] Reiten, I., *Tilting and homologically finite subcategories with application*

to quasi-hereditary algebras, this volume.

[54] Reiten, I., Ringel, C. M., *Infinite dimensional representations of canonical algebras*, Cand. J. Math., to appear.

[55] Rickard, J., Schofield, A., *Cocovers and tilting modules*, Math. Proc. Cambridge Philos. Soc. 106 (1989), no. 1, 1–5.

[56] Riedtmann, C., Schofield, A., *On a simplicial complex associated with tilting modules*, Comment. Math. Helv. 66 (1991), 70–78.

[57] Ringel, C. M., *Infinite-dimensional representations of finite-dimensional hereditary algebras*, Symposia Mathematica, Vol. XXIII (Conf. Abelian groups and their Relationship to the Theory of Modules, INDAM, Rome, 1977), Academic Press, London-New York (1979), 321–412.

[58] Ringel, C. M., Tachikawa, H., *QF-3 rings*, J. Reine Angew. Math. 272 (1974) 49–72.

[59] Smalø, S. O., *Torsion theories and tilting modules*, Bull. London Math. Soc. 16 (1984), 518–522.

[60] Stovicek, J., *All n-cotilting modules are pure-injective*, Proc. Amer. Math. Soc., to appear.

[61] Tonolo, A., *Tilting modules of finite projective dimension: sequentially static and costatic modules*, J. Algebra Appl. 1 (2002), 295–305.

[62] Trlifaj, J., *Infinite dimensional tilting modules and cotorsion pairs*, this volume.

[63] Unger, L., *Combinatorial aspects of the set of tilting modules*, this volume.

[64] Unger, L., *Shellability of simplicial complexes arising in representation theory*, Adv. Math. 144 (1999), no. 2, 211–246.

[65] Wakamatsu, T., *On modules with trivial self-extensions*, J. Algebra 114 (1988) 106–114.

Øyvind Solberg
Institutt for matematiske fag
NTNU
N–7491 Trondheim
Norway

13

Cotilting dualities

Riccardo Colpi and Kent R. Fuller

In the late 1950's K. Morita [31] and G. Azumaya [4] investigated dualities between subcategories of $\mathcal{C}_R$ of $\mathrm{Mod} - R$ and $_S\mathcal{C}$ of $S - \mathrm{Mod}$ that contain the regular representations of the rings and are closed under submodules, epimorphic images and finite direct sums. It was shown that such dualities are *representable* by a bimodule $_SU_R$ in the sense the dualities are isomorphic to the U-dual functors

$$\Delta = \mathrm{Hom}_R(\_, U) : \mathcal{C}_R \rightleftarrows {}_S\mathcal{C} : \mathrm{Hom}_S(\_, U) = \Delta$$

with the evaluation maps $\delta_M : M \to \Delta^2 M$ providing natural isomorphisms

$$\delta : 1_{\mathcal{C}_R} \to \Delta^2 \quad \text{and} \quad \delta : 1_{{}_S\mathcal{C}} \to \Delta^2.$$

Such a duality is known as a *Morita duality* induced by the bimodule $_SU_R$.

Given a bimodule $_SU_R$, a module M is *U-torsionless* if it embeds in a direct product of copies of U, (i.e., belongs to $\mathrm{Cogen}(U)$) or equivalently, if δ_M is a monomorphism. If δ_M is an isomorphism, M is said to be *U-reflexive*.

Morita also proved that a bimodule $_SU_R$ induces such a Morita duality if and only if it is balanced and is an injective cogenerator on both sides if and only if the regular modules R_R and $_SS$ are U-reflexive and the categories of U-reflexive right R- and left S-modules are closed under submodules and epimorphic images. (See [1, Sections 23 and 24], for example.)

A module M is *linearly compact* if for every inverse system of epimorphisms $p_\lambda \colon M \to M_\lambda$, the map $\varprojlim p_\lambda \colon M \to \varprojlim M_\lambda$ is also an epi-morphism. Equivalently, every finitely solvable system $\{x_\alpha, K_\alpha\}_I$ with $x_\alpha \in M$ and $K_\alpha \leqslant M$ is solvable in the sense that if every finite intersec-tion of the cosets $x_\alpha + K_\alpha$ in M is nonempty, then $\cap_I (x_\alpha + M_\alpha) \neq \varnothing$. Several years later B.J. Müller (see [33], for example) proved that if $_SU_R$ induces a Morita duality, then the U-reflexive modules are just the linearly compact modules in $\mathrm{Mod} - R$ and $S - \mathrm{Mod}$.

At about the same time as Morita, J. Dieudonné [19] characterized quasi-Frobenius rings as those rings R such that R-dual functors induce a dual-ity between the categories of finitely generated right and left R-modules, $\mathrm{mod} - R$ and $R - \mathrm{mod}$. Thus a noetherian ring R is quasi-Frobenius if and only if all of its finitely generated modules are R-reflexive.

In the early 1960's J.P. Jans [24] characterized those noetherian rings whose finitely generated R-torsionless modules are R-reflexive as those whose left and right injective dimensions satisfy $\mathrm{inj.dim.}(R_R) \leqslant 1$ and $\mathrm{inj.dim.}(_RR) \leqslant 1$. Shortly thereafter, apparently inspired by Jans's result, E. Matlis [29] proved an early version of a cotilting theorem when he proved that the submodules of finitely generated free modules over a commutative domain D are D-reflexive if and only Q/D is an injective cogenerator if and only if every finitely generated D-torsion module M is Q/D-reflexive, where Q is the field of fractions of D. (Note that here there is a natural isomorphism $\mathrm{Hom}_D(M, Q/D) \cong \mathrm{Ext}_D^1(M, D)$.)

Although they were originally defined for finitely generated modules over artin algebras, as discussed in [30] and [12], the Brenner-Butler notions of a tilting module and the tilting theorem [7],[22],[6] are valid for the entire categories of modules over a pair of rings. However, as in the case of Morita duality vs Morita equivalence, the notion of cotilting modules and cotilting theorems for arbitrary rings must be restricted to smaller categories. In particular, if $_RV$ is a tilting module with $\mathrm{End}(_RV) = S$ over an artin algebra R, then the artin algebra dual yields a bimodule $_SU_R = D(V)$ and a cotilting theorem between $\mathrm{mod} - R$ and $S - \mathrm{mod}$, in the sense of the definition below.

Through out this chapter we let $_SU_R$ be a bimodule, we let Δ represent both $\mathrm{Hom}_R(_-, U)$ and $\mathrm{Hom}_S(_-, U)$, and we denote both $\mathrm{Ext}_R^1(_-, U)$ and $\mathrm{Ext}_S^1(_-, U)$ by Γ.

By an *abelian subcategory* of $\mathrm{Mod} - R$ or $S - \mathrm{Mod}$ we mean a full subcategory that is closed under finite direct sums and contains the kernels and cokernels of all of its homomorphisms. The following definition is from [11].

Definition 0.2. Suppose that $\mathcal{C}_R$ and $_S\mathcal{C}$ are abelian subcategories of $\mathrm{Mod} - R$ and $S - \mathrm{Mod}$ such that $R_R \in \mathcal{C}_R$ and $_SS \in {_S\mathcal{C}}$. Let $_SU_R$ be a bimodule and let

$$\mathcal{T}_R = \mathrm{Ker}\,\Delta \cap \mathcal{C}_R, \quad \mathcal{F}_R = \mathrm{Ker}\,\Gamma \cap \mathcal{C}_R, \quad _S\mathcal{T} = \mathrm{Ker}\,\Delta \cap {_S\mathcal{C}}, \quad _S\mathcal{F} = \mathrm{Ker}\,\Gamma \cap {_S\mathcal{C}}.$$

Then $_SU_R$ induces a *cotilting theorem* between $\mathcal{C}_R$ and $_S\mathcal{C}$ if the following four conditions are satisfied:

(1) $(\mathcal{T}_R, \mathcal{F}_R)$ and $(_S\mathcal{T}, _S\mathcal{F})$ are torsion theories in $\mathcal{C}_R$ and $_S\mathcal{C}$, respectively;

(2) $\Delta: \mathcal{C}_R \to {_S\mathcal{F}}, \quad \Gamma: \mathcal{C}_R \to {_S\mathcal{T}}, \quad \Delta: {_S\mathcal{C}} \to \mathcal{F}_R, \quad \Gamma: {_S\mathcal{C}} \to \mathcal{T}_R$

(3) There are natural transformations $\gamma: \Gamma^2 \to 1_{\mathcal{C}_R}$ and $\gamma: \Gamma^2 \to 1_{_S\mathcal{C}}$ that, together with the evaluation maps $\delta: 1_{\mathcal{C}_R} \to \Delta^2$ and $\delta: 1_{_S\mathcal{C}} \to \Delta^2$, yield exact sequences

$$0 \to \Gamma^2 M \xrightarrow{\gamma_M} M \xrightarrow{\delta_M} \Delta^2 M \to 0$$

and

$$0 \to \Gamma^2 N \xrightarrow{\gamma_N} N \xrightarrow{\delta_N} \Delta^2 N \to 0$$

for each $M \in \mathcal{C}_R$ and each $N \in {_S\mathcal{C}}$.

Observe that if $_SU_R$ induces a *cotilting theorem* between $\mathcal{C}_R$ and $_S\mathcal{C}$, then the restrictions

$$\Delta: \mathcal{F}_R \rightleftarrows {_S\mathcal{F}} : \Delta \quad \text{and} \quad \Gamma: \mathcal{T}_R \rightleftarrows {_S\mathcal{T}} : \Gamma$$

define category equivalences. Also note that (see the definition of an abelian subcategory of $\mathrm{Mod} - R$ or $S - \mathrm{Mod}$ above), since they contain R_R and $_SS$, respectively, $\mathcal{C}_R$ and $_S\mathcal{C}$ contain all finitely presented modules. Morita duality entails a cotilting theorem with $\mathcal{C}_R = \mathcal{F}_R$ and $_S\mathcal{C} = {_S\mathcal{F}}$ the categories of linearly compact modules, and $\mathcal{T}_R$ and $_S\mathcal{T}$ both 0.

The question remains, "what is a cotilting module?" We shall discuss several versions, each of which is the dual of a tilting module if it is finitely generated over an artin algebra. In each version, the $\mathcal{C}'s$ are determined by the $\mathcal{F}'s$ and the naturality of the $\gamma's$ is an issue.

The first version appears in Y. Miyashita's [30] as a noetherian dualization of his tilting modules of projective dimension $r \geqslant 1$. There he showed that if R is right noetherian, S is left noetherian and $_S U_R$ is finitely generated, faithfully balanced, and satisfies inj .dim .$(U) \leqslant r$ and $\mathrm{Ext}^i(U, U) = 0$ for all $i > 0$ on both sides, then $\mathrm{Ext}^i_S(\mathrm{Ext}^i_R(M, U), U) \cong M$ whenever $M \in \mathrm{mod} - R$ with $\mathrm{Ext}^j_R(U, M) = 0$ for all $0 \leqslant j \neq i$. (If $r = 1$, this is just $\Delta^2 M \cong M$, if $\Gamma M = 0$ and $\Gamma^2 M \cong M$, if $\Delta M = 0$.) Moreover, he proved that for arbitrary r, l.gl.dim. $S \leqslant$ r.gl.dim. $R + r$.

1 Generalized Morita Duality and Finitistic Cotilting Modules

If $_S U_R$ induces a Morita duality, then (see [1, Section 24], for example) the categories of U-reflexive modules are closed under extensions, as well as submodules and epimorphic images. In [8] R.R. Colby introduced the notion of a generalized Morita duality.

Definition 1.1. A bimodule $_S U_R$ induces a *generalized Morita duality (GMD)* if $_S U_R$ is faithfully balanced (i.e., U_R, R_R, $_S U$ and $_S S$ are all U-reflexive) and the categories of U-reflexive right R- and left S-modules are closed under submodules and extensions.

Then he proved

Proposition 1.2. *If a bimodule $_S U_R$ induces a generalized Morita duality, then*

1. *If M is a U-reflexive module, then $\Gamma M = 0$, so $\mathrm{Ext}^1_R(U, U) = 0$ and $\mathrm{Ext}^1_S(U, U) = 0$; and*

2. inj .dim .$(U_R) \leqslant 1$ *and* inj .dim .$(_S U) \leqslant 1$.

Thus if $_S U_R$ induces a GMD, it satisfies conditions that are dual to ones that serve to characterize tilting bimodules.

The following definitions are also from Colby's [8]. They have motivated much of the research that we shall discuss here.

Definition 1.3. A faithfully balanced bimodule $_S U_R$ defines a *finitistic generalized Morita duality* if U_R, R_R, $_S U$ and $_S S$ are all noetherian, and the categories of U-reflexive modules in $\text{mod} - R$ and $S - \text{mod}$ are closed under submodules and extensions.

Now we come to the first cotilting theorem for rings (from [8]) other than artin algebras.

Theorem 1.4. *Let* $_SU_R$ *define a finitistic generalized Morita duality. Then* $_SU_R$ *induces a cotilting theorem between* $\text{mod} - R$ *and* $S - \text{mod}$ *with* $\mathcal{F}_R$ *and* $_S\mathcal{F}$ *the* U-*torsionless modules in* $\text{mod} - R$ *and* $S - \text{mod}$. *Moreover, the Grothendieck groups of* $\text{mod} - R$ *and* $S - \text{mod}$ *are isomorphic.*

The modules we now call finitistic cotilting modules were originally called cotilting modules in [8].

Definition 1.5. A module U_R is a *finitistic cotilting module* if it satisfies

1. $\text{Ext}^1_R(U, U) = 0$;

2. $\text{inj.dim.}(U_R) \leqslant 1$; and

3. if $M \in \text{mod} - R$ with $\text{Hom}_R(M, U) = 0 = \text{Ext}^1_R(M, U)$, then $M = 0$.

Also in [8] Colby noted that condition 3. of this definition is implied by

3′. *an injective cogenerator* C_R *admits an exact sequence* $0 \to U_1 \longrightarrow U_0 \longrightarrow C \to 0$ *with* U_0 *and* U_1 *direct summands of direct products of copies of* U.

In addition he established the connection between these two definitions by proving

Theorem 1.6. *Let* $S = \text{End}(U_R)$, *and suppose that* U_R, R_R, $_SU$ *and* $_SS$ *are all noetherian. Then* U_R *is a finitistic cotilting module if and only if* $_SU_R$ *defines a finitistic generalized Morita duality.*

In a later paper [9] Colby proved all but the naturalness of $\gamma \colon \Gamma^2 \to 1_{\mathcal{C}}$ in the following

Theorem 1.7. *Let* $_SU_R$ *define a GMD, and let* $\mathcal{F}_R$ *and* $_S\mathcal{F}$ *denote the reflexive modules in* $\text{Mod} - R$ *and* $S - \text{Mod}$, *and* $\mathcal{C}_R$ *and* $_S\mathcal{C}$ *consist of the epimorphic images of modules in* $\mathcal{F}_R$ *and* $_S\mathcal{F}$, *respectively. Then* $_SU_R$

induces a cotilting theorem between $\mathcal{C}_R$ and $_S\mathcal{C}$ with torsion free classes $\mathcal{F}_R$ and $_S\mathcal{F}$.

Moreover in [9] he established the missing third condition in Proposition 1.2:

> 3. *if $_SU_R$ defines a GMD and $M \in \mathcal{F}$ with $\operatorname{Hom}_R(M,U) = 0 = \operatorname{Ext}^1_R(M,U)$, then $M = 0$.*

That γ is natural was established much later by F. Mantese in [26]. Her methods are key to establishing the naturalness of γ in the cotilting theorems to follow.

The following proposition is from [9] and [10].

Proposition 1.8. *Let U_R be a finitely generated module over an artin algebra and let $S = \operatorname{End}(U_R)$. Then U_R is the artin algebra dual of a tilting module if and only if $_SU_R$ defines a GMD. Moreover, if this is the case, then δ_M is an isomorphism if and only if M is finitely generated and U-torsionless, and γ_M is an isomorphism if and only if M is finitely generated and $\Delta M = 0$.*

It seems that cotilting modules over artin algebras and Morita dualities are the only known examples of generalized Morita dualities. However, in [12] Colby and K.R. Fuller characterized the tilting modules over hereditary noetherian serial rings and proved that any tilting module over a right hereditary noetherian ring is a finitistic cotilting module.

2 Cotilting Modules and Bimodules

In [16] R. Colpi, G. D'Este and A. Tonolo, generalizing the notion of injective cogenerator, made the following

Definition 2.1. *A cotilting module U_R is a module such that $\operatorname{Cogen}(U_R) = \operatorname{Ker}\operatorname{Ext}^1_R(\_, U)$.*

And there they observed that $(\operatorname{Ker}\operatorname{Hom}_R(\_, U), \operatorname{Ker}\operatorname{Ext}^1_R(\_, U))$ is a torsion theory in $\operatorname{Mod} - R$, and they proved (cf., Definition 1.5)

Proposition 2.2. *A module U_R is a cotilting module if and only if it satisfies*

1. inj.dim.$(U_R) \leqslant 1$;

2. $\operatorname{Ext}_R^1(U^\alpha, U) = 0$ *for all cardinal numbers* α;

3. $\operatorname{Ker}\operatorname{Hom}_R(\_, U) \cap \operatorname{Ker}\operatorname{Ext}_R^1(\_, U) = 0$

Subsequently L. Angeleri Hügel, Tonolo and J. Trlifaj [3] proved that condition 3. of this result can be replaced by

> 3'. *an injective cogenerator* C_R *admits an exact sequence* $0 \to U_1 \longrightarrow U_0 \longrightarrow C \to 0$ *with* U_0 *and* U_1 *direct summands of direct products of copies of* U.

In [15] Colpi made the following

Definition 2.3. A faithfully balanced bimodule $_SU_R$ is a *cotilting bimodule* if U_R and $_SU$ are cotilting modules.

And he proved all of the following cotilting theorem, except the facts, later established by Mantese [26], that $\mathcal{C}_R$ and $_S\mathcal{C}$ are abelian subcategories of $\mathrm{Mod}-R$ and $S-\mathrm{Mod}$, i.e., they contain the kernels and cokernels of their homomorphisms, and that the γ's are natural on $\mathcal{C}_R$ and $_S\mathcal{C}$.

Theorem 2.4. *Let* $_SU_R$ *be a cotilting bimodule, let* $\mathcal{F}_R$ *and* $_S\mathcal{F}$ *denote the reflexive modules in* $\mathrm{Mod}-R$ *and* $S-\mathrm{Mod}$, *and let* $\mathcal{C}_R$ *and* $_S\mathcal{C}$ *consist of the cokernels of monomorphisms between modules in* $\mathcal{F}_R$ *and* $_S\mathcal{F}$, *respectively. Then* $_SU_R$ *induces a cotilting theorem between* $\mathcal{C}_R$ *and* $_S\mathcal{C}$ *with torsion free classes* $\mathcal{F}_R$ *and* $_S\mathcal{F}$.

The following notion, a generalization of linear compactness for U-torsionless modules, appears to have originated in [21].

Definition 2.5. If $_SU_R$ is a bimodule, a U-torsionless module M is U-*torsionless linearly compact* if for every inverse system of maps $\{M \xrightarrow{p_\lambda} M_\lambda \mid \lambda \in \Lambda\}$ with each M_λ U-torsionless and each $\operatorname{Coker} p_\lambda \in \operatorname{Ker}\Delta$, it happens that $\operatorname{Coker}(\varprojlim p_\lambda) \in \operatorname{Ker}\Delta$.

In [15] Colpi proved that if $_SU_R$ is a cotilting bimodule, then any U-torsionless linearly compact module is U-reflexive, and subsequently in [17] Colpi and Fuller established the following connection between generalized Morita duality and cotilting bimodules.

Proposition 2.6. *Let* $_SU_R$ *be a cotilting bimodule. Then* $_SU_R$ *induces* a GMD if and only if the U-reflexive modules in $\mathrm{Mod}-R$ and $S-\mathrm{Mod}$ are precisely the U-torsionless linearly compact modules.

Also in [17] they proved that if R is a noetherian serial ring with self-duality induced by a Morita bimodule $_RW_R$, then $_SU_R = \mathrm{Hom}_R(_RV_S, W)$ is a cotilting bimodule whenever $_RV$ is a tilting module with $S = \mathrm{End}(_RV)$, and if R is not artinian, then $_SU_R$ is not finitely generated on either side, so it does not induce a finitistic GMD.

Subsequently, in [14] Colpi proved

Proposition 2.7. *Let $\mathcal{C}_R$ and $_S\mathcal{C}$ as in Theorem 2.4. Then $\mathcal{C}_R$ (respectively, $_S\mathcal{C}$) contains (all the submodules of) the finitely generated modules if and only if $_SU$ (respectively, U_R) is U-torsionless linearly compact.*

Corollary 2.8. *Let $_SU_R$ be a cotilting bimodule and assume that R is right coherent and S is left coherent. Then $_SU_R$ is U-torsionless linearly compact on both sides and R_R and $_SS$ are linearly compact. If, moreover, $_SU_R$ is finitely generated on both sides, then $_SU_R$ induces a cotilting theorem between the abelian subcategories of $\mathrm{Mod}-R$ and $S-\mathrm{Mod}$ consisting of all the submodules of finitely generated modules.*

In [2] Angeleri Hügel considered a kind of cross between a finitistic cotilting module and a cotilting module that she called a *finitely cotilting module*, and a somewhat more general type of module that she called a *Colby module*. She proved a slightly different version of a cotilting theorem for faithfully balanced bimodules of these types, and also proved that a finitely generated module over an artin algebra that is a cotilting module, a finitely cotilting module or a Colby module is the artin algebra dual of a tilting module.

A faithfully balanced bimodule $_SU_R$ that satisfies conditions *1* and *2* of Proposition 2.2 is called a *partial cotilting bimodule*. A. Tonolo proved another cotilting theorem in [32].

Theorem 2.9. *Let $_SU_R$ be a partial cotilting bimodule. Denote by δ^0 and δ^1 respectively the 0-th and the 1-st left derived derived maps of the evaluation map δ. Set $\mathcal{M}_R = \{M \in \mathrm{Mod}-R \mid \delta^0_M \text{ and } \delta^1_M \text{ are isomorphisms}\}$. The class $_S\mathcal{M}$ is defined similarly. Then $\mathcal{M}_R$ and $_S\mathcal{M}$ are abelian subcategories of $\mathrm{Mod}-R$ and $S-\mathrm{Mod}$ respectively, and $_SU_R$ induces a cotilting theorem between them.*

Considering a special kind of cotilting modules. Mantese [27] defined a *hereditary cotilting module* U_R as one such that the torsion theory $(\operatorname{Ker}\Delta, \operatorname{Ker}\Gamma)$ in $\operatorname{Mod}-R$ is hereditary. She characterized them as follows

Proposition 2.10. *Let U_R be a cotilting module. The following are equivalent:*

a) U_R *is a hereditary cotilting module;*

b) $\operatorname{Ext}^1_R(E(U), U) = 0;$

c) Δ^2 *preserves monomorphisms in* $\operatorname{Mod}-R$.

Also there she proved

Theorem 2.11. *Let $_SU_R$ be a cotilting bimodule, hereditary on the right. Then:*

a) $\Delta_S\Gamma_R = 0;$

b) *The class of U_R-reflexive modules is closed under submodules;*

c) *The $_SU$-reflexive modules are precisely the $U-$torsionless linearly compact modules;*

d) *The functor Γ_R is naturally isomorphic to $\operatorname{Hom}_R(\operatorname{Rej}_U(-), E(U)/U).$*

Moreover, $_SU_R$ is hereditary on the left iff $\Delta_R\Gamma_S(U) = 0$ iff $\Gamma_S(U)$ is finitely generated. In this case:

e) *Any finitely cogenerated U-torsionless module is U-reflexive.*

One should note the connection between part (d) of this theorem and the result of Matlis described near the beginning of this chapter.

3 Weak Morita Duality

As discussed earlier, a Morita duality is a duality between full subcategories $\mathcal{C}_R$ of $\operatorname{Mod}-R$ and $_S\mathcal{C}$ of $S-\operatorname{Mod}$ that contain the regular modules R_R and $_SS$ and are closed under submodules, epimorphic images and extensions. Extending the notions of both generalized Morita dualities and finitistic generalized Morita dualities, Colby and

Fuller [13] made the following

Definition 3.1. A *weak Morita duality (WMD)* is a duality between full subcategories $\mathcal{F}_R$ of Mod $-R$ and $_S\mathcal{F}$ of $S-$Mod that contain the regular modules R_R and $_SS$ and are closed under submodules and extensions.

Letting $\mathcal{C}_R$ and $_S\mathcal{C}$ denote the categories of epimorphic images of modules in $\mathcal{F}_R$ and $_S\mathcal{F}$, respectively, they proved the following results regarding weak Morita duality between $\mathcal{F}_R$ and $_S\mathcal{F}$ in [13].

Proposition 3.2. *A weak Morita duality between $\mathcal{F}_R$ and $_S\mathcal{F}$ is induced by a faithfully balanced bimodule $_SU_R$ that satisfies*

1. $\operatorname{Ext}^1_R(U, U) = 0$ *and* $\operatorname{Ext}^1_S(U, U) = 0$;

2. $\operatorname{inj.dim}.(U_R) \leqslant 1$ *and* $\operatorname{inj.dim}.(_SU) \leqslant 1$;

3. $\operatorname{Ker}\Delta \cap \operatorname{Ker}\Gamma \cap \mathcal{C}_R = 0$ *and* $\operatorname{Ker}\Delta \cap \operatorname{Ker}\Gamma \cap {_S\mathcal{C}} = 0$.

Theorem 3.3. *If $_SU_R$ induces a weak Morita duality between $\mathcal{F}_R$ and $_S\mathcal{F}$, then $_SU_R$ induces a cotilting theorem between $\mathcal{C}_R$ and $_S\mathcal{C}$ with torsion free classes $\mathcal{F}_R$ and $_S\mathcal{F}$.*

Also in [13] Colby and Fuller employed the notion of U-torsionless linear compactness to obtain information regarding weak and generalized Morita duality. They proved that if $_SU_R$ induces a weak Morita duality between $\mathcal{F}_R$ and $_S\mathcal{F}$, then the modules in $\mathcal{F}_R$ and $_S\mathcal{F}$ are all U-torsionless linearly compact, and obtained a characterization of WMD's. Also they showed that over an artin algebra, a finitely generated bimodule induces a WMD if and only if it is the artin algebra dual of a tilting module. Then they obtained the following characterization of generalized Morita duality.

Theorem 3.4. *Let $_SU_R$ be a bimodule and let $\mathcal{F}_R$ and $_S\mathcal{F}$ denote the classes of all U-reflexive modules in Mod-R and S-Mod, and let $\mathcal{C}_R$ and $_S\mathcal{C}$ denote the categories of epimorphic images of modules in $\mathcal{F}_R$ and $_S\mathcal{F}$, respectively. Then U induces a generalized Morita duality if and only if*

1. $_SU_R$ *is faithfully balanced,*

2. $\Gamma(M) = 0$ *for all M in $\mathcal{F}_R$ and $\Gamma(N) = 0$ for all N in $_S\mathcal{F}$,*

3. $\operatorname{Ker}\Delta \cap \operatorname{Ker}\Gamma \cap \mathcal{C}_R = 0$ *and* $\operatorname{Ker}\Delta \cap \operatorname{Ker}\Gamma \cap {_S\mathcal{C}} = 0$,

4. $\mathcal{F}_R$ and $_S\mathcal{F}$ consist precisely of the U-torsionless linearly compact modules.

The bimodule $_\mathbb{Z}\mathbb{Z}_\mathbb{Z}$ induces a weak Morita (self) duality between finitely generated free abelian groups (i.e., a finitistic generalized Morita duality), but $\mathbb{Z}$ does not define a generalized Morita duality, nor is it a cotilting bimodule. (See [11, Example 5.8.1]). An interesting problem (in the spirit of Morita duality) would be to determine the U-reflexive modules when $_SU_R$ induces a WMD.

G. D'Este [18] has provided an example of a perfect coherent ring R such that $_RR_R$ is a cotilting bimodule that does not induce a WMD.

4 Pure Injectivity of Cotilting Modules and Reflexivity

A short exact sequence of right R-modules $0 \to K \xrightarrow{f} M \xrightarrow{g} L \to 0$ is *pure exact* if every R-map $X \to L$ with X finitely presented factors through g. Equivalently, exactness of the sequence is preserved by $(\_\otimes_R N)$ for any left R-module N. See [25], for example, for this notion and the following definition.

Definition 4.1. A module U_R is *pure injective (or algebraically compact)* if $\mathrm{Hom}_R(\_,U)$ preserves the exactness of every pure exact sequence.

According to [23, Theorem 9], the endomorphism ring S of a pure injective module is *semiregular* in the sense that $S/J(S)$ is von Neuman regular and idempotents lift modulo $J(S)$.

In [28] Mantese, P. Ruzicka and Tonolo observed that if U_R is a pure injective module, then $\mathrm{Cogen}(U_R)$ is closed under direct limits, and they pointed out that, at the time, all known examples of cotilting modules were pure injective. The following generalizations of linear compactness and reflexivity appeared in [20].

Definition 4.2. A module M is U-linearly compact if $\varprojlim p_\lambda : M \to \varprojlim M_\lambda$ is an epimorphism whenever $p_\lambda : M \to M_\lambda$ is an inverse system of epimorphisms with each $M_\lambda \in \mathrm{Cogen}(U)$.

Definition 4.3. A module M is U-dense relative to a bimodule $_SU_R$ if for each $\alpha \in \Delta^2 M$ and each $f_1, ..., f_n \in \Delta M$, there is an $m \in M$ such that $\alpha(f_i) = f_i(m)$, for $i = 1, ..., n$.

These notions were employed in [28] to prove

Proposition 4.4. *If $_SU_R$ is a cotilting bimodule that is pure injective on both sides, then a U-torsionless module M is U-reflexive if and only if it is both U-linearly compact and U-dense.*

Fortunately, S. Bazzoni [5] has subsequently proved the following theorem. Thus "pure injective on both sides" can be dropped from this characterization of U-reflexivity relative to a cotilting bimodule.

Theorem 4.5. *Every cotilting module is pure injective.*

As an application of Bazzoni's Theorem 4.5 Colpi [14] obtained the following extension of an early theorem of Osofsky [1, Lemma 24.7] on Morita duality.

Proposition 4.6. *If $_SU_R$ is a cotilting bimodule, then no infinite direct sum of non-zero U-reflexive modules is U-reflexive.*

Also in [14] he employed her result to obtain the following further connection between U-reflexivity and different notions of linear compactness.

Proposition 4.7. *Let $_SU_R$ be a cotilting bimodule and assume that R_R and $_SS$ are $U-$torsionless linearly compact. Let M be a U-torsionless module. Then*

1. *M is U-reflexive if and only if it is U-linearly compact;*

2. *every submodule of M is U-reflexive iff M is linearly compact iff ΔM is U-torsionless linearly compact.*

References

[1] F. W. Anderson and K. R. Fuller. *Rings and Categories of Modules.* Springer–Verlag, Inc., New York, Heidelberg, Berlin, second edition, 1992.

[2] L. Angeleri Hügel. Finitely cotilting modules. *Comm. Algebra,* 28:2147–2172, 2000.

[3] L. Angeleri Hügel, A. Tonolo, and J. Trlifaj. Tilting preenvelopes and cotilting precovers. *Algebras and Representation Theory,* 4:155–201, 2002.

[4] G. Azumaya. A duality theory for injective modules. *Amer. J. Math.,* 81:249–278, 1959.

[5] S. Bazzoni. Cotilting modules are pure injective. *Proc. Amer. Math. Soc.*, 131:3665–3672, 2003.

[6] K. Bongartz. Tilted algebras. *Springer–Verlag LNM*, 903:26–38, 1981.

[7] S. Brenner and M. C. R. Butler. Generalizations of the Bernstein–Gelfand–Ponomarev reflection functors. *Springer–Verlag LNM*, 832:103–170, 1980.

[8] R. R. Colby. A generalization of Morita duality and the tilting theorem. *Comm. Alg.*, 17(7):1709–1722, 1989.

[9] R. R. Colby. A cotilting theorem for rings. In *Methods in Module Theory*, pages 33–37. M. Dekker, New York, 1993.

[10] R. R. Colby, R. Colpi, and K. R. Fuller. A note on cotilting modules and generalized Morita duality. In *Rings, Modules, Algebras and Abelian Groups*, 85-88, Lecture Notes in Pure and Appl. Math., 236, Dekker, New York, 2004.

[11] R. R. Colby and K. R. Fuller. *Equivalence and Duality for Module Categories (with tilting and cotilting for rings)*. Cambridge Tracts in Mathematics 161, Cambridge University Press, 2004.

[12] R. R. Colby and K. R. Fuller. Tilting, cotilting, and serially tilted rings. *Comm. Algebra*, 18:1585–1615, 1990.

[13] R. R. Colby and K. R. Fuller. Weak Morita duality. *Comm. Algebra*, 31:1859–1879, 2003.

[14] R. Colpi. Dualities induced by cotilting modules. In *Rings, Modules, Algebras and Abelian Groups*, 89-102, Lecture Notes in Pure and Appl. Math., 236, Dekker, New York, 2004.

[15] R. Colpi. Cotilting bimodules and their dualities. Interactions between ring theory and representations of algebras (Murcia). In *Lecture Notes in Pure and Appl. Math.*, volume 210, pages 81–93. Marcel Dekker, New York, 2000.

[16] R. Colpi, G. D'Este, and A. Tonolo. Quasi–tilting modules and counter equivalences. *J. Algebra*, 191:461–494, 1997.

[17] R. Colpi and K. R. Fuller. Cotilting modules and bimodules. *Pacific J. Math.*, 192:275–291, 2000.

[18] G. D'Este. Reflexive modules are not closed under submodules. In *Lecture Notes in Pure and Appl. Math.*, volume 224, pages 53–64. Marcel Dekker, New York, 2002.

[19] J. Dieudonne. Remarks on quasi-Frobenius rings. *Illinois J. Math.*, 2:346–354, 1958.

[20] J. L. Gómez Pardo. Counterinjective modules and duality. *J. Pure and Appl. Algebra*, 61:165–179, 1989.

[21] J. L. Gómez Pardo, P. A. Guil Asensio, and R. Wisbauer. Morita dualities induced by the M–dual functors. *Comm. Algebra*, 22:5903–5934, 1994.

[22] D. Happel and C. M. Ringel. Tilted algebras. *Trans. Amer. Math. Soc.*, 274:399–443, 1982.

[23] B. Zimmermann Huisgen and W. Zimmermann. Algebraically compact rings and modules. *Math. Z.*, 161:81–93, 1978.

[24] J. P. Jans. Duality in Noetherian rings. *Proc. Amer. Math. Soc.*, 12:829–835, 1961.

[25] C. U. Jensen and H. Lenzing. *Model Theoretic Algebra*, volume 2. Gordon and Breach Science Publishers, New York, 1989.

[26] F. Mantese. Generalizing cotilting dualities. *J. Algebra*, 236:630–644,

2001.

[27] F. Mantese. Hereditary cotilting modules. *J. Algebra*, 238:462–478, 2001.

[28] F. Mantese, P. Ruzicka, and A. Tonolo. Cotilting versus pure-injective modules. *Pacific J. Math.*, 212:321–332, 2003.

[29] E. Matlis. Reflexive domains. *J. Algebra*, 8:1–33, 1967.

[30] Y. Miyashita. Tilting modules of finite projective dimension. *Math. Z.*, 193:113–146, 1986.

[31] K. Morita. Duality for modules and its applications to the theory of rings with minimum condition. *Sci. Rep. Tokyo Kyoiku Daigaku Ser. A*, 6:83–142, 1958.

[32] A. Tonolo. Generalizing Morita duality: a homological approach. *J. Algebra*, 232:282–298, 2000.

[33] W. Xue. *Rings with Morita Duality*. Springer–Verlag LNM vol. 1523, Berlin, Heidelberg, New York, 1992.

Riccardo Colpi
Department of Mathematics
University of Padua
Via Belzoni 7, I-35131 Padova, Italy
colpi@math.unipd.it

Kent Fuller
Department of Mathematics
University of Iowa
Iowa City, IA 52242-1419, USA
kfuller@math.uiowa.edu

14

Representations of finite groups and tilting

Joseph Chuang and Jeremy Rickard

1 A brief introduction to modular representation theory

Let G be a finite group and k, for simplicity, be an algebraically closed field. Then to give a representation of G over a field k is equivalent to giving a module for the group algebra kG. In general, the group algebra is a direct product

$$kG \cong b_0(kG) \times \cdots \times b_n(kG)$$

of connected algebras (i.e., algebras that are not isomorphic to the direct product of smaller algebras), called *blocks*. If the characteristic of k does not divide the order of G, then these blocks are all full matrix algebras over k, but otherwise the blocks may be much more complicated algebras. We will always take $b_0(kG)$ to be the *principal block* (i.e., the unique block that does not annihilate the trivial module k). Many statements about general blocks become less technical for the principal block, so we will often focus on this case for the sake of simplicity.

Recall that to each block $b_i(kG)$ there is associated a p-subgroup D of G, well-defined up to conjugacy, called a *defect group* of the block. This may be defined as a minimal subgroup D of G such that every module for the block is a direct summand of a module induced from D, although there are many other equivalent definitions. In a sense, the defect group controls how complicated the representation theory of the block is. For example, the defect group is trivial (the "defect zero" case) if and only if the block is a simple algebra, and the structure of blocks with cyclic defect group is extremely well understood. The defect groups of the principal block are the Sylow p-subgroups of G, and so are at least

359

as large as the defect groups of any other block. Correspondingly, the principal block typically has a structure at least as complicated as any other block.

In the case that k has characteristic p dividing $|G|$, a theme of modular representation theory since its origins in the work of Brauer has been that the representation theory of G over k is intimately related to that of "p-local" subgroups of G, by which we mean normalizers $N_G(P)$ of non-trivial p-subgroups P of G. Typically, especially if G is a large simple group, the p-local subgroups are far smaller than G, and so to reduce questions about the representations of G to questions about these subgroups is a great advance.

One classical theorem along these lines, relating the blocks of kG to blocks of p-local subgroups is "Brauer correspondence", which, for a fixed p-subgroup D of G, is a natural bijection between the blocks of kG with defect group D and the blocks of $kN_G(D)$ with defect group D. The Brauer correspondent of the principal block of kG is the principal block of $kN_G(D)$, where D is a Sylow p-subgroup of G.

A more recent example of this theme, this time of a still open conjecture, is Alperin's Weight Conjecture [2], which predicts that the number of non-projective simple modules for kG (or the number of simple modules, up to isomorphism, for a non-simple block of kG) should be determined by p-local information. In the case of a block with abelian defect group, this reduces to a very simple statement, although even this special case is still open.

Conjecture 1.1 (Alperin). *Let B be a block of kG with abelian defect group D, and let b be the Brauer correspondent block of $kN_G(D)$. Then B and b have the same number of simple modules.*

2 The abelian defect group conjecture

The simple statement of Alperin's Weight Conjecture for blocks with abelian defect group invites the question of whether there is some structural connection between the blocks B and b that explains why the number of simple modules should be the same. This was given in a slightly later conjecture of Broué [6], called the Abelian Defect Group

Conjecture. In its simplest form, it states:

Conjecture 2.1 (Broué). *Let B be a block of kG with abelian defect group D, and let b be the Brauer correspondent block of $kN_G(D)$. Then the derived categories of B and b are equivalent as triangulated categories.*

Since the Grothendieck group $K_0(B)$ is an invariant of the derived category, as explained in Keller's article in this volume, and the rank of $K_0(B)$ is the number of simple modules of B, this conjecture implies Alperin's Weight Conjecture for blocks with abelian defect group, and also implies that the blocks B and b should share many other invariants. It should be stressed that the blocks B and b are rarely Morita equivalent, and the smaller block b typically has a significantly simpler structure than B, so the derived equivalence predicted by the conjecture gives a lot of useful information about B.

Later in this article we shall see some refinements of the Abelian Defect Group Conjecture.

3 Symmetric algebras

Because the category of representations of a finite group G over a field k is equivalent to the category of modules for a ring kG, the group algebra of G, the general Morita theory for derived categories of module categories, described in Keller's article in this volume, applies to this situation.

However, there are some aspects of the general theory that are simplified because the group algebra is a symmetric algebra. The traditional definition is as follows.

Definition 3.1. A finite-dimensional algebra A over a field k is *symmetric* if there is a linear map $\alpha \colon A \to k$ such that

$$\alpha(xy) = \alpha(yx)$$

for all $x, y \in A$, and such that the kernel of α contains no non-zero left or right ideal of A.

A linear map α satisfying these conditions is called a *symmetrizing form*.

For example, the group algebra kG of a finite group G, or any block of the group algebra, has a symmetrizing form α given by

$$\alpha\left(\sum_{g \in G} \lambda_g g\right) = \lambda_1,$$

the coefficient of the identity element of G.

There are various equivalent forms of the definition, summarized in the following proposition. Most of these are well-known, but a complete proof that they are all equivalent can be found in [106].

Proposition 3.2. *Let A be a finite-dimensional algebra over a field k. The following conditions are equivalent.*

(a) *A is symmetric.*

(b) *A and its dual $A^\vee = \mathrm{Hom}_k(A, k)$ are isomorphic as A-bimodules.*

(c) *$\mathrm{Hom}_k(?, k)$ and $\mathrm{Hom}_A(?, A)$ are isomorphic as functors from the category of right A-modules to the category of left A-modules.*

(c') *$\mathrm{Hom}_k(?, k)$ and $\mathrm{Hom}_A(?, A)$ are isomorphic as functors from the category of left A-modules to the category of right A-modules.*

(d) *For finitely generated projective right A-modules P and finitely generated right A-modules M, there is an isomorphism of k-vector spaces*

$$\mathrm{Hom}_A(P, M) \cong \mathrm{Hom}_A(M, P)^\vee,$$

functorial in both P and M.

(d') *For finitely generated projective left A-modules P and finitely generated left A-modules M, there is an isomorphism of k-vector spaces*

$$\mathrm{Hom}_A(P, M) \cong \mathrm{Hom}_A(M, P)^\vee,$$

functorial in both P and M.

(e) *For objects P and X of the derived category $D(A)$, where P is a perfect complex and X is isomorphic to a bounded complex of finitely generated right A-modules, there is an isomorphism of k-vector spaces*

$$\mathrm{Hom}_{D(A)}(P, X) \cong \mathrm{Hom}_{D(A)}(X, P)^\vee,$$

functorial in both P and X.

(e') *For objects P and X of the derived category $D(A^{op})$, where P is a perfect complex and X is isomorphic to a bounded complex of finitely generated left A-modules, there is an isomorphism of k-vector spaces*

$$\mathrm{Hom}_{D(A^{op})}(P, X) \cong \mathrm{Hom}_{D(A^{op})}(X, P)^{\vee},$$

functorial in both P and X.

Note that condition (d) in the proposition makes it transparent that the property of being symmetric depends only on the module category, so that an algebra Morita equivalent to a symmetric algebra is itself symmetric. Condition (e) does the same for the derived category, so that an algebra derived equivalent to a symmetric algebra is itself symmetric, a fact first proved by different means in [100].

For general algebras A and B over a field, it was shown in [100] that if A and B are derived equivalent, then there is a *two-sided tilting complex*, a bounded complex X of A-B-bimodules, finitely generated and projective as left A-modules and as right B-modules, such that

$$? \otimes_A X : D(A) \to D(B)$$

is an equivalence of derived categories, and that the quasi-inverse equivalence is induced in a similar way by the complex

$$Y \cong \mathrm{Hom}_A(X, A) \cong \mathrm{Hom}_B(X, B)$$

of B-A-bimodules. For general algebras A and B, the functors $\mathrm{Hom}_A(?, A)$ and $\mathrm{Hom}_B(?, B)$ are not isomorphic, so if X is not a two-sided tilting complex, there is no reason to expect that $\mathrm{Hom}_A(X, A)$ and $\mathrm{Hom}_B(X, B)$ should be isomorphic. However, if A and B are symmetric, then conditions (c) and (c') of Proposition 3.2 show that both $\mathrm{Hom}_A(X, A)$ and $\mathrm{Hom}_B(X, B)$ are isomorphic to the vector space dual $X^{\vee}$ of X. Hence the following proposition.

Proposition 3.3. *Let A and B be derived equivalent finite-dimensional symmetric algebras over a field k. Then there is a bounded complex X of A-B-bimodules, finitely generated and projective as left A-modules and as right B-modules, such that*

$$? \otimes_A X : D(A) \to D(B)$$

is an equivalence with quasi-inverse

$$? \otimes_B X^\vee : D(B) \to D(A),$$

In fact, for any complex X of A-B-bimodules, finitely generated and projective as left A-modules and as right B-modules, the functor

$$? \otimes X^\vee : D(B) \to D(A)$$

is both left and right adjoint to the functor

$$? \otimes_A X : D(A) \to D(B),$$

and the units and counits of these adjunctions are induced by natural maps of complexes of A-bimodules

$$X \otimes_B X^\vee \to A$$

and

$$A \to X \otimes_B X^\vee$$

and natural maps of complexes of B-bimodules

$$X^\vee \otimes_A X \to B$$

and

$$B \to X^\vee \otimes_A X,$$

and the condition that X is a two-sided tilting complex is just the condition that these maps are all isomorphisms in the derived categories of A-bimodules and B-bimodules. In fact, since the composition

$$A \to X \otimes_B X^\vee \to A$$

is then an isomorphism in the derived category, it is an isomorphism of bimodules, and so in the category of complexes of bimodules

$$X \otimes_B X^\vee \cong A \oplus Z$$

for some acyclic complex Z. Similarly

$$X^\vee \otimes_A X \cong B \oplus Z'$$

for some acyclic complex Z' of B-bimodules.

The previous proposition perhaps explains why tilting complexes for symmetric algebras seem so numerous. Here is one general construction of such a tilting complex that has been used many times, for example by Okuyama [86].

Proposition 3.4. *Let A be a symmetric finite-dimensional algebra, with indecomposable projective modules $\{P_i : i \in I\}$. For any subset $J \subseteq I$ there is a tilting complex*

$$T = \bigoplus_{i \in I} T_i$$

for A, where

$$T_j = \cdots \to 0 \to Q_j \to P_j \to 0 \to \cdots$$

for $j \in J$, where Q_j is the projective cover of the largest submodule K of P_j such that $\operatorname{Hom}_A(P_i, P_j/K) = 0$ for all $i \in I - J$, and

$$T_i = \cdots \to 0 \to P_i \to 0 \to 0 \to \cdots$$

for $i \in I - J$.

Of course, even if A is a block of a group algebra, the endomorphism algebra of this tilting complex will often not be. However, to the best of our knowledge it is still an open question to decide whether every derived equivalence between blocks of finite groups can be obtained (up to a shift) by a composition of "tilts", or inverses of tilts, of this kind.

Here is one other recent theorem about derived equivalence from [106] that is specific to symmetric algebras.

Theorem 3.5. *Let A be a finite-dimensional symmetric algebra over a field k, and let $X_1, \ldots, X_r$ be objects of the bounded derived category $D^b(\operatorname{mod}(A))$ of finitely generated A-modules such that*

- *$\operatorname{Hom}(X_i, X_j[t]) = 0$ for all $1 \leqslant i, j \leqslant r$ and all $t < 0$.*

- *$\operatorname{Hom}(X_i, X_j) = 0$ if $i \neq j$, and $\operatorname{End}(X_i)$ is a division ring for all i.*

- *$X_1, \ldots, X_r$ generate $D^b(\operatorname{mod}(A))$ as a triangulated category.*

Then there is another (symmetric) algebra B and an equivalence

$$D(A) \to D(B)$$

of derived categories taking the objects $X_1, \ldots, X_r$ to the simple B-modules.

Notice that the conditions are obviously satisfied if $X_1, \ldots, X_r$ are the simple A-modules, even if A is not symmetric. However, there are easy examples to show that the theorem is not in general true for non-symmetric algebras. For example, if A is the algebra of upper-triangular 2×2 matrices over k, and P is the two-dimensional projective module, and S its simple quotient, then

$$X_1 = S, X_2 = P[i]$$

satisfy the conditions of the theorem for any $i > 0$, but if $i > 1$ there can be no equivalence

$$D(A) \to D(B)$$

of derived categories taking the objects X_1, X_2 to the simple B-modules S_1, S_2, since $\mathrm{Hom}(X_i, X_j[1]) = 0$ for $i, j \in \{1, 2\}$ but $\mathrm{Hom}(X_2, X_1[i]) \neq 0$, so that we would have $\mathrm{Ext}^1(S_i, S_j) = 0$ for all $i, j \in \{1, 2\}$, implying that B must be semisimple, whereas $\mathrm{Ext}^i(S_2, S_1[i]) \neq 0$. Similar simple examples show that the theorem is not even true for self-injective algebras [1].

This theorem has been used, in conjunction with "Linckelmann's Theorem" (Theorem 7.1 below), to prove several examples of derived equivalence between blocks of group algebras, in [13] and [31], for example.

4 Characters and derived equivalence

Much explicit calculation related to the representation theory of groups, even modular representation theory, involves characters, which are defined in terms of "ordinary" representations over a field of characteristic zero. There is a standard framework to relate the ordinary and modular representation theory.

We will let $\mathcal{O}$ be a complete discrete valuation ring with an algebraically closed quotient field k of characteristic $p > 0$, and with a field of fractions of characteristic zero with algebraic closure K. The natural surjection $\mathcal{O} \to k$ and the inclusion $\mathcal{O} \to K$ give functors

$$? \otimes_{\mathcal{O}} k \colon \mathrm{mod}(\mathcal{O}G) \to \mathrm{mod}(kG)$$

and

$$? \otimes_{\mathcal{O}} K \colon \mathrm{mod}(\mathcal{O}G) \to \mathrm{mod}(KG)$$

relating the representation theory of a group G over $\mathcal{O}$ to the representation theory over the fields k and K. Thus the ring $\mathcal{O}$ acts as a "bridge" between the ordinary representation theory (and character theory) of a group and the modular representation theory.

The completeness of $\mathcal{O}$ allows us to lift idempotents, and so the indecomposable projective kG-modules lift to indecomposable projective $\mathcal{O}G$-modules, and the block decomposition

$$kG \cong b_0(kG) \times \cdots \times b_n(kG)$$

of kG lifts to a block decomposition

$$\mathcal{O}G \cong b_0(\mathcal{O}G) \times \cdots \times b_n(\mathcal{O}G)$$

of $\mathcal{O}G$, so that to give a block of kG is equivalent to giving a block of $\mathcal{O}G$. A refinement of Broué's Abelian Defect Group Conjecture is then:

Conjecture 4.1 (Broué). *Let B be a block of $\mathcal{O}G$ with abelian defect group D, and let b be the Brauer correspondent block of $\mathcal{O}N_G(D)$. Then the derived categories of B and b are equivalent as triangulated categories.*

This is a refinement, in the sense that it implies the previous version, because if A and B are derived equivalent $\mathcal{O}$-algebras, free as $\mathcal{O}$-modules, and if X is a two-sided tilting complex for A and B, then it is easy to see that $X \otimes_{\mathcal{O}} k$ is a two-sided tilting complex for $A \otimes_{\mathcal{O}} k$ and $B \otimes_{\mathcal{O}} k$, which are therefore also derived equivalent to one another.

The converse is not true. If A and B are $\mathcal{O}$-algebras, free as $\mathcal{O}$-modules, and if $A \otimes_{\mathcal{O}} k$ and $B \otimes_{\mathcal{O}} k$ are derived equivalent, then it is not necessarily true that A and B are derived equivalent. For example, $\mathcal{O}C_p$ and $\mathcal{O}[x]/(x^p)$ are not derived equivalent (being nonisomorphic local $\mathcal{O}$-algebras), even though kC_p and $k[x]/(x^p)$ are isomorphic.

However, we shall see in Section 3 that a further refinement of the Abelian Defect Group Conjecture over k does imply the conjecture over $\mathcal{O}$.

A two-sided tilting complex X for a pair A and B of derived equivalent $\mathcal{O}$-algebras similarly gives rise to a two-sided tilting complex $X \otimes_{\mathcal{O}} K$ for $A \otimes_{\mathcal{O}} K$ and $B \otimes_{\mathcal{O}} K$, which are therefore also derived equivalent. If A and B are blocks of finite groups over $\mathcal{O}$, then these K-algebras are semisimple, with one simple module for each irreducible character of the block. But a derived equivalence for such algebras takes a very simple form: if $U_1, \ldots, U_n$ and $V_1, \ldots, V_n$ are representatives of the isomorphism classes of simple modules for $A \otimes_{\mathcal{O}} K$ and $B \otimes_{\mathcal{O}} K$ respectively, then for some choice of integers t_i and for some permutation σ of $\{1, \ldots, n\}$, the derived equivalence just maps

$$U_i \mapsto V_{\sigma(i)}[t_i],$$

and so the induced map of Grothendieck groups sends

$$[U_i] \mapsto (-1)^{t_i} [V_{\sigma(i)}].$$

Since the Grothendieck group can be identified with the group of virtual characters of the block (i.e., the group of class functions generated by the characters), this is just a "bijection with signs" between the sets of irreducible characters of the two blocks. In other words, if $\mathrm{Ch}(A)$ denotes the group of virtual characters of A, then a derived equivalence between blocks A and B over $\mathcal{O}$ induces an isometry (with respect to the usual inner product of characters)

$$\mathrm{Ch}(A) \to \mathrm{Ch}(B).$$

But more is true. Since, as described in Keller's article in this volume, a derived equivalence restricts to an equivalence between the full subcategories of perfect complexes (bounded complexes of finitely generated projective modules), this isometry restricts to an isomorphism

$$\mathrm{Ch}_{pr}(A) \to \mathrm{Ch}_{pr}(B),$$

where $\mathrm{Ch}_{pr}(A)$ denotes the subgroup of $\mathrm{Ch}(A)$ generated by the characters of projective A-modules, motivating the following definition.

Definition 4.2. A *perfect isometry* between blocks A and B of group algebras over $\mathcal{O}$ is an isometry

$$\mathrm{Ch}(A) \to \mathrm{Ch}(B)$$

that restricts to an isomorphism

$$\mathrm{Ch}_{pr}(A) \to \mathrm{Ch}_{pr}(B).$$

Since the virtual characters that are in $\mathrm{Ch}_{pr}(A)$ can be characterized as those that vanish on p-singular elements of G (i.e., elements whose order is divisible by p), those isometries that are perfect can be described in terms of character values. This is how Broué's original definition [6] was formulated.

The existence of a perfect isometry between two blocks gives some evidence for the existence of a derived equivalence. Stronger evidence is given by a refinement of the notion of a perfect isometry introduced in the same paper of Broué [6]: that of an isotypy. For simplicity we will state a simplified version of the definition that applies to the situation of the Abelian Defect Group Conjecture for principal blocks; the general definition may be found in Broué's paper.

In order to formulate the definition, we note that, given two finite groups G and H, giving a linear map

$$I \colon \mathrm{Ch}(b_0(\mathcal{O}G)) \longrightarrow \mathrm{Ch}(b_0(\mathcal{O}H))$$

is equivalent to giving a class function $\mu_I \in \mathrm{Ch}(\mathcal{O}(G \times H))$, where

$$\mu_I = \sum_{\zeta \in \mathrm{Irr}(b_0(\mathcal{O}G))} \zeta \otimes I(\zeta),$$

$\mathrm{Irr}(b_0(\mathcal{O}G))$ denoting the set of irreducible characters of G belonging to the principal block.

Definition 4.3. Let G and H be finite groups with a common abelian Sylow p-subgroup P. An **isotypy** between the principal blocks of $\mathcal{O}G$ and $\mathcal{O}H$ is a collection of perfect isometries

$$I(Q) \colon \mathrm{Ch}(b_0(C_G(Q))) \longrightarrow \mathrm{Ch}(b_0(C_H(Q))),$$

one for each subgroup $Q \leqslant P$, with the following compatibility condition. Whenever $Q \leqslant P$, $x, y \in P$, $x' \in C_G(x)$ and $y' \in C_H(y)$, where x' and y' are both p-regular elements (i.e., elements whose order is not divisible by p),

$$\mu_{I(Q)}(xx', yy') = \begin{cases} 0 & \text{if } x \text{ and } y \text{ are not conjugate in } G, \\ \mu_{I(Q\langle x \rangle)}(x', y') & \text{if } x = y. \end{cases}$$

Although this definition looks a little complicated, it turns out to be surprisingly easy to work with, and it is often a lot easier to verify the existence of an isotypy than of a more general perfect isometry. Many examples are described in the appendices of Broué's paper [6] and in Rouquier's thesis [111].

5 Splendid equivalences

Given that a perfect isometry is the evidence, at the level of characters, of a derived equivalence, the definition of an isotypy at the end of the last section (a "compatible family" of perfect isometries) raises the question of what the corresponding definition of a "compatible family" of derived equivalences is. This is provided by the notion of a splendid equivalence. We shall for simplicity deal with the case of principal blocks as in the original definition in [103], but note that generalizations to non-principal blocks can be found in the work of Harris, Puig and Linckelmann [25],[28],[94].

We start by introducing the Brauer construction for modules. Let k be a field of characteristic p, and let Q be a subgroup of a finite group G. Then if $M = k[\Omega]$ is a permutation module for G (i.e., M has a basis Ω that is permuted by G), we can form the subspace $k[\Omega^G]$ spanned by the fixed points of the action of G on Ω. However, this construction is not in general functorial, and in fact $k[\Omega^G]$ depends in general, even up to isomorphism, on the choice of permutation basis Ω. If Q is a p-group, however, we can give an alternative description of $k[\Omega^G]$ that makes it clear that in this case it does depend functorially on $k[\Omega]$.

In fact, this functor extends to the whole of the module category to give a functor

$$\mathrm{Br}_Q \colon \mathrm{mod}(kG) \to \mathrm{mod}(kN_G(Q)),$$

the *Brauer construction* with respect to the subgroup Q, where, for a kG-module M,

$$\mathrm{Br}_Q(M) = M^Q / \big(\sum_{R<Q} \mathrm{Tr}_R^Q(Q^R) \big),$$

the quotient of the Q-fixed points of M by the relative traces from all proper subgroups R of Q of the R-fixed points.

Now if G and H are finite groups, then a kG-kH-bimodule M may be regarded as a $k[G \times H]$-module with the action

$$m(g,h) = g^{-1}mh,$$

and if Q is a p-subgroup of both G and H we can apply the Brauer construction with respect to the subgroup

$$\Delta Q = \{(q,q) : q \in Q\} \leqslant G \times H.$$

Since the normalizer $N_{G \times H}(\Delta Q)$ contains $C_G(Q) \times C_H(Q)$ as a subgroup, this can be regarded as a functor from the category of kG-kH-bimodules to the category of $kC_G(Q)$-$kC_H(Q)$-bimodules. We shall denote the image of a bimodule M under this functor as $M(Q)$. For example, if $G = H$ and $M = kG$, regarded as a kG-bimodule, then $M(Q) = kC_G(Q)$.

The Brauer construction is an additive functor, but it is far from being exact. In fact, it is neither left nor right exact, and if Q is non-trivial then it kills the free module, and hence all projective modules. It follows that it does not induce a functor between derived categories in any sensible fashion. To apply it to derived equivalences, then, we need to strengthen the definition of a two-sided tilting complex.

Recall from Section 3 that if X is a two-sided tilting complex for symmetric algebras A and B, then there are natural maps of complexes of A-bimodules

$$X \otimes_B X^{\vee} \to A$$

and

$$A \to X \otimes_B X^{\vee}$$

and natural maps of complexes of B-bimodules

$$X^{\vee} \otimes_A X \to B$$

and

$$B \to X^{\vee} \otimes_A X$$

that are isomorphisms in the appropriate derived categories of bimodules. If they are even isomorphisms in the chain homotopy categories of bimodules, then we say that X is a *split endomorphism two-sided tilting complex*. A similar argument to that in Section 3 then shows that

$$X \otimes_B X^{\vee} \cong A \oplus Z$$

for some contractible complex Z of A-bimodules, and

$$X^{\vee} \otimes_A X \cong B \oplus Z'$$

for some contractible complex Z' of B-bimodules.

If A and B are group algebras over k of groups G and H with a common

p-subgroup Q, the Brauer construction may be applied to these maps to get isomorphisms

$$(X \otimes_B X^{\vee})(Q) \to kC_G(Q),$$

etc., in the chain homotopy category of bimodules for the centralizer of Q. But to get a good interpretation of $(X \otimes_B X^{\vee})(Q)$ we need extra conditions.

Definition 5.1. Let G and H be finite groups with a common abelian p-subgroup P. A *splendid tilting complex* for the principal blocks of kG and kH is a split endomorphism two-sided tilting complex whose terms are all direct sums of direct summands of bimodules of the form

$$kG \otimes_{kR} kH$$

for subgroups R of P. A derived equivalence induced by such a complex is called a *splendid equivalence*.

Considered as a $k[G \times H]$-module, the bimodule $kG \otimes_{kR} kH$ is a permutation module with permutation basis the set of cosets of ΔR in $G \times H$.

In [103] it is proved that if X is such a splendid tilting complex, then $(X \otimes_{kH} X^{\vee})(Q)$ is isomorphic to $X(Q) \otimes_{kC_H(Q)} X(Q)^{\vee}$ for any subgroup Q of P, which is the key to proving:

Theorem 5.2. *Let X be a splendid tilting complex for the principal blocks of group algebras kG and kH, where G and H are finite groups with a common abelian Sylow p-subgroup P. Then $X(Q)$ is a splendid tilting complex for the principal blocks of $kC_G(Q)$ and $kC_H(Q)$ for any subgroup Q of P.*

Thus a splendid tilting complex gives rise to a family of derived equivalences for centralizers, just as an isotypy is a family of perfect isometries for centralizers. However, to apply the Brauer construction sensibly, we need to work over k, not $\mathcal{O}$, since there is no corresponding functorial construction that takes permutation $\mathcal{O}G$-modules to permutation $\mathcal{O}N_G(Q)$-modules for a p-subgroup Q of G. However, we can extend the definition of a splendid tilting complex in a straightforward way to blocks over $\mathcal{O}$.

Definition 5.3. Let G and H be finite groups with a common abelian p-subgroup P. A *splendid tilting complex* for the principal blocks of $\mathcal{O}G$

and $\mathcal{O}H$ is a split endomorphism two-sided tilting complex whose terms are all direct sums of direct summands of bimodules of the form

$$\mathcal{O}G \otimes_{\mathcal{O}R} \mathcal{O}H$$

for subgroups R of P.

Then the second important property of splendid tilting complexes, proved in [103], is the following lifting theorem.

Theorem 5.4. *Let G and H be finite groups with a common abelian p-subgroup P. If X is splendid tilting complex for the principal blocks of kG and kH, then there is a splendid tilting complex $\tilde{X}$ for the principal blocks of $\mathcal{O}G$ and $\mathcal{O}H$, unique up to isomorphism, such that*

$$X \cong \tilde{X} \otimes_{\mathcal{O}} k.$$

Hence a splendid tilting complex X, defined over k, induces a family of splendid tilting complexes $\widetilde{X(Q)}$ defined over $\mathcal{O}$, and so it does induce a family of perfect isometries, which can be shown to be an isotypy [103].

6 Derived equivalence and stable equivalence

Since a block A of a group algebra kG is a self-injective algebra, the module category $\mathrm{mod}(A)$ is a Frobenius category, and so, as described in Keller's article in this volume, its associated stable category $\underline{\mathrm{mod}}(A)$, called the *stable module category* of A, is a triangulated category. In fact, it is closely related to the derived category, as there is an equivalence of triangulated categories

$$D^b(\mathrm{mod}(A))/\mathrm{Perf}(A) \xrightarrow{\sim} \underline{\mathrm{mod}}(A)$$

from the quotient of the bounded derived category of finitely generated modules by the triangulated subcategory of perfect complexes to the stable module category. It follows that if two blocks are derived equivalent then they are stably equivalent, meaning that the stable module categories are equivalent. In fact, more than this is true: there is a *stable equivalence of Morita type*, which means a stable equivalence induced by an exact functor between the module categories [98].

A more general condition that guarantees a stable equivalence of Morita type, analogous to properties of two-sided tilting complexes or split endo-morphism two-sided tilting complexes, is that there should be a bounded complex X of A-B-bimodules, finitely generated and projective as left A-modules and as right B-modules, such that

$$X \otimes_B X^{\vee} \cong A \oplus Z$$

for some perfect complex Z of A-bimodules, and

$$X^{\vee} \otimes_A X \cong B \oplus Z'$$

for some perfect complex Z' of B-bimodules. In fact, if such a complex X exists, it is always possible to choose it to be simply a bimodule (i.e., a complex concentrated in degree zero), but as we shall see, it can be more natural to allow genuine complexes.

Although derived equivalence of blocks implies stable equivalence, the converse is false in general, as described in Section 10, although we know of no counterexamples for blocks with abelian defect group. In the majority of recent cases where derived equivalences of blocks have been verified, there was a stable equivalence already known that was lifted to a derived equivalence.

In fact, Rouquier [115] shows that, at least with some plausible strength-ening of the definition of a splendid tilting complex, the problem of lifting stable equivalences to splendid equivalences is the only obstruction to proving the Abelian Defect Group Conjecture. He gives there a partial converse to Theorem 5.2, providing a method to "glue" local splendid tilting complexes to construct a global complex that induces a stable equivalence. It is because the Brauer construction with respect to a non-trivial p-subgroup Q kills all projective modules that properties of the global complex can only be proved at the level of the stable module category rather than the derived category.

In particular, Rouquier shows, using known facts about splendid equiva-lence for blocks with cyclic defect group [112], that if B is a block whose defect group is abelian of p-rank two (i.e., of the form $C_{p^a} \times C_{p^b}$) or isomorphic to $C_2 \times C_2 \times C_2$, then the block B is stably equivalent to its Brauer correspondent.

7 Lifting stable equivalences

Although the general problem seems very hard, several powerful techniques have been developed recently for lifting stable equivalences of Morita type to derived equivalences (or splendid equivalences). These are mostly based on the following simple but extremely useful theorem [64], now commonly known as "Linckelmann's Theorem".

Theorem 7.1 (Linckelmann). *Let A and B be finite-dimensional self-injective algebras over a field k with no projective simple modules (for example, non-semisimple blocks), and suppose that*

$$F \colon \underline{\mathrm{mod}}(A) \xrightarrow{\sim} \underline{\mathrm{mod}}(B)$$

is a stable equivalence of Morita type preserving simple modules (i.e., such that $S \in \underline{\mathrm{mod}}(A)$ is isomorphic to a simple A-module if and only if $F(S) \in \underline{\mathrm{mod}}(B)$ is isomorphic to a simple B-module). Then F is induced (up to isomorphism) by an equivalence

$$\mathrm{mod}(A) \xrightarrow{\sim} \mathrm{mod}(B)$$

of module categories.

Okuyama [86] was the first to systematically exploit this to lift stable equivalences to derived equivalences, and the method has been used extensively since then.

The idea is as follows. Suppose that B is a block with Brauer correspondent b. Typically B is a complicated block of a large group, but b is a much simpler block of a much smaller group, and explicit calculations may be possible with the representations of b that are far beyond reach for the representations of B. Thus we would like to be able to work as far as possible with b and not with B. To quote one striking example, the Monster sporadic group M has an abelian Sylow 11-subgroup P of order 121, and it is known, for example by the results of Rouquier discussed in the previous section, that in characteristic 11 the principal blocks of M and $N_M(P)$ are stably equivalent, and of course they are conjectured to be derived equivalent. The order of M is

80801742479451287588645990496171075700575436800000000,

whereas the order of $N_M(P)$ is

72600.

Suppose that we know a stable equivalence

$$F: \underline{\mathrm{mod}}(B) \xrightarrow{\sim} \underline{\mathrm{mod}}(b)$$

of Morita type between B and b. Suppose further that we can construct a derived equivalence between b and another algebra C which induces a stable equivalence such that the composition

$$\underline{\mathrm{mod}}(B) \xrightarrow{\sim} \underline{\mathrm{mod}}(b) \xrightarrow{\sim} \underline{\mathrm{mod}}(C)$$

takes the simple B-modules to the simple C-modules. Then by Linckelmann's Theorem the algebras B and C are Morita equivalent, and so, since b and C are derived equivalent, it follows that B and b are derived equivalent. Notice that all we need to know about B to apply this method is the images of the simple B-modules under the stable equivalence F (we admit that even this is far from known in the example of the Monster group described above).

Okuyama [86],[84] and others have used this method, constructing the derived equivalent algebra C using a sequence of (one-sided) tilting complexes of the kind described in Section 3. In all these cases he was also able to prove that the derived equivalences of blocks that he produced were actually splendid equivalences. Theorem 3.5 is designed to produce derived equivalences where the images of simple modules are known, and Chuang [13], Holloway [31] and others have used this theorem in conjunction with Linckelmann's Theorem to prove instances of the Abelian Defect Group Conjecture.

8 Clifford theory

Clifford theory relates the representation theory of a group $\widetilde{G}$ with a normal subgroup G to that of G and the quotient group $\widetilde{G}/G$.

Here we shall give just a taste of Clifford theory in the context of derived equivalences, focusing on the easiest situation, when the order of $\widetilde{G}/G$ is not divisible by p, and describing applications to the Abelian Defect Group Conjecture. The main idea is to lift derived equivalences between blocks of normal subgroups of two groups with isomorphic quotient groups to derived equivalences between blocks of the two groups themselves.

To be more precise, let G and H be normal subgroups of finite groups

$\widetilde{G}$ and $\widetilde{H}$ with isomorphic quotients $\widetilde{G}/G = \widetilde{H}/H = E$ of order not divisible by p. Let A and B be blocks of $\mathcal{O}G$ and $\mathcal{O}H$ invariant under conjugation by elements of E, so that $\widetilde{A} = \mathcal{O}\widetilde{G}.A$ and $\widetilde{B} = \mathcal{O}\widetilde{H}.B$ are products of blocks of $\mathcal{O}\widetilde{G}$ and of $\mathcal{O}\widetilde{H}$.

We say that a two-sided tilting complex X for A and B is *E-liftable* if X, regarded as a complex of $\mathcal{O}[G \times H]$-modules, extends to a complex of $\mathcal{O}\Delta$-modules, where $\Delta = \{(\widetilde{g}, \widetilde{h}) \in \widetilde{G} \times \widetilde{H} \mid \widetilde{g}G = \widetilde{h}H \in E\}$. Of course if an extension exists, it may not be unique.

Andrei Marcus [69] proves a lifting criterion for derived equivalences.

Theorem 8.1 (Marcus). *Suppose X is an E-liftable two-sided tilting complex for A and B. Then $\widetilde{X} = \mathrm{Ind}_{\Delta}^{\widetilde{G} \times \widetilde{H}}(X)$ is a two-sided tilting complex for $\widetilde{A}$ and $\widetilde{B}$. Moreover if X is splendid then $\widetilde{X}$ is splendid as well.*

This is useful for verifying cases of the Abelian Defect Group Conjecture, as we now explain, restricting to the case of principal blocks for simplicity's sake. Let $\widetilde{G}$ be a group with normal subgroup G, and let P be an abelian Sylow p-subgroup of $\widetilde{G}$ contained in G. By the Frattini argument we have $\widetilde{G}/G = N_{\widetilde{G}}(P)/N_G(P) =: E$.

Suppose that the Abelian Defect Group Conjecture is true for the principal block of $\mathcal{O}G$ and that moreover there exists an E-liftable (splendid) two-sided tilting complex X for the principal blocks of $\mathcal{O}G$ and $\mathcal{O}N_G(P)$. Then a summand of the complex $\widetilde{X}$ of $\mathcal{O}[\widetilde{G} \times N_{\widetilde{G}}(P)]$-modules provided by Marcus's theorem is a (splendid) two-sided tilting complex for the principal blocks of $\mathcal{O}\widetilde{G}$ and $\mathcal{O}N_{\widetilde{G}}(P)$, confirming the truth of the Abelian Defect Group Conjecture for the principal block of $\mathcal{O}\widetilde{G}$.

One of the most impressive applications of this Clifford theory is a reduction theorem for the Abelian Defect Group Conjecture, due to Andrei Marcus [69], which parallels some of the work of Fong and Harris [21] on perfect isometries. The starting point for the approach taken by Fong and Harris is a structure theorem for finite groups with abelian Sylow p-subgroups they deduced from the classification of finite simple groups. Marcus uses this together with Theorem 8.1 to prove that the the Abelian Defect Group Conjecture, at least for principal blocks, essentially reduces to the case of simple groups.

Theorem 8.2 (Marcus). *Suppose that for any simple group G with* *abelian Sylow p-subgroup P and any subgroup E of $Aut(G)/G$ of order not divisible by p there exists an E-liftable two-sided tilting complex for the principal blocks of $\mathcal{O}G$ and $\mathcal{O}N_G(P)$. Then the Abelian Defect Group Conjecture is true for all principal blocks of all finite groups (with abelian Sylow p-subgroups).*

Because of the nature of Fong and Harris's structure theorem, in order to verify the Abelian Defect Group Conjecture for the principal block of $\mathcal{O}L$ for a particular group L, it in fact suffices to establish the hypothesis of Theorem 8.2 for all simple groups G isomorphic to a subnormal subgroup of L.

In some situations it may be easier to prove the Abelian Defect Group Conjecture for a group $\widetilde{G}$ than for a normal subgroup G. So it makes sense to have a counterpart to Theorem 8.1, giving conditions for 'descending' from $\widetilde{G}$ to G. A hint for how to do this comes from the observation that the complex $\widetilde{X}$ provides something stronger than an equivalence of derived module categories: the direct sum decompositions $\widetilde{A} = \oplus_{e \in E} Ae$ and $\widetilde{B} = \oplus_{e \in E} Be$ give $\widetilde{A}$ and $\widetilde{B}$ the structure of group-graded algebras, and $\widetilde{X}$ is an complex of E-graded bimodules inducing an equivalence of derived categories of E-graded modules. Andrei Marcus [70] proves the following converse to Theorem 8.1.

Theorem 8.3 (Marcus). *Suppose $\widetilde{Y}$ is a complex of E-graded $\mathcal{O}[\widetilde{G} \times \widetilde{H}]$-modules inducing an equivalence of the derived categories of E-graded modules over $\widetilde{A}$ and $\widetilde{B}$. Then the 1-component $Y = \widetilde{Y}_1$ of $\widetilde{Y}$ is a two-sided tilting complex for A and B.*

Marcus [79] uses this to deduce the truth of the Abelian Defect Group Conjecture for alternating groups from the known result for symmetric groups.

9 Cases for which the Abelian Defect Group Conjecture has been verified

In the list below, B will denote a block of $\mathcal{O}G$ with abelian defect group D, and b will be its Brauer correspondent, a block of $\mathcal{O}N_G(D)$, so that the conjecture predicts that B and b should have equivalent derived categories. Actually, in all these cases the derived equivalences are known to be induced by splendid two-sided tilting complexes.

The webpage

www.maths.bris.ac.uk/~majcr/adgc/

constructed by Jeremy Rickard and Naoko Kunugi contains a list of known examples where the Abelian Defect Group Conjecture has been verified, and is occasionally updated as new examples are found. It was used to help compile the list below.

(1) General families of blocks.

- All blocks with cyclic defect group.

   - Rickard [98], Linckelmann [60], and Rouquier [112], [114].

- All blocks of p-solvable groups.

   - Dade [17], Puig [89], Harris and Linckelmann [25]. In this case, the blocks B and b are Morita equivalent.

- All blocks with defect group $C_2 \times C_2$.

   - Rickard [103], Linckelmann [61], [62] and Rouquier [115].

- All principal blocks with defect group $C_3 \times C_3$.

   - Koshitani and Kunugi [47]. This depends on the classification of finite simple groups, and uses results of Okuyama and others mentioned below for specific groups, and general methods due to Marcus [69].

(2) Symmetric groups and related groups.

- All blocks of symmetric groups with defect groups of order p^2.

   - Chuang [12].

- All blocks of symmetric groups with abelian defect groups.

   - Rickard [97], Chuang and Kessar [14], Chuang and Rouquier [15].

- The principal block of the alternating group A_5, $p = 2$. (Defect group $C_2 \times C_2$.)

   - Rickard [103].

- The principal blocks of the alternating groups A_6, A_7, and A_8, $p = 3$. (Defect group $C_3 \times C_3$.)

- Okuyama [86].

• All blocks of alternating groups with abelian defect groups.

  - Marcus [79]. This is deduced from the case of symmetric groups using Clifford theory.

(3) Sporadic groups and related groups

  • The principal blocks of the Mathieu groups M_{11}, M_{22}, M_{23}, $p = 3$. (Defect group $C_3 \times C_3$.)

    - Okuyama [86].

  • The principal block of the Higman-Sims group HS, $p = 3$. (Defect group $C_3 \times C_3$.)

    - Okuyama [86].

  • The principal block of the Janko group J_1, $p = 2$. (Defect group $C_2 \times C_2 \times C_2$.)

    - Gollan and Okuyama [24].

  • The principal block of the Hall-Janko group $HJ = J_2$, $p = 5$. (Defect group $C_5 \times C_5$.)

    - Holloway [31].

  • The non-principal block of the O'Nan group $O'N$ with defect group $C_3 \times C_3$.

    - Koshitani, Kunugi and Waki [48].

  • The non-principal block of the Higman-Sims group HS with defect group $C_3 \times C_3$.

    - Holm [34], Koshitani, Kunugi and Waki [48].

  • The non-principal block of the Held group He with defect group $C_3 \times C_3$.

    - Koshitani, Kunugi and Waki [49].

  • The non-principal block of the Suzuki group Suz with defect group $C_3 \times C_3$.

    - Koshitani, Kunugi and Waki [49].

- The non-principal block of the double cover $2.J_2$ of the Hall-Janko group with defect group $C_5 \times C_5$.

  - Holloway [31].

(4) Groups of Lie type in defining characteristic

- The principal block of $PSL_2(p^n)$ in characteristic p. (Defect group $C_p{}^n$.)

  - Okuyama [84] for the general case; earlier Chuang [13] for the case $n = 2$ and Rouquier [112] for the case $p^n = 8$.

- The non-principal block with full defect of $SL_2(p^2)$ in characteristic p. (Defect group $C_p \times C_p$.)

  - Holloway [30]

(5) Groups of Lie type in non-defining characteristic

- The principal block in characteristic p of the group $G = \mathbf{G}^F$ of rational points of a connected reductive group $\mathbf{G}$ defined over the field $\mathbf{F}_q$ of q elements, where p is a prime dividing $q - 1$ but not dividing the order of the Weyl group of G.

  - This follows from result of Puig [90]. In this case the blocks B and b are Morita equivalent.

  - This general theorem includes, for $p = 3$, some cases where G has Sylow 3-subgroup $C_3 \times C_3$, which were needed for the theorem of Koshitani and Kunugi. Namely: $G = Sp_4(q)$, where $q = 4$ or 7 (mod 9); $G = PSU_4(q^2)$ or $PSU_5(q^2)$, where $q = 4$ or 7 (mod 9) (see also Koshitani and Miyachi [52]). Of course, Puig's theorem also deals with the case $q = 1$ (mod 9), but then the Sylow p-subgroup is larger than $C_3 \times C_3$.

- The principal blocks of the Ree groups $R(3^{2n+1}) = {}^2G_2(3^{2n+1})$, $p = 2$. (Defect group $C_2 \times C_2 \times C_2$.)

  - It follows from work of Landrock and Michler [59] that the principal blocks of $R(3^{2n+1})$ are all Morita equivalent (for different values of n). This reduces the conjecture to the case $n = 0$, which is known, since $R(3) = SL_2(8).3$.

- The principal blocks of $SU_3(q^2)$, where q is a prime power, and

$p > 3$ is a prime dividing $q + 1$. (Defect group $C_r \times C_r$, where $r = p^a$ is the largest power of p dividing $q + 1$.)

- Kunugi and Waki [58].

• The principal blocks of $PSL_3(q)$, $p = 3$, where $q = 4$ or 7 (mod 9). (Defect group $C_3 \times C_3$.)

- Kunugi [54] proves that these blocks are all Morita equivalent. The conjecture then follows since Okuyama [86] proved the case of $PSL_3(4)$.

• The principal blocks of $Sp_4(q)$, $p = 3$, where $q = 2$ or 5 (mod 9). (Defect group $C_3 \times C_3$.)

- It follows from work of Okuyama and Waki [88] that these blocks are all Morita equivalent. The conjecture then follows, because $Sp_4(2)$ is isomorphic to the symmetric group S_6.

• The principal blocks of $Sp_4(q)$, where q is a prime power and p is an odd prime dividing $q + 1$. (Defect group $C_r \times C_r$, where $r = p^a$ is the largest power of p dividing $q + 1$.)

- Kunugi, Okuyama, and Waki [57] for the general case; earlier Holloway [31] for the case $q = 4$, $p = 5$.

• The principal blocks of $PSU_3(q^2)$, $p = 3$, where $q = 2$ or 5 (mod 9). (Defect group $C_3 \times C_3$.)

- Koshitani and Kunugi [46] prove that the blocks B and b are Morita equivalent.

• The principal blocks of $GL_4(q)$, $p = 3$, where $q = 2$ or 5 (mod 9). (Defect group $C_3 \times C_3$.)

- Koshitani and Miyachi [51] prove that these blocks are all Morita equivalent. The conjecture then follows since Okuyama [86] proved the case of $GL_4(2)$, which is isomorphic to the alternating group A_8.

• The principal blocks of $GL_5(q)$, $p = 3$, where $q = 2$ or 5 (mod 9). (Defect group $C_3 \times C_3$.)

- Koshitani and Miyachi [51] prove that the blocks B and b are Morita equivalent.

- The principal blocks of $G_2(q)$, where q is a prime power and $p > 3$ is a prime dividing $q + 1$. (Defect group $C_r \times C_r$, where $r = p^a$ is the largest power of p dividing $q + 1$.)

  - Okuyama [85] in general; earlier Usami and Yoshida in the case $q = 4$, $p = 5$.

- Unipotent blocks of weight 2 of general linear groups in non-defining characteristic p. (Defect group $C_r \times C_r$, for $r = p^a$ some power of p.)

  - Hida and Miyachi [29], Turner [121].

10 Nonabelian defect groups

As yet there is no general conjecture for blocks of finite groups with arbitrary defect group that extends the Abelian Defect Group Conjecture. The most naive attempt at generalization fails because a block with a nonabelian defect group may have more simple modules than its Brauer correspondent block. For example, the principal block of S_4 in characteristic 2 has two simple modules while its Brauer correspondent, the principal block of the dihedral group D_8, has only one.

Even in cases where a block and its Brauer correspondent do have the same number of simple modules, they may still not be derived equivalent. Consider for example a finite group G with a (not necessarily Abelian) p-Sylow subgroup P with the property that P has trivial intersection with any other p-Sylow subgroup. In this situation, Alperin's weight conjecture predicts that the principal blocks of kG and $kN_G(P)$ have the same number of simple modules. Moreover it is known that induction and restriction of modules induce a stable equivalence of Morita type between the blocks. However this cannot in general be improved to a derived equivalence; one of the smaller counterexamples is $G = Sz(8)$ in characteristic 2 [16, 108].

Nonetheless there are interesting examples of derived (or even splendid) equivalences between blocks with nonabelian defect groups, a few of which are listed below.

(1) Two blocks of symmetric groups are splendidly equivalent if and only if they have isomorphic defect groups [15]. The smallest

384   *J. Chuang and J. Rickard*

example involving nonabelian defect groups is a splendid equivalence between the principal blocks of $\mathcal{O}S_4$ and $\mathcal{O}S_5$, $p = 2$. Here the defect group is a dihedral group of order 8.

(2) Alvis-Curtis duality [10]. This is a self-equivalence of the derived category for any finite reductive group in non-defining characteristic.

(3) The principal blocks of $G_2(q)$, $p = 3$, where $q = 2$ or 5 (mod 9), are all Morita equivalent [129]. The defect groups are extraspecial of order 27 and exponent 3.

(4) The principal blocks of $PGL_3(q)$ and $PGU_3(q^2)$, $p = 3$, where q is a prime power such that $q + 1$ is divisible by 3, are splendidly equivalent [128, 56]. The defect groups are extraspecial of order 27 and exponent 3.

References

[1] Salah Al-Nofayee. *t-structures and derived equivalence for self-injective algebras*. PhD thesis, University of Bristol, 2004.
[2] J. L. Alperin. Weights for finite groups. In *The Arcata Conference on Representations of Finite Groups (Arcata, Calif., 1986)*, volume 47 of *Proc. Sympos. Pure Math.*, pages 369–379. Amer. Math. Soc., Providence, RI, 1987.
[3] M. Broué. Rickard equivalences and block theory. In *Groups '93 Galway/St. Andrews, Vol. 1 (Galway, 1993)*, pages 58–79. Cambridge Univ. Press, Cambridge, 1995.
[4] Michel Broué. Blocs, isométries parfaites, catégories dérivées. *C. R. Acad. Sci. Paris Sér. I Math.*, 307(1):13–18, 1988.
[5] Michel Broué. Isométries de caractères et équivalences de Morita ou dérivées. *Inst. Hautes Études Sci. Publ. Math.*, (71):45–63, 1990.
[6] Michel Broué. Isométries parfaites, types de blocs, catégories dérivées. *Astérisque*, (181-182):61–92, 1990.
[7] Michel Broué. Equivalences of blocks of group algebras. In *Finite-dimensional algebras and related topics (Ottawa, ON, 1992)*, pages 1–26. Kluwer Acad. Publ., Dordrecht, 1994.
[8] Michel Broué, Gunter Malle, and Jean Michel. Generic blocks of finite reductive groups. *Astérisque*, (212):7–92, 1993. Représentations unipotentes génériques et blocs des groupes réductifs finis.
[9] Marc Cabanes and Claudine Picaronny. Types of blocks with dihedral or quaternion defect groups. *J. Fac. Sci. Univ. Tokyo Sect. IA Math.*, 39(1):141–161, 1992.
[10] Marc Cabanes and Jeremy Rickard. Alvis-Curtis duality as an equivalence of derived categories. In *Modular representation theory of finite groups (Charlottesville, VA, 1998)*, pages 157–174. de Gruyter, Berlin, 2001.

[11] Pierre Cartier. La théorie des blocs et les groupes génériques. *Astérisque*, (227):Exp. No. 781, 4, 171–208, 1995. Séminaire Bourbaki, Vol. 1993/94.

[12] Joseph Chuang. The derived categories of some blocks of symmetric groups and a conjecture of Broué. *J. Algebra*, 217(1):114–155, 1999.

[13] Joseph Chuang. Derived equivalence in $sl_2(p^2)$. *Trans. Amer. Math. Soc.*, 353(7):2897–2913 (electronic), 2001.

[14] Joseph Chuang and Radha Kessar. Symmetric groups, wreath products, Morita equivalences, and Broué's abelian defect group conjecture. *Bull. London Math. Soc.*, 34(2):174–184, 2002.

[15] Joseph Chuang and Raphaël Rouquier. Derived equivalences for symmetric groups and sl(2)-categorification. math.RT/0407205.

[16] Gerald Cliff. On centers of 2-blocks of Suzuki groups. *J. Algebra*, 226(1):74–90, 2000.

[17] Everett C. Dade. A correspondence of characters. In *The Santa Cruz Conference on Finite Groups (Univ. California, Santa Cruz, Calif., 1979)*, pages 401–403. Amer. Math. Soc., Providence, R.I., 1980.

[18] Michel Enguehard. Isométries parfaites entre blocs de groupes symétriques. *Astérisque*, (181-182):157–171, 1990.

[19] Michel Enguehard. Isométries parfaites entre blocs de groupes linéaires ou unitaires. *Math. Z.*, 206(1):1–24, 1991.

[20] Karin Erdmann, Stuart Martin, and Joanna Scopes. Morita equivalence for blocks of the Schur algebras. *J. Algebra*, 168(1):340–347, 1994.

[21] Paul Fong and Morton E. Harris. On perfect isometries and isotypies in finite groups. *Invent. Math.*, 114(1):139–191, 1993.

[22] Paul Fong and Morton E. Harris. On perfect isometries and isotypies in alternating groups. *Trans. Amer. Math. Soc.*, 349(9):3469–3516, 1997.

[23] Murray Gerstenhaber and Mary E. Schaps. The modular version of Maschke's theorem for normal abelian p-Sylows. *J. Pure Appl. Algebra*, 108(3):257–264, 1996.

[24] Holger Gollan and Tetsuro Okuyama. Derived equivalences for the smallest Janko group.

[25] M. E. Harris and M. Linckelmann. Splendid derived equivalences for blocks of finite p-solvable groups. *J. London Math. Soc. (2)*, 62(1):85–96, 2000.

[26] M. E. Harris and M. Linckelmann. On the Glauberman and Watanabe correspondences for blocks of finite p-solvable groups. *Trans. Amer. Math. Soc.*, 354(9):3435–3453 (electronic), 2002.

[27] Morton E. Harris. Categorically equivalent and isotypic blocks. *J. Algebra*, 166(2):232–244, 1994.

[28] Morton E. Harris. Splendid derived equivalences for blocks of finite groups. *J. London Math. Soc. (2)*, 60(1):71–82, 1999.

[29] Akihiko Hida and Hyoue Miyachi. On the principal blocks of finite general linear groups in non-defining characteristic. *Sūrikaisekikenkyūsho Kōkyūroku*, (1140):127–130, 2000.

[30] Miles Holloway. *Derived equivalences for group algebras*. PhD thesis, Bristol, 2001.

[31] Miles Holloway. Broué's conjecture for the Hall-Janko group and its double cover. *Proc. London Math. Soc. (3)*, 86(1):109–130, 2003.

[32] Thorsten Holm. Derived equivalences and Hochschild cohomology for blocks with quaternion defect groups. In *Darstellungstheorietage Jena 1996*, pages 75–86. Akad. Gemein. Wiss. Erfurt, Erfurt, 1996.

[33] Thorsten Holm. Derived equivalent tame blocks. *J. Algebra*, 194(1):178–200, 1997.

[34] Thorsten Holm. Derived categories, derived equivalences and representation theory. In *Proceedings of the Summer School on Representation Theory of Algebras, Finite and Reductive Groups (Cluj-Napoca, 1997)*, pages 33–66, Cluj, 1998. "Babeş-Bolyai" Univ.

[35] Thorsten Holm. Derived equivalence classification of algebras of dihedral, semidihedral, and quaternion type. *J. Algebra*, 211(1):159–205, 1999.

[36] Hiroshi Horimoto. On a correspondence between blocks of finite groups induced from the Isaacs character correspondence. *Hokkaido Math. J.*, 30(1):65–74, 2001.

[37] Michael Kauer and Klaus W. Roggenkamp. Higher-dimensional orders, graph-orders, and derived equivalences. *J. Pure Appl. Algebra*, 155(2-3):181–202, 2001.

[38] Bernhard Keller. A remark on tilting theory and DG algebras. *Manuscripta Math.*, 79(3-4):247–252, 1993.

[39] Bernhard Keller. Deriving DG categories. *Ann. Sci. École Norm. Sup. (4)*, 27(1):63–102, 1994.

[40] Bernhard Keller. Tilting theory and differential graded algebras. In *Finite-dimensional algebras and related topics (Ottawa, ON, 1992)*, pages 183–190. Kluwer Acad. Publ., Dordrecht, 1994.

[41] Bernhard Keller. Introduction to abelian and derived categories. In *Representations of reductive groups*, pages 41–61. Cambridge Univ. Press, Cambridge, 1998.

[42] Bernhard Keller. Bimodule complexes via strong homotopy actions. *Algebr. Represent. Theory*, 3(4):357–376, 2000. Special issue dedicated to Klaus Roggenkamp on the occasion of his 60th birthday.

[43] Radha Kessar. Shintani descent and perfect isometries for blocks of finite general linear groups. *J. Algebra*, 276(2):493–501, 2004.

[44] Radha Kessar and Markus Linckelmann. On perfect isometries for tame blocks. *Bull. London Math. Soc.*, 34(1):46–54, 2002.

[45] Steffen König and Alexander Zimmermann. *Derived equivalences for group rings*. Springer-Verlag, Berlin, 1998. With contributions by Bernhard Keller, Markus Linckelmann, Jeremy Rickard and Raphaël Rouquier.

[46] Shigeo Koshitani and Naoko Kunugi. The principal 3-blocks of the 3-dimensional projective special unitary groups in non-defining characteristic. *J. Reine Angew. Math.*, 539:1–27, 2001.

[47] Shigeo Koshitani and Naoko Kunugi. Broué's conjecture holds for principal 3-blocks with elementary abelian defect group of order 9. *J. Algebra*, 248(2):575–604, 2002.

[48] Shigeo Koshitani, Naoko Kunugi, and Katsushi Waki. Broué's conjecture for non-principal 3-blocks of finite groups. *J. Pure Appl. Algebra*, 173(2):177–211, 2002.

[49] Shigeo Koshitani, Naoko Kunugi, and Katsushi Waki. Broué's abelian defect group conjecture for the Held group and the sporadic Suzuki group. *J. Algebra*, 279(2):638–666, 2004.

[50] Shigeo Koshitani and Gerhard O. Michler. Glauberman correspondence of p-blocks of finite groups. *J. Algebra*, 243(2):504–517, 2001.

[51] Shigeo Koshitani and Hyoue Miyachi. The principal 3-blocks of four- and

five-dimensional projective special linear groups in non-defining characteristic. *J. Algebra*, 226(2):788–806, 2000.

[52] Shigeo Koshitani and Hyoue Miyachi. Donovan conjecture and Loewy length for principal 3-blocks of finite groups with elementary abelian Sylow 3-subgroup of order 9. *Comm. Algebra*, 29(10):4509–4522, 2001.

[53] Naoko Kunugi. Derived equivalences in symmetric groups. *Sūrikaisekikenkyūsho Kōkyūroku*, (1140):131–135, 2000. Research on the cohomology theory of finite groups (Japanese) (Kyoto, 1999).

[54] Naoko Kunugi. Morita equivalent 3-blocks of the 3-dimensional projective special linear groups. *Proc. London Math. Soc. (3)*, 80(3):575–589, 2000.

[55] Naoko Kunugi. Derived equivalences for blocks of finite groups. In *Proceedings of the 33rd Symposium on Ring Theory and Representation Theory (Shimane, 2000)*, pages 55–59. Tokyo Univ. Agric. Technol., Tokyo, 2001.

[56] Naoko Kunugi and Tetsuro Okuyama. Examples of splendid equivalent blocks with non-abelian defect groups. Talk given at Mathematisches Forschungsinstitut Oberwolfach, 2003.

[57] Naoko Kunugi, Tetsuro Okuyama, and Katsushi Waki. Constructions of derived equivalences using Okuyama's method. Talk given at Colloque "Catégories dérivées et groupes finis", CIRM, Luminy, 22-26 octobre 2001.

[58] Naoko Kunugi and Katsushi Waki. Derived equivalences for the 3-dimensional special unitary groups in non-defining characteristic. *J. Algebra*, 240(1):251–267, 2001.

[59] Peter Landrock and Gerhard O. Michler. Principal 2-blocks of the simple groups of Ree type. *Trans. Amer. Math. Soc.*, 260(1):83–111, 1980.

[60] Markus Linckelmann. Derived equivalence for cyclic blocks over a P-adic ring. *Math. Z.*, 207(2):293–304, 1991.

[61] Markus Linckelmann. A derived equivalence for blocks with dihedral defect groups. *J. Algebra*, 164(1):244–255, 1994.

[62] Markus Linckelmann. The source algebras of blocks with a Klein four defect group. *J. Algebra*, 167(3):821–854, 1994.

[63] Markus Linckelmann. On stable and derived equivalences of blocks and algebras. *An. Ştiinţ. Univ. Ovidius Constanţa Ser. Mat.*, 4(1):74–97, 1996. Representation theory of groups, algebras, and orders (Constanţa, 1995).

[64] Markus Linckelmann. Stable equivalences of Morita type for self-injective algebras and p-groups. *Math. Z.*, 223(1):87–100, 1996.

[65] Markus Linckelmann. On derived equivalences and local structure of blocks of finite groups. *Turkish J. Math.*, 22(1):93–107, 1998.

[66] Markus Linckelmann. On stable equivalences of Morita type. In *Derived equivalences for group rings*, pages 221–232. Springer, Berlin, 1998.

[67] Markus Linckelmann. On splendid derived and stable equivalences between blocks of finite groups. *J. Algebra*, 242(2):819–843, 2001.

[68] Andrei Marcus. Graded equivalences and Broué's conjecture. *An. Ştiinţ. Univ. Ovidius Constanţa Ser. Mat.*, 4(2):107–126, 1996. Representation theory of groups, algebras, and orders (Constanţa, 1995).

[69] Andrei Marcus. On equivalences between blocks of group algebras: reduction to the simple components. *J. Algebra*, 184(2):372–396, 1996.

[70] Andrei Marcus. Equivalences induced by graded bimodules. *Comm.*

Algebra, 26(3):713–731, 1998.

[71] Andrei Marcus. Equivariant Morita equivalences between blocks of group algebras. *Mathematica*, 40(63)(2):227–234, 1998.

[72] Andrei Marcus. Homology of fully graded algebras, Morita and derived equivalences. *J. Pure Appl. Algebra*, 133(1-2):209–218, 1998. Ring theory (Miskolc, 1996).

[73] Andrei Marcus. Stable equivalences, crossed products and symmetric algebras. *Mathematica*, 40(63)(1):123–129, 1998.

[74] Andrei Marcus. Derived equivalences and Dade's invariant conjecture. *J. Algebra*, 221(2):513–527, 1999.

[75] Andrei Marcus. Blocks of normal subgroups and Morita equivalences. *Ital. J. Pure Appl. Math.*, (7):77–84, 2000. International Algebra Conference (Iaşi, 1998).

[76] Andrei Marcus. Clifford theory of bimodules. *Comm. Algebra*, 28(4):2029–2041, 2000.

[77] Andrei Marcus. Twisted group algebras, normal subgroups and derived equivalences. *Algebr. Represent. Theory*, 4(1):25–54, 2001. Special issue dedicated to Klaus Roggenkamp on the occasion of his 60th birthday.

[78] Andrei Marcus. Tilting complexes for group graded algebras. *J. Group Theory*, 6(2):175–193, 2003.

[79] Andrei Marcus. Broué's conjecture for alternating groups. *Proc. Amer. Math. Soc.*, 132(1):7–14, 2004.

[80] Andrei Marcus and Ciprian Modoi. Graded endomorphism rings and equivalences. *Comm. Algebra*, 31(7):3219–3249, 2003.

[81] Chedva Mejer and Mary Schaps. Separable deformations of blocks with abelian normal defect group and of derived equivalent global blocks. In *Representation theory of algebras (Cocoyoc, 1994)*, pages 505–517. Amer. Math. Soc., Providence, RI, 1996.

[82] Hyohe Miyachi. Morita equivalences for general linear groups in non-defining characteristic. In *Proceedings of the 33rd Symposium on Ring Theory and Representation Theory (Shimane, 2000)*, pages 95–105. Tokyo Univ. Agric. Technol., Tokyo, 2001.

[83] Miyako Nakabayashi. Principal 3-blocks with extra-special 3-defect groups of order 27 and exponent 3. *Natur. Sci. Rep. Ochanomizu Univ.*, 50(1):1–8, 1999.

[84] Tetsuro Okuyama. Derived equivalences in $SL_2(q)$.

[85] Tetsuro Okuyama. Derived equivalent blocks in finite chevalley groups of small rank. Talk given at Conference on Frobenius Algebras and Related Topics, Torun, Poland, 2003.

[86] Tetsuro Okuyama. Some examples of derived equivalent blocks of finite groups.

[87] Tetsuro Okuyama. Remarks on splendid tilting complexes. In Shigeo Koshitani, editor, *Representation theory of finite groups and related topics*, volume 1149 of *RIMS Kokyuroku (Proceedings of Research Insititute for Mathematical Sciences*, pages 53–59, 2000.

[88] Tetsuro Okuyama and Katsushi Waki. Decomposition numbers of sp(4, q). *J. Algebra*, 199(2):544–555, 1998.

[89] Lluís Puig. Local block theory in p-solvable groups. In *The Santa Cruz Conference on Finite Groups (Univ. California, Santa Cruz, Calif., 1979)*, volume 37 of *Proc. Sympos. Pure Math.*, pages 385–388. Amer. Math. Soc., Providence, R.I., 1980.

[90] Lluís Puig. Algèbres de source de certains blocs des groupes de Chevalley. *Astérisque*, (181-182):9, 221–236, 1990.

[91] Lluís Puig. Une correspondance de modules pour les blocs à groupes de défaut abéliens. *Geom. Dedicata*, 37(1):9–43, 1991.

[92] Lluís Puig. On Joanna Scopes' criterion of equivalence for blocks of symmetric groups. *Algebra Colloq.*, 1(1):25–55, 1994.

[93] Lluis Puig. A survey on the local structure of Morita and Rickard equivalences between Brauer blocks. In *Representation theory of finite groups (Columbus, OH, 1995)*, pages 101–126. de Gruyter, Berlin, 1997.

[94] Lluís Puig. *On the local structure of Morita and Rickard equivalences between Brauer blocks.* Birkhäuser Verlag, Basel, 1999.

[95] Lluís Puig and Yoko Usami. Perfect isometries for blocks with abelian defect groups and Klein four inertial quotients. *J. Algebra*, 160(1):192–225, 1993.

[96] Lluís Puig and Yoko Usami. Perfect isometries for blocks with abelian defect groups and cyclic inertial quotients of order 4. *J. Algebra*, 172(1):205–213, 1995.

[97] Jeremy Rickard. MRSI talk, Berkeley, November 1990.

[98] Jeremy Rickard. Derived categories and stable equivalence. *J. Pure Appl. Algebra*, 61(3):303–317, 1989.

[99] Jeremy Rickard. Morita theory for derived categories. *J. London Math. Soc. (2)*, 39(3):436–456, 1989.

[100] Jeremy Rickard. Derived equivalences as derived functors. *J. London Math. Soc. (2)*, 43(1):37–48, 1991.

[101] Jeremy Rickard. Lifting theorems for tilting complexes. *J. Algebra*, 142(2):383–393, 1991.

[102] Jeremy Rickard. Translation functors and equivalences of derived categories for blocks of algebraic groups. In *Finite-dimensional algebras and related topics (Ottawa, ON, 1992)*, pages 255–264. Kluwer Acad. Publ., Dordrecht, 1994.

[103] Jeremy Rickard. Splendid equivalences: derived categories and permutation modules. *Proc. London Math. Soc. (3)*, 72(2):331–358, 1996.

[104] Jeremy Rickard. The abelian defect group conjecture. In *Proceedings of the International Congress of Mathematicians, Vol. II (Berlin, 1998)*, number Extra Vol. II, pages 121–128 (electronic), 1998.

[105] Jeremy Rickard. Bousfield localization for representation theorists. In *Infinite length modules (Bielefeld, 1998)*, Trends Math., pages 273–283. Birkhäuser, Basel, 2000.

[106] Jeremy Rickard. Equivalences of derived categories for symmetric algebras. *J. Algebra*, 257(2):460–481, 2002.

[107] Jeremy Rickard and Mary Schaps. Folded tilting complexes for Brauer tree algebras. *Adv. Math.*, 171(2):169–182, 2002.

[108] Geoffrey R. Robinson. A note on perfect isometries. *J. Algebra*, 226(1):71–73, 2000.

[109] K. W. Roggenkamp. Derived equivalences of group rings. *An. Ştiinţ. Univ. Ovidius Constanţa Ser. Mat.*, 2:137–151, 1994. XIth National Conference of Algebra (Constanţa, 1994).

[110] Raphaël Rouquier. Gluing p-permutation modules.

[111] Raphaël Rouquier. Isométries parfaites dans les blocs à défaut abélien des groupes symétriques et sporadiques. *J. Algebra*, 168(2):648–694, 1994.

[112] Raphaël Rouquier. From stable equivalences to Rickard equivalences for blocks with cyclic defect. In *Groups '93 Galway/St. Andrews, Vol. 2*, pages 512–523. Cambridge Univ. Press, Cambridge, 1995.

[113] Raphaël Rouquier. Some examples of Rickard complexes. *An. Ştiinţ. Univ. Ovidius Constanţa Ser. Mat.*, 4(2):169–173, 1996. Representation theory of groups, algebras, and orders (Constanţa, 1995).

[114] Raphaël Rouquier. The derived category of blocks with cyclic defect groups. In *Derived equivalences for group rings*, pages 199–220. Springer, Berlin, 1998.

[115] Raphaël Rouquier. Block theory via stable and Rickard equivalences. In *Modular representation theory of finite groups (Charlottesville, VA, 1998)*, pages 101–146. de Gruyter, Berlin, 2001.

[116] Raphaël Rouquier. Complexes de chaînes étales et courbes de Deligne-Lusztig. *J. Algebra*, 257(2):482–508, 2002.

[117] Raphaël Rouquier and Alexander Zimmermann. Picard groups for derived module categories. *Proc. London Math. Soc. (3)*, 87(1):197–225, 2003.

[118] Lucia Sanus. Perfect isometries and the Isaacs correspondence. *Osaka J. Math.*, 40(2):313–326, 2003.

[119] Masato Sawabe and Atumi Watanabe. On the principal blocks of finite groups with abelian Sylow p-subgroups. *J. Algebra*, 237(2):719–734, 2001.

[120] Joanna Scopes. Cartan matrices and Morita equivalence for blocks of the symmetric groups. *J. Algebra*, 142(2):441–455, 1991.

[121] W. Turner. Equivalent blocks of finite general linear groups in non-describing characteristic. *J. Algebra*, 247(1):244–267, 2002.

[122] Yoko Usami. Perfect isometries for blocks with abelian defect groups and dihedral inertial quotients of order 6. *J. Algebra*, 172(1):113–125, 1995.

[123] Yoko Usami. Perfect isometries for blocks with abelian defect groups and the inertial quotients isomorphic to D_6, $Z_4 \times Z_2$ or $Z_3 \times Z_3$. *Sūrikaisekikenkyūsho Kōkyūroku*, (896):13–25, 1995. Research into algebraic combinatorics (Japanese) (Kyoto, 1993).

[124] Yoko Usami. Perfect isometries and isotypies for blocks with abelian defect groups and the inertial quotients isomorphic to $Z_3 \times Z_3$. *J. Algebra*, 182(1):140–164, 1996.

[125] Yoko Usami. Perfect isometries and isotypies for blocks with abelian defect groups and the inertial quotients isomorphic to $Z_4 \times Z_2$. *J. Algebra*, 181(3):727–759, 1996.

[126] Yoko Usami. Perfect isometries for principal blocks with abelian defect groups and elementary abelian 2-inertial quotients. *Sūrikaisekikenkyūsho Kōkyūroku*, (991):36–43, 1997. Group theory and combinatorial mathematics (Japanese) (Kyoto, 1996).

[127] Yoko Usami. Perfect isometries for principal blocks with abelian defect groups and elementary abelian 2-inertial quotients. *J. Algebra*, 196(2):646–681, 1997.

[128] Yoko Usami. Principal blocks with extra-special defect groups of order 27. *Sūrikaisekikenkyūsho Kōkyūroku*, (1149):98–113, 2000. Representation theory of finite groups and related topics (Japanese) (Kyoto, 1998).

[129] Yoko Usami and Miyako Nakabayashi. Morita equivalent principal 3-blocks of the Chevalley groups $G_2(q)$. *Proc. London Math. Soc. (3)*,

86(2):397–434, 2003.

[130] Atumi Watanabe. The Glauberman character correspondence and perfect isometries for blocks of finite groups. *J. Algebra*, 216(2):548–565, 1999.

[131] Atumi Watanabe. The Isaacs character correspondence and isotypies between blocks of finite groups. In *Groups and combinatorics—in memory of Michio Suzuki*, volume 32 of *Adv. Stud. Pure Math.*, pages 437–448. Math. Soc. Japan, Tokyo, 2001.

[132] Atumi Watanabe. Morita equivalence of Isaacs correspondence for blocks of finite groups. *Arch. Math. (Basel)*, 82(6):488–494, 2004.

[133] Alexander Zimmermann. Derived equivalences of orders. In *Representation theory of algebras (Cocoyoc, 1994)*, pages 721–749. Amer. Math. Soc., Providence, RI, 1996.

[134] Alexander Zimmermann. Two sided tilting complexes for Green orders and Brauer tree algebras. *J. Algebra*, 187(2):446–473, 1997.

Joseph Chuang and Jeremy Rickard
University of Bristol
School of Mathematics
University Walk
Bristol BS8 1TW
ENGLAND

15
Morita theory in stable homotopy theory

Brooke Shipley

Abstract

We discuss an analogue of Morita theory for ring spectra, a thickening
of the category of rings inspired by stable homotopy theory. This follows
work by Rickard and Keller on Morita theory for derived categories. We
also discuss two results for derived equivalences of DGAs which show
they differ from derived equivalences of rings.

1 Introduction

Although the usual paradigm in algebraic topology is to translate topological problems into algebraic ones, here we discuss the translation of
algebra into topology. Specifically, we discuss an analogue of Morita
theory for a thickening of the category of rings inspired by stable homotopy theory. Here rings in the classical sense correspond to ordinary
cohomology theories, whereas "rings up to homotopy" correspond to
generalized cohomology theories. Although these generalized rings have
a considerable history behind them, only recent progress has allowed the
wholesale transport of algebraic methods into this domain.

The topological analogue of Morita theory is very similar to the
following two algebraic versions. To emphasize this similarity we
delay discussion of the technical terminology used in these algebraic
statements. We include the third condition below since it is the most
familiar criterion for classical Morita equivalences, but we concentrate

393

on the equivalence of the first two conditions in the other contexts.

Theorem 1.1. *Two rings R and R' are* Morita equivalent *if the following equivalent conditions hold.*

(1) *The categories of right modules over R and R' are equivalent.*

(2) *There is a finitely generated projective (strong) generator M in Mod-R' such that the endomorphism ring $(_,R')(M, M)$ is isomorphic to R.*

(3) *There is an R-R' bimodule N such that $- \otimes_R N \colon \text{Mod-}R \longrightarrow \text{Mod-}R'$ is an equivalence of categories.*

Next we state a reformulation from [8] of Rickard's characterization of equivalences of derived categories [36, 37]; see also [18, 9.2] and [21].

Theorem 1.2. *Two rings R and R' are* derived equivalent *if the following equivalent conditions hold.*

(1) *The unbounded derived categories of R and R' are triangulated equivalent.*

(2) *There is a compact generator M of $\mathcal{D}(R')$ such that the graded endomorphism ring in the derived category, $\mathcal{D}(R')(M, M)_*$, is isomorphic to R (concentrated in degree zero).*

If M satisfies the conditions in (2), it is called a tilting complex.

For the analogue in stable homotopy theory, one must make the appropriate changes in terminology. The real difference lies in the meaning of "ring" and "equivalence"; we devote a section below to defining each of these terms. In Section 2 we introduce "abelian groups up to homotopy" or *spectra* and the associated "rings up to homotopy" or *ring spectra*. Although spectra are the main object of study in stable homotopy theory and have been studied for almost forty years, only recent work has made the definitions of ring spectra easily accessible. In Section 3 we consider a notion of "up to homotopy" equivalences of categories, or *Quillen equivalences*. Here "homotopy" is determined by defining *Quillen model structures* on the relevant categories. This extra structure allows one to apply standard techniques of homotopy theory in non-standard settings.

Theorem 1.3 ([43], [10]). *The following two statements are equivalent for ring spectra R and R'.*

(1) *The Quillen model categories of R-module spectra and R′-module spectra are Quillen equivalent.*

(2) *There is a compact generator M in* Ho(R'-modules) *such that the derived endomorphism ring spectrum* $\overline{\mathrm{Hom}}_{R'}(M, M)$ *is weakly equivalent to R.*

The proofs of these three statements basically have the same format. In each situation, one can prove that (1) implies (2) by noting that the image of R under the given equivalence has the properties required of M. For (2) implies (1), the conditions on M are exactly what is needed to show that the appropriate analogue of $(_,R')(M, -)$ induces the necessary equivalence. For example, in the classical statement one asks that M is a finitely generated projective module to ensure that $(_,R')(M, -)$ preserves sums (and is exact). In a triangulated category $\mathcal{T}$, M is *compact* if $\mathcal{T}(M, -)$ preserves sums. (For example, in $\mathcal{D}(R)$ a bounded complex of finitely generated projectives is compact [3] and conversely [18, 5.3], [29].) Also, if M is compact then it is a *(weak) generator* of $\mathcal{T}$ if it detects trivial objects; that is, an object X of $\mathcal{T}$ is trivial if and only if there are no graded maps from M to X, $\mathcal{T}(M, X)_* = 0$. (For triangulated categories with infinite coproducts this is shown to be equivalent to the more common definition of a generator in [43, 2.2.1].) See [41] for a more detailed survey developing the three theorems above.

The topological analogue of Morita theory has had many applications in stable homotopy theory, see Remark 3.5, but here we discuss some algebraic applications. In the last two sections of this paper we discuss results for derived equivalences of rings and differential graded algebras. The first of these results shows that any derived equivalence of rings lifts to the stronger "up to homotopy" equivalence mentioned above; see Theorem 4.1. As a consequence, all possible homotopy invariants are preserved by derived equivalences of rings, including algebraic K-theory; see Corollary 4.2. The next results show that this is not the case for differential graded algebras. Example 4.5 produces two DGAs whose derived categories are equivalent as triangulated categories but which have non-isomorphic K-theories. In Section 5 we also give an explicit example of a derived equivalence of differential graded algebras that does not arise from an algebraic tilting complex. Instead there is a topological tilting spectrum which comes from considering DGAs as examples of ring spectra.

Acknowledgments: This paper is based on a talk I gave at the MSRI con-

ference "Commutative Algebra: Interactions with Homological Algebra
and Representation Theory" in February of 2003. I'd like to thank John
Greenlees for the invitation to speak at that conference and Henning
Krause for the invitation to write these notes. I'd also like to thank Dan
Dugger and Stefan Schwede for the enjoyable collaborations which led
to this work.

2 Spectral Algebra

Before considering "rings up to homotopy" we must consider "abelian
groups up to homotopy". The analogue of an abelian group here is
a *spectrum*. Each spectrum corresponds to a generalized cohomology
theory. For example, the Eilenberg-Mac Lane spectrum HA is associated with ordinary cohomology with coefficients in the abelian group A.
Another well-known cohomology theory is complex K-theory; $K^*(X)$
basically classifies the complex vector bundles on X and the associated
spectrum is denoted K. The analogue of a ring is then a generalized
cohomology theory with a product. Both HA and K are ring spectra;
the associated products are the cup product and the product induced
by the tensor product of vector bundles.

At a first approximation, a spectrum is a sequence of pointed spaces.
Each of these spaces represents one degree of the cohomology theory;
for example $H^n(X; A)$, the nth ordinary cohomology of a space X, is
isomorphic to homotopy classes of maps from X with a disjoint base
point added to $K(A, n)$, denoted $[X_+, K(A, n)]$. Here $K(A, n)$ is the
Eilenberg-Mac Lane space whose homotopy type is determined by having homotopy concentrated in degree n, $\pi_n K(A, n) = A$; for example,
$K(\mathbb{Z}, 1)$ is the circle S^1. Complex K-theory is represented by the infinite
unitary group U in odd degrees and by the classifying space cross the
integers, $BU \times \mathbb{Z}$, in even degrees.

One needs additional structure on a sequence of pointed spaces
though to make sure the associated homology theory satisfies the
Eilenberg-Steenrod axioms for a homology theory. (Note that since
we are considering generalized homology theories here we remove the
dimension axiom.) To introduce this structure we need the following
pointed version of the Cartesian product of spaces. The smash product
should be thought of as an analogue of the tensor product in algebra;

here the base point acts like a zero element.

Definition 2.1. The *smash product* $X \wedge Y$ of two pointed spaces X and Y with base points pt_X and pt_Y is given by $X \wedge Y = X \times Y/(X \times pt_Y) \cup (pt_X \times Y)$. The *suspension* of a pointed space X, ΣX is defined by $S^1 \wedge X$.

Finally we give the formal definition of a spectrum.

Definition 2.2. A *spectrum* Y is a sequence of pointed spaces $(Y_1, Y_2, \cdots, Y_n, \cdots)$ with structure maps $\Sigma Y_n \longrightarrow Y_{n+1}$. A map of spectra $f: Y \longrightarrow Z$ is given by a sequence of maps $f_n: Y_n \longrightarrow Z_n$ which commute with the structure maps.

Given any pointed space X, there is an associated *suspension spectrum* $\Sigma^\infty X$ given by $(X, \Sigma X, \Sigma^2 X, \cdots)$. A particularly important example is the sphere spectrum, $\mathbb{S} = \Sigma^\infty S^0$ given by $(S^0, S^1, S^2, \cdots)$ since the suspension of the n-sphere is the $n + 1$-sphere.

We should mention here that in general the association of a cohomology theory to a spectrum Y is not as simple as the formula given for HA and K. We explain this point in Remark 3.2.

As mentioned above, a ring spectrum is a spectrum associated to a generalized cohomology theory with a product. Ordinary cohomology theory with coefficients in a ring R has such a product given by the cup product $H^p(X; R) \otimes H^q(X; R) \longrightarrow H^{p+q}(X; R)$. This product is induced by a compatible family of maps on the associated spaces: $K(R, p) \wedge K(R, q) \longrightarrow K(R, p + q)$. Basically then, a ring spectrum R is a spectrum with compatibly associative and unital products $R_p \wedge R_q \longrightarrow R_{p+q}$. The unital condition here ensures that these products interact in a compatible way with the suspension structure as well. Unfortunately, although this simple outline is a good beginning for the definition of ring spectra, no one has actually been able to finish it in a way that captures the objects we actually care about.

One wants to define a smash product on the category of spectra which acts like a tensor product, that is, is a symmetric monoidal product. Then the ring spectra would be spectra R with an associative and unital product map $R \wedge R \longrightarrow R$. The problem here is that one must choose a sequence of spaces to define a spectrum $R \wedge R$ from among the two-dimensional array of spaces $R_p \wedge R_q$. Boardman gave the first approximation to defining such a smash product; his version is only commutative and associative up to homotopy [2]; see also [1, III]. Using this smash product one can instead consider A_∞ or E_∞ ring spectra, which are associative or commutative rings up to all higher homotopies [28]. These definitions are cumbersome in comparison to the algebraic analogues though.

In 1991, Gaunce Lewis published a paper which seemed to imply that these definitions were as good as could be [23]. He showed that no smash product exists which satisfies five reasonable axioms. One of his axioms was that the smash product is strictly commutative and associative; the other axioms asked for reasonable relationships between the smash product for spectra and the smash product for spaces.

Luckily, it turns out that Lewis' axioms were just a bit too strong. In the last few years, several new ways of defining categories of spectra with a good smash product have been discovered. In these categories the smash product acts like a tensor product and the usual algebraic definitions capture the correct notion of ring, algebra and module spectra. This has made it possible to really do algebra in the setting of stable homotopy theory.

Rather than give an overview of the different categories and products that have been defined, we concentrate on the one version, symmetric spectra, which has proved most useful for comparisons with algebra. The definitions for symmetric spectra are also the closest to the simple approximate definitions above. Because the different models for spectra all agree 'up to homotopy', in a way we discuss in Remark 3.4, one can always choose to use whichever category is most convenient to the problem at hand.

Definition 2.3 ([17]). A *symmetric sequence* is a sequence of pointed spaces, $(X_1, X_2, \cdots)$, with an action of Σ_n, the nth symmetric group, on X_n, the nth space. For two symmetric sequences X and Y, define $X \otimes Y$ as the sequence with $(X \otimes Y)_n = \bigvee_{p+q=n} \Sigma_n \wedge_{\Sigma_p \times \Sigma_q} (X_p \wedge Y_q)$.

The suspension spectra introduced above are examples of symmetric sequences with the symmetric action taking place on the suspension coordinates. That is, $\Sigma^n X = S^1 \wedge S^1 \cdots S^1 \wedge X$ and the copies of S^1 are permuted. In particular, the sequence of spheres, $\mathbb{S}$, is a commutative ring under this product; specifically, there is an associative, commutative and unital map $\mathbb{S} \otimes \mathbb{S} \longrightarrow \mathbb{S}$ induced by the $\Sigma_p \times \Sigma_q$-equivariant maps $S^p \wedge S^q \longrightarrow S^{p+q}$. Here we define $S^n = (S^1)^{\wedge n} = S^1 \wedge S^1 \cdots S^1$.

Definition 2.4 ([17]). A *symmetric spectrum* is a module over the commutative ring $\mathbb{S}$ in the category of symmetric sequences. This module structure on a symmetric sequence X is determined by an associative and unital map $\alpha_X \colon \mathbb{S} \otimes X \longrightarrow X$. Unraveling this further, a symmetric spectrum X is a symmetric sequence $(X_1, X_2, \cdots)$ with compatible $\Sigma_p \times \Sigma_q$-equivariant maps $S^p \wedge X_q \longrightarrow X_{p+q}$. The smash product $X \wedge Y = X \otimes_{\mathbb{S}} Y$ is the coequalizer of the two maps $1 \otimes \alpha_Y, \alpha_X \otimes 1$ from $X \otimes \mathbb{S} \otimes Y$ to $X \otimes Y$. Similarly, a *symmetric ring spectrum* is an $\mathbb{S}$-algebra; specifically, a symmetric ring spectrum R is a symmetric sequence with compatible unit map $\eta \colon \mathbb{S} \longrightarrow R$ and associative multiplication map $\mu \colon R \otimes R \longrightarrow R$.

The components of the multiplication map for a symmetric ring spectrum R are $\Sigma_p \times \Sigma_q$ equivariant maps $R_p \wedge R_q \longrightarrow R_{p+q}$. For the rest of this article we shorten "symmetric ring spectrum" to just "ring spectrum"; in particular, note that a *commutative* symmetric ring spectrum would have the added condition that $\mu \tau = \mu$ where τ is the twist map on $R \wedge R$. For R a (commutative) ring spectrum, the definitions of R-modules and R-algebras follow similarly.

The Eilenberg-Mac Lane spectrum HR for R any classical ring plays an important role in this paper. Since we want HR to be a ring spectrum (commutative if R is), we need to be careful about our choice of the spaces $K(R, n)$. It is easiest to define $K(R, n)$ as a simplicial set and then take its geometric realization if we want to work with topological spaces; see [14, 27]. (Actually, throughout this paper "space" can be taken to mean either simplicial set or topological space.) Define the simplicial set S^1 to be $\Delta[1]/\partial\Delta[1]$, the one-simplex with its two endpoints identified. Then define $HR_n = K(R, n)$ to be the simplicial set which in level k is the free R-module with basis the non-basepoint k-simplices of $S^n = (S^1)^{\wedge n}$ [17, 1.2.5].

3 Quillen model categories

In algebra one considers derived categories; given an abelian category $\mathcal{A}$, the *derived category* of $\mathcal{A}$, $\mathcal{D}(\mathcal{A})$, is the localization obtained from the category $\mathcal{C}h(\mathcal{A})$ of (unbounded) chain complexes in $\mathcal{A}$ by inverting the *quasi-isomorphisms*, or maps which induce isomorphisms in homology.

The analogues in homotopy theory are homotopy categories; beginning with a category $\mathcal{C}$ and a notion of *weak equivalence* the *homotopy category* Ho($\mathcal{C}$) is obtained by inverting the weak equivalences. By requiring more structure on $\mathcal{C}$, namely a *Quillen model structure*, one can avoid the set theoretic difficulties which exist when inverting a general class of morphisms. This extra structure also enables the application of standard techniques of homotopy theory and captures more homotopical information than the homotopy category alone; see Section 5 for an example of two non-Quillen equivalent model categories with equivalent homotopy categories.

A *Quillen model category* is a category $\mathcal{C}$ with three distinguished types of maps called *weak equivalences, cofibrations* and *fibrations* which satisfy the five axioms below [35]. A modern variation on these axioms appears in [16], and [11] is a very good, short introduction to model categories. It may be useful when reading these axioms to keep a couple of archetypal examples in mind. The category of bounded below chain complexes of R-modules $\mathcal{C}h_+(R)$, is a Quillen model category with weak equivalences the quasi-isomorphisms, cofibrations the injections with levelwise projective cokernels and fibrations the surjections. Similarly, $\mathcal{C}h_+(R)$ with the same weak equivalences, cofibrations the injections and fibrations the surjections with levelwise injective kernels is also a Quillen model category. These structures can both be extended to the category of unbounded chain complexes but the cofibrations in the first *(projective)* case and the fibrations in the second *(injective)* case are less explicitly defined [16]. The homotopy categories associated to these model categories are equal (as triangulated categories) to the corresponding (bounded or unbounded) variants of the derived category of R; see [49, 10.3.2].

Axioms 3.1 ([35]). *Axioms for a Quillen model category.*

(1) $\mathcal{C}$ *admits all finite limits and colimits.*

(2) *If two out of three of f, g and gf are weak equivalences, so is the third.*

(3) *Cofibrations, fibrations and weak equivalences are closed under retract.*

(4) *Any map f may be factored in two ways: $f = pi$ with i a cofibration and p a fibration and a weak equivalence (a trivial fibration),*

and f = qj with j a cofibration and weak equivalence (a trivial
cofibration) and q a fibration.

(5) *Given a commuting square*

    with A $\longrightarrow$ B a cofibration and X $\longrightarrow$ Y a fibration, then a morphism B $\longrightarrow$ X making both triangles commute exists if either of the two vertical maps is a weak equivalence.

Remark 3.2. To give some feel for working with a model category, note
that in the projective model category on $\mathcal{C}h_+(R)$ the first factorization
mentioned in (4) above applied to a map $0 \longrightarrow M$ (with M an R-
module concentrated in degree zero) produces a projective resolution of
M. Similarly, the second factorization produces injective resolutions in
the injective model category on $\mathcal{C}h_+(R)$.

Also, the formula mentioned above for associating a cohomology theory
to a spectrum works only for fibrant spectra. An object in a model
category is *fibrant* if the map $Y \longrightarrow *$ is a fibration, where $*$ is the
terminal object. In the standard model category structures on both
symmetric spectra and the category of spectra in Definition 2.2, it turns
out that Y is fibrant if and only if $Y_n \longrightarrow \mathrm{map}_*(S^1, Y_{n+1}) = \Omega Y_{n+1}$ is a
weak equivalence of spaces where this map is adjoint to the map $\Sigma Y_n \longrightarrow$
Y_{n+1}. Thus, for Y fibrant $[X, Y_n] \cong [X, \Omega Y_{n+1}] \cong [\Sigma X, Y_{n+1}]$; this
corresponds to the fact that for any cohomology theory the cohomology
group in degree $n+1$ of ΣX agrees with the nth cohomology group of X.
A weakly equivalent *fibrant replacement* always exists due to axiom (4);
one factors the map $Y \longrightarrow *$ as $Y \longrightarrow Y^f \longrightarrow *$ with $Y \longrightarrow Y^f$ a weak
equivalence and $Y^f \longrightarrow *$ a fibration. Then, for a general spectrum Y,
the associated cohomology theory is given by homotopy classes of maps
into the levels of Y^f.

We also have the following notion of equivalence between model
categories.

Definition 3.3 ([35],[16]). A *Quillen adjunction* between two Quillen
model categories $\mathcal{C}$ and $\mathcal{D}$ is given by an adjoint pair of functors $L :$
$\mathcal{C} \rightleftarrows \mathcal{D} : R$ where the left adjoint, L, preserves cofibrations and the right

adjoint, R, preserves fibrations. It follows that L also preserves trivial cofibrations and R preserves trivial fibrations. Under these conditions the adjoint functors induce adjoint derived functors on the homotopy categories $\bar{L} : \mathrm{Ho}\,\mathcal{C} \rightleftarrows \mathrm{Ho}\,\mathcal{D} : \bar{R}$. This pair is a *Quillen equivalence* if $\bar{L}$ and $\bar{R}$ form an equivalence of categories between $\mathrm{Ho}\,\mathcal{C}$ and $\mathrm{Ho}\,\mathcal{D}$. Two model categories are said to be *Quillen equivalent* if there is a string of Quillen equivalences between them.

We call all of the homotopical information associated to a Quillen model structure a *homotopy theory*. For example, the identity functors between the projective and injective model categories defined above on $\mathcal{C}h_+(R)$ form a Quillen equivalence. That is, the projective and injective model categories define the same homotopy theory, which encompasses the homological algebra associated with R.

Remark 3.4. The category of spectra in Definition 2.2 [4], the category of symmetric spectra [17] and all of the other new categories of spectra [12, 24, 25, 26] have Quillen equivalent model structures [26, 39]. Moreover, the associated categories of rings, modules and algebras are also Quillen equivalent. This is the sense referred to in Section 2 in which these models all agree 'up to homotopy'; since Quillen equivalent model structures define the same homotopy theory, any homotopically invariant statement about one model applies to the others as well.

These models for spectra actually define stable homotopy theories; that is, theories where suspension is invertible in the homotopy category. Suspension corresponds to shifting the spaces in a spectrum down by one; this is easiest to see with suspension spectra. One can show that up to homotopy this functor is invertible with inverse given by shifting up by one. The category $\mathcal{C}h(R)$ is another example of a stable homotopy theory; again suspension and desuspension are homotopic to shifting. The homotopy category associated to a stable homotopy theory is a triangulated category [16, 7.1.6]; the triangles are given by the homotopy fiber sequences (which can be shown to agree with the homotopy cofiber sequences). In particular, Quillen equivalences between stable model categories induce triangulated equivalences on the homotopy categories.

We end this section with two remarks related to Theorem 1.3 and Quillen equivalences.

Remark 3.5. There have been many applications of variations of Theorem 1.3 in stable homotopy theory. A variation on this theorem shows that any stable model category with a compact generator is Quillen equivalent to modules over a ring spectrum [43]. This shows that many homotopical settings can be translated into settings in spectral algebra. Applications of this variation include various characterizations of the stable homotopy theory of spectra [44, 46, 40]. Since any rational ring spectrum is an $H\mathbb{Q}$-algebra, combining this theorem with Theorem 4.3 gives algebraic models for any rational stable homotopy theory [48]. This is used in [47, 15] to form practical algebraic models for rational torus-equivariant spectra for tori of any dimension.

Remark 3.6. In [43] condition (1) of Theorem 1.3 actually requires that the functors involved in the Quillen equivalences preserve the enrichment of these module categories over spectra. In [10] we show that this requirement is not actually necessary by using ideas similar to those in [6, 7].

4 Differential graded algebras

In the rest of this paper we explore some results which lie between the Morita theory for derived categories of rings and the Morita theory for ring spectra.

First, we consider the overlap between Theorems 1.2 and 1.3. Both of these theorems apply to rings since there is an Eilenberg-Mac Lane ring spectrum associated to any ring. Then we extend the association between spectra and classical algebraic objects to differential graded rings, modules and algebras and their associated homotopy theories. This shows that the category of DGAs lies between rings and ring spectra. Finally, we discuss the Morita theory of DGAs.

Since any Quillen equivalence induces an equivalence on the homotopy (or derived) categories, Theorem 1.3 may seem to give a stronger result for ring spectra than Theorem 1.2 does for rings. Theorem 1.3, however, has a stronger hypothesis on the endomorphism ring of the generator since the weak equivalence type of a ring spectrum is a finer invariant in general than the homotopy ring. (Here the graded ring of homotopy groups of $\overline{\mathrm{Hom}}_{R'}(M, M)$ is isomorphic to the graded ring of endomorphisms of M in $\mathrm{Ho}(R'\text{-modules})$.) When that homotopy is concentrated in degree zero though, the weak equivalence type of a ring spectrum is determined by its homotopy. Thus, for an Eilenberg-Mac Lane spectrum the conditions on the generator in the two theorems are equivalent. From this one can show that for rings the notions of Quillen equivalence and derived equivalence actually agree.

Theorem 4.1 ([8]). *For two rings R and R' the following are equivalent.*

(1) *The derived categories $\mathcal{D}(R)$ and $\mathcal{D}(R')$ are triangulated equivalent.*

(2) $Ch(R)$ *and* $Ch(R')$ *are Quillen equivalent model categories.*

(3) *There is a compact generator M in $\mathcal{D}(R')$ such that the graded endomorphism ring in the derived category $\mathcal{D}(R')(M, M)_*$ is isomorphic to R.*

This is one case where the homotopy (or derived) category actually determines the whole homotopy theory. For this reason, one expects a derived equivalence between two rings to induce an isomorphism on any homotopy invariant of rings. For example, Hochschild homology and cyclic homology have been shown to be invariants of derived equivalences [36, 37, 19]; see also [20]. One can also show that a Quillen equivalence preserves algebraic K-theory. We have the following result. For regular rings this can also be derived from Neeman's work on the K-theory of abelian categories [30, 31, 32, 33, 34]; this result also appears in [5] with flatness conditions.

Corollary 4.2 ([8]). *If $\mathcal{D}(R)$ and $\mathcal{D}(R')$ are triangulated equivalent, then $K_*(R) \cong K_*(R')$.*

Just as the Eilenberg-Mac Lane ring spectrum HR represents ordinary cohomology with coefficients in R, there is a ring spectrum HA associated to hypercohomology with coefficients in a DGA A. Here H is defined as a composite of several functors, so we do not define it explicitly for DGAs. In fact H induces a Quillen equivalence between $H\mathbb{Z}$-algebras and DGAs. The following statement collects the various related results. For the relevant Quillen model structures, see [17, 42].

Theorem 4.3.   (1) [43, 38] *For any classical ring R, the model categories of $Ch(R)$ and HR-modules are Quillen equivalent.*

(2) [48] *The model categories of differential graded algebras and H$\mathbb{Z}$-algebras are Quillen equivalent.*

(3) [48] *For any DGA A, the model categories of differential graded A-modules and HA-modules are Quillen equivalent.*

Using this theorem to consider DGAs as lying between rings and ring spectra, one might conjecture that there is an intermediary Morita theory for DGAs similar to Theorem 4.1 for rings. The homology ring, however, does not determine the quasi-isomorphism type for a DGA; for example, a non-formal DGA is not quasi-isomorphic to its homology (see Section 5 for an example). Thus, a Morita theorem for DGAs must require a condition on the quasi-isomorphism type of the derived endomorphism DGA of a generator rather than just its homology. (Here the derived endomorphism DGA is an analogue of the derived endomorphism ring spectrum.) This leads to the following question.

Question 4.4. *For two DGAs A and B, are the following equivalent?*

(1) $\mathcal{D}(A)$ *is triangulated equivalent to* $\mathcal{D}(B)$.

(2) *The model categories of differential graded modules over A and B are Quillen equivalent.*

(3) *There is a compact generator M in* $\mathcal{D}(A)$ *whose derived endomorphism DGA is quasi-isomorphic to B.*

If a generator M exists which satisfies (3), then one can show that the model categories of differential graded modules over A and B are Quillen equivalent and hence $\mathcal{D}(A)$ and $\mathcal{D}(B)$ are triangulated equivalent. In fact this follows from Theorems 4.3 and 1.3 by considering the DGAs as examples of ring spectra and replacing differential graded modules by module spectra. Hence, indeed (3) implies (2) implies (1). The other implications fail though. In Example 4.5 below we give an example of a derived equivalence with no underlying Quillen equivalence; so (1) does not imply (2). In Section 5 below we'll also give an example of a Quillen equivalence between DGAs with no generator satisfying (3); so (2) does not imply (3). This example arises from considering DGAs as special examples of ring spectra.

Example 4.5. There are two DGAs A and B whose derived categories are equivalent even though the associated model categories of differential graded modules are not Quillen equivalent. In particular, Corollary 4.2

does not extend to DGAs because A and B have non-isomorphic K-theories. This example is based on Marco Schlichting's work in [45] which shows that for $p > 3$ the stable module categories over $(\mathbb{Z}/p)[\epsilon]/\epsilon^2$ and $\mathbb{Z}/p^2$ are triangulated equivalent but the associated K-theories are not isomorphic.

Proposition 4.6 ([9]). *The model category underlying each of these stable module categories is Quillen equivalent to a category of differential graded modules over a DGA.*

$$\mathrm{Stmod}((\mathbb{Z}/p)[\epsilon]/\epsilon^2) \simeq_{Quillen} \mathrm{d.\,g.\,Mod}\text{-}A$$

$$\mathrm{Stmod}(\mathbb{Z}/p^2) \simeq_{Quillen} \mathrm{d.\,g.\,Mod}\text{-}B$$

For the dual numbers, the DGA A is $\mathbb{Z}/p[x, x^{-1}]$, a polynomial algebra on a class in degree one and its inverse with trivial differential. Let $k = \mathbb{Z}[x, x^{-1}]$, a similarly graded polynomial algebra over $\mathbb{Z}$. For $\mathbb{Z}/p^2$, the DGA B is generated over k by a class e in degree one with the relations that $e^2 = 0$, $ex + xe = x^2$, $de = p$ and $dx = 0$. Here A and B are the endomorphism DGAs of the Tate resolution of a generator $(\mathbb{Z}/p)$ of the respective stable module categories [9]; see [18, 4.3] for a related general statement.

Because the stable module categories are triangulated equivalent, it follows that $\mathcal{D}(A)$ and $\mathcal{D}(B)$ are also equivalent. Since Quillen equivalences induce isomorphisms in K-theory [8], Schlichting's work showing the stable module categories' K-theories are non-isomorphic implies that the model categories underlying the stable module categories are not Quillen equivalent. Hence the differential graded modules over A and B are not Quillen equivalent, either.

5 Two topologically equivalent DGAs

As mentioned above, in this section we discuss an example from [10] of two DGAs A and B with Quillen equivalent categories of differential graded modules where there is no generator satisfying the properties listed in Question 4.4. This example arises from replacing the DGAs by their associated $H\mathbb{Z}$-algebras HA and HB. One can then forget the $H\mathbb{Z}$-algebra structure and consider them as ring spectra, or $\mathbb{S}$-algebras. Since some structure has been forgotten it is reasonable to expect that there are more maps and equivalences between HA and HB as $\mathbb{S}$-algebras than between the DGAs A and B. This leads to the following definition.

Definition 5.1. Two DGAs A and B are *topologically equivalent* if their associated $H\mathbb{Z}$-algebras HA and HB are equivalent as ring spectra ($\mathbb{S}$-algebras).

Quasi-isomorphic DGAs are topologically equivalent as well, but the converse does not hold. Below we give an example of two DGAs A and A' which are topologically equivalent but not quasi-isomorphic. Moreover we show that these two DGAs have equivalent derived categories even though there is no compact generator in $\mathcal{D}(A')$ whose derived endomorphism DGA is quasi-isomorphic to A. In fact, any two topologically equivalent DGAs have equivalent derived categories.

Theorem 5.2. *If A and B are topologically equivalent DGAs then $\mathcal{D}(A)$ is triangulated equivalent to $\mathcal{D}(B.)$. Moreover, the associated categories of differential graded modules are Quillen equivalent.*

This is based on the following.

Corollary 5.3. *If $R \longrightarrow R'$ is a weak equivalence of ring spectra, then the model categories of R-module spectra and R'-module spectra are Quillen equivalent.*

This follows from Theorem 1.3 by taking R' as the compact generator required in condition (2) since $\overline{\mathrm{Hom}}_{R'}(R', R') = R'$ is weakly equivalent to R. Although this is stated as a corollary here, it is actually an ingredient in the proof of Theorem 1.3.

Proof of Theorem 5.2. By Theorem 4.3 differential graded modules over A and module spectra over HA are Quillen equivalent and the same is true for B and HB. Then Corollary 5.3 provides the bridge between these pairs showing that HA-module spectra and HB-module spectra are Quillen equivalent since HA and HB are weakly equivalent. Since Quillen equivalences between stable model categories induce triangulated equivalences on the homotopy (or derived) categories, the first statement follows from the second. $\qquad\qquad\square$

Now we introduce the two DGAs from [10] which are topologically equivalent but not quasi-isomorphic. The first DGA we consider is a truncated polynomial ring over the integers on a class in degree one, $A = \mathbb{Z}[e]/(e^4)$, with $d(e) = 2$. Note that A is not graded commutative, since e^2 is not trivial. Since $d(e^2) = 0$ and $d(e^3) = 2e^2$, the homology of A is an exterior algebra over $\mathbb{Z}/2$ on a class in degree 2. Let $A' = H_*A = \Lambda_{\mathbb{Z}/2}(\alpha_2)$, with trivial differential.

One can show that A and A' are not quasi-isomorphic. In fact, A is not quasi-isomorphic to any DGA over $\mathbb{Z}/2$. In particular, there are no maps from $\mathbb{Z}/2$, thought of as a DGA concentrated in degree zero, to A even up to homotopy; that is, $[\mathbb{Z}/2, A] = 0$ in the homotopy category of associative DGAs. On the other hand, any DGA over $\mathbb{Z}/2$ would have a unit map to it from $\mathbb{Z}/2$. To calculate that $[\mathbb{Z}/2, A] = 0$, one replaces $\mathbb{Z}/2$ by a quasi-isomorphic free associative DGA over $\mathbb{Z}$ and shows that there are no maps from it to A. In low degrees, one possibility for this replacement begins with a generator x in degree one with $dx = 2$, a generator y in degree three with $dy = x^2$ and other generators in degree four or higher.

This also shows that there is no compact generator M in $\mathcal{D}(A')$ whose derived endomorphism DGA is quasi-isomorphic to A. Since A' is a DGA over $\mathbb{Z}/2$, all of the homomorphism groups between differential graded A' modules are naturally $\mathbb{Z}/2$ vector spaces. It follows that the derived endomorphism DGA of any differential graded A' module would also be a DGA over $\mathbb{Z}/2$, so this endomorphism DGA cannot be quasi-isomorphic to A. Once we show that A and A' are topologically equivalent, A and A' give a counter example to the equivalence of conditions (2) and (3) in Question 4.4 by Theorem 5.2.

We use topological Hochschild cohomology, THH^*, to show that the associated ring spectra HA and HA' are weakly equivalent, or A and A' are topologically equivalent. This is the analogue of Hochschild cohomology, HH^*, for spectral algebra. For a ring R and an R-bimodule M, DGAs with non-zero homology $H_0 = R$ and $H_n = M$ are classified by $HH_{\mathbb{Z}}^{n+2}(R; M)$. Similarly, ring spectra with non-zero homotopy $\pi_0 = R$ and $\pi_n = M$ are classified by $THH_{\mathbb{S}}^{n+2}(HR; HM)$ [22].

For R a k-algebra and M an R-bimodule over k, $HH_k^*(R; M)$, is calculated as the derived bimodule homomorphisms from R to M over the derived tensor product $R\overline{\otimes}_k R^{op}$. In the case where k is a field, it is not necessary to derive the tensor product $R \otimes_k R^{op}$ so this extra wrinkle is often suppressed. As an example, one can calculate that $\mathbb{Z}/2\overline{\otimes}_{\mathbb{Z}}\mathbb{Z}/2$ is quasi-isomorphic to the exterior algebra $\Lambda_{\mathbb{Z}/2}(x_1)$, so $HH_{\mathbb{Z}}^*(\mathbb{Z}/2;\mathbb{Z}/2) = Ext_{\Lambda_{\mathbb{Z}/2}(x_1)}(\mathbb{Z}/2,\mathbb{Z}/2) = \mathbb{Z}/2[\sigma_2]$.

THH^* is defined similarly, by considering maps of bimodule spectra. In particular, $THH_{H\mathbb{Z}}^*(HR; HM) = HH_{\mathbb{Z}}^*(R, M)$, although THH^* can be considered over any ring spectrum. Here we need to consider $THH_{\mathbb{S}}^*(H\mathbb{Z}/2, H\mathbb{Z}/2)$; by [13], this is $\Gamma_{\mathbb{Z}/2}[\tau_2]$, a divided power algebra over $\mathbb{Z}/2$ which is isomorphic to an exterior algebra over $\mathbb{Z}/2$ on classes e_i for $i \geqslant 1$ with e_i in degree 2^i.

The unit map $\mathbb{S} \longrightarrow H\mathbb{Z}$ induces a map

$$\Phi \colon HH_{\mathbb{Z}}^*(\mathbb{Z}/2, \mathbb{Z}/2) \longrightarrow THH_{\mathbb{S}}^*(H\mathbb{Z}/2, H\mathbb{Z}/2).$$

The two elements in HH^2 classify the two rings $\mathbb{Z}/4$ and $\mathbb{Z}/2 \oplus \mathbb{Z}/2$. Since the associated Eilenberg-Mac Lane ring spectra are also distinct, Φ must be injective in degree two; hence $\Phi(\sigma) = \tau$. Since A and A' are not quasi-isomorphic, one can check that σ^2 and 0 in HH^4 correspond respectively to A and A'. Since $\tau^2 = 0$ and Φ is a ring homomorphism, $\Phi(\sigma^2) = 0$. So HA and HA' correspond to the same homotopy type as ring spectra; that is, HA and HA' are weakly equivalent and A and A' are topologically equivalent.

References

[1] Adams, J. Frank, *Stable homotopy and generalised homology*, University of Chicago Press, Chicago, 1974.

[2] Boardman, J.M., *Stable homotopy theory*, preprint, Warwick University, 1964.

[3] M. Bökstedt and A. Neeman, *Homotopy limits in triangulated categories* Compositio Math. **86** (1993), 209-234.

[4] A. K. Bousfield and E. M. Friedlander, *Homotopy theory of Γ-spaces, spectra, and bisimplicial sets*. Geometric applications of homotopy theory (Proc. Conf., Evanston, Ill., 1977), II, pp. 80–130, Lecture Notes in Math., **658**, Springer, Berlin, 1978.

[5] J. Daplayrat-Glutron, *Equivalence dérivées et K-théorie (d'après Thomason-Trobaugh)*, Mémoire de DEA, Université Denis Diderot - Paris7, 1999, http://www.math.jussieu.fr/~keller/

[6] D. Dugger, *Universal homotopy theories*, Adv. Math. **164** (2001), no. 1, 144-176.

[7] D. Dugger, *Replacing model categories with simplicial ones*, Trans. Amer. Math. Soc., **353** (2001), no. 12, 5003-5027.

[8] D. Dugger and B. Shipley, *K-theory and derived equivalences*, Duke J. Math., **124** (2004), no. 3, 587-617.

[9] D. Dugger and B. Shipley, *Stable module categories and differential graded rings*, in preparation.

410   *B. Shipley*

[10] D. Dugger and B. Shipley, *Topological equivalences of DGAs*, in preparation.

[11] W. G. Dwyer and J. Spalinski, *Homotopy theories and model categories*, Handbook of algebraic topology (Amsterdam), North-Holland, Amsterdam, 1995, pp. 73–126.

[12] A. D. Elmendorf, I. Kriz, M. A. Mandell, and J. P. May, *Rings, modules, and algebras in stable homotopy theory. With an appendix by M. Cole*, Mathematical Surveys and Monographs, **47**, American Mathematical Society, Providence, RI, 1997.

[13] Vincent Franjou, Jean Lannes, and Lionel Schwartz, *Autour de la cohomologie de Mac Lane des corps finis* Invent. Math. **115** (1994), no. 3, 513–538.

[14] Paul G. Goerss and John F. Jardine, Simplicial homotopy theory. Progress in Mathematics, **174**. Birkhauser Verlag, Basel, 1999.

[15] J. P. C. Greenlees and B. Shipley, *Algebraic models for rational torus-equivariant cohomology theories*, preprint.

[16] M. Hovey, *Model categories*, Mathematical Surveys and Monographs, **63**, American Mathematical Society, Providence, RI, 1999.

[17] M. Hovey, B. Shipley, and J. Smith, *Symmetric spectra*, J. Amer. Math. Soc. **13** (2000), 149–208.

[18] B. Keller, *Deriving DG categories*, Ann. Sci. École Norm. Sup. (4) **27** (1994), 63–102.

[19] B. Keller, *Invariance and localization for cyclic homology of DG algebras*, J. Pure Appl. Algebra **123** (1998), 223-273.

[20] B. Keller, *Hochschild cohomology and derived Picard groups*, J. Pure Appl. Algebra, **190** (2004), 177-196.

[21] B. Keller, *Derived categories and tilting*, this volume.

[22] A. Lazarev, *Homotopy Theory of A_∞ Ring Spectra and Applications to MU-Modules*, K-Theory **24** (3) (2001), 243-281.

[23] Lewis, L.G., Jr., *Is there a convenient category of spectra?*, J. Pure and Appl. Alg. **73** (1991), 233-246.

[24] M. Lydakis, *Smash products and Γ-spaces*, Math. Proc. Cambridge Philos. Soc. **126** (1999), 311–328.

[25] M. Lydakis, *Simplicial functors and stable homotopy theory*, Preprint, Universität Bielefeld, 1998.

[26] M. A. Mandell, J. P. May, S. Schwede and B. Shipley, *Model categories of diagram spectra*, Proc. London Math. Soc., **82** (2001), 441-512.

[27] J. P. May, *Simplicial Objects in Algebraic Topology*, Van Nostrand Mathematical Studies, no. 11, Van Nostrand, Princeton, NJ, 1967.

[28] J. P. May, E_∞ ring spaces and E_∞ ring spectra. With contributions by Frank Quinn, Nigel Ray, and Jorgen Tornehave. Lecture Notes in Mathematics, **577** Springer-Verlag, Berlin-New York, 1977. 268 pp.

[29] A. Neeman, *The connection between the K-theory localisation theorem of Thomason, Trobaugh and Yao, and the smashing subcategories of Bousfield and Ravenel*, Ann. Sci. Ec. Norm. Sup. **25** (1992), 547-566.

[30] A. Neeman, *K-theory for triangulated categories I(A): Homological functors*, Asian J. Math. Vol 1, no. 2 (1997), 330–417.

[31] A. Neeman, *K-theory for triangulated categories. I(B). Homological functors*. Asian J. Math. **1** (1997), no. 3, 435–529.

[32] A. Neeman, *K-theory for triangulated categories. II. The subtlety of the theory and potential pitfalls*. Asian J. Math. **2** (1998), no. 1, 1–125.

[33] A. Neeman, *K-theory for triangulated categories. III(A). The theorem of the heart.* Asian J. Math. 2 (1998), no. 3, 495–589.

[34] A. Neeman, *K-theory for triangulated categories. III(B). The theorem of the heart.* Asian J. Math. 3 (1999), no. 3, 557–608.

[35] D. G. Quillen, *Homotopical algebra,* Lecture Notes in Mathematics, **43**, Springer-Verlag, 1967.

[36] J. Rickard, *Morita theory for derived categories,* J. London Math. Soc. (2) **39** (1989), 436–456.

[37] J. Rickard, *Derived equivalences as derived functors,* J. London Math. Soc. (2) **43** (1991), 37–48.

[38] A. Robinson, *The extraordinary derived category,* Math. Z. **196** (1987), no.2, 231–238.

[39] S. Schwede, *S-modules and symmetric spectra,* Math. Ann. **319** (2001), 517-532.

[40] S. Schwede, *The stable homotopy category has a unique model at the prime 2,* Advances in Mathematics **164** (2001), 24-40.

[41] S. Schwede, *Morita theory in abelian, derived and stable model categories* in Structured Ring Spectra, 33–86, London Mathematical Society Lecture Notes 315, edited by Andrew Baker and Birgit Richter, Cambridge Univ. Press, Cambridge, 2004.

[42] S. Schwede and B. Shipley, *Algebras and modules in monoidal model categories,* Proc. London Math. Soc. **80** (2000), 491-511.

[43] S. Schwede and B. Shipley, *Stable model categories are categories of modules,* Topology **42** (2003), 103-153.

[44] S. Schwede and B. Shipley, *A uniqueness theorem for stable homotopy theory,* Mathematische Zeitschrift **239** (2002), 803-828.

[45] M. Schlichting, *A note on K-theory and triangulated categories,* Invent. Math. **150** (2002), no. 1, 111–116.

[46] B. Shipley, *Monoidal uniqueness of stable homotopy theory,* Advances in Mathematics **160** (2001), 217-240.

[47] B. Shipley, *An algebraic model for rational S^1-equivariant homotopy theory,* Quart. J. Math **53** (2002), 87–110.

[48] B. Shipley, *HZ-algebra spectra are differential graded algebras,* Preprint 2002.

[49] Charles A. Weibel, An introduction to homological algebra. Cambridge Studies in Advanced Mathematics, **38**. Cambridge University Press, Cambridge, 1994.

Brooke Shipley
Department of Mathematics
University of Illinois at Chicago
Chicago, IL
USA

Appendix

Some remarks concerning tilting modules and tilted algebras. Origin. Relevance. Future.

Claus Michael Ringel

The project to produce a Handbook of Tilting Theory was discussed during the Fraueninsel Conference *20 Years of Tilting Theory*, in November 2002. A need was felt to make available surveys on the basic properties of tilting modules, tilting complexes and tilting functors, to collect outlines of the relationship to similar constructions in algebra and geometry, as well as reports on the growing number of generalizations. At the time the Handbook was conceived, there was a general consensus about the overall frame of tilting theory, with the tilted algebra as the core, surrounded by a lot of additional considerations and with many applications in algebra and geometry. One was still looking forward to further generalizations (say something like "pre-semi-tilting procedures for near-rings"), but the core of tilting theory seemed to be in a final shape. The Handbook was supposed to provide a full account of the theory as it was known at that time. The editors of this Handbook have to be highly praised for what they have achieved. But the omissions which were necessary in order to bound the size of the volume clearly indicate that there should be a second volume.

Part 1 will provide an outline of this core of tilting theory. Part 2 will then be devoted to topics where tilting modules and tilted algebras have shown to be relevant. I have to apologize that these parts will repeat some of the considerations of various chapters of the Handbook, but such a condensed version may be helpful as a sort of guideline. Both Parts 1 and 2 contain historical annotations and reminiscences. The final Part 3 will be a short report on some striking recent developments which are motivated by the cluster theory of Fomin and Zelevinsky. In particular, we will guide the reader to the basic properties of cluster tilted algebras, to the relationship between tilted algebras and cluster tilted algebras, but also to the cluster categories which provide a universal setting for all the related tilted and cluster tilted algebras. In addition, we will focus the attention to the complex of cluster tilting objects and exhibit a quite elementary description of this complex. In Part 1 some problems concerning tilting modules and tilted algebras are raised and one may jump directly to Part 3 in order to see in which way these questions have been answered by the cluster tilting theory. We stress that it should be possible to look at the Parts 2 and 3 independently.

1 Basic Setting

The setting to be exhibited is the following: We start with a hereditary artin algebra A and a tilting A-module T. It is the endomorphism ring $B = \operatorname{End}(T)$, called a *tilted* algebra, which attracts the attention. The main interest lies in the comparison of the categories $\operatorname{mod} A$ and $\operatorname{mod} B$ (for any ring R, let us denote by $\operatorname{mod} R$ the category of all R-modules of finite length).

1.1 The representation types of the hereditary categories one starts with

We may assume that A is connected (this means that 0 and 1 are the only central idempotents), and we may distinguish whether A is representation-finite, tame, or wild; for hereditary algebras, this distinction is well understood: the corresponding quiver (or better species) is a Dynkin diagram, a Euclidean diagram, or a wild diagram, respectively. There is a parallel class of algebras: if we start with a canonical algebra A instead of a finite-dimensional hereditary algebra (or, equivalently, with a weighted projective line, or a so called "exceptional curve" in the species case), there is a corresponding tilting procedure. Again the representation theory distinguishes three different cases: A may be domestic, tubular, or wild. Now two of the six cases coincide: the algebras obtained from the domestic canonical algebras via tilting are precisely those which can be obtained from a Euclidean algebra via tilting. Thus, there are 5 possibilities which are best displayed as the following "T": the upper horizontal line refers to the hereditary artin algebras, the middle vertical line to the canonical algebras.

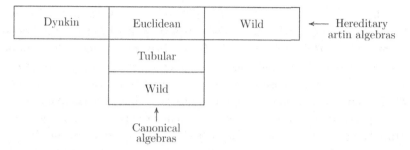

↑
Canonical
algebras

There is a common frame for the five different classes: start with an artin algebra A, such that the bounded derived category $D^b(\operatorname{mod} A)$ is equivalent to the bounded derived category $D^b(\mathcal{H})$ of a hereditary abelian category $\mathcal{H}$. Let T be a tilting object in $D^b(\operatorname{mod} A)$ and B its endomorphism ring. Then B has been called a *quasi-tilted* algebra by Happel-Reiten-Smalø, and according to Happel and Happel-Reiten these categories $D^b(\operatorname{mod} A)$ are just the derived categories of artin algebras which are hereditary or canonical. In the "T" displayed above, the upper horizontal line concerns the derived categories with a slice, the middle vertical line those with a separating tubular family. More information can be found in Chapter 6 by Lenzing.

Most of the further considerations will be formulated for tilted algebras only. However usually there do exist corresponding results for all the quasi-tilted algebras. To restrict the attention to the tilted algebras has to be seen as an expression just of laziness, and does not correspond to the high esteem which I have for the remaining algebras (and the class of quasi-tilted algebras in general).

1.2 The functors $\operatorname{Hom}_A(T,-)$ and $\operatorname{Ext}_A^1(T,-)$

Thus, let us fix again a hereditary artin algebra A and let D be the standard duality of $\operatorname{mod} A$ (if k is the center of A, then $D = \operatorname{Hom}_k(-,k)$; note that k is semisimple). Thus DA is an injective cogenerator in $\operatorname{mod} A$. We consider a tilting A-module T, and let $B = \operatorname{End}(T)$. The first feature which comes to mind and which was the observation by Brenner and Butler which started the game[1] , is the following: the func-

---

[1] Of course, we are aware that examples of tilting modules and tilting functors had been studied before. Examples to be mentioned are first the Coxeter functors introduced by Gelfand and Ponomarev in their paper on the four-subspace-problem (1970), then the BGP-reflection functors (Bernstein, Gelfand and Ponomarev,

tor $\mathrm{Hom}_A(T, -)$ yields an equivalence between the category[2] $\mathcal{T}$ of all
A-modules generated by T and the category $\mathcal{Y}$ of all B-modules cogenerated by the B-module $\mathrm{Hom}_A(T, DA)$. Now the dimension vectors of
the indecomposable A-modules in $\mathcal{T}$ generate the Grothendieck group
$K_0(A)$. If one tries to use $\mathrm{Hom}_A(T, -)$ in order to identify the Grothendieck groups $K_0(A)$ and $K_0(B)$, one observes that the positivity cones
overlap, but differ: the new axes which define the positive cone for B are
"tilted" against those for A. This was the reason for Brenner and Butler
to call it a tilting procedure. But there is a second "tilting" phenomenon
which concerns the corresponding torsion pairs[3] . In order to introduce
these torsion pairs, we have to look not only at the functor $\mathrm{Hom}_A(T, -)$,
but also at $\mathrm{Ext}_A^1(T, -)$. The latter functor yields an equivalence between
the category $\mathcal{F}$ of all A-modules M with $\mathrm{Hom}_A(T, M) = 0$ and the category $\mathcal{X}$ of all B-modules N with $T \otimes_B N = 0$. Now the pair $(\mathcal{F}, \mathcal{T})$
is a torsion pair in the category of A-modules, and the pair $(\mathcal{Y}, \mathcal{X})$ is a
torsion pair in the category of B-modules:

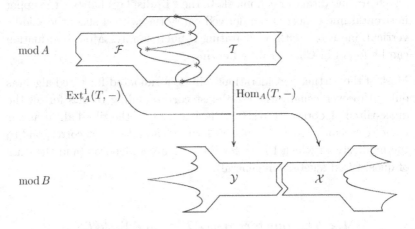

1973), their generalisation by Auslander, Platzeck and Reiten (1979), now called
the APR-tilting functors, and also a lot of additional ad-hoc constructions used
around the globe, all of which turn out to be special tilting functors. But the
proper start of tilting theory is clearly the Brenner-Butler paper (1980). The
axiomatic approach of Brenner and Butler was considered at that time as quite
unusual and surprising in a theory which still was in an experimental stage. But
it soon turned out to be a milestone in the development of representation theory.

2 Subcategories like $\mathcal{T}$ and $\mathcal{F}$ will play a role everywhere in this appendix. In case
we want to stress that they are defined using the tilting module T, we will write
$\mathcal{T}(T)$ instead of $\mathcal{T}$, and so on.

3 In contrast to the usual convention in dealing with a torsion pair or a "torsion
theory", we name first the torsion-free class, then the torsion class: this fits to
the rule that in a rough thought, maps go from left to right, and a torsion pair
concerns regions with "no maps backwards".

and one encounters the amazing fact that under the pair of functors $\mathrm{Hom}_A(T, -)$ and $\mathrm{Ext}^1_A(T, -)$ the torsion-free class of a torsion pair is flipped over the torsion class in order to form a new torsion pair in reversed order[4] . The stars * indicate a possible distribution of the indecomposable direct summands T_i of T, and one should keep in mind that for any i, the Auslander-Reiten translate τT_i of T_i belongs to $\mathcal{T}$ (though it may be zero). We have said that the modules in $\mathcal{T}$ are those generated by T, but similarly the modules in $\mathcal{F}$ are those cogenerated by τT.

According to Happel[5] , the category $\mathrm{mod}\, B$ should be seen as being embedded into the derived category $D^b(\mathrm{mod}\, A)$

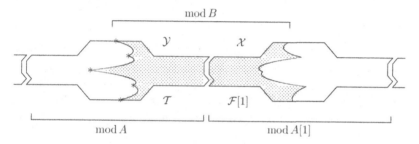

Under this embedding, $\mathcal{Y} = \mathcal{T}$ is the intersection of $\mathrm{mod}\, B$ with $\mathrm{mod}\, A$, whereas $\mathcal{X} = \mathcal{F}[1]$ is the intersection of $\mathrm{mod}\, B$ with $\mathrm{mod}\, A[1]$ (the shift functor in a triangulated category will always be denoted by $[1]$). This embedding functor $\mathrm{mod}\, B \longrightarrow D^b(\mathrm{mod}\, A)$ extends to an equivalence of $D^b(\mathrm{mod}\, B)$ and $D^b(\mathrm{mod}\, A)$, and this equivalence is one of the essential features of tilting theory.

Looking at the torsion pair $(\mathcal{Y}, \mathcal{X})$, there is a sort of asymmetry due to the fact that $\mathcal{Y}$ is always sincere (this means that every simple module

---

4 The discovery of this phenomenon was based on a detailed examination of many examples (and contributions by Dieter Vossieck, then a student at Bielefeld, should be acknowledged). At that time only the equivalence of $\mathcal{T}(T)$ and $\mathcal{Y}(T)$ was well understood. The obvious question was to relate the remaining indecomposable B-modules (those in $\mathcal{X}(T)$) to suitable A-modules. As Dieter Happel recalls, the first examples leading to a full understanding of the whole tilting process were tilting modules for the $\mathbb{E}_6$-quiver with subspace orientation.

5 When he propagated this in 1984, it was the first clue that the use of derived categories may be of interest when dealing with questions in the representation theory of finite dimensional algebras. The derived categories had been introduced by Grothendieck in order to construct derived functors when dealing with abelian categories which have neither sufficiently many projective nor sufficiently many injective objects, and at that time they were considered as useless in case there are enough projectives and enough injectives, as in the cases $\mathrm{mod}\, A$ and $\mathrm{mod}\, B$.

occurs as a composition factor of some module in $\mathcal{Y}$), whereas $\mathcal{X}$ does not have to be sincere (this happens if $\mathcal{Y}$ contains an indecomposable injective module). As a remedy, one should divide $\mathcal{Y}$ further as follows: $\mathcal{Y}$ contains the slice module $S = \mathrm{Hom}_A(T, DA)$, let $\mathcal{S} = \mathrm{add}\, S$, and denote by $\mathcal{Y}'$ the class of all B-modules in $\mathcal{Y}$ without an indecomposable direct summand in $\mathcal{S}$. It is the triple $(\mathcal{Y}', \mathcal{S}, \mathcal{X})$, which really should be kept in mind:

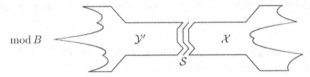

with all the indecomposables lying in one of the classes $\mathcal{Y}', \mathcal{S}, \mathcal{X}$ and with no maps backwards (the only maps from $\mathcal{S}$ to $\mathcal{Y}'$, from $\mathcal{X}$ to $\mathcal{S}$, as well as from $\mathcal{X}$ to $\mathcal{Y}'$ are the zero maps). Also note that any indecomposable projective B-module belongs to $\mathcal{Y}'$ or $\mathcal{S}$, any indecomposable injective module to $\mathcal{S}$ or $\mathcal{X}$. The module class $\mathcal{S}$ is a slice (as explained in Chapter 3 by Brüstle) and any slice is obtained in this way. The modules in $\mathcal{Y}'$ are those cogenerated by τS, the modules in $\mathcal{X}$ are those generated by $\tau^{-1}S$.

Here is an example. Start with the path algebra A of a quiver of Euclidean type $\widetilde{\mathbb{A}}_{22}$ having one sink and one source. Let $B = \mathrm{End}(T)$, where T is the direct sum of the simple projective, the simple injective and the two indecomposable regular modules of length 3 (this is a tilting module), then the quiver of B is the same as the quiver of A, but B is an algebra with radical square zero. Thus B is given by a square with two zero relations.

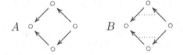

The category $\mathrm{mod}\, B$ looks as follows:

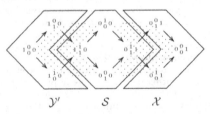

The separation of mod B into the three classes $\mathcal{Y}', \mathcal{S}, \mathcal{X}$ can be phrased in the language of **cotorsion pairs**. Cotorsion pairs are very well related to tilting theory (see the Chapters 8 and 11 by Reiten and Trlifaj), but still have to be rated more as a sort of insider tip. We recall the definition: the pair $(\mathcal{V}, \mathcal{W})$ of full subcategories of mod A is said to be a *cotorsion pair* provided $\mathcal{V}$ is the class of all A-modules V with $\mathrm{Ext}_A^1(V, W) = 0$ for all W in $\mathcal{W}$, and $\mathcal{W}$ is the class of all A-modules W such that $\mathrm{Ext}_A^1(V, W) = 0$ for all V in $\mathcal{V}$. The cotorsion pair is said to be *split*, provided every indecomposable A-module belongs to $\mathcal{V}$ or $\mathcal{W}$. Usually some indecomposables will belong to both classes, they are said to form the *heart*. In our case the following holds: *The pair* $\big(\mathrm{add}(\mathcal{Y}', \mathcal{S}), \mathrm{add}(\mathcal{S}, \mathcal{X})\big)$ *forms a split cotorsion pair with heart* $\mathcal{S}$.

We also see that the modules in $\mathcal{Y}'$ and in $\mathcal{S}$ have projective dimension at most 1, those in $\mathcal{S}$ and in $\mathcal{X}$ have injective dimension at most 1. As a consequence, if X, Y are indecomposable modules with $\mathrm{Ext}_B^2(X, Y) \neq 0$, then X belongs to $\mathcal{X}$ and Y belongs to $\mathcal{Y}'$.

Let me add a remark even if it may be considered to be superfluous — its relevance should become clear in the last part of this appendix. If we feel that the subcategory $\mathcal{Y}'$ has the same importance as $\mathcal{X}$ (thus that it is of interest), then we should specify an equivalent subcategory, say T' of mod A and an equivalence $T' \longrightarrow \mathcal{Y}'$. Such an equivalence is given by the functor

$$\mathrm{Hom}_A(\tau^{-1}T, -)\colon \mathrm{mod}\, A \longrightarrow \mathrm{mod}\, B$$

or, equivalently, by $\mathrm{Hom}_A(T, \tau-)$, since τ^{-1} is left adjoint to τ. This functor vanishes on $\mathcal{F}$ as well as on T, and it yields an equivalence between the subcategory T' of all A-modules generated by $\tau^{-1}T$ and the subcategory $\mathcal{Y}'$ of mod B. Note that the functor can also be written in the form $D\,\mathrm{Ext}_A^1(-, T)$, due to the Auslander-Reiten formula $D\,\mathrm{Ext}^1(M, T) \simeq \mathrm{Hom}_A(T, \tau M)$. In this way, we see that we deal with equivalences

$$D\,\mathrm{Ext}_A^1(-, T)\colon T' \longrightarrow \mathcal{Y}' \qquad \text{and} \qquad \mathrm{Ext}_A^1(T, -)\colon \mathcal{F} \longrightarrow \mathcal{X},$$

which are sort of dual to each other.

It seems to be worthwhile to have a short look at the rather trivial case when no modules are lost, so that the tilting procedure is a kind of rearrangement of module classes. *The following assertions are equivalent:*

(i) *The tilting module T is a slice module.*

(ii) *The endomorphism ring B is hereditary.*

(iii) *The torsion pair $(\mathcal{F}, \mathcal{T})$ splits.*

(iv) $\operatorname{Ext}_A^1(\tau T, T) = 0.$

The equivalence of these assertions are well-known, but not too easy to trace. Some implications are quite obvious, for example that (ii) and (iii) are implied by (i). Let us show that (ii) implies (i): Since T is a tilting module, the B-module $T' = \operatorname{Hom}_A(T, DA) \simeq \operatorname{Hom}_k(T, k)$ is a slice module in mod B. Since the B-module T' is a tilting module and $A = \operatorname{End}_B(T')$, we can use this tilting module in order to tilt from mod B to mod A. Since B is hereditary, we obtain in mod A the slice module $\operatorname{Hom}_B(T', DB) \simeq \operatorname{Hom}_k(T', k) \simeq T$. This shows (i). The equivalences of (ii) and (iv), as well as of (iii) and (iv), can be seen as consequences of more general considerations which will be presented later.

If T is not a slice module, so that B has global dimension equal to 2, then the algebras A and B play quite a different role: the first difference is of course the fact that A is hereditary, whereas B is not. Second, there are the two torsion pairs $(\mathcal{F}, \mathcal{T})$ in mod A and $(\mathcal{Y}, \mathcal{X})$ in mod B - the second one is a split torsion pair, the first one not. This means that we loose modules going from mod A to mod B via tilting. Apparently, no one cared about the missing modules, at least until quite recently. There are two reasons: First of all, we know (see Chapter 3 by Brüstle), that the study of indecomposable modules over a representation-finite algebra is reduced via covering theory to the study of representation-finite tilted algebras. Such an algebra B may be of the form $B = \operatorname{End}_A(T)$, where T is a tilting A-module, with A representation-infinite. Here we describe the B-modules in terms of A and we are only interested in the finitely many indecomposable A-modules which belong to $\mathcal{F}$ or $\mathcal{T}$, the remaining A-modules seem to be of no interest, we do not miss them. But there is a second reason: the fashionable reference to derived categories is used to appease anyone, who still mourns about the missing modules. They are lost indeed as modules, but they survive as complexes: since the derived categories of A and B are equivalent, corresponding to any indecomposable A-module, there is an object in the derived category which is given by a complex of B-modules. However, I have to admit that I prefer modules to complexes, whenever possible — thus I was delighted, when the lost modules were actually found, as described in Part 3 of this appendix.

1.3 The simplicial complex Σ_A

We will always denote by $n = n(A)$ the number of isomorphism classes of simple A-modules. The interest in tilting A-modules directly leads to a corresponding interest in their direct summands. These are the modules without self-extensions and are called *partial tilting modules*. In particular, one may consider the indecomposable ones: an indecomposable A-module without self-extensions is said to be *exceptional* (or a "stone", or a "brick without self-extensions", or a "Schurian module without self-extensions"). But there is also an interest in the partial tilting modules with precisely $n-1$ isomorphism classes of indecomposable direct summands, the so-called *almost complete* partial tilting modules. If $\overline{T}$ is an almost complete partial tilting module and X is indecomposable with $\overline{T} \oplus X$ a tilting module, then X (or its isomorphism class) is called a *complement* for $\overline{T}$. It is of interest that any almost complete partial tilting module $\overline{T}$ has either 1 or 2 complements, and it has 2 if and only if $\overline{T}$ is sincere. Recall that a module is said to be *basic*, provided it is a direct sum of pairwise non-isomorphic indecomposable modules. The isomorphism classes of basic partial tilting modules form a simplicial complex Σ_A, with vertex set the set of isomorphism classes of exceptional modules (the vertices of a simplex being its indecomposable direct summands), see Chapter 10 by Unger. Note that this simplicial complex is of pure dimension $n - 1$. The assertion concerning the complements shows that it is a pseudomanifold with boundary. The boundary consists of all the non-sincere almost complete partial tilting modules.

As an example, consider the path algebra A of the quiver $\circ \leftarrow \circ \leftarrow \circ$. The simplicial complex Σ_A has the following shape:

Some questions concerning the simplicial complex Σ_A remained open: What happens under a change of orientation? What happens under a tilting functor? Is there a way to get rid of the boundary? Here we are again in a situation where a remedy is provided by the derived categories: If we construct the analogous simplicial complex of tilting complexes in $D^b(\text{mod } A)$, then one obtains a pseudo manifold without boundary, but this is quite a large simplicial complex!

If $\overline{T}$ is an almost complete partial tilting module, and X and Y are non-isomorphic complements for $\overline{T}$, then either $\text{Ext}_A^1(Y,X) \neq 0$ or $\text{Ext}_A^1(X,Y) \neq 0$ (but not both). If $\text{Ext}_A^1(Y,X) \neq 0$ (what we may assume), then there exists an exact sequence $0 \longrightarrow X \longrightarrow T' \longrightarrow Y \longrightarrow 0$ with $T' \in \text{add } \overline{T}$, and one may write $\overline{T} \oplus X < \overline{T} \oplus Y$. In this way, one gets a partial ordering on the set of isomorphism classes of basic tilting modules. One may consider the switch between the tilting modules $\overline{T} \oplus X$ and $\overline{T} \oplus Y$ as an exchange process which stops at the boundary. We will see in Part 3 that it is possible to define an exchange procedure across the boundary of Σ_A, and that this can be arranged in such a way that one obtains an interesting small extension of the simplicial complex Σ_A.

1.4 The BGP-reflection functors and the structure of tilted algebras

Bernstein, Gelfand and Ponomarev have defined reflection functors in order to be able to compare representations of quivers with different orientation, but also in order to construct inductively indecomposable representations. A BGP-reflection functor furnishes a quite small change of the given module category. Let us consider this in more detail. Let Q be a quiver and i a sink of Q. We denote by $\sigma_i Q$ the quiver obtained from Q by changing the orientation of all arrows ending in i, thus i becomes a source in $\sigma_i Q$. Let $S(i)$ be the simple kQ-module corresponding to the vertex i, and $S'(i)$ the simple $k\sigma_i Q$-module corresponding to the vertex i. The BGP-reflection functor $\sigma_i\colon \text{mod } kQ \longrightarrow \text{mod } k\sigma_i Q$ provides an equivalence between the categories

$$\text{mod } kQ/\langle \text{add } S(i)\rangle \longrightarrow \text{mod } k\sigma_i Q/\langle \text{add } S'(i)\rangle.$$

In general, given rings R, R' one may look for a simple R-module S and a simple R'-module S' such that the categories

$$\text{mod } R/\langle \text{add } S\rangle \longrightarrow \text{mod } R'/\langle \text{add } S'\rangle$$

are equivalent. In this case, let us say that R, R' are *nearly Morita-equivalent*. As we have seen, for Q a quiver with a sink i, the path algebras kQ and $k\sigma_i Q$ are nearly Morita-equivalent. I am not aware that other pairs of nearly Morita-equivalent rings have been considered until very recently, but Part 3 will provide a wealth of examples.

Unfortunately, the BGP-reflection functors are defined only for sinks and for sources of the given quiver. This has to be considered as a real deficiency, since there is no similar restriction in Lie theory. Indeed, in Lie theory the use of reflections for all the vertices is an important tool. A lot of efforts have been made in representation theory in order to overcome this deficiency, see for example the work of Kac on the dimension vectors of the indecomposable representations of a quiver.

A final question should be raised here. There is a very nice homological characterization of the quasi-tilted algebras by Happel-Reiten-Smalø [28]: these are the artin algebras of global dimension at most 2, such that any indecomposable module has projective dimension at most 1 or injective dimension at most 1. But it seems that a corresponding characterization of the subclass of tilted algebras is still missing. Also, in case we consider tilted k-algebras, where k is an algebraically closed field, the possible quivers and their relations are not known.

2 Connections

The relevance of tilting theory relies on the many different connections it has not only to other areas of representation theory, but also to algebra and geometry in general. Let me give some indications. If nothing else is said, A will denote a hereditary artin algebra, T a tilting A-module and B its endomorphism ring.

2.1 Homology

Already the definition (the vanishing of Ext^1) refers to homology. We have formulated above that the first feature which comes to mind is the functor $\mathrm{Hom}_A(T, -)$. But actually all the tilting theory concerns the study of the corresponding derived functors $\mathrm{Ext}_A^i(T, -)$, or better, of the right derived functor $R\,\mathrm{Hom}_A(T, -)$.

The best setting to deal with these functors are the corresponding derived categories $D^b(\text{mod}\,A)$ and $D^b(\text{mod}\,B)$, they combine to the right derived functor $R\,\text{Hom}_A(T, -)$, and this functor is an equivalence, as Happel has shown. Tilting modules T in general were defined in such a way that $R\,\text{Hom}_A(T, -)$ is still an equivalence. The culmination of this development was Rickard's characterization of rings with equivalent derived categories: such equivalences are always given by "tilting complexes". A detailed account can be found in Chapter 5 by Keller.

Tilting theory can be exhibited well by using spectral sequences. In Bongartz's presentation of tilting theory one finds the formulation: *Well-read mathematicians tend to understand tilting theory using spectral-sequences* (which is usually interpreted as a critical comment about the earlier papers). But it seems that the first general account of this approach is only now available: the contribution of Brenner and Butler (see Chapter 4) in this volume. A much earlier one by Vossieck should have been his Bielefeld Ph.D. thesis in 1984, but he never handed it in.

2.2 Geometry and invariant theory.

The Bielefeld interest in tilting modules was first not motivated by homological, but by geometrical questions. Happel's Ph.D. thesis had focused the attention to quiver representations with an open orbit (thus to all the partial tilting modules). In particular, he showed that the number $s(V)$ of isomorphism classes of indecomposable direct summands of a representation V with open orbit is bounded by the number of simple modules. In this way, the study of open orbits in quiver varieties was a (later hidden) step in the development of tilting theory. When studying open orbits, we are in the setting of what Sato and Kimura [45] call **prehomogeneous vector spaces**. On the one hand, the geometry of the complement of the open orbit is of interest, on the other hand one is interested in the structure of the ring of semi-invariants.

Let k be an algebraically closed field and Q a finite quiver (with vertex set Q_0 and arrow set Q_1), and we may assume that Q has no oriented cyclic path, thus the path algebra kQ is just a basic hereditary finite-dimensional k-algebra. For any arrow α in Q_1, denote by $t\alpha$ its tail and by $h\alpha$ its head, and fix some dimension vector $\mathbf{d}$. Let us consider representations V of Q with a fixed dimension vector $\mathbf{d}$, we may assume

$V(x) = k^{\mathbf{d}(x)}$; thus the set of these representations forms the affine space

$$\mathcal{R}(Q, \mathbf{d}) = \bigoplus_{\alpha \in Q_1} \mathrm{Hom}_k(k^{\mathbf{d}(t\alpha)}, k^{\mathbf{d}(h\alpha)}).$$

The group $\mathrm{GL}(Q, \mathbf{d}) = \prod_{x \in Q_0} \mathrm{GL}(\mathbf{d}(x))$ operates on this space via a sort of conjugation, and the orbits under this action are just the isomorphism classes of representations. One of the results of Happel [26] asserts that given a sincere representation V with open orbit, then $|Q_0| - s(V)$ is the number of isomorphism classes of representations W with $\mathbf{dim}\, V = \mathbf{dim}\, W$ and $\dim\, \mathrm{Ext}_A^1(W, W) = 1$ (in particular, there are only finitely many such isomorphism classes; we also see that $|Q_0| \geqslant s(V)$).

Consider now the ring $\mathrm{SI}(Q, \mathbf{d})$ of semi-invariants on $\mathcal{R}(Q, \mathbf{d})$; by definition these are the invariants of the subgroup $\mathrm{SL}(Q, \mathbf{d}) = \prod_{x \in Q_0} \mathrm{SL}(\mathbf{d}(x))$ of $\mathrm{GL}(Q, \mathbf{d})$. Given two representations V, W of Q, one may look at the map:

$$d_W^V: \bigoplus_{x \in Q_0} \mathrm{Hom}_k(V(x), W(x)) \longrightarrow \bigoplus_{\alpha \in Q_1} \mathrm{Hom}_k(V(t\alpha), W(h\alpha)),$$

sending $(f(x))_x$ to $(f(h\alpha)V(\alpha) - W(\alpha)f(t\alpha))_\alpha$. Its kernel is just $\mathrm{Hom}_{kQ}(V, W)$, its cokernel $\mathrm{Ext}_{kQ}^1(V, W)$. In case d_W^V is a square matrix, one may consider its determinant. According to Schofield [46], this is a way of producing semi-invariants. Namely, the Grothendieck group $K_0(kQ)$ carries a (usually non-symmetric) bilinear form $\langle -, - \rangle$ with $\langle \mathbf{dim}\, V, \mathbf{dim}\, W \rangle = \dim_k \mathrm{Hom}_{kQ}(V, W) - \dim_k \mathrm{Ext}_{kQ}^1(V, W)$, thus d_W^V is a square matrix if and only if $\langle \mathbf{dim}\, V, \mathbf{dim}\, W \rangle = 0$. So, if $\mathbf{d} \in \mathbb{N}_0^{Q_0}$ and if we select a representation W such that $\langle \mathbf{d}, \mathbf{dim}\, W \rangle = 0$, then $c_W(V) = \det d_W^V$ yields a semi-invariant c_W in $\mathrm{SI}(Q, \alpha)$. Derksen and Weyman (and also Schofield and Van den Bergh) have shown that these semi-invariants form a generating set for $\mathrm{SI}(Q, \mathbf{d})$. In fact, it is sufficient to consider only indecomposable representations W, thus exceptional kQ-modules.

2.3 Lie theory

It is a well-accepted fact that the representation theory of hereditary artinian rings has a strong relation to Lie algebras and quantum groups (actually one should say: a strong relation to Lie algebras via quantum groups). Such a relationship was first observed by Gabriel when he showed that the representation-finite connected quivers are just those with underlying graph being of the form $\mathbb{A}_n, \mathbb{D}_n, \mathbb{E}_6, \mathbb{E}_7, \mathbb{E}_8$ and that in these cases the indecomposables correspond bijectively to the positive roots. According to Kac, this extends to arbitrary finite quivers without oriented cycles: the dimension vectors of the indecomposable representations are just the positive roots of the corresponding Kac-Moody Lie-algebra. It is now known that it is even possible to reconstruct this Lie algebra using the representation theory of hereditary artin algebras, via Hall algebras. Here one encounters the problem of specifying the subring of a Hall algebra, generated by the simple modules. Schofield induction (to be discussed below) shows that all the exceptional modules belong to this subring.

It seems to be appropriate to discuss the role of the necessary choices. Let me start with a semisimple finite-dimensional complex Lie-algebra. First, there is the choice of a Cartan subalgebra, it yields the root system of the Lie-algebra. Second, the choice of a root basis yields a triangular decomposition (and the set of positive roots), this is needed in order to define a Borel subalgebra and the corresponding category $\mathcal{O}$. Finally, the choice of a total ordering of the root basis (or, better, of an orientation of the edges of the Dynkin diagram) allows to work with a Coxeter element in the Weyl group. Of course, one knows that all these choices are inessential, when dealing with a finite dimensional Lie-algebra. The situation is more subtle if we deal with arbitrary Kac-Moody Lie-algebras: different orderings of the root basis may yield Coxeter elements which are not conjugate – the first case is $\widetilde{\mathbb{A}}_3$, where one has to distinguish between $\widetilde{\mathbb{A}}_{3,1}$ and $\widetilde{\mathbb{A}}_{2,2}$.

On the other hand, when we start with a representation-finite hereditary artin algebra A, no choice at all is needed in order to write down its Dynkin diagram: it is intrinsically given as the Ext-quiver of the simple A-modules, and we obtain in this way a Dynkin diagram with orientation. A change of orientation corresponds to module categories with quite distinct properties (as already the algebras of type $\mathbb{A}_3$ show). This difference is still preserved when one looks at the corresponding Hall algebras, and it comes as a big surprise that only a small twist of its multiplication is needed in order to get an algebra which is independent of the orientation.

2.4 The combinatorics of root systems

It is necessary to dig deeper into root systems since they play an important role for dealing with A-modules. Of interest is the corresponding quadratic form, and the reflections which preserve the root system (but not necessarily the positivity of roots), and their compositions, in particular the Coxeter transformations and the BGP-reflection functors. We will return to the reflection functors when we deal with generalizations of Morita equivalences, but also in Part 3.

As we have mentioned the relationship between the representation theory of a hereditary artin algebra A and root systems is furnished by the dimension vectors: We consider the Grothendieck group $K_0(A)$ (of all finite length A-modules modulo exact sequences). Given an A-module M, we denote by $\mathbf{dim}\, M$ the corresponding element in $K_0(A)$; this is what is called its *dimension vector*. The dimension vectors of the indecomposable A-modules are the positive roots of the root system in question. A positive root $\mathbf{d}$ is said to be a *Schur root* provided there exists an indecomposable A-module M with $\mathbf{dim}\, M = \mathbf{d}$ and $\operatorname{End}_A(M)$ a division ring. The dimension vectors of the exceptional modules are Schur roots, they are just the real (or Weyl) Schur roots. In case A is representation-finite then all the positive roots are Schur roots, also for $n(A) = 2$ all the real roots are Schur roots. But in all other cases, the set of real Schur roots depends on the choice of orientation. For example, consider the following three orientations of $\widetilde{\mathbb{D}}_4$:

The two dimension vectors on the left are Schur roots, whereas the right one is not a Schur root.

In order to present the dimension vectors of the indecomposable A-modules, one may depict the Grothendieck group $K_0(A)$; a very convenient way seems to be to work with homogeneous coordinates, say with the projective space of $K_0(A) \otimes_{\mathbb{Z}} \mathbb{R}$. It is the merit of Derksen and Weyman [22] of having popularized this presentation well: they managed to get it to a cover of the Notices of the American Mathematical Society [21]. One such example has been shown in Part 1, when we presented the simplicial complex Σ_A, whith A the path algebra of the linearly oriented quiver of type $\mathbb{A}_3$. In general, dealing with hereditary A, one is interested in the position of the Schur roots. Our main concern are the real Schur roots as the dimension vectors $\dim E$ of the exceptional modules E. They are best presented by marking the corresponding *exceptional lines:* look for orthogonal exceptional pairs E_1, E_2 (this means: E_1, E_2 are exceptional modules with $\operatorname{Hom}_A(E_1, E_2) = \operatorname{Hom}_A(E_2, E_1) = \operatorname{Ext}^1_A(E_2, E_1) = 0$) and draw the line segment from $\dim E_1$ to $\dim E_2$. The discussion of Schofield induction below will explain the importance of these exceptional lines.

As we know, a tilting A-module T has precisely $n = n(A)$ isomorphism classes of indecomposable direct summands, say $T_1, \ldots, T_n$ and the dimension vectors $\dim T_1, \ldots, \dim T_n$ are linearly independent. We may consider the cone $C(T)$ in $K_0(A) \otimes \mathbb{R}$ generated by $\dim T_1, \ldots, \dim T_n$. These cones are of special interest. Namely, if a dimension vector $\mathbf{d}$ belongs to $C(T)$, then there is a unique isomorphism class of modules M with $\dim M = \mathbf{d}$ such that $\operatorname{End}_A(M)$ is of minimal dimension, and such a module M has no self-extensions. On the other hand, the dimension vector of any module M without self-extensions lies in such a cone (since these modules, the partial tilting modules, are the direct summands of tilting modules). The set of these cones forms a fan as they are considered in toric geometry (see for example the books of Fulton and Oda).

As Hille [29] has pointed out, one should use the geometry of these cones in order to introduce the following notion: If $T = \bigoplus_{i=1}^n T_i$ is a basic tilting module with indecomposable modules T_i of length $|T_i|$, he calls $\prod_{i=1}^n |T_i|^{-1}$ the *volume* of T. It follows that

$$\sum_T v(T) \leqslant 1$$

(where the summation extends over all isomorphism classes of basic tilting A-modules), with equality if and only if A is representation-finite or tame. This yields interesting equalities: For example, the preprojective tilting modules of the Kronecker quiver yield

$$\frac{1}{1} \cdot \frac{1}{3} + \frac{1}{3} \cdot \frac{1}{5} + \frac{1}{5} \cdot \frac{1}{7} + \cdots = \frac{1}{2}.$$

One may refine these considerations by replacing the length $|T_i|$ by the k-dimension of T_i, at the same time replacing A by all the Morita equivalent algebras. In this way one produces power series identities in $n(A)$ variables which should be of general interest, for example

$$\frac{1}{x} \cdot \frac{1}{2x+y} + \frac{1}{2x+y} \cdot \frac{1}{3x+2y} + \frac{1}{3x+2y} \cdot \frac{1}{4x+3y} + \cdots = \frac{1}{2xy}.$$

Namely, let x and y denote the k-dimension of the simple projective, or simple injective A-module, respectively. The indecomposable preprojective A-module P_t of length $2t-1$ is of dimension $tx + (t-1)y$. This means that the the tilting module $P_t \oplus P_{t+1}$ contributes the summand $\frac{1}{tx+(t-1)y} \cdot \frac{1}{(t+1)x+ty}$. The sequence of tilting modules $P_1 \oplus P_2$, $P_2 \oplus P_3, \ldots$ yields the various summands on the left side.

2.5 Combinatorial structure of modules

A lot of tilting theory is devoted to combinatorial considerations. The combinatorial invariants just discussed concern the Grothendieck group. But also the exceptional modules themselves have a combinatorial flavor: they are "tree modules" [42]. As we have mentioned, the orbit of a tilting module is open in the corresponding module variety, and this holds true with respect to all the usual topologies, in particular, the Zariski topology, but also the usual real topology in case the base field is $\mathbb{R}$ or $\mathbb{C}$. This means that a slight change of the coefficients in any realization of T using matrices will not change the isomorphism class. Now in general to be able to change the coefficients slightly, will not allow to prescribe a finite set (for example $\{0,1\}$) of coefficients which one may like to use: the corresponding matrices may just belong to the complement of the orbit. However, in case we deal with the path algebra of a quiver, the exceptional modules have this nice property: there always exists a realization of E using matrices with coefficients only 0 and 1. A stronger statement holds true: If E has dimension d, then there is a matrix realization which uses precisely $d-1$ coefficients equal to 1, and all the remaining ones are 0 (note that in order to be indecomposable, we need at least $d-1$ non-zero coefficients; thus we assert that really the minimal possible number of non-zero coefficients can be achieved).

2.6 Numerical linear algebra

Here we refer to the previous consideration: The relevance of 0-1-matrices in numerical linear algebra is well-known. Thus linear algebra problems, which can be rewritten as dealing with partial tilting modules, are very suitable for numerical algorithms, because of two reasons: one can restrict to 0-1-matrices and the matrices to be considered involve only very few non-zero entries.

2.7 Module theory

Of course, tilting theory is part of module theory. It provides a very useful collection of non-trivial examples for many central notions in ring and module theory. The importance of modules without self-extensions has been realized a long time ago, for example one may refer to the lecture notes of Tachikawa from 1973. Different names are in use for such modules such as "splitters".

It seems that the tilting theory exhibited for the first time a wide range of torsion pairs, with many different features: there are the splitting torsion pairs, which one finds in the module category of any tilted algebra, as well as the various non-split torsion pairs in the category $\operatorname{mod} A$ itself. As we have mentioned in Part 1, tilting theory also gives rise to non-trivial examples of cotorsion pairs. And there are corresponding approximations, but also filtrations with prescribed factors. Questions concerning subcategories of module categories are considered in many of the contributions in this Handbook, in particular in Chapter 8 by Reiten, but also in the Chapters 9, 11 and 12 by Donkin, Trlifaj and Solberg, respectively.

We also should mention the use of **perpendicular categories**. Starting with an exceptional A-module E, the category $E^{\perp}$ of all A-modules M with $\operatorname{Hom}_A(E, M) = 0 = \operatorname{Ext}^1_A(E, M)$ is again a module category, say $E^{\perp} \simeq \operatorname{mod} A'$, where A' is a hereditary artin algebra with $n(A') = n(A) - 1$. These perpendicular categories are an important tool for inductive arguments and they can be considered as a kind of localisation.

Another notion should be illuminated here: recall that a left R-module M is said to have the **double centralizer property** (or to be balanced), provided the following holds: If we denote by S the endomorphism ring of $_RM$, say operating on the right on M, we obtain a right S-module M_S, and we may now look at the endomorphism ring R' of M_S. Clearly, there is a canonical ring homomorphism $R \longrightarrow R'$ (sending $r \in R$ to the left multiplication by r on M), and now we require that this map is surjective (in case M is a faithful R-module, so that the map $R \longrightarrow R'$ is injective, this means that we can identify R and R': the ring R is determined by the categorical properties of M, namely its endomorphism ring S, and the operation of S on the underlying abelian group of M). Modules with the double centralizer property are very important in ring and module theory. Tilting modules satisfy the double centralizer property and this is used in many different ways.

Of special interest is also the following **subquotient realization** of mod A. All the modules in $\mathcal{T}$ are generated by T, all the modules in $\mathcal{F}$ are cogenerated by τT. It follows that *for any A-module M, there exists an A-module X with submodules $X''' \subseteq X'' \subseteq X'$ such that X'' is a direct sum of copies of T, whereas X/X'' is a direct sum of copies of τT and such that $M = X'/X'''$* (it then follows that X''/X''' is the torsion submodule of M and X'/X'' its torsion-free factor module). In particular, we see that $(\mathcal{F}, \mathcal{T})$ is a split torsion pair if and only if $\operatorname{Ext}_A^1(\tau T, T) = 0$ (the equivalence of conditions (iii) und (iv) in Part 1). This is one of the results which stresses the importance of the bimodule $\operatorname{Ext}_A^1(\tau T, T)$. Note that the extensions considered when we look at $\operatorname{Ext}_A^1(\tau T, T)$ are opposite to those the Auslander-Reiten translation τ is famous for (namely the Auslander-Reiten sequences, they correspond to elements of $\operatorname{Ext}_A^1(T_i, \tau T_i)$, where T_i is a non-projective indecomposable direct summand of T). We will return to the bimodule $\operatorname{Ext}_A^1(\tau T, T)$ in Part 3.

2.8 Morita equivalence

Tilting theory is a powerful generalization of Morita equivalence. This can already be demonstrated very well by the reflection functors. When Gabriel showed that the representations of a Dynkin quiver correspond to the positive roots and thus only in an inessential way on the given orientation, this was considered as a big surprise. The BGP-reflection functors explain in which way the representation theory of a quiver is independent of the orientation: one can change the orientation of all the arrows in a sink or a source, and use reflection functors in order to obtain a bijection between the indecomposables. Already for the quivers of type $\mathbb{A}_n$ with $n \geqslant 3$, we get interesting examples, relating say a serial algebra (using one of the two orientations with just one sink and one source) to a non-serial one.

The reflection functors are still near to classical Morita theory, since no modules are really lost: here, we only deal with a kind of rearrangement of the categories in question. We deal with split torsion pairs $(\mathcal{F}, \mathcal{T})$ in $\mathrm{mod}\, A$ and $(\mathcal{Y}, \mathcal{X})$ in $\mathrm{mod}\, B$, with $\mathcal{F}$ equivalent to $\mathcal{X}$ and $\mathcal{T}$ equivalent to $\mathcal{Y}$. Let us call two hereditary artin algebras *similar* provided they can be obtained from each other by a sequence of reflection functors. In case we consider the path algebra of a quiver which is a tree, then any change of orientation leads to a similar algebra. But already for the cycle with 4 vertices and 4 arrows, there are two similarity classes, namely the quiver $\tilde{\mathbb{A}}_{3,1}$ with a path of length 3, and the quivers of type $\tilde{\mathbb{A}}_{2,2}$.

One property of the reflection functors should be mentioned (since it will be used in Part 3). Assume that i is a sink for A (this means that the corresponding simple A-module $S(i)$ is projective). Let $S'(i)$ be the corresponding simple $\sigma_i A$-module (it is injective). If M is any A-module, then $S(i)$ is not a composition factor of M if and only if $\mathrm{Ext}^1_{\sigma_i A}(S'(i), \sigma_i M) = 0$. This is a situation, where the reflection functor yields a universal extension; for similar situations, let me refer to [39].

The general tilting process is further away from classical Morita theory, due to the fact that the torsion pair $(\mathcal{F}, \mathcal{T})$ in $\mathrm{mod}\, A$ is no longer split.

2.9 Duality theory

Tilting theory is usually formulated as dealing with equivalences of subcategories (for example, that $\mathrm{Hom}_A(T, -)\colon \mathcal{T} \longrightarrow \mathcal{Y}$ is an equivalence). However, one may also consider it as a duality theory, by composing the equivalences obtained with the duality functor D, thus obtaining a duality between subcategories of the category $\mathrm{mod}\, A$ and subcategories of the category $\mathrm{mod}\, B^{\mathrm{op}}$. The new formulations obtained in this way actually look more symmetrical, thus may be preferable. Of course, as long as we deal with finitely generated modules, there is no mathematical difference. This changes, as soon as one takes into consideration also modules which are not of finite length.

But the interpretation of tilting processes as dualities is always of interest, also when dealing with modules of finite length: In [2], Auslander considers (for R a left and right noetherian ring) the class $\mathcal{W}(R)$ of all left R-modules of the form $\mathrm{Ext}^1_R(N_R, R_R)$, where N_R is a finitely generated right R-module, and he asserts that it would be of interest to know whether this class is always closed under submodules. A first example of a ring R with $\mathcal{W}(R)$ not being closed under submodules has been exhibited by Huang [30], namely the path algebra $R = kQ$ of the quiver Q of type $\mathbb{A}_3$ with 2 sources. Let us consider the general case when $R = A$ is a hereditary artin algebra. The canonical injective cogenerator $T = DA$ is a tilting module, thus $\operatorname{Ext}^1_A(T, -)$ is a full and dense functor from $\operatorname{mod} A$ onto $\mathcal{X}(T)$. The composition of functors

$$\operatorname{mod} A^{\mathrm{op}} \xrightarrow{D} \operatorname{mod} A \xrightarrow{\operatorname{Ext}^1(T,-)} \operatorname{mod} A$$

is the functor $\operatorname{Ext}^1_A(-, A_A)$, thus we see that $\mathcal{W}(A) = \mathcal{X}(T)$. On the other hand, $\mathcal{T}(T)$ are the injective A-modules, they are mapped under $\operatorname{Hom}_A(T, -)$ to the class $\mathcal{Y}(T)$, and these are the projective A-modules. It follows that $\mathcal{X}(T)$ is the class of all A-modules without an indecomposable projective direct summand. As a consequence, $\mathcal{W}(A) = \mathcal{X}(T)$ *is closed under submodules if and only if A is a Nakayama algebra.* (It is an easy exercise to show that $\mathcal{X}(T)$ is closed under submodules if and only if the injective envelope of any simple projective module is projective, thus if and only if A is a Nakayama algebra).

It should be stressed that Morita himself seemed to be more interested in dualities than in equivalences. What is called Morita theory was popularized by P.M.Cohn and H. Bass, but apparently was considered by Morita as a minor addition to his duality theory. When Gabriel heard about tilting theory, he immediately interpreted it as a non-commutative analog of Roos duality.

The use of general tilting modules as a source for dualities has been shown to be very fruitful in the representation theory of algebraic groups, of Lie algebras and of quantum groups. This is explained in detail in Chapter 9 by Donkin. As a typical special case one should have the classical **Schur-Weyl duality** in mind, which relates the representation theory of the general linear groups and that of the symmetric groups, see Chapter 9 by Donkin, but also [36].

In the realm of commutative complete local noetherian rings, Auslander and Reiten [4] considered **Cohen-Macaulay rings** with dualizing module W. They showed that W *is the only basic cotilting module.* On the basis of this result, they introduced the notion of a dualizing module for arbitrary artin algebras.

2.10 Schofield induction

This is an inductive procedure for constructing all exceptional modules starting with the simple ones, by forming exact sequences of the following kind: Assume we deal with a hereditary k-algebra, where k is algebraically closed, and let E_1, E_2 be orthogonal exceptional modules with dim $\operatorname{Ext}_A^1(E_1, E_2) = t$ and $\operatorname{Ext}_A^1(E_2, E_1) = 0$. Then, for every pair (a_1, a_2) of positive natural numbers satisfying $a_1^2 + a_2^2 - ta_1a_2 = 1$, there exists (up to equivalence) a unique non-split exact sequence of the form

$$0 \longrightarrow E_2^{a_2} \longrightarrow E \longrightarrow E_1^{a_1} \longrightarrow 0$$

(call it a *Schofield sequence*). Note that the middle term of such a Schofield sequence is exceptional again, and it is an amazing fact that starting with the simple A-modules without self-extension, all the exceptional A-modules are obtained in this way. Even a stronger assertion is true: If E is an exceptional module with support of cardinality s (this means that E has precisely s isomorphism classes of composition factors), then there are precisely $s - 1$ Schofield sequences with E as middle term. What is the relation to tilting theory? Starting with E one obtains the Schofield sequences by using the various indecomposable direct summands of its Bongartz complement as an A/I-module, where I is the annihilator of E: the $s - 1$ summands yield the $s - 1$ sequences [41].

2.11 Exceptional sequences, mutations

Note that a tilted algebra is always directed: the indecomposable summands of a tilting module $E_1, \ldots, E_m$ can be ordered in such a way that $\operatorname{Hom}_A(E_i, E_j) = 0$ for $i > j$. We may call such a sequence $(E_1, \ldots, E_m)$ a tilting sequence, and there is the following generalization which is of interest in its own (and which was considered by the Rudakov school [44]): Call $(E_1, \ldots, E_m)$ an *exceptional sequence* provided all the modules E_i are exceptional A-modules and $\operatorname{Hom}_A(E_i, E_j) = 0$ and $\operatorname{Ext}_A^1(E_i, E_j) = 0$ for $i > j$. There are many obvious examples of exceptional sequences which are not tilting sequences, the most important one being sequences of simple modules in case the Ext-quiver of the simple modules is directed. Now one may be afraid that this generalization could yield too many additional sequences, but this is not the case. In general most of the exceptional sequences are tilting sequences! An exceptional sequence

$(E_1, \ldots, E_m)$ is said to be complete provided $m = n(A)$ (the number of simple A-modules). There is a braid group action on the set of complete exceptional sequences, and this action is transitive [18, 40]. This means that all the exceptional sequences can be obtained from each other by what one calls "mutations". As a consequence, one obtains the following: If $(E_1, \ldots, E_n)$ is a complete exceptional sequence, then there is a permutation π such that $\operatorname{End}_A(E_i) = \operatorname{End}_A(S_{\pi(i)})$, where $S_1, \ldots, S_n$ are the simple A-modules. In particular, this means that for any tilted algebra B, the radical factor algebras of A and of B are Morita equivalent.

An exceptional module E defines also **partial reflection functors** [39] as follows: consider the following full subcategories of $\operatorname{mod} A$. Let $\mathcal{M}^E$ be given by all modules M with $\operatorname{Ext}^1_A(E, M) = 0$ such that no non-zero direct summand of M is cogenerated by E; dually, let $\mathcal{M}_E$ be given by all modules M with $\operatorname{Ext}^1_A(M, E) = 0$ such that no non-zero direct summand of M is generated by E; also, let $\mathcal{M}^{-E}$ be given by all M with $\operatorname{Hom}_A(M, E) = 0$ and $\mathcal{M}_{-E}$ by all M with $\operatorname{Hom}_A(E, M) = 0$. For any module M, let $\sigma^{-E}(M)$ be the intersection of the kernels of maps $M \longrightarrow E$ and $\sigma_{-E}(M) = M/t_E M$, where $t_E M$ is the sum of the images of maps $M \longrightarrow E$. In this way, we obtain equivalences

$$\sigma^{-E} \colon \mathcal{M}^E/\langle E \rangle \longrightarrow \mathcal{M}^{-E}, \quad \text{and} \quad \sigma_{-E} \colon \mathcal{M}_E/\langle E \rangle \longrightarrow \mathcal{M}_{-E}.$$

Here $\langle E \rangle$ is the ideal of all maps which factor through $\operatorname{add} E$. The reverse functors σ^E and σ_E are provided by forming universal extensions by copies of E (from above or below, respectively). Note that on the level of dimension vectors these partial functors $\sigma = \sigma^E, \sigma^{-E}, \sigma_E, \sigma_{-E}$ yield the usual reflection formula:

$$\dim \sigma(M) = \dim M - \frac{2\langle \dim M, \dim E \rangle}{\langle \dim E, \dim E \rangle} \dim E.$$

2.12 Slices

An artin algebra B is a tilted algebra if and only if $\operatorname{mod} B$ has a slice. Thus the existence of slices characterizes the tilted algebras. The necessity to explain the importance of slices has to be mentioned as a (further) impetus for the development of tilting theory. In my 1979 Ottawa lectures, I tried to describe several module categories explicitly. At that time, the knitting of preprojective components was one of the main tools, and I used slices in such components in order to guess what later turned out to be tilting functors, namely functorial constructions using pushouts and pullbacks. The obvious question about a possible theoretical foundation was raised by several participants, but it could be answered only a year later at the Puebla conference. Under minor restrictions (for example, the existence of a sincere indecomposable module) preprojective components will contain slice modules and these are tilting modules with a hereditary endomorphism ring! This concerns the concealed algebras to be mentioned below, but also all the representation-directed algebras. Namely, using covering theory, the problem of describing the structure of the indecomposable modules over a representation-finite algebra is reduced to the representation-directed algebras with a sincere indecomposable module, and such an algebra is a tilted algebra, since it obviously has a slice module.

In dealing with an artin algebra of finite representation type, and looking at its Auslander-Reiten quiver, one may ask for sectional subquivers say of Euclidean types. Given such a subquiver Γ, applying several times τ or τ^{-1} (and obtaining in this way "parallel" subquivers), one has to reach a projective, or an injective vertex, respectively. Actually, Bautista and Brenner have shown that the number of parallel subquivers is bounded, the bound is called the **replication number.** If one is interested in algebras with optimal replication numbers, one only has to look at representation-finite tilted algebras of Euclidean type. Note that given a hereditary algebra A of Euclidean type and a tilting A-module T, then $B = \text{End}_A(T)$ is representation-finite if and only if T has both preprojective and preinjective indecomposable direct summands.

It is natural to look inside preprojective and preinjective components for slices. In 1979 one did not envision that there could exist even regular components with a slice module. But any connected wild hereditary algebra with at least three simple modules has a regular tilting module T, and then the **connecting component** of $B = \text{End}_A(T)$ is regular. One should be aware that the category $\text{mod}\,B$ looks quite amazing: the connecting component (which is a regular component in this case) connects two wild subcategories, like a tunnel between two busy regions. Inside the tunnel, there are well-defined paths for the traffic, and the traffic goes in just one direction.

$$* * *$$

Tilting modules can be used to study **specific classes of artin algebras.** Some examples have been mentioned already. We have noted that all the representation finite k-algebras, with k algebraically closed, can be described using tilted algebras (the condition on k is needed in order to be able to use covering theory). We obtain in this way very detailed information on the structure of the indecomposables. One of the first uses of tilting theory concerned the representation-finite tree algebras, see Chapter 3 by Brüstle.

2.13 Concealed algebras

By definition, B is a *concealed* algebra, provided $B = \text{End}_A(T)$, where T is a preprojective A-module with A hereditary. The tame concealed k-algebras B where k is algebraically closed, have been classified by Happel and Vossieck, and Bongartz has shown in which way they can be used in order to determine whether a k-algebra is representation-finite.

2.14 Representations of posets

The representation theory of posets always has been considered as an important tool when studying questions in representation theory in general: there are quite a lot of reduction techniques which lead to a vector space with a bunch of subspaces, but the study of a vector space with a bunch of subspaces with some inclusions prescribed, really concerns the representation theory of the corresponding poset. On the other hand, the representation theory of finite posets is very similar to the representation theory of some quite well-behaved algebras, and the relationship is often given by tilting modules. For example, when dealing with a disjoint union of chains, then we deal with the subspace representations of a star quiver Q (the quiver Q is obtained from a finite set of linearly oriented quivers of type $\mathbb{A}$, with all the sinks identified to one vertex, the center of the star). If c is the center of the star quiver Q, then the subspace representations are the torsion-free modules of the (split) torsion pair $(\mathcal{Y}, \mathcal{X})$, with $\mathcal{X}$ being the representations V of Q such that $V_c = 0$. We also may consider the opposite quiver Q^{op} and the (again split) torsion pair $(\mathcal{F}, \mathcal{T})$, where now $\mathcal{F}$ are the representations V of Q^{op} with $V_c = 0$. The two orientations used here are obtained by a sequence of reflections, and the two split torsion pairs $(\mathcal{F}, \mathcal{G}), (\mathcal{Y}, \mathcal{X})$ are given by a tilting module which is a slice module:

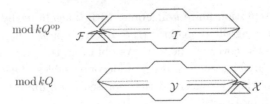

2.15 The Crawley-Boevey-Kerner functors

If R is an artin algebra and W an R-module, let us write $\langle \tau^\bullet W \rangle$ for
the ideal of $\mod R$ of all maps which factor through a direct sum of
modules of the form $\tau^z W$ with $z \in \mathbb{Z}$. We say that the module categories
$\mod R$ and $\mod R'$ are *almost equivalent* provided there is an R-module
W and an R'-module W' such that the categories $\mod R / \langle \tau^\bullet W \rangle$ and
$\mod R' / \langle \tau^\bullet W' \rangle$ are equivalent. The Crawley-Boevey-Kerner functors
were introduced in order to show the following: *If k is a field and Q and
Q' are connected wild quivers, then the categories $\mod kQ$ and $\mod kQ'$
are almost equivalent.* The proof uses tilting modules, and the result may
be rated as one of the most spectacular applications of tilting theory.
Thus it is worthwhile to outline the essential ingredients. This will be
done below.

Here are some. remarks concerning almost equivalent categories. It is
trivial that the module categories of all representation-finite artin algebras are almost equivalent. If k is a field, and Q, Q' are tame connected
quivers, then $\mod kQ$ and $\mod kQ'$ are almost equivalent only if Q and
Q' have the same type ($\tilde{\mathbb{A}}_{pq}, \tilde{\mathbb{D}}_n, \tilde{\mathbb{E}}_6, \tilde{\mathbb{E}}_7, \tilde{\mathbb{E}}_8$). Let us return to wild quivers Q, Q' and a Crawley-Boevey-Kerner equivalence

$$\eta \colon \mod kQ / \langle \tau^\bullet W \rangle \longrightarrow \mod kQ' / \langle \tau^\bullet W' \rangle,$$

with finite length modules W, W'. Consider the case of an uncountable
base field k, so that there are uncountably many isomorphism classes
of indecomposable modules for $R = kQ$ as well as for $R' = kQ'$. The
ideals $\langle \tau^\bullet W \rangle$ and $\langle \tau^\bullet W' \rangle$ are given by the maps which factor through
a countable set of objects, thus nearly all the indecomposable modules
remain indecomposable in $\mod kQ / \langle \tau^\bullet W \rangle$ and $\mod kQ' / \langle \tau^\bullet W' \rangle$, and
non-isomorphic ones (which are not sent to zero) remain non-isomorphic.
In addition, one should note that the equivalence η is really constructive
(not set-theoretical rubbish), with no unfair choices whatsoever. This
will be clear from the further discussion.

Nearly all quivers are wild. For example, if we consider the m-subspace quivers $Q(m)$, then one knows that $Q(m)$ is wild provided $m \geqslant 5$. Let us concentrate on a comparison of the wild quivers $Q(6)$ and $Q(5)$. To assert that Q, Q' are wild quivers means that there are full embeddings $\mod kQ \longrightarrow \mod kQ'$ and $\mod kQ' \longrightarrow \mod kQ$. But the Crawley-Boevey-Kerner theorem provides a completely new interpretation of what "wildness" is about. The definition of "wildness" itself is considered as quite odd, since it means in particular that there is a full embedding of $\mod kQ(6)$ into $\mod kQ(5)$. One may reformulate the wildness assertion as follows: any complication which occurs for 6 subspaces can be achieved (in some sense) already for 5 subspaces. But similar results are known in mathematics, since one is aware of other categories which allow to realize all kinds of categories as a subcategory. Also, "wildness" may be interpreted as a kind of fractal behaviour: inside the category $\mod kQ(5)$ we find proper full subcategories which are equivalent to $\mod kQ(5)$, again a quite frequent behaviour. These realization results are concerned with small parts of say $\mod kQ(5)$; one looks at full subcategories of the category $\mod kQ$ which have desired properties, but one does not try to control the complement. This is in sharp contrast to the Crawley-Boevey-Kerner property which provides a **global** relation between $\mod kQ(5)$ and $\mod kQ(6)$, actually, between the module categories of any two wild connected quivers. In this way we see that there is a kind of homogeneity property of wild abelian length categories which had not been anticipated before.

The Crawley-Boevey-Kerner result may be considered as a sort of Schröder-Bernstein property for abelian length categories. Recall that the Schröder-Bernstein theorem asserts that if two sets S, S' can be embedded into each other, then there is a bijection $S \longrightarrow S'$. For any kind of mathematical structure with a notion of embedding, one may ask whether two objects are isomorphic in case they can be embedded into each other. Such a property is very rare, even if we replace the isomorphism requirement by some weaker requirement. But this is what is asserted by the Crawley-Boevey-Kerner property.

Let us outline the construction of η. We start with a connected wild hereditary artin algebra A, and a regular exceptional module E which is quasi-simple (this means that the Auslander-Reiten sequence ending in E has indecomposable middle term, call it $\mu(E)$), such a module exists provided $n(A) \geqslant 3$. Denote by $E^{\perp}$ the category of all A-modules M such that $\operatorname{Hom}_A(E, M) = 0 = \operatorname{Ext}_A^1(E, M)$. One knows (Geigle-Lenzing,

Strausemi simple) that $E^\perp$ is equivalent to the category $\operatorname{mod} C$, where C is a connected wild hereditary algebra C and $n(C) = n(A) - 1$. The aim is to compare the categories $\operatorname{mod} C$ and $\operatorname{mod} A$, they are shown to be almost equivalent.

It is easy to see that the module $\mu(E)$ belongs to $E^\perp$, thus it can be regarded as a C-module. Since $E^\perp = \operatorname{mod} C$, there is a projective generator T' in $E^\perp$ with $\operatorname{End}_A(T') = C$. Claim: $T' \oplus E$ *is a tilting module.* For the proof we only have to check that $\operatorname{Ext}_A^1(T', E) = 0$. Since T' is projective in $E^\perp$, it follows that $\operatorname{Ext}^1(T', \mu(E)) = 0$. However, there is a surjective map $\mu(E) \longrightarrow E$ and this induces a surjective map $\operatorname{Ext}^1(T', \mu(E)) \longrightarrow \operatorname{Ext}^1(T', E)$.

As we know, the tilting module $T = \overline{T} \oplus E$ defines a torsion pair $(\mathcal{F}, \mathcal{T})$, with $\mathcal{T}$ the A-modules generated by T. Let us denote by $\tau_T M = t\tau_A M$ the torsion submodule[6] of $\tau_A M$. The functor η is now defined as follows:

$$\eta(M) = \lim_{t \to \infty} \tau_A^t \tau_T^{-2t} \tau_C^t(M).$$

One has to observe that the limit actually stabilizes: for large t, there is no difference whether we consider t or $t + 1$. The functor η is full, the image is just the full subcategory of all regular A-modules. There is a non-trivial kernel: a map is sent to zero if and only if it belongs to $\langle \tau^\bullet W \rangle$, where $W = C \oplus \mu(E) \oplus DC$. Also, let $W' = A \oplus DA$. Then η is an equivalence

$$\eta \colon \operatorname{mod} C/\langle \tau^\bullet W \rangle \longrightarrow \operatorname{mod} A/\langle \tau^\bullet W' \rangle.$$

One may wonder how special the assumptions on A and C are. Let us say that A *dominates* C provided there exists a regular exceptional module E which is quasi-simple with $\operatorname{mod} C$ equivalent to $E^\perp$. Given any two wild connected quivers Q, Q', there is a sequence of wild connected quivers $Q = Q_0, \ldots, Q_t = Q'$ such that kQ_i either dominates or is dominated by kQ_{i-1}, for $1 \leqslant i \leqslant t$. This implies that the module categories of all wild path algebras are almost equivalent.

The equivalence η can be constructed also in a different way [35], using

---

6 The notation shall indicate that this functor τ_T has to be considered as an Auslander-Reiten translation: it is the relative Auslander-Reiten translation in the subcategory $\mathcal{T}$. And there is the equivalence $\mathcal{T} \simeq \mathcal{Y}$, where $\mathcal{Y}$ is a full subcategory of $\operatorname{mod} B$, with $B = \operatorname{End}_A(T)$. Since $\mathcal{Y}$ is closed under τ in $\operatorname{mod} B$, the functor τ_T corresponds to the Auslander-Reiten translation τ_B in $\operatorname{mod} B$.

partial reflection functors. Let $E(i) = \tau^i E$, for all $i \in \mathbb{Z}$. Note that for any regular A-module M, one knows that

$$\mathrm{Hom}_A(M, E(-t)) = 0 = \mathrm{Hom}_A(E(t), M) \qquad \text{for} \qquad t \gg 0,$$

according to Baer and Kerner. Thus, if we choose t sufficiently large, we can apply the partial reflection functors $\sigma^{E(-t)}$ and $\sigma_{E(t)}$ to M. The module obtained from M has the form

$$\frac{E(-t)\,E(-t) \cdots E(-t)}{\overline{\rule{0pt}{1em}\quad M \quad}}$$
$$\overline{E(t) \quad E(t) \quad \cdots \quad E(t)}$$

and belongs to

$$\mathcal{M}^{E(-t)} \cap \mathcal{M}_{E(t)} \quad \subseteq \quad \mathcal{M}^{-E(-t+1)} \cap \mathcal{M}_{-E(t-1)}.$$

Thus we can proceed, applying now $\sigma^{E(-t+1)}$ and $\sigma_{E(t-1)}$. We use induction, the last partial reflection functors to be applied are the functors $\sigma^{E(-1)}$, $\sigma_{E(1)}$, and then finally σ^E. In this way we obtain a module in

$$\mathcal{M}_{E(1)} \cap \mathcal{M}^E = E^\perp$$

as required. It has the following structure:

$$
\begin{array}{c}
\overline{E \quad\cdots\cdots\cdots\cdots\cdots\cdots \quad E} \\
\overline{E(-1) \quad\cdots\cdots\cdots\cdots\cdots \quad E(-1)} \\
\vdots \\
\overline{E(-t) \cdots E(-t)} \\
\overline{\quad M \quad} \\
\overline{E(t) \quad\cdots\quad E(t)} \\
\vdots \\
\overline{E(1) \quad\cdots\cdots\cdots\cdots\quad E(1)}
\end{array}
$$

2.16 The shrinking functors for the tubular algebras

Again these are tilting functors (here, A no longer is a hereditary artin algebra, but say a canonical algebra - we are still in the realm of the "T" displayed in Part 1, now even in its center), and such functors belong to the origin of the development. If one looks at the Brenner-Butler tilting paper, the main examples considered there were of this kind. So one of the first applications of tilting theory was to show the similarity of the module categories of various tubular algebras. And this is also the setting which later helped to describe in detail the module category of a tubular algebra: one uses the shrinking functors in order to construct all the regular tubular families, as soon as one is known to exist.

2.17 Self-injective algebras

Up to coverings and (in characteristic 2) deformations, the trivial extensions of the tilted algebras of Dynkin type (those related to the left arm of the "T" displayed in Part 1) yield all the representation-finite self-injective algebras (recall that the trivial extension of an algebra R is the semi-direct product $R \ltimes DR$ of R with the dual module DR). In private conversation, such a result was conjectured by Tachikawa already in 1978, and it was the main force for the investigations of him and Wakamatsu, which he presented at the Ottawa conference in 1979. There he also dealt with the trivial extension of a tilted algebra of Euclidean type (the module category has two tubular families). This motivated Hughes-Waschbüsch to introduce the concept of a repetitive algebra. But it is also part of one of the typical quarrels between Zürich and the rest of the world: with Gabriel hiding the Hughes-Waschbüsch manuscript from Bretscher-Läser-Riedtmann (asking a secretary to seal the envelope with the manuscript and to open it only several months later...), so that they could proceed "independently".

The representation theory of artin algebras came into limelight when Dynkin diagrams popped up for representation-finite algebras. And this occurred twice, first for hereditary artin algebras in the work of Gabriel (as the Ext-quiver), but then also for self-injective algebras in the work of Riedtmann (as the tree class of the stable Auslander-Reiten quiver). The link between these two classes of rings is furnished by tilted algebras and their trivial extensions. As far as I know, it is Tachikawa who deserves the credit for this important insight.

The reference to trivial extensions of tilted algebras actually closes a circle in our considerations, due to another famous theorem of Happel. We have started with the fact that tilting functors provide derived equivalences. Thus the derived category of a tilted algebra can be identified with the derived category of a hereditary artin algebra. For all artin algebras R of finite global dimension (in particular our algebras A and B), there is a an equivalence between $D^b(\text{mod } R)$ and the stable module category of the repetitive algebra $\widehat{R}$. But $\widehat{R}$ is just a $\mathbb{Z}$-covering of the trivial extension of R.

2.18 Artin algebras with Gorenstein dimension at most 1

We have mentioned that the two classes of algebras: the selfinjective ones and the hereditary ones, look very different, but nevertheless they have some common behaviour. Auslander and Reiten [4] have singled out an important property which they share, they are Gorenstein algebras of Gorenstein dimension at most 1. An artin algebra A is called *Gorenstein*[7] provided $_ADA$ has finite projective dimension and $_AA$ has finite injective dimension. For Gorenstein algebras, proj-dim $_ADA$ = inj-dim $_AA$, and this number is called the *Gorenstein dimension* of A. It is not known whether the finiteness of the projective dimension of $_ADA$ implies the finiteness of the injective dimension of $_AA$. It is conjectured that this is the case: this is the Gorenstein symmetry conjecture, and this conjecture is equivalent to the conjecture that the small finitistic dimension of A is finite. The artin algebras of Gorenstein dimension 0 are the selfinjective algebras. An artin algebra has Gorenstein dimension at most 1 if and only if DA is a tilting module (of projective dimension at most 1).

If A is a Gorenstein algebra of Gorenstein dimension at most 1, then there is a strict separation of the indecomposable modules: an A-module M of finite projective dimension or finite injective dimension satisfies both proj-dim $M \leqslant 1$, inj-dim $M \leqslant 1$. (The proof is easy: Assume proj-dim $M \leqslant m$, thus the m-th syzygy module $\Omega_m(M)$ is projective. Now for any short exact sequence $0 \longrightarrow X \longrightarrow Y \longrightarrow Z \longrightarrow 0$, it is clear that inj-dim $X \leqslant 1$, inj-dim $Y \leqslant 1$ imply inj-dim $Z \leqslant 1$. One applies this inductively to the exact sequences $\Omega_i(M) \longrightarrow P_i \longrightarrow \Omega_{i-1}(M)$, where P_i is projective, starting with $i = m$ and ending with $i = 0$. This shows that inj-dim$(M) \leqslant 1$. The dual argument shows that a module of finite injective dimension has projective dimension at most 1.) As a consequence, if A is not hereditary, then the global dimension of A

---

7 This definition is one of the many possibilities to generalize the notion of a commutative Gorenstein ring to a non-commutative setting. Note that a commutative artin algebra R is a Gorenstein ring if and only if R is selfinjective. Of course, a commutative connected artin algebra R is a local ring, and a local ring has a non-zero module of finite projective dimension only in case R is selfinjective.

is infinite. Also, if P is an indecomposable projective A-module, then either its radical is projective or else the top of P is a simple module which has infinite projective and infinite injective dimension.

Until very recently, the interest in artin algebras of Gorenstein dimension at most 1 has been quite moderate, the main reason being a lack of tempting examples: of algebras which are neither selfinjective nor hereditary. But now there is a wealth of such examples, as we will see in Part 3.

* * *

We hope that we have convinced the reader that the use of tilting modules and tilted algebras lies at the heart of nearly all the major developments in the representation theory of artin algebras in the last 25 years. In this report we usually restrict to tilting modules in the narrow sense (as being finite length modules of projective dimension at most 1). In fact, most of the topics mentioned are related to tilting A-modules T, where A is a hereditary artin algebra (so that there is no need to stress the condition proj-dim $T \leqslant 1$). However, the following two sections will widen the viewpoint, taking into account also various generalizations.

2.19 Representations of semisimple complex Lie algebras and algebraic groups

The highest weight categories which arise in the representation theory of semisimple complex Lie algebras and algebraic groups can be analyzed very well using quasi-hereditary artin algebras as introduced by Cline-Parshal-Scott. One of the main features of such a quasi-hereditary artin algebra is its characteristic module, this is a tilting module (of finite projective dimension). Actually, the experts use a different convention, calling its indecomposable direct summands "tilting modules", see Chapter 9 by Donkin. If T is the characteristic module, then add T consists of the A-modules which have both a standard filtration and a costandard filtration, and it leads to a duality theory which seems to be of great interest.

2.20 The homological conjectures

The homological conjectures are one of the central themes of module theory, so clearly they deserve special interest. They go back to mathematicians like Nakayama, Eilenberg, Auslander, Bass, but also Rosenberg, Zelinsky, Buchsbaum and Nunke should be mentioned, and were formulated between 1940 and 1960. Unfortunately, there are no written accounts about the origin, but we may refer to surveys by Happel, Smalø and Zimmermann-Huisgen. The modern development in representation theory of artin algebras was directed towards a solution of the Brauer-Thrall conjectures, and there was for a long time a reluctance to work on the homological conjectures. The investigations concerning the various representation types have produced a lot of information on special classes of algebras, but for these algebras the homological conjectures are usually true for trivial reasons. As Happel has pointed out, the lack of knowledge of non-trivial examples may very well mean that counter-examples could exist. Here is a short discussion of this topic, in as far as modules without self-extensions are concerned.

Let me start with the Nakayama conjecture which according to B. Müller can be phrased as follows: If R is an artin algebra and M is a generator and cogenerator for mod R with $\mathrm{Ext}_R^i(M, M) = 0$ for all $i \geqslant 1$, then M has to be projective. Auslander and Reiten [3] proposed in 1975 that the same conclusion should hold even if M is not necessarily a cogenerator (this is called the "generalized Nakayama conjecture"). This incorporates a conjecture due to Tachikawa (1973): If R is self-injective and M is an R-module with $\mathrm{Ext}_R^i(M, M) = 0$ for all $i > 0$, then M is projective. The relationship of the generalized Nakayama conjecture with tilting theory was noted by Auslander and Reiten [3, 4]. Then there is the conjecture on the finiteness of the number of complements of an almost complete partial tilting module, due to Happel and Unger. And there is a conjecture made by Beligiannis and Reiten [5], called the Wakamatsu tilting conjecture (because it deals with Wakamatsu tilting modules, see Chapter 8 by Reiten): If T is a Wakamatsu tilting module of finite projective dimension, then T is a tilting module. The Wakamatsu tilting conjecture implies the generalized Nakayama conjecture (apparently, this was first observed by Buan) and also the Gorenstein symmetry conjecture, see [5]. In a joint paper, Mantese and Reiten [37] showed that it is implied by the finitistic dimension conjecture, and that it implies the conjecture on a finite number of complements, which according to Buan and Solberg is known to imply the generalized Nakayama conjecture. There is also the equivalence of the generalized Nakayama conjecture with projective almost complete partial tilting modules having only a finite number of complements (Happel-Unger, Buan-Solberg, both papers are in the Geiranger proceedings). Coelho, Happel and Unger proved that the finitistic dimension conjecture implies the conjecture on a finite number of complements.

Further relationship of tilting theory with the finitistic dimension conjectures is discussed in detail in Chapter 11 by Trlifaj and in Chapter 12 by Solberg. But also other results presented in the Handbook have to be seen in this light. We know from Auslander and Reiten, that the finitistic dimension of an artin algebra R is finite, in case the subcategory of all modules of finite projective dimension is contravariantly finite in mod R. This has been the motivation to look at the latter condition carefully (see for example Chapter 10 by Unger).

With respect to applications outside of ring and module theory, many more topics could be mentioned. We have tried to stay on a basic level, whereas there are a lot of mathematical objects which are derived from representation theoretical data and this leads to a fruitful interplay (dealing with questions on quantum groups, with the shellability of simplicial complexes, or with continued fraction expansions of real numbers): There are many unexpected connections to analysis, to number theory, to combinatorics — and again, it is usually the tilting theory which plays an important role.

3 The new cluster tilting approach

Let me repeat: at the time the Handbook was conceived, there was a common feeling that the tilted algebras (as the core of tilting theory) were understood well and that this part of the theory had reached a sort of final shape. But in the meantime this has turned out to be wrong: the tilted algebras have to be seen as factor algebras of the so called cluster tilted algebras, and it may very well be, that in future the cluster tilted algebras and the cluster categories will topple the tilted algebras. The impetus for introducing and studying cluster tilted algebras came from outside, in a completely unexpected way. We will mention below some of the main steps of this development. But first let me jump directly to the relevant construction.

3.1 The cluster tilted algebras

We return to the basic setting, the hereditary artin algebra A, the tilting A-module T and its endomorphism ring B. Consider the semi-direct ring extension

$$\widetilde{B} = B \ltimes \mathrm{Ext}^2_B(DB, B).$$

This is called the *cluster tilted algebra* corresponding to B. Since this is the relevant definition, let me say a little more about this construction[8] : $\widetilde{B}$ has B as a subring, and there is an ideal J of $\widetilde{B}$ with $J^2 = 0$, such that $\widetilde{B} = B \oplus J$ as additive groups and J is as a B-B-bimodule isomorphic to $\mathrm{Ext}^2_B(DB, B)$; in order to construct $\widetilde{B}$ one may take $B \oplus \mathrm{Ext}^2_B(DB, B)$, with componentwise addition, and one uses $(b, x)(b', x') = (bb', bx' + xb')$, for $b, b' \in B$ and $x, x' \in \mathrm{Ext}^2_B(DB, B)$ as multiplication. The definition shows that $\widetilde{B}$ can be considered as a $\mathbb{Z}$-graded (or also $\mathbb{Z}/2$-graded) algebra, with $(\widetilde{B})_0 = B$ and $(\widetilde{B})_1 = J$.

We consider again the example of B given by a square with two zero relations. Here $\mathrm{Ext}^2_B(DB, B)$ is 8-dimensional and $\widetilde{B}$ is a 16-dimensional algebra:

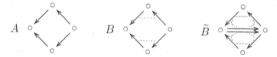

Non-isomorphic tilted algebras B may yield isomorphic cluster tilted algebras $\widetilde{B}$. Here are all the tilted algebras which lead to the cluster tilted algebra just considered:

It is quite easy to write down the **quiver of a cluster-tilted algebra.** Here, we assume that we deal with k-algebras, where k is an algebraically

---

8 One may wonder what properties the semi-direct product $R \ltimes \mathrm{Ext}^2_R(DR, R)$ for any artin algebra R has in general (at least in case R has global dimension at most 2); it seems that this question has not yet been studied.

closed field. We get the quiver of $\widetilde{B}$ from the quiver with relations of B by just replacing the dotted arrows[9] by solid arrows in opposite direction [1]. The reason is the following: Let us denote by N the radical of B. Then $N \oplus J$ is the radical of $\widetilde{B} = B \ltimes J$, and $N^2 \oplus (NJ + JN)$ is equal to the square of the radical of $\widetilde{B}$. This shows that the additional arrows for $\widetilde{B}$ correspond to $J/(NJ + JN)$. Note that $J/(NJ + JN)$ is the top of the B-B-bimodule J. Now the top of the bimodule $\operatorname{Ext}^2_B(DB, B)$ is $\operatorname{Ext}^2_B(\operatorname{soc} {}_B DB, \operatorname{top} {}_B B)$, since B has global dimension at most 2. It is well-known that $\operatorname{Ext}^2_B(\operatorname{soc} {}_B DB, \operatorname{top} {}_B B)$ describes the relations of the algebra B, and we see in this way that relations for B correspond to the additional arrows for $\widetilde{B}$. Since $\operatorname{rad} \widetilde{B} = \operatorname{rad} B \oplus J$ and J is an ideal of $\widetilde{B}$ with $J^2 = 0$, we also see: If $(\operatorname{rad} B)^t = 0$, then $(\operatorname{rad} \widetilde{B})^{2t} = 0$. The quiver of any tilted algebra is directed, thus $(\operatorname{rad} B)^{n(B)} = 0$, therefore $(\operatorname{rad} \widetilde{B})^{2n(B)} = 0$.

The recipe for obtaining the quiver of $\widetilde{B}$ shows that there are always oriented cyclic paths (unless B is hereditary). However, such a path is always of length at least 3. Namely, since the quiver of B has no loops, there cannot be any relation for B starting and ending at the same vertex. Thus, the quiver of $\widetilde{B}$ cannot have a loop [8]. Also, Happel ([27], Lemma IV.1.11) has shown that for simple B-modules S, S' with $\operatorname{Ext}^1_B(S, S') \neq 0$ one has $\operatorname{Ext}^2_B(S, S') = 0$. This means that the quiver of $\widetilde{B}$ cannot have a pair of arrows in opposite direction [11].

It should be of interest whether knowledge about the quiver with relations of a cluster tilted algebra $\widetilde{B}$ can provide new insight into the structure of the tilted algebras themselves. There is a lot of ongoing research on cluster tilted algebras, let us single out just one result. Assume that we deal with k-algebras, where k is algebraically closed. Then: *Any cluster tilted k-algebra of finite representation type is uniquely determined by its quiver* [12]. This means: in the case of finite representation type, the quiver determines the relations! What happens in general is still under investigation.

If A is a hereditary artin algebra and T a tilting A-module with endomorphism ring B, we have introduced the corresponding cluster tilted algebras as the algebra $\widetilde{B} = B \ltimes J$, with B-B-bimodule $J = \operatorname{Ext}^2_B(DB, B)$.

---

9 Actually, the usual convention for indicating relations is to draw dotted lines, not dotted arrows. However, these dotted lines are to be seen as being directed, since the corresponding relations are linear combinations of paths with fixed starting point and fixed end point.

The original definition of $\tilde{B}$ by Buan, Marsh and Reiten [10] used another description of J, namely $J = \mathrm{Ext}^1_A(T, \tau^{-1}T)$, and it was observed by Assem, Brüstle and Schiffler [1] that the bimodules $\mathrm{Ext}^1_A(T, \tau^{-1}T)$ and $\mathrm{Ext}^2_B(DB, B)$ are isomorphic[10] (using this Ext^2-bimodule has the advantage that it refers only to the algebra B itself, and not to T). It was Zhu Bin [51] who stressed that cluster tilted algebras should be explored as semi-direct ring extensions.

Since this isomorphism is quite essential, let me sketch an elementary proof, without reference to derived categories. Let V be the universal extension of τT by copies of T from above, thus there is an exact sequence

$$(*) \qquad\qquad 0 \longrightarrow \tau T \longrightarrow V \longrightarrow T^m \longrightarrow 0$$

for some m, and $\mathrm{Ext}^1_A(T, V) = 0$. Applying $\mathrm{Hom}_A(-, T)$ to $(*)$ shows that $\mathrm{Ext}^1_A(V, T) \simeq \mathrm{Ext}^1_A(\tau T, T)$. Applying $\mathrm{Hom}_A(T, -)$ to $(*)$ yields the exact sequence

$$0 \longrightarrow \mathrm{Hom}_A(T, V) \longrightarrow \mathrm{Hom}_A(T, T^m) \longrightarrow \mathrm{Ext}^1_A(T, \tau T) \longrightarrow 0.$$

This is an exact sequence of B-modules and $\mathrm{Hom}_A(T, T^m)$ is a free B-module, thus we see that $\mathrm{Hom}_A(T, V)$ is a syzygy module for the B-module $\mathrm{Ext}^1_A(T, \tau T)$. But the latter means that

$$\mathrm{Ext}^2_B(\mathrm{Ext}^1_A(T, \tau T), {}_B B) \simeq \mathrm{Ext}^1_B(\mathrm{Hom}_A(T, V), {}_B B).$$

The left hand side is nothing else than $\mathrm{Ext}^2_B(DB, B)$, since the B-module DB and $\mathrm{Ext}^1_A(T, \tau T)$ differ only by projective-injective direct summands. The right hand side $\mathrm{Ext}^1_B(\mathrm{Hom}_A(T, V), \mathrm{Hom}_A(T, T))$ is the image of $\mathrm{Ext}^1_A(V, T)$ under the (exact) equivalence $\mathrm{Hom}_A(T, -) \colon \mathcal{T} \longrightarrow \mathcal{Y}$ (here we use that V belongs to $\mathcal{T}$). This completes the proof[11] .

Now let us deal with the **representations** of $\tilde{B}$. The $\tilde{B}$-modules can

---

10 In addition, we should remark that $\mathrm{Ext}^1_A(T, \tau^{-1}T)$ can be identified with $\mathrm{Ext}^1_A(\tau T, T)$ (as B-B-bimodules). The reason is the fact that the functor τ^{-1} is left adjoint to τ, for A hereditary, thus $\mathrm{Ext}^1_A(T, \tau^{-1}T) \simeq D\mathrm{Hom}_A(\tau^{-1}T, \tau T) \simeq D\mathrm{Hom}_A(T, \tau^2 T) \simeq \mathrm{Ext}^1_A(\tau T, T)$. The importance of the bimodule $\mathrm{Ext}^1_A(\tau T, T)$ has been stressed already in section 2.7; I like to call it the "magic" bimodule for such a tilting process. All the bimodule isomorphisms mentioned here should be of interest when dealing with the magic bimodule J. In particular, when working with injective $\tilde{B}$-modules, it seems to be convenient to know that $DJ \simeq \mathrm{Hom}_A(T, \tau^2 T)$.

11 Note that the isomorphy of $\mathrm{Ext}^2_B(DB, B)$ and $\mathrm{Ext}^1_A(\tau T, T)$ yields a proof for the implication (ii) $\Rightarrow$ (iv) mentioned in Part 1. Since we know that B has global dimension at most 2, the vanishing of $\mathrm{Ext}^2_B(DB, B)$ implies that $\mathrm{Ext}^2_B(X, Y) = 0$ for all B-modules X, Y, thus we also see that (iv) $\Rightarrow$ (ii).

be described as follows: they are pairs of the form (M, γ), where M is a B-module, and $\gamma \colon J \otimes_B M \longrightarrow M$ is a B-linear map. As we know, in $\mathrm{mod}\, B$ there is the splitting torsion pair $(\mathcal{Y}, \mathcal{X})$ and it turns out that $J \otimes_B X = 0$ for $X \in \mathcal{X}$, and that $J \otimes_B Y$ belongs to $\mathcal{X}$ for all $Y \in \mathcal{Y}$ (for the definition of the module classes $\mathcal{X}, \mathcal{Y}$, but also for $\mathcal{Y}'$ and $\mathcal{S}$ we refer to section 1.2). Let us consider a pair (M, γ) in $\mathrm{mod}\, \widetilde{B}$ and write $M = Y \oplus S \oplus X$, with $Y \in \mathcal{Y}'$, $S \in \mathcal{S}$, and $X \in \mathcal{X}$. Then the image of γ is contained in $\mathcal{Y}'$ and $Y \oplus S$ is contained in the kernel of γ (in particular, $(S, 0)$ is a direct summand of (M, γ)).

Note that $(\mathcal{Y}, \mathcal{X})$ still is a torsion pair in $\mathrm{mod}\, \widetilde{B}$ (a module $(X \oplus Y, \gamma)$ with $X \in \mathcal{X}$ and $Y \in \mathcal{Y}$ has $(X, 0)$ as torsion submodule, has $(Y, 0)$ as its torsion-free factor module, and the map γ is the obstruction for the torsion submodule to split off). Let us draw the attention to a special feature of this torsion pair $(\mathcal{Y}, \mathcal{X})$ in $\mathrm{mod}\, \widetilde{B}$: there exists an ideal, namely J, such that the modules annihilated by J are just the modules in $\mathrm{add}(\mathcal{X}, \mathcal{Y})$.

Buan, Marsh and Reiten [10] have shown that the category $\mathrm{mod}\, \widetilde{B}$ can be described in terms of $\mathrm{mod}\, A$ (via the corresponding cluster category). Let us present such a description in detail. We will use that $J = \mathrm{Ext}^1_A(T, \tau^{-1}T)$ (as explained above). The algebra $\widetilde{B}$ has as $\mathbb{Z}$-covering the following (infinite dimensional) matrix algebra:

$$
B_\infty = \begin{bmatrix} \ddots & & & \\ & B & J & \\ & & B & J \\ & & & B \end{bmatrix} \, \begin{matrix} \\ \\ \\ \ddots \end{matrix}
$$

with B on the main diagonal, J directly above the main diagonal, and zeros elsewhere (note that this algebra has no unit element in case $B \neq 0$). It turns out that it is sufficient to determine the representations of the convex subalgebras of the form $B_2 = \begin{bmatrix} B & J \\ 0 & B \end{bmatrix}$. We can write B_2-modules as columns $\begin{bmatrix} N \\ N' \end{bmatrix}$ and use matrix multiplication, provided we have specified a map $\gamma \colon J \otimes N' \longrightarrow N$. In the example considered (B a square, with two zero relations), the algebras B_∞ and B_2 are as follows:

In order to exhibit all the B_2-modules, we use the functor $\Phi\colon \operatorname{mod} A \longrightarrow \operatorname{mod} B_2$ given by

$$\Phi(M) = \begin{bmatrix} \operatorname{Ext}^1_A(T, M) \\ \operatorname{Hom}_A(\tau^{-1}T, M) \end{bmatrix},$$

with $\gamma\colon \operatorname{Ext}^1_A(T, \tau^{-1}T) \otimes \operatorname{Hom}_A(\tau^{-1}T, M) \longrightarrow \operatorname{Ext}^1_A(T, M)$ being the canonical map of forming induced exact sequences (this is just the Yoneda multiplication)[12] . Now Φ itself is not faithful, since obviously T is sent to zero[13] . However, it induces a fully faithful functor (which again will be denoted by Φ):

$$\Phi\colon \operatorname{mod} A/\langle T\rangle \longrightarrow \operatorname{mod} B_2,$$

where $\operatorname{mod} A/\langle T\rangle$ denotes the factor category of $\operatorname{mod} A$ modulo the ideal of all maps which factor through $\operatorname{add} T$. The image of the functor Φ is given by

$$\begin{bmatrix} \mathcal{X} \\ 0 \end{bmatrix} \bigsqcup \begin{bmatrix} 0 \\ \mathcal{Y}' \end{bmatrix}.$$

In general, given module classes $\mathcal{K}, \mathcal{L}$ in $\operatorname{mod} R$, we write $\mathcal{K}\bigsqcup\mathcal{L}$ for the class of all R-modules M with a submodule K in $\mathcal{K}$ such that

---

12 The reader should recall that the functors $\operatorname{Ext}^1_A(T, -)$ and $\operatorname{Hom}_A(\tau^{-1}T, -)$ have been mentioned already in Part 1. These are the functors which provide the equivalences $\mathcal{F} \simeq \mathcal{X}$ and $\mathcal{T}' \simeq \mathcal{Y}'$, respectively.

13 The comparison with the Buan-Marsh-Reiten paper [10] shows a slight deviation: The functor they use vanishes on the modules τT and not on T (and if we denote by T_i an indecomposable direct summand of T, then the image of T_i becomes an indecomposable projective $\tilde{B}$-module). Instead of looking at the functor Φ, we could have worked with $\Phi'(M) = \begin{bmatrix} \operatorname{Ext}^1_A(\tau T, M) \\ \operatorname{Hom}_A(T, M) \end{bmatrix}$, again taking for γ the canonical map. This functor Φ' vanishes on τT. On the level of cluster categories, the constructions corresponding to Φ and Φ' differ only by the Auslander-Reiten translation in the cluster category, and this is an auto-equivalence of the cluster category. But as functors $\operatorname{mod} A \longrightarrow \operatorname{mod} B_2$, the two functors Φ, Φ' are quite different. Our preference for the functor Φ has the following reason: the functor Φ kills precisely $n = n(A)$ indecomposable A-modules, thus the number of indecomposable $\tilde{B}$-modules which are not contained in the image of Φ is also n, and these modules form a slice. This looks quite pretty: the category $\operatorname{mod} \tilde{B}$ is divided into the image of the functor Φ and one additional slice.

M/K belongs to $\mathcal{L}$. Thus, we assert that the image of Φ is the class of the $\tilde{B}$-modules $\left[\begin{smallmatrix} N \\ N' \end{smallmatrix}\right]$ with $N \in \mathcal{X}$ and $N' \in \mathcal{Y}'$. (In order to see that $\text{Hom}_A(\tau^{-1}T, M) \in \mathcal{Y}'$, first note that $\text{Hom}_A(\tau^{-1}T, M) = \text{Hom}_A(T, \tau M)$, thus this is a B-module in $\mathcal{Y}$. We further have $\text{Hom}_A(T, \tau M) = \text{Hom}_A(T, t\tau M)$, where $t\tau M$ is the torsion submodule of τM. If we assume that $\text{Hom}_A(T, t\tau M)$ has an indecomposable submodule in $\mathcal{S}$, say $\text{Hom}_A(T, Q)$, where Q is an indecomposable injective A-module, then we obtain a non-zero map $Q \longrightarrow t\tau M \subseteq \tau M$, since $\text{Hom}_A(T, -)$ is fully faithful on $\mathcal{T}$. However, the image of this map is injective (since A is hereditary) and τM is indecomposable, thus τM is injective, which is impossible).

We want to draw a rough sketch of the shape of $\text{mod}\,B_2$, in the same spirit as we have drawn a picture of $\text{mod}\,B$ in Part 1:

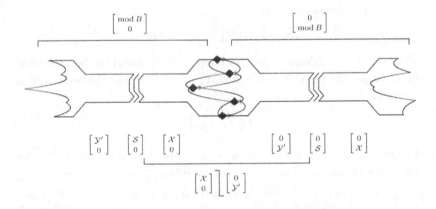

As we have mentioned, the middle part $\left[\begin{smallmatrix} \mathcal{X} \\ 0 \end{smallmatrix}\right] \bigsqcup \left[\begin{smallmatrix} 0 \\ \mathcal{Y}' \end{smallmatrix}\right]$ (starting with $\left[\begin{smallmatrix} \mathcal{X} \\ 0 \end{smallmatrix}\right]$ and ending with $\left[\begin{smallmatrix} 0 \\ \mathcal{Y}' \end{smallmatrix}\right]$) is the image of the functor Φ, thus this part of the category $\text{mod}\,B_2$ is equivalent to $\text{mod}\,A/\langle T \rangle$. Note that this means that there are some small "holes" in this part, they are indicated by black lozenges; these holes correspond to the position in the Auslander-Reiten quiver of A which are given by the indecomposable direct summands T_i of T (and are directly to the left of the small stars).

It follows that $\text{mod}\,\tilde{B}$ has the form:

$\mathrm{mod}\,\tilde{B}$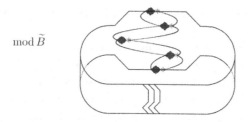

Here, we have used the covering functor $\Pi\colon \mathrm{mod}\,B_\infty \longrightarrow \mathrm{mod}\,\tilde{B}$ (or better its restriction to $\mathrm{mod}\,B_2$): under this functor the subcategories $\left[\,{}^{\mathrm{mod}\,B}_{0}\,\right]$ and $\left[\,{}^{0}_{\mathrm{mod}\,B}\,\right]$ are canonically identified. In particular, a fundamental domain for the covering functor is given by the module classes $\left[\,{}^{\mathcal{X}}_{0}\,\right] \bigsqcup \left[\,{}^{0}_{\mathcal{Y}'}\,\right]$ and $\left[\,{}^{0}_{\mathcal{S}}\,\right]$.

This shows that $\mathrm{mod}\,\tilde{B}$ decomposes into the modules in $\mathcal{X}\,\rfloor\,\mathcal{Y}'$ (these are the $\tilde{B}$-modules N with a submodule $X \subseteq N$ in $\mathcal{X}$, such that N/X belongs to $\mathcal{Y}'$) on the one hand, and the modules in $\mathcal{S}$ on the other hand. Under the functor Φ, $\mathrm{mod}\,A/\langle T\rangle$ is embedded into $\mathrm{mod}\,\tilde{B}$ with image the module class $\mathcal{X}\,\rfloor\,\mathcal{Y}'$. This is a controlled embedding (as defined in [43]), with control class $\mathcal{S}$.

The functor

$$\mathrm{mod}\,A \xrightarrow{\Phi} \mathrm{mod}\,B_2 \xrightarrow{\Pi} \mathrm{mod}\,\tilde{B}$$

has the following interesting property: only finitely many indecomposables are killed by the functor (the indecomposable direct summands of T) and there are only finitely many indecomposables (actually, the same number) which are not in the image of the functor (the indecomposable modules in $\mathcal{S}$). Otherwise, it yields a bijection between indecomposables.

It should be noted that some of the strange phenomena of tilted algebras disappear when passing to cluster tilted algebras. For example, the tunnel effect mentioned above changes as follows: there still is the tunnel, but no longer does it connect two separate regions; it now is a sort of by-pass for a single region. On the other hand, we should stress that the pictures which we have presented and which emphasise the existence of cyclic paths in $\mathrm{mod}\,\tilde{B}$ are misleading in the special case when T is a slice module: in this case, $J = \mathrm{Ext}^2_B(DB, B) = 0$, thus $\tilde{B} = B$ is again hereditary.

The cluster tilting theory has produced a lot of surprising results —

C. M. Ringel

it even answered some question which one did not dare to ask. For example, dealing with certain classes of algebras such as special biserial ones, one observes that sometimes there do exist indecomposable direct summands X of the radical of an indecomposable projective module P, such that the Auslander-Reiten translate τX is a direct summand of the socle factor module of an indecomposable injective module I. Thus, in the Auslander-Reiten quiver of $\tilde{B}$, there are non-sectional paths of length 4 from I to P

$$I \longrightarrow \tau X \longrightarrow X' \longrightarrow X \longrightarrow P.$$

Is this configuration of interest? I did not think so before I was introduced to cluster tilting theory, but according to [10], this configuration is a very typical one when dealing with cluster tilted algebras.

As an illustration, we show what happens in the non-regular components of our example B_∞ (where B is the square with two zero relations). The upper line exhibits the part of the quiver of B_∞ which is needed as support for the modules shown below:

For both components, the dashed boundary lines have to be identified. In this way, the right picture with the vertical identification yields what is called a tube, the left picture gives a kind of horizontal hose. In contrast to the tube with its mouth, the hose extends in both directions indefinitely. The big circles indicate the position of the modules T_i in the corresponding components of $\operatorname{mod} A$, these are the modules which are killed by the functor Φ. In both components we find non-sectional paths of length 4 from an indecomposable injective B_∞-module I to an indecomposable projective B_∞-module P such that the simple modules $\operatorname{soc} I$ and $\operatorname{top} P$ are identified under the covering functor Π.

We also want to use this example in order to illustrate the fact that the image of Φ in $\mathrm{mod}\,\tilde{B}$ is complemented by a slice $\mathcal{S}$:

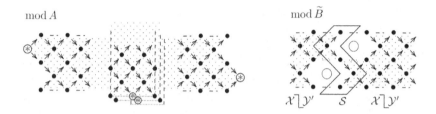

When looking at the non-sectional paths from I to P of length 4, where I is an indecomposable injective $\tilde{B}$-module, P an indecomposable projective $\tilde{B}$-module such that $S = \mathrm{soc}\,I \simeq \mathrm{top}\,P$, one should be aware that the usual interest lies in paths from P to I. Namely, there is the so called "hammock" for the simple module S, dealing with pairs of maps of the form $P \longrightarrow M \longrightarrow I$ with composition having image S (and M indecomposable).

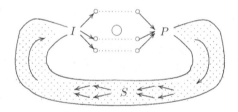

Taking into account not only the hammock, but also the non-sectional paths of length 4 from I to P leads to a kind of organized round trip. Since the simple module S has no self-extension, it is the only indecomposable module M such that $\mathrm{Hom}_{\tilde{B}}(P', M) = 0$, for any indecomposable projective $\tilde{B}$-module $P' \not\simeq P$. We will return to this hammock configuration (P, S, I) later.

Readers familiar with the literature will agree that despite of the large number of papers devoted to questions in the representation theory of artin algebras, only few classes of artin algebras are known where there is a clear description of the module categories[14] . The new developments

---

14 Say in the same way as the module categories of hereditary artin algebras are described. We consider here algebras which may be wild, thus we have to be cautious of what to expect from a "clear description".

outlined here show that the cluster tilted algebras are such a class: As for the hereditary artin algebras, the description of the module category is again given by the root system of a Kac-Moody Lie algebra.

Keller and Reiten [34] have shown, that *cluster tilted algebras are Gorenstein algebras of Gorenstein dimension at most 1.* This is a very remarkable assertion! The proof uses in an essential way cluster categories, and provides further classes of Gorenstein algebras of Gorenstein dimension at most 1. Note that the cluster tilted algebra $\widetilde{B}$ is hereditary only in case $\widetilde{B} = B$, thus only for T a slice module. There are examples where $\widetilde{B}$ is self-injective (for example for $B = kQ/\langle\rho\rangle$ with Q the linearly directed $\mathbb{A}_3$-quiver and ρ the path of length 2). In general, $\widetilde{B}$ will be neither hereditary nor self-injective.

3.2 The complex Σ'_A

We have mentioned in Part 1 that the simplicial complex Σ_A of tilting modules always has a non-empty boundary (for $n(A) \geqslant 2$). Now the cluster theory provides a recipe for embedding this simplicial complex in a slightly larger one without boundary. Let me introduce here this complex Σ'_A directly in terms of mod A, using a variation of the work of Marsh, Reineke und Zelevinsky [38][15] . It is obtained from Σ_A by just adding $n = n(A)$ vertices, and of course further simplices. Recall that a *Serre subcategory*[16] $\mathcal{U}$ of an abelian category is a subcategory which is closed under submodules, factor modules and extensions; thus in case we deal with a length category such as mod A, then $\mathcal{U}$ is specified by the simple modules contained in $\mathcal{U}$ (an object belongs to $\mathcal{U}$ if and only if its composition factors lie in $\mathcal{U}$). In particular, for a simple A-module S, let us denote by $(-S)$ the subcategory of all A-modules which do not have S as a composition factor. Any Serre subcategory is the intersection of such subcategories.

Here is the definition of Σ'_A: As simplices take the pairs $(M,\mathcal{U})$ where $\mathcal{U}$ is a Serre subcategory of mod A and M is (the isomorphism class of) a basic module in $\mathcal{U}$ without self-extensions; write $(M',\mathcal{U}') \leqslant (M,\mathcal{U})$

---

15 The title of the paper refers to "associahedra": in the case of the path algebra of a quiver of type $\mathbb{A}_n$, the dual of the simplicial complex Σ'_A is an associahedron (or Stasheff polytope). For quivers of type $\mathbb{B}_n$ and $\mathbb{C}_n$ one obtains a Bott-Taubes cyclohedron.

16 The Serre subcategories are nothing else then the subcategories of the form mod A/AeA, where e in an idempotent of A.

provided M' is a direct summand of M and $\mathcal{U}' \supseteq \mathcal{U}$ (note the reversed order!). Clearly[17], Σ_A can be considered as a subcomplex of Σ'_A, namely as the set of all pairs $(M, \operatorname{mod} A)$.

There are two kinds of vertices of Σ'_A, namely those of the form $(E, \operatorname{mod} A)$ with E an exceptional A-module (these are the vertices belonging to Σ_A), and those of the form $(0, (-S))$ with S simple. It is fair to say that the latter ones are indexed by the "negative simple roots"; of course these are the vertices which do not belong to Σ_A. Given a simplex $(M, \mathcal{U})$, its vertices are the elements $(E, \operatorname{mod} A)$, where E is an indecomposable direct summand of M, and the elements $(0, (-S))$, where $\mathcal{U} \subseteq (-S)$. The $(n-1)$-simplices are those of the form $(M, \mathcal{U})$, where M is a basic tilting module in $\mathcal{U}$. The vertices outside Σ_A belong to one $(n-1)$-simplex, namely to $(0, \{0\})$. The $(n-2)$-simplices are of the form $(M, \mathcal{U})$, where M is an almost complete partial tilting module for $\mathcal{U}$. If it is sincere in $\mathcal{U}$, there are precisely two complements in $\mathcal{U}$. If it is not sincere in $\mathcal{U}$, then there is only one complement in $\mathcal{U}$, but there also is a simple module S such that X belongs to $(-S)$, thus X is a tilting module for $\mathcal{U} \cap (-S)$. This shows that any $(n-2)$-simplex belongs to precisely two $(n-1)$-simplices.

As an example, we consider again the path algebra A of the quiver $\circ \leftarrow \circ \leftarrow \circ$. The simplicial complex Σ'_A is a 2-sphere and looks as follows (considering the 2-sphere as the 1-point compactification of the real plane):

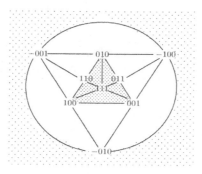

Here, the vertex $(0, (-S))$ is labeled as $-\dim S$. We have shaded the

---

17 In the same way, we may identify the set of simplicies of the form $(M, \mathcal{U})$ with $\mathcal{U}$ fixed, as $\Sigma_{A/AeA}$, where $\mathcal{U} = \operatorname{mod} A/AeA$. In this way, we see that Σ'_A can be considered as a union of all the simplicial complexes $\Sigma_{A/AeA}$.

subcomplex Σ_A (the triangle in the middle) as well as the $(n-1)$-simplex $(0, \{0\})$ (the outside).

Consider now a reflection functor σ_i, where i is a sink, say. We obtain an embedding of $\Sigma_{\sigma_i A}$ into Σ'_A as follows: There are the exceptional $\sigma_i A$-modules of the form $\sigma_i E$ with E an exceptional A-module, different from the simple A-module $S(i)$ concentrated at the vertex i, and in between these modules $\sigma_i E$ the simplex structure is the same as in between the modules E. In addition, there is the simple $\sigma_i A$-module $S'(i)$ again concentrated at i. Now we know that E has no composition factor $S(i)$ if and only if $\mathrm{Ext}^1_{\sigma_i A}(S'(i), \sigma_i M) = 0$. This shows that the simplex structure of Σ_A involving $(0, (-S(i))$ and vertices of the form $(E, \mathrm{mod}\, A)$ is the same as the simplex structure of $\Sigma_{\sigma_i A}$ in the vicinity of $(S'(i), \mathrm{mod}\, \sigma_i A)$.

We may consider the simplicial complex Σ'_A as a subset of the real n-dimensional space $K_0(A) \otimes \mathbb{R}$, where $n = n(A)$, namely as a part of the corresponding unit $(n-1)$-sphere, with all the $(n-1)$-simplices defined by n linear inequalities. In case A is representation-finite, we deal with the $(n-1)$-sphere itself, otherwise with a proper subset. For example, in the case of the path algebra A of the quiver $\circ \leftarrow \circ \leftarrow \circ$, the inequalities are $\varphi_1 \geqslant 0$, $\varphi_2 \geqslant 0$, $\varphi_3 \geqslant 0$, where $\varphi_1, \varphi_2, \varphi_3$ are the linear forms inserted in the corresponding triangle:

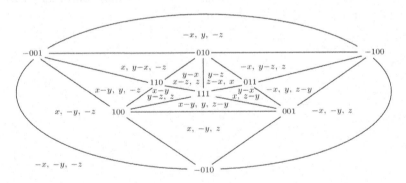

In general, any $(n-1)$-simplex $(M, \mathcal{U})$ is equipped with n linear forms $\varphi_1, \ldots, \varphi_n$ on $K_0(A)$ such that the following holds: an A-module N without self-extensions belongs to $\mathrm{add}\, M$ if and only if $\varphi_i(\mathbf{dim}\, N) \geqslant 0$.

In the same way as Σ_A, also Σ'_A can be identified with a fan in $K_0(A) \otimes \mathbb{R}$. For any simplex $(M, \mathcal{U})$ with vertices $(E, \mathrm{mod}\, A)$ and $(0, (-S))$, where

E are indecomposable direct summands of M, and S simple modules which do not belong to $\mathcal{U}$, take the cone $C(M,\mathcal{U})$ generated by the vectors $\dim E$ and $-\dim S$.

3.3 The cluster categories

We have exhibited the cluster tilted algebras without reference to cluster categories, in order to show the elementary nature of these concepts. But a genuine understanding of cluster tilted algebras as well as of Σ'_A is not possible in this way. Starting with a hereditary artin algebra A, let us introduce now the corresponding cluster category $\mathcal{C}_A$. We have to stress that this procedure reverses the historical development[18] : the cluster categories were introduced first, and the cluster tilted algebras only later. The aim of the definition of cluster categories was to illuminate the combinatorics behind the so called cluster algebras, in particular the combinatorics of the cluster complex.

Let me say a little how cluster tilted algebras were found. Everything started with the introduction of "cluster algebras" by Fomin and Zelevinksy [23]: these are certain subrings of rational function fields, thus commutative integral domains. At first sight, one would not guess any substantial relationship to non-commutative artin algebras. But it turned out that the Dynkin diagrams, as well as the general Cartan data, play an important role for cluster algebras too. As it holds true for the hereditary artin algebras, it is the corresponding root system, which is of interest. This is a parallel situation, although not completely. For the cluster algebras one needs to understand not only the positive roots, but the *almost positive* roots: this set includes besides the positive roots also the negative simple roots. As far as we know, the set of almost positive roots had not been considered before[19] . The first link between cluster

---

18 In the words of Fomin and Zelevinsky [25], this Part 3 altogether is completely revisionistic.

19 Lie theory is based on the existence of perfect symmetries — partial structures (such as the set of positive roots) which allow only broken symmetries tend to be accepted just as necessary working tools. The set of almost positive roots seems to be as odd as that of the positive ones: it depends on the same choices, but does not even enjoy the plus-minus merit of being half of a neat entity. This must have been the mental reasons that the intrinsic beauty of the cluster complex was realized only very recently. But let me stress here that the cluster complex seems to depend not only on the choice of a root basis, but on the ordering of the basis (or better, on the similarity class): with a difference already for the types $\tilde{A}_{2,2}$ and $\tilde{A}_{3,1}$.

theory and tilting theory was given by Marsh, Reineke, Zelevinksy in [38] when they constructed the complex Σ'_A. Buan, Marsh, Reineke, Reiten, Todorov [8] have shown in which way the representation theory of hereditary artin algebras can be used in order to construct a category $\mathcal{C}_A$ (the cluster category) which is related to the set of almost positive roots[20] in the same way as the module category of a hereditary artin algebra is related to the corresponding set of positive roots.

As we have seen, a tilted algebra B should be regarded as the factor algebra of its cluster tilted algebra $\tilde{B}$, if we want to take into account also the missing modules. But mod $\tilde{B}$ has to be considered as the factor category of some triangulated category $\mathcal{C}_A$, the corresponding cluster category. Looking at $\mathcal{C}_A$, we obtain a common ancestor of all the algebras tilted from algebras in the similarity class of A. In the setting of the pictures shown, the corresponding cluster category has the form

$\mathcal{C}_A$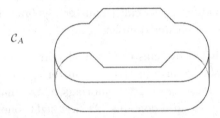

The cluster category $\mathcal{C}_A$ should be considered as a universal kind of category belonging to the similarity class of the hereditary artin algebra A in order to obtain all the module categories mod $\tilde{B}$, where $\tilde{B}$ is a cluster tilted algebra of type similar to A.

What one does is the following: start with the derived category $D^b(\text{mod } A)$ of the hereditary artin algebra A, with shift functor [1],

---

20 A slight unease should be mentioned: as we will see, there is an embedding of mod A into the cluster category which preserves indecomposability and reflects isomorphy (but it is not a full embedding), thus this part of the cluster category corresponds to the positive roots. There are precisely $n = n(A)$ additional indecomposable objects: they should correspond to the negative simple roots, but actually the construction relates them to the negative of the dimension vectors of the indecomposable projectives. Thus, the number of additional objects is correct, and there is even a natural bijection between the additional indecomposable objects and the simple modules, thus the simple roots. But in this interpretation one may hesitate to say that "one has added the negative simple roots" (except in case any vertex is a sink or a source). On the other hand, in our presentation of the cluster complex we have used as additional vertices the elements $(0, (-S))$, and they really look like "negative simple roots". Thus, we hope that this provides a better feeling.

and take as $\mathcal{C}_A$ the orbit category with respect to the functor $\tau_D^{-1}[1]$ (we write τ_D for the Auslander-Reiten translation in the derived category, and τ_C for the Auslander-Reiten translation in $\mathcal{C}_A$). As a fundamental domain for the action of this functor one can take the disjoint union of mod A (this yields all the positive roots) and the shifts of the projective A-modules by [1] (this yields $n = n(A)$ additional indecomposable objects). It should be mentioned that Keller [33] has shown that $\mathcal{C}_A$ is a triangulated category; this is now the basis of many considerations dealing with cluster categories and cluster tilted algebras. Now if we take a tilting module T in mod A, we may look at the endomorphism ring $\tilde{B}$ of T in $\mathcal{C}_A$ (or better: the endomorphism ring of the image of T under the canonical functors mod $A \subseteq D^b(\text{mod } A) \longrightarrow \mathcal{C}_A$), and obtain a cluster tilted algebra[21] as considered above. The definition immediately yields that $\tilde{B} = B \ltimes J$, where $J = \text{Hom}_{D^b(\text{mod } A)}(T, \tau_D^{-1}T[1]) = \text{Ext}_A^1(T, \tau^{-1}T)$. The decisive property is that there is a canonical equivalence of categories[22]

$$\mathcal{C}_A/\langle T\rangle \longrightarrow \text{mod } \tilde{B}.$$

In particular, we see that the triangulated category $\mathcal{C}_A$ has many factor categories which are abelian[23] .

What happens when we form the factor category $\mathcal{C}_A/\langle T\rangle$? Consider an indecomposable direct summand E of the tilting A-module T as an object in the cluster category $\mathcal{C}_A$ and the meshes starting and ending in E:

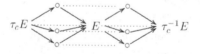

---

21 This is the way, the cluster tilted algebras were introduced and studied by Buan, Marsh and Reiten [10].

22 Instead of $\mathcal{C}_A/\langle T\rangle$, one may also take the equivalent category $\mathcal{C}_A/\langle \tau_C T\rangle$. The latter is of interest if one wants the indecomposable summands of T in $\mathcal{C}_A$ to become indecomposable projective objects.

23 We have mentioned that the cluster theory brought many surprises. Here is another one: One knows for a long time many examples of abelian categories $\mathcal{A}$ with an object M such that the category $\mathcal{A}/\langle M\rangle$ (obtained by setting zero all maps which factor through add M) becomes a triangulated category: just take $\mathcal{A} = \text{mod } R$, where R is a self-injective artin algebra R and $M = {}_R R$. The category mod $R/\langle {}_R R\rangle = \underline{\text{mod}}R$ is the stable module category of R. But we are not aware that non-trivial examples were known of a triangulated category $\mathcal{D}$ with an object N such that $\mathcal{D}/\langle N\rangle$ becomes abelian. Cluster tilting theory is just about this!

In the category $C_A/\langle T\rangle$, the object E becomes zero, whereas both $\tau_C E$ and $\tau_C^{-1}E$ remain non-zero. In fact, $\tau_C^{-1}E$ becomes a projective object and $\tau_C E$ becomes an injective object: We obtain in this way in $\operatorname{mod}\widetilde{B} = C_A/\langle T\rangle$ an indecomposable projective module $P = \tau_C^{-1}E$ and an indecomposable injective module $I = \tau_C E$, such that $\operatorname{top}P \simeq \operatorname{soc}I$. This explains the round trip phenomenon for $\widetilde{B}$ mentioned above: there is the hammock corresponding to the simple $\widetilde{B}$-module $\operatorname{top}P \simeq \operatorname{soc}I$, starting from $I = \tau_C E$, and ending in $P = \tau_C^{-1}E$. And either $\operatorname{rad}P$ is projective (and $I/\operatorname{soc}I$ injective) or else there are non-sectional paths of length 4 from I to P.

There is a decisive symmetry condition[24] in the cluster category $C = C_A$:

$$\operatorname{Hom}_C(X, Y[1]) \simeq D\operatorname{Hom}_C(Y, X[1]).$$

This is easy to see: since we form the orbit category with respect to $\tau_D^{-1}[1]$, this functor becomes the identity functor in C, and therefore the Auslander-Reiten functor τ_C and the shift functor $[1]$ in C coincide. On the other hand, the Auslander-Reiten (or Serre duality) formula for C asserts that $\operatorname{Hom}_C(X, Y[1]) \simeq D\operatorname{Hom}_C(Y, \tau_C X)$. A triangulated category is said to be d-Calabi-Yau provided the shift functor $[d]$ is a Serre (or Nakayama) functor, thus provided there is a functorial isomorphism

$$\operatorname{Hom}(X, -) \simeq D\operatorname{Hom}(-, X[d])$$

(for a discussion of this property, see for example [33]). As we see, *the cluster category is 2-Calabi-Yau.*

The cluster category has Auslander-Reiten sequences. One component Γ_0 of the Auslander-Reiten quiver of C_A has only finitely many τ_C-orbits, namely the component containing the indecomposable projective (as well as the indecomposable injective) A-modules. The remaining components of the Auslander-Reiten quiver of C_A have tree class $\mathbb{A}_\infty$.

In a cluster category $C = C_A$, an object is said to be a *cluster-tilting object*[25] provided first $\operatorname{Hom}_C(T, T[1]) = 0$, and second, that T is maximal

---

24 If we write $\operatorname{Ext}^1(X, Y) = \operatorname{Hom}_C(X, Y[1])$, then this symmetry condition reads that $\operatorname{Ext}^1(X, Y)$ and $\operatorname{Ext}^1(Y, X)$ are dual to each other, in particular they have the same dimension.

25 It has to be stressed that the notion of a "cluster-tilting object" in a cluster category does not conform to the tilting notions used otherwise in this Handbook! If T is such a cluster-tilting object, then it may be that $\operatorname{Hom}_C(T, T[i]) \neq 0$ in $C = C_A$ for some $i \geqslant 2$. Observe that in a 2-Calabi-Yau category such as C_A, we have $\operatorname{Hom}_C(X, X[2]) \neq 0$, for any non-zero object X.

with this property in the following sense: if $\mathrm{Hom}_{\mathcal{C}}(T\oplus X,(T\oplus X)[1])=0$, then X is in $\mathrm{add}\,T$. If T is a tilting A-module, then one can show quite easily that T, considered as an object of $\mathcal{C}_A$ is a cluster-tilting object.

Let us consider the hereditary artin algebras in one similarity class and the reflection functors between them. One may identify the correspond-ing cluster categories using the reflection functors, as was pointed out by Bin Zhu [52]. In this way, one can compare the tilting modules of all the hereditary artin algebras in one similarity class. It turns out that *the cluster-tilting objects in $\mathcal{C}_A$ are just the tilting modules for the various artin algebras obtained from A by using reflection functors* [8]. In order to see this, let T be a cluster-tilting object in $\mathcal{C}_A$. Let Γ_0 be the component of the Auslander-Reiten quiver of $\mathcal{C}_A$ which contains the indecomposable projective A-modules. If no indecomposable direct summand of T belongs to Γ_0, then T can be considered as an A-module, and it is a regular tilting A-module. On the other hand, if there is an indecomposable direct summand of T, say T_1, which belongs to Γ_0, then let $\mathcal{S}$ be the class of all indecomposable objects X in Γ_0 with a path from X to T_1 in Γ_0, and such that any path from X to T_1 in Γ_0 is sectional. Then no indecomposable direct summand of T belongs to $\tau_C\mathcal{S}$. We may identify the factor category $\mathcal{C}_A/\langle \tau_C\mathcal{S}\rangle$ with $\mathrm{mod}\,A'$ for some hereditary artin algebra A', and consider T as an A'-module (the object T_1, consid-ered as an A'-module, is projective and faithful). Clearly, A' is obtained from A by a sequence of BGP-reflection functors.

Also, the usual procedure of going from a tilting module to another one by exchanging just one indecomposable direct summand gets more regular. Of course, there is the notion of an almost complete partial cluster-tilting object and of a complement, parallel to the corresponding notions of an almost complete partial tilting module and its comple-ments. Here we get: *Any almost complete partial cluster-tilting object $\overline{T}$ has precisely two complements* [8]. We indicate the proof: We can as-sume that $\overline{T}$ is an A-module. If $\overline{T}$ is sincere, then we know that there are two complements for $\overline{T}$ considered as an almost complete partial tilting A-module. If $\overline{T}$ is not sincere, then there is only one complement for $\overline{T}$ considered as an almost complete partial tilting A-module. But there is also one (and obviously only one) indecomposable projective module P with $\mathrm{Hom}_A(P,\overline{T})=0$, and the τ_C-shift of P in the cluster category is the second complement we are looking for!

An important point seems to be the following: *The simplicial complex* *of partial cluster-tilting objects in the cluster category $\mathcal{C}_A$ is nothing else than Σ'_A,* with the following identification: If T is a basic partial cluster-tilting object in $\mathcal{C}_A$, we can write T as the direct sum of a module M in $\operatorname{mod} A$ and objects of the form $\tau_{\mathcal{C}} P(i)$, with $P(i)$ indecomposable projective in $\operatorname{mod} A$, and i in some index set Θ. Then M corresponds in Σ'_A to the pair $(M, \mathcal{U})$, where $\mathcal{U} = \bigcap_{i \in \Theta}(-S(i))$. The reason is very simple: $\operatorname{Hom}_{\mathcal{C}}(\tau_{\mathcal{C}} P(i), M[1]) \simeq \operatorname{Hom}_{\mathcal{C}}(P(i), M) = \operatorname{Hom}_A(P(i), M)$, with $\mathcal{C} = \mathcal{C}_A$.

The complex Σ'_A should be viewed as a convenient index scheme[26] for the set of cluster tilted algebras obtained from the hereditary artin algebras in the similarity class of A. Any maximal simplex of σ'_A is a cluster-tilting object in $\mathcal{C}_A$, and thus we can attach to it its endomorphism ring. Let us redraw the complex Σ'_A for the path algebra A of the quiver $\circ \leftarrow \circ \leftarrow \circ$, so that the different vertices and triangles are better seen:

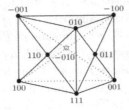

There are two kinds of vertices, having either 4 or 5 neighbours. The vertices with 5 neighbours form two triangles (the bottom and the top triangle), and these are the cluster-tilting objects with endomorphism ring of infinite global dimension. The remaining triangles yield hereditary endomorphism rings and again, there are two kinds: The quiver of the endomorphism ring may have one sink and one source, these rings are given by the six triangles which have an edge in common with the bottom or the top triangle. Else, the endomorphism ring is hereditary and the radical square is zero: these rings correspond to the remaining six triangles:

---

26 But we should also mention the following: The set of isomorphism classes of basic cluster-tilting objects in $\mathcal{C}_A$ is no longer partially ordered. In fact, given an almost complete partial cluster-tilting object $\overline{T}$ and its two complements X and Y, there are triangles $X \longrightarrow T' \longrightarrow Y \longrightarrow$ and $Y \longrightarrow T'' \longrightarrow X \longrightarrow$ with $T', T'' \in \operatorname{add} \overline{T}$.

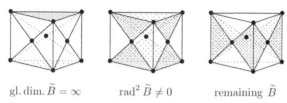

$$\text{gl. dim.}\ \tilde{B} = \infty \qquad \text{rad}^2\,\tilde{B} \neq 0 \qquad \text{remaining}\ \tilde{B}$$

Consider an almost complete partial cluster-tilting object $\overline{T}$ in $\mathcal{C} = \mathcal{C}_A$. As we have mentioned, there are precisely two complements for $\overline{T}$, say E and E'. Let $T = \overline{T} \oplus E$, and $T' = \overline{T} \oplus E'$. Thus, there are given two cluster-tilted algebras $\tilde{B} = \mathrm{End}_{\mathcal{C}}(T)$, and $\widetilde{B'} = \mathrm{End}_{\mathcal{C}}(T')$, we may call them *adjacent*, this corresponds to the position of T and T' in the complex Σ'_A. We can identify $\mathcal{C}_A/\langle T \rangle$ with mod $\tilde{B}$, and $\mathcal{C}_A/\langle T' \rangle$ with mod $\widetilde{B'}$ We saw that E as an indecomposable direct summand of T yields an indecomposable projective $\tilde{B}$-module $P = \tau_{\mathcal{C}}^{-1} E$ and an indecomposable injective $\tilde{B}$-module $I = \tau_{\mathcal{C}} E$, such that soc $I \simeq$ top P. Since $\overline{T} \oplus E'$ is a cluster-tilting object, it is not difficult to show that $\mathrm{Hom}_{\tilde{B}}(P', E') = 0$ for any indecomposable projective $\tilde{B}$-module $P' \not\simeq P$. But this implies that E' is identified under the equivalence of $\mathcal{C}_A/\langle T \rangle$ and mod $\tilde{B}$ with the simple $\tilde{B}$-module which is the socle of I and the top of P. In the same way, we see that E is a simple $\widetilde{B'}$-module, namely the top of the $\widetilde{B'}$-module $P = \tau_{\mathcal{C}}^{-1} E'$ and the socle of the indecomposable injective $\widetilde{B'}$-module $I = \tau_{\mathcal{C}} E'$. Thus, there is the following sequence of identifications:

$$\text{mod}\ \tilde{B}/\langle \text{add}\, E' \rangle \simeq \mathcal{C}_A/\langle \text{add}(T \oplus E') \rangle$$
$$= \mathcal{C}_A/\langle \text{add}(T' \oplus E) \rangle \simeq \text{mod}\ \widetilde{B'}/\langle \text{add}\, E \rangle.$$

Altogether this means that *artin algebras $\tilde{B}$ and $\widetilde{B'}$ which are adjacent, are nearly Morita equivalent* [10]. We had promised to the reader, that we will return to the hammock configuration (P, S, I), where S is a simple $\tilde{B}$-module, $P = P(S)$ its projective cover, and $I = I(S)$ its injective envelope: but this is the present setting. Using the cluster category notation, we can write $P = \tau_{\mathcal{C}}^{-1} E$, $I = \tau_{\mathcal{C}} E$, and then $S = E'$, where E, E' are complements to an almost complete partial cluster-tilting object $\overline{T}$. When we form the category mod $\tilde{B}/\langle \text{add}\, S \rangle$, the killing of the simple $\tilde{B}$-module S creates a hole in mod $\tilde{B}$. From the hammock $\mathrm{Hom}(P, -)$ in mod $\tilde{B}$ the following parts survive:

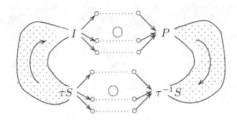

Note that the new hole is of the same nature as the hole between I and P (which was created when we started from the cluster category $\mathcal{C}$, killing the object E). Indeed, one may fill alternatively one of the two holes and obtains $\operatorname{mod} \widetilde{B}$, or $\operatorname{mod} \widetilde{B}'$, respectively.

Altogether, we see: A cluster category $\mathcal{C} = \mathcal{C}_A$ has a lot of nice factor categories which are abelian (the module categories $\operatorname{mod} \widetilde{B}$), and one should regard $\mathcal{C}$ as being obtained from patching together the various factor categories in the same way as manifolds are built up from open subsets by specifying the identification maps of two such subsets along what will become their intersection. The patching process for the categories $\operatorname{mod} \widetilde{B}$ is done via the nearly Morita equivalences for adjacent tilting objects[27] .

The reader will have noticed that this exchange process for adjacent algebras generalizes the BGP-reflection functors (and the APR-tilting functors) to vertices which are not sinks or sources. Indeed, for $\widetilde{B} = \operatorname{End}_{\mathcal{C}}(\overline{T} \oplus E)$, and $\widetilde{B}' = \operatorname{End}_{\mathcal{C}}(\overline{T} \oplus E')$, the indecomposable direct summand E of $\overline{T} \oplus E$ corresponds to a vertex of the quiver of $\widetilde{B}$, and similarly, E' corresponds to a vertex of the quiver of $\widetilde{B}'$. In the BGP and the APR setting, one of the modules E, E' is simple projective, the other one is simple injective — here now E and E' are arbitrary exceptional modules[28] .

This concludes our attempt to report about some of the new results in tilting theory which are based on cluster categories. Let us summarize the importance of this development. First of all, the cluster tilted algebras provide a nice depository for storing the modules which are lost when we pass from hereditary artin algebras to tilted algebras; there is

---

27　It seems that there is not yet any kind of axiomatic approach to this new patching
　　process.

28　A direct description of this reflection process seems to be still missing. It will
　　require a proper understanding of all the cluster tilted algebras $\widetilde{B}$ with $n(\widetilde{B}) = 3$.
　　A lot is already known about such algebras, see [11].

a magic bimodule which controls the situation. We obtain in this way a wealth of algebras whose module categories are described by the root system of a Kac-Moody Lie-algebra. These new algebras are no longer hereditary, but are still of Gorenstein dimension at most 1. For the class of cluster tilted algebras, there is a reflection process at any vertex of the quiver, not only at sinks and sources. This is a powerful generalization of the APR-tilting functors (thus also of the BGP-reflection functors), and adjacent cluster tilted algebras are nearly Morita equivalent. The index set for this reflection process is the simplicial complex Σ'_A and the introduction of this simplicial complex solved also another riddle of tilting theory: it provides a neat way of enlarging the simplicial complex of tilting A-modules in order to get rid of its boundary. We have mentioned in Part 1 that both the missing modules problem as well as the boundary problem concern the module category, but disappear on the level of derived categories. Thus it is not too surprising that derived categories play a role: as it has turned out, the cluster categories, as suitable orbit categories of the corresponding derived categories, are the decisive new objects. These are again triangulated categories, and are to be considered as the universal structure behind all the tilted and cluster tilted algebras obtained from a single hereditary artin algebra A (and the hereditary artin algebras similar to A).

3.4 Appendix: Cluster algebras

Finally we should speak about the source of all these developments, the introduction of cluster algebras by Fomin and Zelevinsky. But we are hesitant, for two reasons: first, there is our complete lack of proper expertise, but also it means that we leave the playground of tilting theory. Thus this will be just an appendix to the appendix. The relationship between cluster algebras on the one hand, and the representation theory of hereditary artin algebras and cluster tilted algebras on the other hand is fascinating, but also very subtle[29] . At first, one observed certain analogies and coincidences. Then there was an experimental period, with

---

29 Since this report is written for the Handbook of Tilting Theory, we are only concerned with the relationship of the cluster algebras to tilting theory. There is a second relationship to the representation theory of artin algebras, namely to Hall algebras, as found by Caldero and Chapoton [15], and Caldero-Keller [16, 17] , see also Hubery [31]. And there are numerous interactions between cluster theory and many different parts of mathematics. But all this lies beyond the scope of this volume.

many surprising findings (for example, that the Happel-Vossieck list of tame concealed algebras corresponds perfectly to the Seven list of minimal infinite cluster algebras [47], as explained in [14]). In the meantime, many applications of cluster-tilted algebras to cluster algebras have been found [13, 9], and the use of Hall algebra methods provides a conceptual understanding of this relationship [15, 16, 17, 31].

Here is at least a short indication what cluster algebras are. As we said already, the cluster algebras are (commutative) integral domains. The cluster algebras we are interested in (those related to hereditary artin algebras)[30] are finitely generated (this means finitely generated "over nothing", say over $\mathbb{Z}$), thus they can be considered as subrings of a finitely generated function field $\mathbb{Q}(x_1, \ldots, x_n)$ over the rational numbers $\mathbb{Q}$. This is the way they usually are presented in the literature (but the finite generation is often not stressed). In fact, one of the main theorems of cluster theory asserts that we deal with subrings of the ring of Laurent polynomials $\mathbb{Z}[x_1^{\pm 1}, \ldots, x_n^{\pm 1}]$ (this is the subring of all elements of the form $\frac{p}{q}$ where p is in the polynomial ring $\mathbb{Z}[x_1, \ldots, x_n]$ and q is a monomial in the variables $x_1, \ldots, x_n$).

Since we deal with a noetherian integral domain, the reader may expect to be confronted with problems in algebraic geometry, or, since we work over $\mathbb{Z}$ with those of arithmetical geometry. But this was not the primary interest. Instead, the cluster theory belongs in some sense to algebraic combinatorics, and the starting question concerns the existence of a nice $\mathbb{Z}$-basis of such a cluster algebra, say similar to all the assertions about canonical bases in Lie theory.

What are clusters? Recall that a cluster algebra is a subring of $\mathbb{Z}[x_1^{\pm 1}, \ldots, x_n^{\pm 1}]$. What one is looking for is a convenient $\mathbb{Z}$-basis of the cluster algebra. One may assume that the elements of the basis are written in the form $\frac{p}{q}$, where $p \in \mathbb{Z}[x_1, \ldots, x_n]$ is not divisible by the variables $x_1, \ldots, x_n$ and $q = x_1^{d_1} \cdots x_n^{d_n}$ with exponents $d_i \in \mathbb{Z}$; the Laurent monomial q is said to be the *denominator*[31] of $\frac{p}{q}$ and one may call $\dim q = (d_1, \ldots, d_n)$ its dimension vector. There seems to be an inductive procedure to produce at least a part of a $\mathbb{Z}$-basis by first obtaining the "cluster variables", and then forming monomials of the cluster variables belonging to a fixed cluster. At least, this works for the cluster

---

30 these are the so-called acylic cluster algebras [6].
31 Note that this means that the variable x_i itself will be rewritten in the form $1/(x_i^{-1})$; its denominator is $q = x_i^{-1}$.

algebras of finite type and in this case one actually obtains a complete $\mathbb{Z}$-basis. One of the main topics discussed in cluster theory concerns the shape of the cluster variables in general.

Consider the case of the path algebra of a finite quiver Q without oriented cycles. According to Caldero-Keller [17], the simplicial complex Σ'_A with $A = kQ$ can be identified with the cluster complex corresponding to Q. Under this correspondence, the cluster variables correspond to the exceptional A-modules and the elements of the form $(0, (-S))$. When we introduced the simplicial complex Σ'_A, the maximal simplices were labeled $(M, \mathcal{U})$ with M a basic tilting module in a Serre subcategory $\mathcal{U}$ of mod A. Recall that such an $(n-1)$-simplex $(M, \mathcal{U})$ in Σ'_A is equipped with n linear forms $p_1, \ldots, p_n$ on $K_0(A)$ such that an A-module N without self-extensions belongs to add M if and only if $\varphi_i(\mathbf{dim}\, N) \geqslant 0$, for $1 \leqslant i \leqslant n$. And there is the parallel assertion: A cluster monomial with denominator q belongs to the cluster corresponding to $(M, \mathcal{U})$ if and only if $\varphi_i(\mathbf{dim}\, q) \geqslant 0$, for $1 \leqslant i \leqslant n$.

Here are the cluster variables for the cluster algebra of type $\mathbb{A}_3$, inserted as the vertices of the cluster complex Σ'_A:

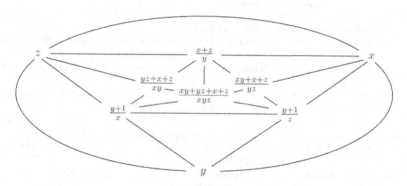

Acknowledgment. We were able to provide only a small glimpse of what cluster tilting theory is about — it is a theory in fast progress, but also with many problems still open. The report would not have been possible without the constant support of many experts. In particular, we have to mention Idun Reiten who went through numerous raw versions. The author has also found a lot of help by Thomas Brüstle, Philippe Caldero, Dieter Happel, Otto Kerner and Robert Marsh, who answered all kinds of questions. And he is indebted to Aslak Buan, Philipp Fahr,

Rolf Farnsteiner, Lutz Hille, Bernhard Keller and Markus Reineke for
helpful comments concerning an early draft of this report.

References. In order to avoid a too long list of references, we tried to restrict to the new developments. We hope that further papers mentioned
throughout our presentation can be identified well using the appropriate
chapters of this Handbook as well as standard lists of references. But
also concerning the cluster approach, there are many more papers of
interest and most are still preprints (see the arXiv). Of special interest
should be the survey by Buan and Marsh [7].

References

[1] Assem, Brüstle and Schiffler: Cluster-tilted algebras as trivial extensions. arXiv: math.RT/0601537

[2] Auslander: Comments on the functor Ext. Topology 8 (1969), 151-166.

[3] Auslander, Reiten: On a generalized version of the Nakayama conjecture. Proc. Amer.Math.Soc. 52 (1975), 69-74.

[4] Auslander, Reiten: Applications of contravariantly finite subcategories.
Advances Math. 86 (1991), 111-152.

[5] Beligiannis, Reiten: Homological and homotopical aspects of torsion
theories. Memoirs Amer.Math.Soc. (To appear).

[6] Berenstein, Fomin, Zelevinsky: Cluster algebras III: Upper bounds and
double Bruhat cells. arXiv: math.RT/0305434

[7] Buan, Marsh: Cluster-tilting theory. Proceedings of the ICRA meeting,
Mexico 2004. Contemporary Mathematics. (To appear).

[8] Buan, Marsh, Reineke, Reiten, Todorov: Tilting theory and cluster combinatorics. Advances Math. (To appear). arXiv: math.RT/0402054

[9] Buan, Marsh, Reiten, Todorov: Clusters and seeds in acyclic cluster
algebras. arXiv: math.RT/0510359

[10] Buan, Marsh, Reiten: Cluster tilted algebras. Trans. Amer. Math. Soc.
(To appear). arXiv: math.RT/0402075

[11] Buan, Marsh, Reiten: Cluster mutation via quiver representations.
arXiv: math.RT/0412077.

[12] Buan, Marsh, Reiten: Cluster-tilted algebras of finite representation
type. arXiv: math.RT/0509198

[13] Buan, Reiten: Cluster algebras associated with extended Dynkin quivers. arXiv: math.RT/0507113

[14] Buan, Reiten, Seven: Tame concealed algebras and cluster algebras of
minimal infinite type. arXiv: math.RT/0512137

[15] Caldero, Chapoton: Cluster algebras as Hall algebras of quiver representations. arXiv: math.RT/0410187

[16] Caldero, Keller: From triangulated categories to cluster algebras. I.
arXiv: math.RT/0506018

[17] Caldero, Keller: From triangulated categories to cluster algebras. II. arXiv: math.RT/0510251.

[18] Crawley-Boevey: Exceptional sequences of representations of quivers in 'Representations of algebras', Proc. Ottawa 1992, eds V. Dlab and H. Lenzing, Canadian Math. Soc. Conf. Proc. 14 (Amer. Math. Soc., 1993), 117-124.

[19] Crawley-Boevey, Kerner: A functor between categories of regular modules for wild hereditary algebras. Math. Ann., 298 (1994), 481-487.

[20] Derksen, Weyman: Semi-invariants of quivers and saturation for Littlewood-Richardson coefficients, J. Amer. Math. Soc. 13 (2000), no. 3, 456-479.

[21] Derksen, Weyman: Quiver Representations. Notices of the AMS 52, no. 2.

[22] Derksen, Weyman: On the canonical decomposition of quiver representations, Compositio Math. (To appear).

[23] Fomin, Zelevinsky: Cluster algebras I: Foundations. J. Amer. Math. Soc. 15 (2002), 497-529.

[24] Fomin, Zelevinsky: Cluster algebras II. Finite type classification. Invent. Math. 154 (2003), 63-121.

[25] Fomin, Zelevinsky: Cluster algebras. Notes for the CDM-03 conference. arXiv: math.RT/0311493

[26] Happel: Relative invariants and subgeneric orbits of quivers of finite and tame type. Spinger LNM 903 (1981), 116-124.

[27] Happel: Triangulated categories in the representation theory of finite dimensional algebras. London Math.Soc. Lecture Note Series 119 (1988)

[28] Happel, Reiten, Smalø: Tilting in abelian categories and quasitilted algebras. Memoirs Amer. Math. Soc. 575 (1995).

[29] Hille: The volume of a tilting module. Lecture. Bielefeld 2005, see www.math.uni-bielefeld.de/birep/select/hille_6.pdf.

[30] Huang: Generalized tilting modules with finite injective dimension. arXiv: math.RA/0602572

[31] Hubery: Acyclic cluster algebras via Ringel-Hall algebras. See wwwmath.uni-paderborn.de/hubery/Cluster.pdf

[32] Hughes, Waschbüsch: Trivial extensions of tilted algebras. Proc. London Math. Soc. 46 (1983), 347-364.

[33] Keller: On triangulated orbit categories. arXiv: math.RT/0503240.

[34] Keller, Reiten: Cluster-tilted algebras are Gorenstein and stably Calabi-Yau. arXiv: math.RT/0512471

[35] Kerner, Takane: Universal filtrations for modules in perpendicular categories. In Algebras and Modules II (ed. Reiten, Smalø, Solberg). Amer.Math. Soc. 1998. 347-364.

[36] König, Slungard, Xi: Double centralizer properties, dominant dimension, and tilting modules. J. Algebra 240 (2001), 393-412.

[37] Mantese, Reiten: Wakamatsu tilting modules. J Algebra 278 (2004), 532-552.

[38] Marsh, Reineke, Zelevinsky: Generalized associahedra via quiver representations. Trans. Amer. Math. Soc. 355 (2003), 4171-4186

[39] Ringel: Reflection functors for hereditary algebras. J. London Math. Soc. (2) 21 (1980), 465-479.

[40] Ringel: The braid group action on the set of exceptional sequences of a hereditary algebra. In: Abelian Group Theory and Related Topics.

Contemp. Math. 171 (1994), 339-352.

[41] Ringel: Exceptional objects in hereditary categories. Proceedings Constantza Conference. An. St. Univ. Ovidius Constantza Vol. 4 (1996), f. 2, 150-158.

[42] Ringel: Exceptional modules are tree modules. Lin. Alg. Appl. 275-276 (1998) 471-493.

[43] Ringel: Combinatorial Representation Theory: History and Future. In: Representations of Algebras. Vol. I (ed. D.Happel, Y.B.Zhang). BNU Press. (2002) 122-144.

[44] Rudakov: Helices and vector bundles. London Math. Soc. Lecture Note Series. 148.

[45] Sato, Kimura: A classification of irreducible prehomogeneous vector spaces and their relative invariants. Nagoya Math. J. 65 (1977), 1-155.

[46] Schofield: Semi-invariants of quivers. J. London Math. Soc. 43 (1991), 383-395.

[47] Seven: Recognizing Cluster Algebras of Finite type. arXiv: math.CO/0406545

[48] Tachikawa, Wakamatsu: Tilting functors and stable equivalences for selfinjective algebras. J. Algebra (1987).

[49] Zelevinsky: From Littlewood-Richardson coefficients to cluster algebras in three lectures. In: Fomin (ed): Symmetric Functions 2001. 253-273.

[50] Zelevinsky: Cluster algebras: Notes for the 2004 IMCC. arXiv: math.RT/0407414.

[51] Zhu, Bin: Equivalences between cluster categories. arXiv: math.RT/0511382

[52] Zhu, Bin: Applications of BGP-reflection functors: isomorphisms of cluster algebras. arXiv: math.RT/0511384

Claus Michael Ringel

Fakultät für Mathematik, Universität Bielefeld

POBox 100 131, D-33501 Bielefeld, Germany

Email: ringel@math.uni-bielefeld.de

Printed in the United States
By Bookmasters